O9-BSA-841

STUDENT'S SOLUTIONS MANUAL TO ACCOMPANY

SEARS, ZEMANSKY, AND YOUNG

University Physics

SEVENTH EDITION

STUDENT'S SOLUTIONS MANUAL TO ACCOMPANY

SEARS, ZEMANSKY, AND YOUNG

University Physics

SEVENTH EDITION

A. LEWIS FORD

TEXAS A & M UNIVERSITY

ADDISON-WESLEY PUBLISHING COMPANY

Reading, Massachusetts Menlo Park, California
Don Mills, Ontario Wokingham, England Amsterdam Sydney
Singapore Tokyo Madrid Bogotá Santiago San Juan

Reproduced by Addison-Wesley from camera-ready copy supplied by the author.

ISBN 0-201-06686-6
8 9 10 AL 9594939291

PREFACE

This student's solution manual contains detailed solutions for 334 of the 1016 Exercises and 207 of the 622 Problems in the 7th edition of University Physics. The Exercises and Problems included here were selected solely from the odd-numbered Exercises and Problems in the text, for which the answers are tabulated in the back of the textbook. The Exercises and Problems to be included were not selected at random, but rather have been carefully selected so as to include at least one representative example of each problem type. This solution manual greatly expands the set of worked-out examples that go along with the presentation of the physics laws and concepts in the text. The remaining 682 Exercises and 415 Problems constitute an ample set of problems for the students to tackle on their own. In addition, there are 172 Challenge Problems, for which no solutions are given here.

This solution manual is written for student use. A primary function of the manual is to provide the student with models to follow in working physics problems. The problems are worked out in the manual in the manner and style in which the students should carry out their own problem solutions.

The author will gratefully receive comments as to style, points of physics, errors, or anything else relating to the manual.

Texas A and M University
College Station, TX 77843
December, 1986

A. L. F.

CONTENTS

Chapter			Page
15	Exercises	1, 3, 9, 13, 15, 17, 21	124
	Problems	25, 27, 31, 35	
16	Exercises	3, 7, 11, 13, 15, 17	130
	Problems	23, 25, 29, 31	
17	Exercises	3, 7, 9, 13	135
	Problems	17, 19, 21	
18	Exercises	3, 5, 9, 11, 15, 17	139
	Problems	19, 23, 25	
19	Exercises	1, 5, 7, 11, 13, 17	144
	Problems	21, 23	
20	Exercises	3, 5, 9, 11, 13	151
	Problems	17, 21	
21	Exercises	3, 5, 9, 15, 17	155
	Problems	19, 23, 25	
22	Exercises	3, 5, 7, 11	159
	Problems	13, 15	
23	Exercises	3, 7, 9, 11	162
	Problems	13, 17, 19	
24	Exercises	1, 5, 7, 13, 15, 17	169
	Problems	19, 23, 25	
25	Exercises	5, 7, 9, 13, 17, 19, 23	177
	Problems	27, 29, 31, 35	
26	Exercises	1, 5, 7, 9, 15, 17, 19, 21, 25, 27	185
	Problems	29, 33, 37, 39, 41, 45	
27	Exercises	1, 7, 9, 13, 15, 19	197
	Problems	21, 23, 27, 31, 33	
28	Exercises	1, 7, 11, 13, 17, 19, 23, 25	204
	Problems	31, 33, 35, 37, 39	
29	Exercises	3, 7, 9, 11, 13, 15, 17, 21, 25	210
	Problems	29, 33, 35, 37, 39, 45, 47, 49	

Chapter			Page
30	Exercises	1, 3, 7, 9, 13, 19, 23, 25, 27	223
	Problems	29, 31, 37, 41, 43, 45	
31	Exercises	1, 3, 5, 9, 11, 15, 21, 25	232
	Problems	27, 31, 33, 37, 39, 41	
32	Exercises	1, 5, 7, 11, 13, 15	240
	Problems	21, 23, 27, 31, 33	
33	Exercises	3, 5, 7, 9, 13, 19, 21	248
	Problems	23, 27, 33, 35, 37	
34	Exercises	1, 9, 11, 13, 15, 17, 19	253
	Problems	23, 27, 29, 31	
35	Exercises	3, 7, 9, 11	260
	Problems	17, 19, 21	
36	Exercises	1, 5, 9, 11, 13, 17, 21	263
	Problems	27, 31, 33, 35, 39	
37	Exercises	1, 5, 9, 11, 15, 17	268
	Problems	19, 21, 23, 27	
38	Exercises	3, 5, 9, 11, 13, 15, 19, 23, 27, 31, 33, 37	273
	Problems	39, 43, 45, 47, 49, 53, 57	
39	Exercises	1, 5, 7, 11, 13, 15, 19, 21	285
	Problems	27, 29, 31, 37, 39	
40	Exercises	1, 3, 5, 7, 13, 15, 21, 25	291
	Problems	27, 29, 33, 35, 37, 41	
41	Exercises	1, 5, 9, 11, 15, 19, 23	297
	Problems	25, 27, 29, 35, 37	
42	Exercises	1, 5, 9, 11, 13, 15	302
	Problems	19, 21, 25, 27	
43	Exercises	1, 7, 9	309
	Problems	13, 15	
44	Exercises	1, 5, 7, 9, 13, 17, 19, 21	312
	Problems	25, 29	

CHAPTER 1

Exercises 3, 5, 9, 11, 19, 23, 25, 29, 31, 33, 37, 41, 43

Problems 45, 47, 49

Exercises

1-3

a) $1450\ \text{mi}\cdot\text{hr}^{-1} = (1450\ \text{mi}\cdot\text{hr}^{-1})\left(\frac{1\ \text{hr}}{60\ \text{min}}\right)\left(\frac{1\ \text{min}}{60\ \text{s}}\right) = \underline{0.403\ \text{mi}\cdot\text{s}^{-1}}$

b) $0.403\ \text{mi}\cdot\text{s}^{-1} = (0.403\ \text{mi}\cdot\text{s}^{-1})\left(\frac{1.609\ \text{km}}{1\ \text{mi}}\right)\left(\frac{10^3\ \text{m}}{1\ \text{km}}\right) = \underline{648\ \text{m}\cdot\text{s}^{-1}}$

1-5

$2.0\ \text{L} = 2.0\ \text{L}\left(\frac{10^3\ \text{cm}^3}{1\ \text{L}}\right)\left(\frac{1\ \text{in}}{2.54\ \text{cm}}\right)^3 = 122\ \text{in}^3$

But the answer should be expressed as 120 in^3 to show only two significant figures, since 2.0 L has only two significant figures.

1-9

a) $1.42\times10^9\ \text{cycles}\cdot\text{s}^{-1} \Rightarrow \frac{1}{1.42\times10^9\ \text{cycles}\cdot\text{s}^{-1}} = \underline{7.04\times10^{-10}\ \text{s}}$ for 1 cycle

b) $(1.42\times10^9\ \text{cycles}\cdot\text{s}^{-1})\left(\frac{3600\ \text{s}}{1\ \text{hr}}\right) = \underline{5.11\times10^{12}\ \text{cycles}\cdot\text{hr}^{-1}}$

c) First calculate the number of cycles in one year:

$(5.11\times10^{12}\ \text{cycles}\cdot\text{hr}^{-1})\left(\frac{24\ \text{hr}}{1\ \text{da}}\right)\left(\frac{365\ \text{da}}{1\ \text{yr}}\right) = 4.48\times10^{16}\ \text{cycles}\cdot\text{yr}^{-1}$

10 billion years $= 10^{10}$ yr, so the number of cycles in this time is

$(4.48\times10^{16})(10^{10})\ \text{cycles} = \underline{4.48\times10^{26}\ \text{cycles}}$

d) 1 second error in 10^5 yr

4600 million years $= 4.60\times10^9$ yr

The error thus would be $\left(\frac{1\ \text{s}}{10^5\ \text{yr}}\right)(4.60\times10^9\ \text{yr}) = 4.60\times10^4\ \text{s}\left(\frac{1\ \text{hr}}{3600\ \text{s}}\right) = \underline{12.8\ \text{hr}}$

1-11

$\pi\times10^7\ \text{s} = 3.142\times10^7\ \text{s}$

Actually, $1\ \text{yr} = 1\ \text{yr}\left(\frac{365\ \text{da}}{1\ \text{yr}}\right)\left(\frac{24\ \text{hr}}{1\ \text{da}}\right)\left(\frac{3600\ \text{s}}{1\ \text{hr}}\right) = 3.154\times10^7\ \text{s}$.

The fractional error is thus $\frac{3.154\times10^7\ \text{s} - 3.142\times10^7\ \text{s}}{3.154\times10^7\ \text{s}} = \underline{0.0038}$, or 0.38%

1-19

I estimate that my scalp's area is about that of a 12-in diameter circle.

$$12\,\text{in} = (12\,\text{in})\left(\frac{2.54\,\text{cm}}{1\,\text{in}}\right) = 30\,\text{cm} = 300\,\text{mm}$$

The estimated area is thus $A = \pi r^2$ with $r = \frac{d}{2} = 150\,\text{mm}$:

$$A = \pi(150\,\text{mm})^2 = 7\times10^4\,\text{mm}^2.$$

I further <u>estimate</u> that on my scalp there are 5 hairs per mm^2.

The number of hairs on my head is thus estimated to be about $(5\,\text{hairs}\cdot\text{mm}^{-2})(7\times10^4\,\text{mm}^2) = 3.5\times10^5$ hairs, or about <u>350,000 hairs</u>.

1-23

This requires several estimates, and I can only hope to make plausible ones.

I will take the speed of the cars to be $50\,\text{mi}\cdot\text{hr}^{-1}$. Then a 50 mi long string of cars traveling in each lane will pass through in one hour.

I further estimate that the length of a car is 12 ft and that there is a 5 car length gap behind each car. Thus each car plus its gap occupies about 72 ft.

The number of cars traveling in one lane that passes through in one hour is

$$\frac{(50\,\text{mi})\left(\frac{5280\,\text{ft}}{1\,\text{mi}}\right)}{72\,\text{ft}\cdot\text{car}^{-1}} = 3700\text{ cars}.$$

Since it is a two-lane tunnel the total number of cars that pass through in one hour is <u>7400 cars</u>.

1-25

In each case I will give a sketch that shows the directions of the x and y components and of the resultant vector.

a)

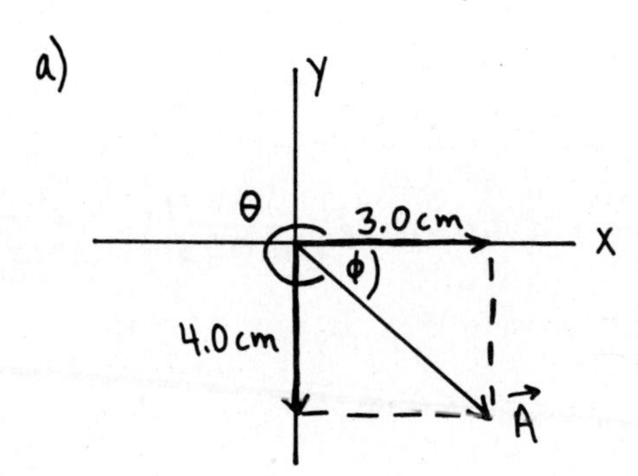

$$A = \sqrt{A_x^2 + A_y^2} = \sqrt{(3.0\,\text{cm})^2 + (4.0\,\text{cm})^2} = \underline{5.0\,\text{cm}}$$

$$\tan\phi = \frac{4.0\,\text{cm}}{3.0\,\text{cm}} = 1.33 \Rightarrow \phi = 53.1^\circ$$

$\theta = 360^\circ - \phi = \underline{307^\circ}$, where this angle is measured counterclockwise from the positive x-axis.

1-25 (cont)

b)

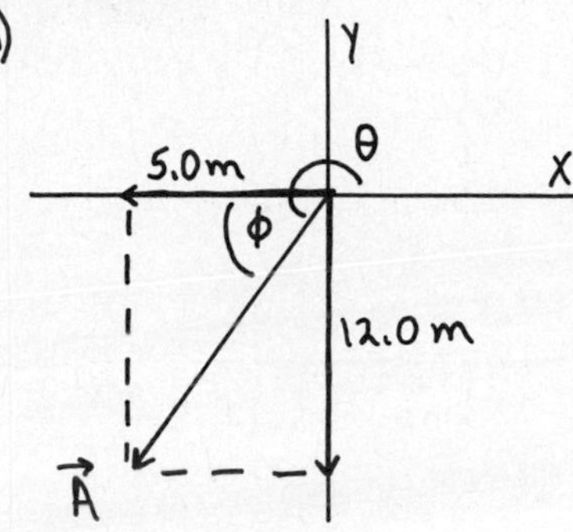

$$A = \sqrt{(5.0\,m)^2 + (12.0\,m)^2} = \underline{13.0\,m}$$

$$\tan\phi = \frac{12.0\,m}{5.0\,m} = 2.4 \Rightarrow \phi = 67.4°$$

$\theta = 180° + \phi = \underline{247°}$, measured counterclockwise from the positive x-axis.

c)

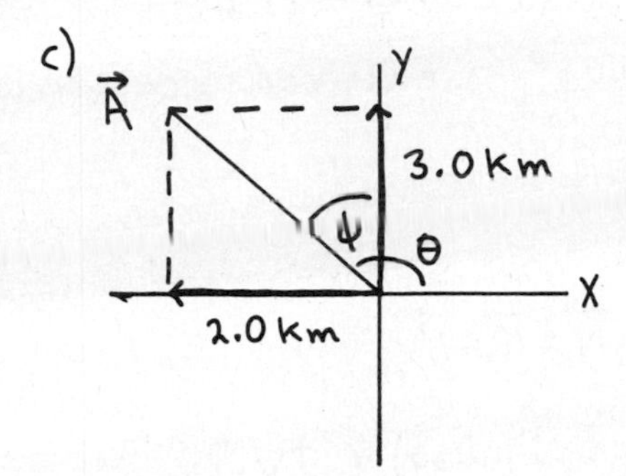

$$A = \sqrt{(2.0\,km)^2 + (3.0\,km)^2} = \underline{3.6\,km}$$

$$\tan\phi = \frac{2.0\,km}{3.0\,km} = 0.67 \Rightarrow \phi = 33.7°$$

$\theta = 90° + \phi = \underline{124°}$, measured counterclockwise from the positive x-axis.

1-29

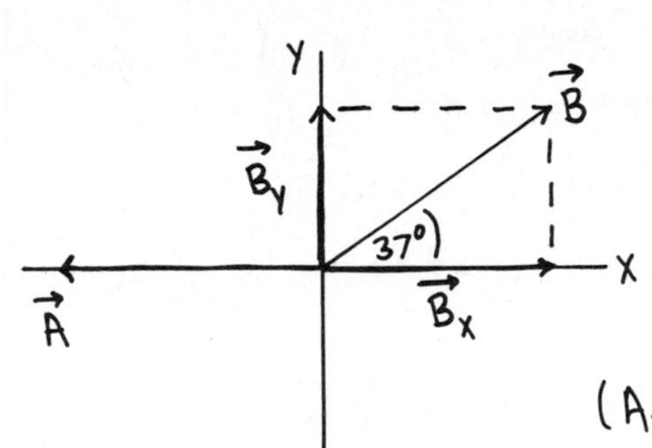

$$B_x = B\cos 37° = (20\,m)(0.799) = 16.0\,m$$
$$B_y = B\sin 37° = (20\,m)(0.602) = 12.0\,m$$

(Note that both B_x and B_y are positive.)

$$A_x = -A = -7.0\,m$$

(A_x is negative since $\vec{A}$ is in the negative x-direction.)

$$A_y = 0$$

a) We can use the above components of $\vec{A}$ and $\vec{B}$ to find the components of the vector $\vec{C}$ that is the sum of $\vec{A}$ and $\vec{B}$: $\vec{C} = \vec{A} + \vec{B}$.

Thus $C_x = A_x + B_x = -7.0\,m + 16.0\,m = 9.0\,m$

$C_y = A_y + B_y = 0 + 12.0\,m = 12.0\,m$

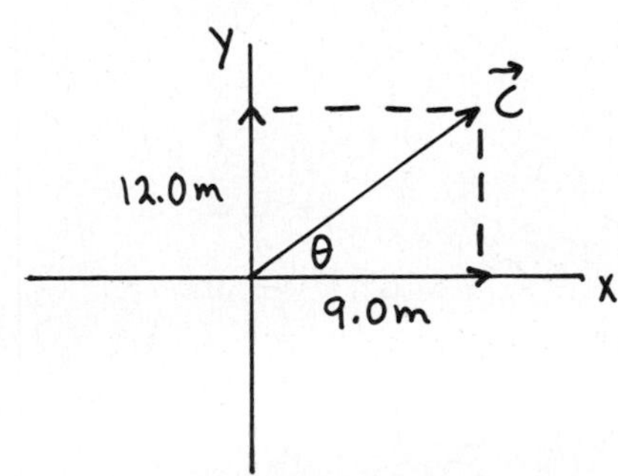

$$C = \sqrt{C_x^2 + C_y^2} = \sqrt{(9.0\,m)^2 + (12.0\,m)^2} = \underline{15.0\,m}$$

$\tan\theta = \frac{12.0\,m}{9.0\,m} = 1.33 \Rightarrow \underline{\theta = 53.1°}$, measured counterclockwise from the positive x axis.

1-29 (cont)

b) Let $\vec{D} = \vec{A} - \vec{B}$, then $D_x = A_x - B_x$ and $D_y = A_y - B_y$.

$$D_x = -7.0\,m - 16.0\,m = -23.0\,m$$
$$D_y = 0 - 12.0\,m = -12.0\,m$$

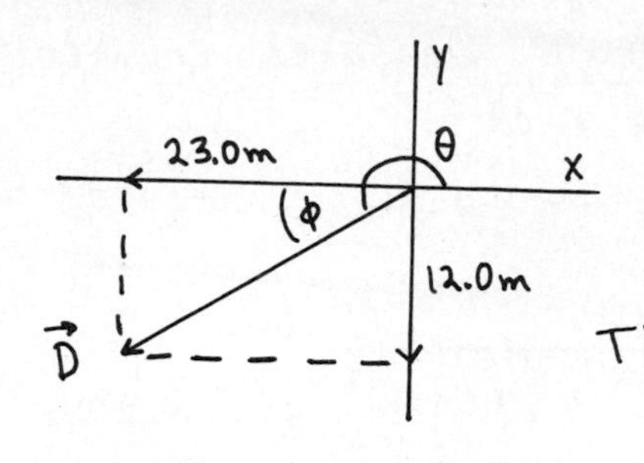

$$D = \sqrt{D_x^2 + D_y^2} = \sqrt{(-23.0m)^2 + (-12.0m)^2} = 25.9m$$

$$\tan\phi = \frac{12.0\,m}{23.0\,m} = 0.522 \Rightarrow \phi = 27.6°$$

Thus $\theta = 180° + \phi = \underline{208°}$, measured counterclockwise from the x-axis.

1-31

Let $\vec{R}$ be the resultant; $\vec{R} = \vec{M} + \vec{N} \Rightarrow \vec{N} = \vec{R} - \vec{M}$

a)

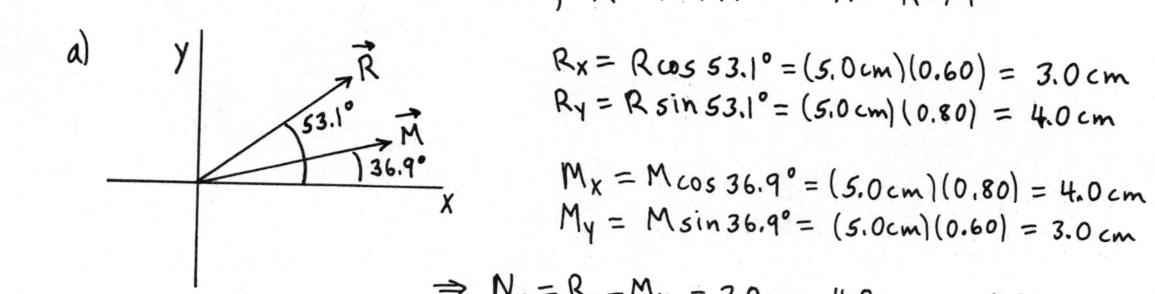

$$R_x = R\cos 53.1° = (5.0cm)(0.60) = 3.0cm$$
$$R_y = R\sin 53.1° = (5.0cm)(0.80) = 4.0cm$$

$$M_x = M\cos 36.9° = (5.0cm)(0.80) = 4.0cm$$
$$M_y = M\sin 36.9° = (5.0cm)(0.60) = 3.0cm$$

$$\Rightarrow N_x = R_x - M_x = 3.0cm - 4.0cm = \underline{-1.0\,cm}$$
$$N_y = R_y - M_y = 4.0\,cm - 3.0\,cm = \underline{+1.0\,cm}$$

b)

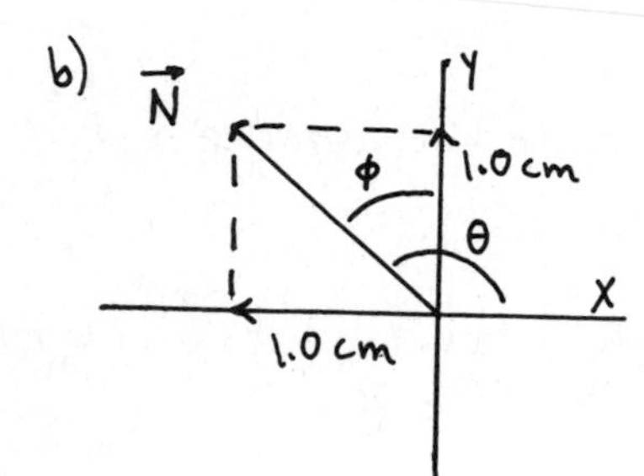

$$N = \sqrt{N_x^2 + N_y^2} = \sqrt{(-1.0cm)^2 + (1.0cm)^2} = \underline{1.41cm}$$

$$\tan\phi = \frac{1.0\,cm}{1.0\,cm} = 1.0 \Rightarrow \phi = 45°$$

$\theta = 90° + \phi = \underline{135°}$, measured counterclockwise from the positive x-axis.

1-33

a) $A = \sqrt{A_x^2 + A_y^2} = \sqrt{2^2 + 3^2} = \underline{3.61}$

$B = \sqrt{B_x^2 + B_y^2} = \sqrt{1^2 + (-2)^2} = \underline{2.24}$

b) $\vec{C} = \vec{A} + \vec{B}$; $\vec{C} = (2+1)\hat{i} + (3-2)\hat{j}$

$$\underline{\vec{C} = 3\hat{i} + \hat{j}}$$

1-33 (cont)

c)

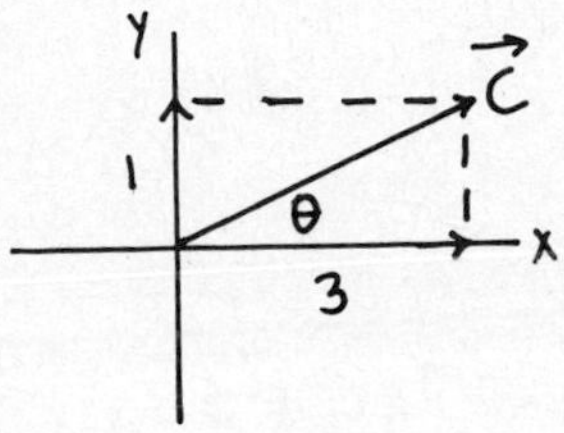

$$C = \sqrt{C_x^2 + C_y^2}$$

$$C = \sqrt{3^2 + 1^2} = \underline{3.16}$$

$$\tan\theta = \tfrac{1}{3} = 0.333$$

$$\theta = \underline{18.4^\circ}$$

d) $\vec{D} = \vec{A} - \vec{B}$; $\vec{D} = (2-1)\vec{i} + (3-(-2))\vec{j}$

$$\underline{\vec{D} = \vec{i} + 5\vec{j}}$$

e)

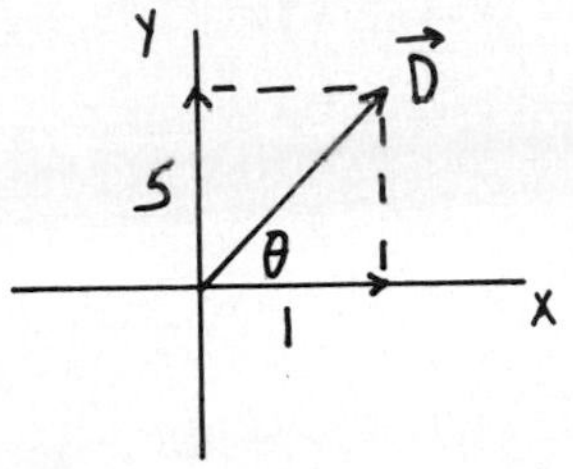

$$D = \sqrt{D_x^2 + D_y^2}$$

$$D = \sqrt{5^2 + 1^2} = \underline{5.10}$$

$$\tan\theta = \tfrac{5}{1} = 5$$

$$\theta = \underline{78.7^\circ}$$

1-37

Use the method of Example 1-5.

$\vec{A}\cdot\vec{B} = AB\cos\theta$

$\vec{A}\cdot\vec{B} = A_xB_x + A_yB_y$

$\Rightarrow$ $AB\cos\theta = A_xB_x + A_yB_y$

$$\boxed{\cos\theta = \frac{A_xB_x + A_yB_y}{AB}}$$

$\vec{A} = -2\vec{i} + 5\vec{j} \Rightarrow A = \sqrt{A_x^2 + A_y^2} = \sqrt{(-2)^2 + 5^2} = 5.39$

$\vec{B} = 3\vec{i} - \vec{j} \Rightarrow B = \sqrt{B_x^2 + B_y^2} = \sqrt{3^2 + (-1)^2} = 3.16$

$$\Rightarrow \cos\theta = \frac{(-2)(3) + (5)(-1)}{(5.39)(3.16)} = \frac{-11}{(5.39)(3.16)} = -0.646 \Rightarrow \theta = \underline{130^\circ}$$

1-41

a) $\vec{A} = 2\vec{i} - 5\vec{j}$; $\vec{B} = 5\vec{i} + 2\vec{j}$

$\vec{A}\cdot\vec{B} = (2)(5) + (-5)(2) = \underline{0}$

$\vec{A}\times\vec{B} = (2\vec{i} - 5\vec{j})\times(5\vec{i} + 2\vec{j})$; $\vec{i}\times\vec{i} = \vec{j}\times\vec{j} = 0$; $\vec{i}\times\vec{j} = \vec{k}$; $\vec{j}\times\vec{i} = -\vec{k}$

Thus $\vec{A}\times\vec{B} = 10\underbrace{\vec{i}\times\vec{i}}_{0} + 4\underbrace{\vec{i}\times\vec{j}}_{\vec{k}} - 25\underbrace{\vec{j}\times\vec{i}}_{-\vec{k}} - 10\underbrace{\vec{j}\times\vec{j}}_{0}$

$$\vec{A}\times\vec{B} = 29\vec{k}$$

1-41 (cont)

b) $\vec{A} = 2\vec{i} - 5\vec{j}$; $\vec{B} = 4\vec{i} - 10\vec{j}$

$\vec{A}\cdot\vec{B} = (2)(4) + (-5)(-10) = \underline{58}$

$$\vec{A}\times\vec{B} = (2\vec{i} - 5\vec{j})\times(4\vec{i} - 10\vec{j}) = 8\underbrace{\vec{i}\times\vec{i}}_{0} - 20\underbrace{\vec{i}\times\vec{j}}_{\vec{k}} - 20\underbrace{\vec{j}\times\vec{i}}_{-\vec{k}} + 50\underbrace{\vec{j}\times\vec{j}}_{0}$$

$\vec{A}\times\vec{B} = -20\vec{k} + 20\vec{k} = \underline{0}$

Note: $\vec{B} = 2\vec{A}$. Thus $\vec{A}$ and $\vec{B}$ are parallel, and we know from this that their cross product is zero.

1-43

a)

$$\vec{C} = \begin{vmatrix} \vec{i} & \vec{j} & \vec{k} \\ A_x & A_y & A_z \\ B_x & B_y & B_z \end{vmatrix}$$

Expand the determinant by the first row:

$$\vec{C} = \begin{vmatrix} \vec{i} & \vec{j} & \vec{k} \\ A_x & A_y & A_z \\ B_x & B_y & B_z \end{vmatrix} = \begin{vmatrix} A_y & A_z \\ B_y & B_z \end{vmatrix}\vec{i} - \begin{vmatrix} A_x & A_z \\ B_x & B_z \end{vmatrix}\vec{j} + \begin{vmatrix} A_x & A_y \\ B_x & B_y \end{vmatrix}\vec{k}$$

$$\vec{C} = (A_yB_z - A_zB_y)\vec{i} - (A_xB_z - A_zB_x)\vec{j} + (A_xB_y - A_yB_x)\vec{k}$$

$$\vec{C} = (A_yB_z - A_zB_y)\vec{i} + (A_zB_x - A_xB_z)\vec{j} + (A_xB_y - A_yB_x)\vec{k}$$

Comparing this to $\vec{C} = C_x\vec{i} + C_y\vec{j} + C_z\vec{k}$ implies

$C_x = A_yB_z - A_zB_y$, $C_y = A_zB_x - A_xB_z$, $C_z = A_xB_y - A_yB_x$, which is eq. (1-29).

b) First compute $(\vec{A}\times\vec{B})\cdot\vec{C}$ in terms of components:

$$\vec{A}\times\vec{B} = (A_yB_z - A_zB_y)\vec{i} + (A_zB_x - A_xB_z)\vec{j} + (A_xB_y - A_yB_x)\vec{k}$$

Thus $(\vec{A}\times\vec{B})\cdot\vec{C} = (A_yB_z - A_zB_y)C_x + (A_zB_x - A_xB_z)C_y + (A_xB_y - A_yB_x)C_z$.

Now show that this equals the determinant:
Expand the determinant along the 3rd row:

$$\begin{vmatrix} A_x & A_y & A_z \\ B_x & B_y & B_z \\ C_x & C_y & C_z \end{vmatrix} = C_x\begin{vmatrix} A_y & A_z \\ B_y & B_z \end{vmatrix} - C_y\begin{vmatrix} A_x & A_z \\ B_x & B_z \end{vmatrix} + C_z\begin{vmatrix} B_x & B_y \\ C_x & C_y \end{vmatrix}$$

1-43 (cont)

$$= C_x (A_y B_z - A_z B_y) - C_y (A_x B_z - A_z B_x) + C_z (B_x C_y - B_y C_x)$$

$$= (A_y B_z - A_z B_y) C_x + (A_z B_x - A_x B_z) C_y + (B_x C_y - B_y C_x) C_z,$$

which does equal what we got for $(\vec{A} \times \vec{B}) \cdot \vec{C}$.

Problems

1-45

Use a coordinate system where east is in the +x-direction and north is in the +y-direction.

Let $\vec{D}$ be the unknown 4th displacement.
Then the resultant displacement is $\vec{R} = \vec{A} + \vec{B} + \vec{C} + \vec{D}$. And since he ends up back where he started, $\vec{R} = 0$.

$$0 = \vec{A} + \vec{B} + \vec{C} + \vec{D} \Rightarrow \vec{D} = -(\vec{A} + \vec{B} + \vec{C}); \quad D_x = -(A_x + B_x + C_x)$$
$$D_y = -(A_y + B_y + C_y)$$

Y, (N) — $\vec{B}$ 50m, 30° — $\vec{A}$ 100m — X (E) — (W) — $\vec{C}$ 150m, 45° — (S)

$A_x = 100\text{ m}$
$A_y = 0$

$B_x = -(50\text{m}) \sin 30° = -25\text{ m}$
$B_y = +(50\text{m}) \cos 30° = 43.3\text{m}$

$C_x = -(150\text{m}) \sin 45° = -106\text{m}$
$C_y = -(150\text{m}) \cos 45° = -106\text{m}$

$$D_x = -(A_x + B_x + C_x) = -(100\text{m} - 25\text{ m} - 106\text{m}) = +31\text{ m}$$
$$D_y = -(A_y + B_y + C_y) = -(0 + 43.3\text{m} - 106\text{m}) = +63\text{ m}$$

(N) Y — $\vec{D}$ — 63m — ϕ — 31m — X (E) — (W) — (S)

$$D = \sqrt{D_x^2 + D_y^2} = \sqrt{(31\text{m})^2 + (63\text{m})^2} = \underline{70\text{m}}$$

$$\tan \phi = \frac{31\text{m}}{63\text{ m}} = 0.492 \Rightarrow \phi = 26.2°$$

The direction of $\vec{D}$ is <u>26.2° E of N</u>.

(Or, equivalently, 63.8° N of E.)

1-47

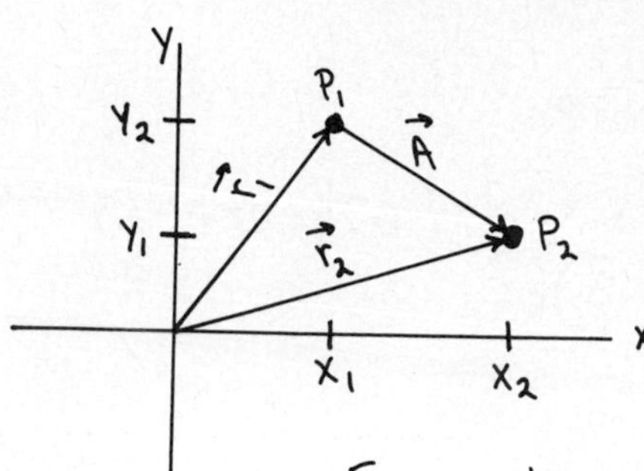

Let the vectors $\vec{r}_1$ and $\vec{r}_2$ be the position vectors for P_1 and P_2; $\vec{r}_1$ and $\vec{r}_2$ go from the origin to P_1 and P_2.

Then $\vec{r}_1 = x_1\vec{i} + y_1\vec{j}$

$\vec{r}_2 = x_2\vec{i} + y_2\vec{j}$

From the sketch, $\vec{r}_2 = \vec{r}_1 + \vec{A} \Rightarrow \vec{A} = \vec{r}_2 - \vec{r}_1$

$$\vec{A} = (x_2\vec{i} + y_2\vec{j}) - (x_1\vec{i} + y_1\vec{j}) = (x_2 - x_1)\vec{i} + (y_2 - y_1)\vec{j}$$

Comparing this expression to $\vec{A} = A_x\vec{i} + A_y\vec{j}$ gives

$A_x = x_2 - x_1$ and $A_y = y_2 - y_1$, as was to be shown.

Magnitude of $\vec{A}$:

$$A = \sqrt{A_x^2 + A_y^2} = \sqrt{(x_2 - x_1)^2 + (y_2 - y_1)^2}$$

Direction of $\vec{A}$:

$$\tan\theta = \frac{A_y}{A_x} = \frac{y_2 - y_1}{x_2 - x_1} \Rightarrow \theta = \tan^{-1}\left(\frac{y_2 - y_1}{x_2 - x_1}\right),$$ where θ is measured counterclockwise from the x-axis.

1-49

a) Choose the x-axis to be in the direction of $\vec{A}$; this makes the analysis much simpler.

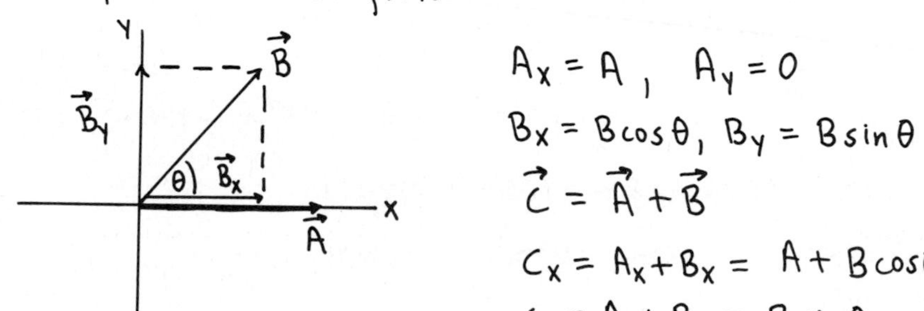

$A_x = A$, $A_y = 0$

$B_x = B\cos\theta$, $B_y = B\sin\theta$

$\vec{C} = \vec{A} + \vec{B}$

$C_x = A_x + B_x = A + B\cos\theta$

$C_y = A_y + B_y = B\sin\theta$

$$C = \sqrt{C_x^2 + C_y^2} = \sqrt{(A + B\cos\theta)^2 + (B\sin\theta)^2} = \sqrt{A^2 + 2AB\cos\theta + B^2\cos^2\theta + B^2\sin^2\theta}$$

$$C = \sqrt{A^2 + 2AB\cos\theta + B^2(\sin^2\theta + \cos^2\theta)}$$

But $\sin^2\theta + \cos^2\theta = 1 \Rightarrow C = \sqrt{A^2 + 2AB\cos\theta + B^2}$, as was to be shown.

b) Let $B = A$ in the expression derived in (a)

$$\Rightarrow C = \sqrt{A^2 + 2A^2\cos\theta + A^2} = \sqrt{2A^2(1 + \cos\theta)} = A\sqrt{2(1 + \cos\theta)}$$

1-49 (cont)

$$C = A \Rightarrow \sqrt{2(1+\cos\theta)} = 1$$

$$2(1+\cos\theta) = 1 \Rightarrow 1+\cos\theta = \tfrac{1}{2} \Rightarrow \cos\theta = -\tfrac{1}{2}$$

This occurs for $\theta = \underline{120^\circ}$ or $\underline{240^\circ}$.

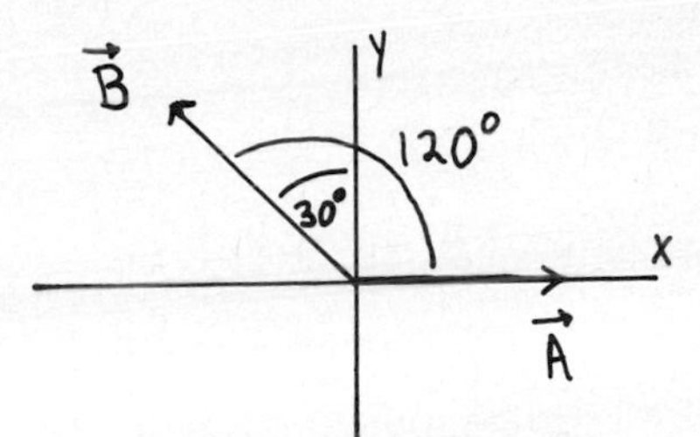

or

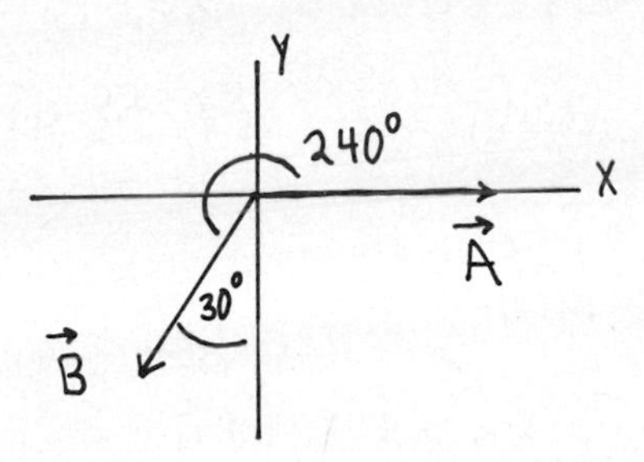

c) $\vec{D} = \vec{A} - \vec{B} \Rightarrow D_x = A_x - B_x = A - B\cos\theta$

$D_y = A_y - B_y = -B\sin\theta$

$$D = \sqrt{D_x^2 + D_y^2} = \sqrt{(A - B\cos\theta)^2 + (-B\sin\theta)^2} = \sqrt{A^2 - 2AB\cos\theta + B^2\cos^2\theta + B^2\sin^2\theta}$$

$$D = \sqrt{A^2 - 2AB\cos\theta + B^2\underbrace{(\sin^2\theta + \cos^2\theta)}_{=1}}$$

$$\Rightarrow D = \sqrt{A^2 - 2AB\cos\theta + B^2}$$

d) $B = A \Rightarrow \quad D = \sqrt{A^2 - 2A^2\cos\theta + A^2} = A\sqrt{2(1-\cos\theta)}$

$D = A \Rightarrow 2(1-\cos\theta) = 1$

$\cos\theta = \tfrac{1}{2}$; this occurs for $\theta = \underline{60^\circ}$ or $\underline{300^\circ}$

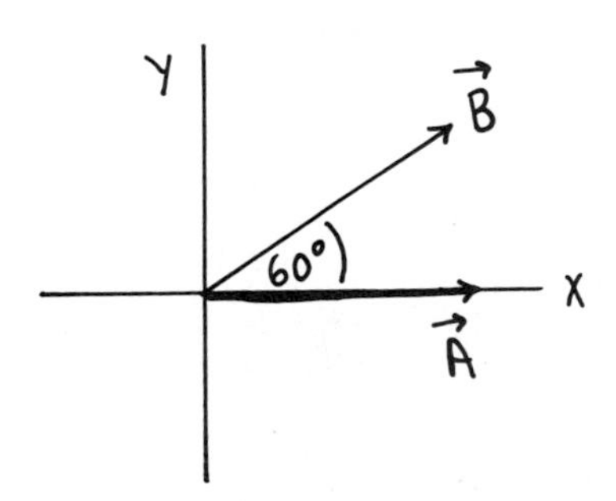

or

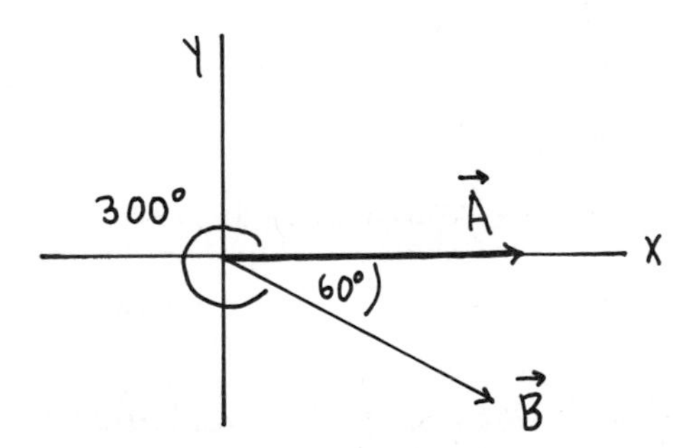

CHAPTER 2

Exercises 3, 5, 7, 13, 15, 19, 21, 23, 27, 29, 31

Problems 37, 39, 41, 45, 47, 49

Exercises

2-3

$$x = at + bt^2$$

$$v_{av} = \frac{\Delta x}{\Delta t} = \frac{x_2 - x_1}{t_2 - t_1}$$

$t_1 = 0,\ t_2 = 1\,s$

at $t_1 = 0$, $x_1 = 0$

at $t_2 = 1\,s$, $x_2 = (8\,cm\cdot s^{-1})(1\,s) + (-3\,cm\cdot s^{-2})(1\,s)^2 = 8\,cm - 3\,cm = 5\,cm$

$$v_{av} = \frac{5\,cm - 0}{1\,s - 0} = \underline{5.0\,cm\cdot s^{-1}}$$

$t_1 = 0,\ t_2 = 4\,s$

at $t_2 = 4\,s$, $x_2 = (8\,cm\cdot s^{-1})(4\,s) + (-3\,cm\cdot s^{-2})(4\,s)^2 = 32\,cm - 48\,cm = -16\,cm$

$$v_{av} = \frac{-16\,cm - 0}{4\,s - 0} = \underline{-4.0\,cm\cdot s^{-1}}$$

2-5

$$x = bt^2$$

$$v = \frac{dx}{dt} = 2bt$$

At $t = 3\,s$, $v = 2(10\,cm\cdot s^{-2})(3\,s) = \underline{60\,cm\cdot s^{-1}}$.

This may also be calculated using the equivalent expression $v = \lim\limits_{\Delta t \to 0} \frac{\Delta x}{\Delta t}$.

Let $t_1 = 3\,s$, $t_2 = 3\,s + \Delta t \Rightarrow t_2 - t_1 = \Delta t$.

$x_1 = bt_1^2 = (10\,cm\cdot s^{-2})(3\,s)^2 = 90\,cm$

$x_2 = bt_2^2 = (10\,cm\cdot s^{-2})(3\,s + \Delta t)^2 = 90\,cm + (60\,cm\cdot s^{-1})\Delta t + (10\,cm\cdot s^{-2})(\Delta t)^2$

$\Rightarrow x_2 - x_1 = \cancel{90\,cm} + (60\,cm\cdot s^{-1})\Delta t + (10\,cm\cdot s^{-2})(\Delta t)^2 - \cancel{90\,cm} = (60\,cm\cdot s^{-1})\Delta t + (10\,cm\cdot s^{-2})(\Delta t)^2$

$$v = \lim_{\Delta t \to 0}\left[\frac{(60\,cm\cdot s^{-1})\Delta t + (10\,cm\cdot s^{-2})(\Delta t)^2}{\Delta t}\right] = \lim_{\Delta t \to 0}\left[60\,cm\cdot s^{-1} + (10\,cm\cdot s^{-2})\Delta t\right]$$

$v = 60\,cm\cdot s^{-1}$, in agreement with the above.

2-7

a) $a_{av} = \frac{\Delta v}{\Delta t}$

time interval	Δv	a_{av}
$0 \to 2s$	0	0
$2s \to 4s$	$2\ m \cdot s^{-1}$	$1.0\ m \cdot s^{-2}$
$4s \to 6s$	$3\ m \cdot s^{-1}$	$1.5\ m \cdot s^{-2}$
$6s \to 8s$	$5\ m \cdot s^{-1}$	$2.5\ m \cdot s^{-2}$
$8s \to 10s$	$5\ m \cdot s^{-1}$	$2.5\ m \cdot s^{-2}$
$10s \to 12s$	$5\ m \cdot s^{-1}$	$2.5\ m \cdot s^{-2}$
$12s \to 14s$	$2\ m \cdot s^{-1}$	$1.0\ m \cdot s^{-2}$
$14s \to 16s$	0	0

a is not constant over the entire 0 to 16 s interval. It is constant from 6 s to 12 s.

b)

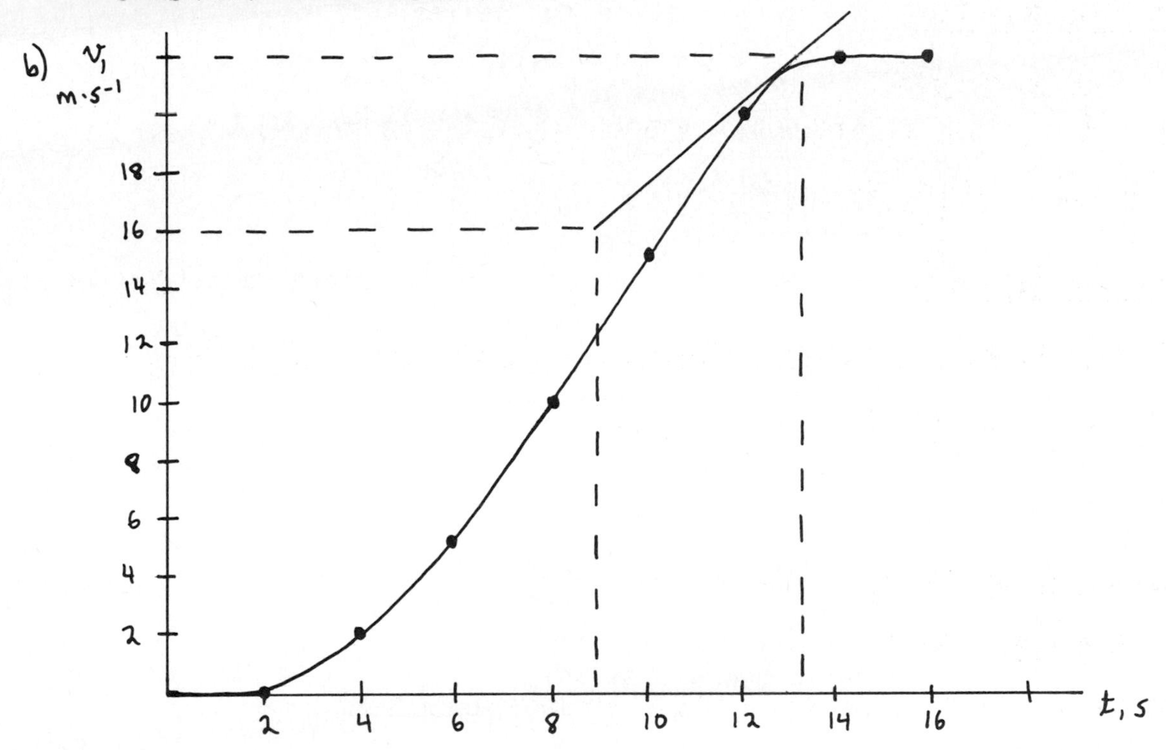

I have sketched a smooth curve through the points.

At $t = 8s$ the graph is a straight line. The slope is $\frac{v(t=10s) - v(t=6s)}{10s - 6s} = \frac{15\ m \cdot s^{-1} - 5\ m \cdot s^{-1}}{4s} = 2.5\ m \cdot s^{-2}$. Thus at $t = 8s$, $\underline{a = 2.5\ m \cdot s^{-2}}$.

At $t = 13s$ I have sketched an approximate tangent to the curve. The slope of this tangent line is $\frac{22.0\ m \cdot s^{-1} - 16\ m \cdot s^{-1}}{13.1\ s - 8.8\ s} = 1.3\ m \cdot s^{-2}$. Thus at $t = 13s$, $\underline{a = 1.3\ m \cdot s^{-2}}$.
(This is an approximate calculation, the actual value you get can vary somewhat, depending on how you draw the smooth curve and your eyeball estimate of the

2-7 (cont)

tangent to the curve.)

At $t = 15\,s$ the graph is a horizontal straight line, with zero slope. Hence at $t = 15\,s$, $\underline{a = 0}$.

2-13

Take the direction in which the car is moving to be positive.

a) $v_0 = ?$
$v = 15\,m \cdot s^{-1}$
$x - x_0 = 60\,m$
$t = 6\,s$

The desired equation is one that contains no a. We can obtain such an equation by combining the two eqs. $v = v_0 + at$ and $x - x_0 = v_0 t + \frac{1}{2} a t^2$.

$$v = v_0 + at \Rightarrow a = \frac{v - v_0}{t}$$

Use this in the other eq. to replace a

$$\Rightarrow x - x_0 = v_0 t + \frac{1}{2}\left(\frac{v - v_0}{t}\right) t^2$$

$$x - x_0 = v_0 t + \frac{1}{2} v t - \frac{1}{2} v_0 t = \frac{1}{2}(v_0 + v) t$$

$$\boxed{x - x_0 = \left(\frac{v_0 + v}{2}\right) t}$$

We can hereafter use this eq. as a fourth constant acceleration eq.

Note that by comparison to $x - x_0 = v_{av} t$, we have $v_{av} = \left(\frac{v_0 + v}{2}\right)$ for motion with constant acceleration.

Now back to the problem:

$$x - x_0 = \left(\frac{v_0 + v}{2}\right) t$$

Solve for $v_0 \Rightarrow v_0 = \frac{2(x - x_0)}{t} - v = \frac{2(60\,m)}{6\,s} - 15\,m \cdot s^{-1} = 20\,m \cdot s^{-1} - 15\,m \cdot s^{-1} = \underline{5\,m \cdot s^{-1}}$

b) $a = ?$

$v = v_0 + at \Rightarrow a = \frac{v - v_0}{t} = \frac{15\,m \cdot s^{-1} - 5\,m \cdot s^{-1}}{6\,s} = \underline{1.67\,m \cdot s^{-2}}$

2-15

a) $a(t)$ is the slope of the tangent to the v versus t curve.

At $\underline{t = 3\,s}$ the v versus t curve is a horizontal straight line $\Rightarrow \underline{a = 0}$.

At $\underline{t = 7\,s}$ the curve is a straight-line segment with slope $\frac{45\,m \cdot s^{-1} - 20\,m \cdot s^{-1}}{9\,s - 5\,s} = \underline{6.25\,m \cdot s^{-2}}$. Thus $a = 6.25\,m \cdot s^{-2}$.

2-15 (cont)

At $t = 11s$ the curve is again a straight-line segment, now with slope $\frac{0 - 45\,m\cdot s^{-1}}{13s - 9s} = -11.2\,m\cdot s^{-2}$. Thus $a = -11.2\,m\cdot s^{-2}$.

b) For the interval $t=0$ to $t=5s$ the acceleration is constant and equal to zero.
For the interval $t=5s$ to $t=9s$ the acceleration is constant and equal to $6.25\,m\cdot s^{-2}$.
For the interval $t=9s$ to $t=13s$ the acceleration is constant and equal to $-11.2\,m\cdot s^{-2}$.

During the first 5 seconds the acceleration is constant, so the constant acceleration kinematic formulas can be used.

$v_0 = 20\,m\cdot s^{-1}$
$a = 0$
$t = 5s$
$x - x_0 = ?$

$x - x_0 = v_0 t + \frac{1}{2}at^2$ (with $\frac{1}{2}at^2 \rightarrow 0$)

$x - x_0 = (20\,m\cdot s^{-1})(5s) = 100\,m$, this is the distance the officer travels in the first 5 seconds.

During the interval $t = 5s$ to $9s$ the acceleration is again constant. The constant acceleration formulas can be applied for this 4 second interval. (Note that the acceleration is not constant for $t=0$ to $t=9s$.)

$v_0 = 20\,m\cdot s^{-1}$ (at $t = 5s$)
$a = 6.25\,m\cdot s^{-2}$
$t = 4s$
$x_0 = 100\,m$ (at $t = 5s$)
$x - x_0 = ?$

$$x - x_0 = v_0 t + \frac{1}{2}at^2$$
$$x - x_0 = (20\,m\cdot s^{-1})(4s) + \frac{1}{2}(6.25\,m\cdot s^{-2})(4s)^2$$
$$x - x_0 = 80\,m + 50\,m = 130\,m$$

Thus $x = x_0 + 130\,m = 230\,m$.

At $t = 9s$ the officer is at $x = 230\,m$, so she has traveled 230m in the first 9 seconds.

During the interval $t=9s$ to $t = 13s$ the acceleration is again constant. The constant acceleration formulas can be applied for this 4 second interval, but not for the whole $t=0$ to $t=13s$ interval.

$v_0 = 45\,m\cdot s^{-1}$ (at $t = 9s$)
$a = -11.2\,m\cdot s^{-2}$
$t = 4s$
$x_0 = 230\,m$ (at $t = 9s$)
$x - x_0 = ?$

$$x - x_0 = v_0 t + \frac{1}{2}at^2$$
$$x - x_0 = (45\,m\cdot s^{-1})(4s) + \frac{1}{2}(-11.2\,m\cdot s^{-2})(4s)^2$$
$$x - x_0 = 180\,m - 89.6\,m = 90.4\,m$$

Thus $x = x_0 + 90.4\,m = 320\,m$.

At $t = 13s$ the officer is at $t = 320\,m$, so she has traveled 320 m in the first 13 seconds.

2-19

a) The maximum speed occurs at the end of the initial acceleration period.

$a = 10\,m\cdot s^{-2}$

$t = 10\,min = 600\,s$

$v = ?$

$v_0 = 0$

$$v = v_0 + at$$

$$v = 0 + (10\,m\cdot s^{-2})(600\,s) = \underline{6000\,m\cdot s^{-1}}$$

b) The motion consists of three constant acceleration intervals. In the middle part of the trip $a=0$ and $v = 6000\,m\cdot s^{-1}$, but we can't directly find the distance traveled during this part of the trip because we don't know the time. Instead, find the distance traveled in the first ($a = 10\,m\cdot s^{-2}$) part of the trip and in the last ($a = -10\,m\cdot s^{-2}$) part of the trip. Subtract these distances from the total distance of 4.0×10^5 km to find the distance traveled in the middle ($a=0$) part of the trip.

first segment

$x - x_0 = ?$

$t = 10\,min = 600\,s$

$a = +10\,m\cdot s^{-2}$

$v_0 = 0$

$v = 6000\,m\cdot s^{-1}$

$$x - x_0 = \cancel{v_0 t}^{\,0} + \tfrac{1}{2}at^2$$

$$x - x_0 = \tfrac{1}{2}(10\,m\cdot s^{-2})(600\,s)^2 = 1.8 \times 10^6\,m = 1.8 \times 10^3\,km$$

third segment

$x - x_0 = ?$

$t = 10\,min = 600\,s$

$a = -10\,m\cdot s^{-2}$

$v_0 = 6000\,m\cdot s^{-1}$

$v = 0$

$$x - x_0 = v_0 t + \tfrac{1}{2}at^2$$

$$x - x_0 = (6000\,m\cdot s^{-1})(600\,s) + \tfrac{1}{2}(-10\,m\cdot s^{-2})(600\,s)^2$$

$$x - x_0 = 3.6 \times 10^6\,m - 1.8 \times 10^6\,m = 1.8 \times 10^6\,m = 1.8 \times 10^3\,km$$

Therefore the distance traveled at constant speed is

$$4.0 \times 10^5\,km - 2(1.8 \times 10^3\,km) = 3.964 \times 10^5\,km$$

The fraction is $\dfrac{3.964 \times 10^5\,km}{4.00 \times 10^5\,km} = 0.991.$

c) Find the time for the constant speed segment:

$x - x_0 = 3.964 \times 10^5\,km = 3.964 \times 10^8\,m$

$v = 6000\,m\cdot s^{-1}$

$a = 0$

$$t = \frac{x - x_0}{v_0} = \frac{3.964 \times 10^8\,m}{6000\,m\cdot s^{-1}} = 6.61 \times 10^4\,\cancel{s}\left(\frac{1\,min}{60\,\cancel{s}}\right) = 1.10 \times 10^3\,min$$

The total time for the whole trip is thus

$$10\,min + 1100\,min + 10\,min = 1120\,min, \text{ or } \underline{18.7\,hr.}$$

2-21

$a = (1.2\,\text{m}\cdot\text{s}^{-3})\,t - (0.12\,\text{m}\cdot\text{s}^{-4})\,t^2$; at $t=0$, $x=0$ and $v=0$.

a) $v_2 - v_1 = \int_{t_1}^{t_2} a\,dt$

Let $t_1 = 0$ and t_2 be any later time t.

$$v - v_0 = \int_0^t a\,dt = \int_0^t \left[(1.2\,\text{m}\cdot\text{s}^{-3})t - (0.12\,\text{m}\cdot\text{s}^{-4})t^2\right] dt$$

$$v - v_0 = (1.2\,\text{m}\cdot\text{s}^{-3}) \int_0^t t\,dt - (0.12\,\text{m}\cdot\text{s}^{-4}) \int_0^t t^2\,dt$$

$$v - v_0 = (1.2\,\text{m}\cdot\text{s}^{-3})\left(\tfrac{1}{2}t^2\right) - (0.12\,\text{m}\cdot\text{s}^{-4})\left(\tfrac{1}{3}t^3\right) = (0.6\,\text{m}\cdot\text{s}^{-3})t^2 - (0.04\,\text{m}\cdot\text{s}^{-4})t^3$$

$$v_0 = 0 \Rightarrow \boxed{v(t) = (0.6\,\text{m}\cdot\text{s}^{-3})t^2 - (0.04\,\text{m}\cdot\text{s}^{-4})\,t^3}$$

Now that know $v(t)$ can do the integral for $x(t)$:

$x_2 - x_1 = \int_{t_1}^{t_2} v\,dt$

Let $t_1 = 0$ and t_2 be any later time t.

$$x - x_0 = \int_0^t v\,dt = \int_0^t \left[(0.6\,\text{m}\cdot\text{s}^{-3})t^2 - (0.04\,\text{m}\cdot\text{s}^{-4})t^3\right] dt$$

$$x - x_0 = (0.6\,\text{m}\cdot\text{s}^{-3}) \int_0^t t^2\,dt - (0.04\,\text{m}\cdot\text{s}^{-4}) \int_0^t t^3\,dt$$

$$x - x_0 = (0.6\,\text{m}\cdot\text{s}^{-3})\left(\tfrac{1}{3}t^3\right) - (0.04\,\text{m}\cdot\text{s}^{-4})\left(\tfrac{1}{4}t^4\right) = (0.2\,\text{m}\cdot\text{s}^{-3})t^3 - (0.01\,\text{m}\cdot\text{s}^{-4})t^4$$

$$x_0 = 0 \Rightarrow \boxed{x(t) = (0.2\,\text{m}\cdot\text{s}^{-3})t^3 - (0.01\,\text{m}\cdot\text{s}^{-4})t^4}$$

b) At the time t when $v(t)$ is maximum, $\frac{dv}{dt} = 0$. (Since $a = \frac{dv}{dt}$, the maximum velocity is when $a = 0$. For earlier times a is positive so v is still increasing. For later times a is negative and v is decreasing.)

Use this equation to solve for t at which $v(t)$ is maximum:

$$a = \frac{dv}{dt} = 0 \Rightarrow (1.2\,\text{m}\cdot\text{s}^{-3})t - (0.12\,\text{m}\cdot\text{s}^{-4})t^2 = 0$$

$$t = \frac{1.2\,\text{m}\cdot\text{s}^{-3}}{0.12\,\text{m}\cdot\text{s}^{-4}} = 10.0\,\text{s}$$

Now calculate v at $t = 10\,\text{s}$, from the $v(t)$ calculated in part (a):

$$v = (0.6\,\text{m}\cdot\text{s}^{-3})(10\,\text{s})^2 - (0.04\,\text{m}\cdot\text{s}^{-4})(10\,\text{s})^3$$

$$v = 60\,\text{m}\cdot\text{s}^{-1} - 40\,\text{m}\cdot\text{s}^{-1} = \underline{20\,\text{m}\cdot\text{s}^{-1}}$$

2-23

a) Take upward to be positive. At the maximum height $v=0$. So, taking the final position to be at the maximum height and the initial position to be just after the ball has left the thrower's hand:

$y-y_0 = 20\text{ m}$

$a = -9.8\text{ m}\cdot\text{s}^{-2}$

$v = 0$

$v_0 = ?$

$$\cancel{v^2}^{\,0} = v_0^2 + 2a(y-y_0)$$

$$v_0 = \sqrt{-2a(y-y_0)} = \sqrt{-2(-9.8\text{ m}\cdot\text{s}^{-2})(20\text{ m})} = \underline{19.8\text{ m}\cdot\text{s}^{-1}}$$

b) Take upward to be positive, the initial time to be just after the ball is thrown, and the final time to be when the ball returns to this initial position on its way back down. Then $y-y_0=0$.

$y-y_0 = 0$

$v_0 = 19.8\text{ m}\cdot\text{s}^{-1}$

$a = -9.8\text{ m}\cdot\text{s}^{-2}$

$t = ?$

$$\cancel{y-y_0}^{\,0} = v_0 t + \tfrac{1}{2}at^2$$

$$0 = v_0 t + \tfrac{1}{2}at^2$$

$$t = -\frac{2v_0}{a} = -\frac{2(19.8\text{ m}\cdot\text{s}^{-1})}{-9.8\text{ m}\cdot\text{s}^{-2}} = \underline{4.04\text{ s}}$$

Alternatively, calculate the time to go up to maximum height:

$v = 0$ (at max. height)

$v_0 = 19.8\text{ m}\cdot\text{s}^{-1}$

$a = -9.8\text{ m}\cdot\text{s}^{-2}$

$t = ?$

$$v = v_0 + at$$

$$t = \frac{v-v_0}{a} = \frac{0-19.8\text{ m}\cdot\text{s}^{-1}}{-9.8\text{ m}\cdot\text{s}^{-2}} = 2.02\text{ s}$$

Then calculate the time to come from the maximum height back down:

$v_0 = 0$ (at max height)

$y-y_0 = -20\text{ m}$

$a = -9.8\text{ m}\cdot\text{s}^{-2}$

$t = ?$

$$y-y_0 = \cancel{v_0 t}^{\,0} + \tfrac{1}{2}at^2$$

$$t = \sqrt{\frac{2(y-y_0)}{a}} = \sqrt{\frac{2(-20\text{ m})}{-9.8\text{ m}\cdot\text{s}^{-2}}} = 2.02\text{ s}$$

The total time in the air is then $2.02\text{ s} + 2.02\text{ s} = \underline{4.04\text{ s}}$.

2-27

Take upward to be positive.

a) Consider the motion from the initial point to 5 s later:

$y-y_0 = -160\text{ ft}$ (ball is <u>below</u> its starting point)

$t = 5\text{ s}$

$a = -32\text{ ft}\cdot\text{s}^{-2}$ (note units)

$v_0 = ?$

$$y-y_0 = v_0 t + \tfrac{1}{2}at^2$$

$$v_0 = \frac{y-y_0}{t} - \tfrac{1}{2}at$$

$$v_0 = \frac{-160\text{ ft}}{5\text{ s}} - \tfrac{1}{2}(-32\text{ ft}\cdot\text{s}^{-2})(5\text{ s})$$

$$v_0 = -32\text{ ft}\cdot\text{s}^{-1} + 80\text{ ft}\cdot\text{s}^{-1} = \underline{48\text{ ft}\cdot\text{s}^{-1}}$$

2-27 (cont)

b) Consider the motion from the starting point to the maximum height: At the maximum height $v=0$.

$v=0$

$a=-32\,ft\cdot s^{-2}$

$v_0 = 48\,ft\cdot s^{-1}$ (from part (a))

$y-y_0 = ?$

$$v^2 = v_0^2 + 2a(y-y_0)$$

$$y-y_0 = \frac{v^2-v_0^2}{2a} = \frac{0^2-(48\,ft\cdot s^{-1})^2}{2(-32\,ft\cdot s^{-2})}$$

$$y-y_0 = \underline{+36\,ft}$$

c) $v = \underline{0}$

d) $a = \underline{32\,ft\cdot s^{-2}}$; direction is <u>downward</u>.

e) $v = ?$

$y-y_0 = -64\,ft$

$v_0 = 48\,ft\cdot s^{-1}$

$a = -32\,ft\cdot s^{-2}$

$$v^2 = v_0^2 + 2a(y-y_0)$$

$$v = \pm\sqrt{v_0^2+2a(y-y_0)} = \pm\sqrt{(48\,ft\cdot s^{-1})^2+2(-32\,ft\cdot s^{-2})(-64\,ft)}$$

$$v = \pm 80\,ft\cdot s^{-1}$$

Our physical sense tells us that at this point the ball must be traveling downward, so $v = \underline{-80\,ft\cdot s^{-1}}$.

2-29

a) $v = (1000\,mi\cdot hr^{-1})\left(\frac{5280\,ft}{1\,mi}\right)\left(\frac{1\,hr}{3600\,s}\right) = 1467\,ft\cdot s^{-1}$

$t = 1.8\,s$

$v_0 = 0$

$a = ?$

$$v = v_0 + at \Rightarrow a = \frac{v-v_0}{t} = \frac{1467\,ft\cdot s^{-1} - 0}{1.8\,s} = \underline{815\,ft\cdot s^{-2}}$$

b) $g = 32\,ft\cdot s^{-2}$

$$\Rightarrow \frac{a}{g} = \frac{815\,ft\cdot s^{-2}}{32\,ft\cdot s^{-2}} = \underline{25.5}$$

c) $x-x_0 = \cancel{v_0 t}^{\,0} + \frac{1}{2}at^2 = \frac{1}{2}(815\,ft\cdot s^{-2})(1.8\,s)^2 = \underline{1320\,ft}$

d) We will calculate the acceleration, assuming it to be constant.

$t = 1.4\,s$

$v_0 = (632\,mi\cdot hr^{-1})\left(\frac{5280\,ft}{1\,mi}\right)\left(\frac{1\,hr}{3600\,s}\right) = 927\,ft\cdot s^{-1}$

$v = 0$

$a = ?$

$$v = v_0 + at \Rightarrow a = \frac{v-v_0}{t} = \frac{0-927\,ft\cdot s^{-1}}{1.4\,s} = -662\,ft\cdot s^{-2}$$

$$\frac{a}{g} = \frac{-662\,ft\cdot s^{-2}}{32\,ft\cdot s^{-2}} = -20.7 \Rightarrow \underline{a = -20.7g}$$

2-29 (cont)

If the acceleration while the sled is stopping is constant the acceleration is only 20.7g. But if the acceleration is not constant it is certainly possible that at some point the instantaneous acceleration could be as large as 40g.

2-31

a) v_{MS} = velocity of the man relative to the sidewalk

v_{SE} = velocity of the sidewalk relative to the earth

v_{ME} = velocity of the man relative to the earth

These velocities are related by $v_{ME} = v_{MS} + v_{SE}$.

The man travels 150m relative to the earth, so it is v_{ME} that we need for our calculations. Let the direction in which the sidewalk is moving be positive.

Then $v_{MS} = +2\,m\cdot s^{-1}$, $v_{SE} = +1\,m\cdot s^{-1}$

$$v_{ME} = +2\,m\cdot s^{-1} + 1\,m\cdot s^{-1} = +3\,m\cdot s^{-1}$$

$x - x_0 = 150\,m$

$v_0 = v = 3\,m\cdot s^{-1}$ } The velocities are constant.

$a = 0$

$t = ?$

$$x - x_0 = v_0 t + \tfrac{1}{2}at^2 \quad (a = 0)$$

$$t = \frac{x - x_0}{v_0} = \frac{150\,m}{3\,m\cdot s^{-1}} = \underline{50\,s}$$

b) Since the man walks with a speed greater than that of the sidewalk, he must be going "upstream" against the motion of the sidewalk.

v_{ME} ←, ← v_{MS}, v_{SE} →, ←150m→

Let the direction the man is walking be positive. $v_{MS} = +2\,m\cdot s^{-1}$, $v_{SE} = -1\,m\cdot s^{-1}$

$$v_{ME} = v_{MS} + v_{SE} = +2\,m\cdot s^{-1} - 1\,m\cdot s^{-1} = +1\,m\cdot s^{-1}$$

$x - x_0 = 150\,m$

$v_0 = v = 1\,m\cdot s^{-1}$

$a = 0$

$t = ?$

$$x - x_0 = v_0 t + \tfrac{1}{2}at^2 \quad (a = 0)$$

$$t = \frac{x - x_0}{v_0} = \frac{150\,m}{1\,m\cdot s^{-1}} = \underline{150\,s}$$

Problems

2-37

a) At $t=0$ the auto and truck are at the same position. "The auto overtakes the truck" means that after some time, call it T, the auto and truck will have undergone the same displacement d.

truck	auto
$a = 0$ (constant v)	$a = 2\,m\cdot s^{-2}$
$v_0 = v = 10\,m\cdot s^{-1}$	$v_0 = 0$ (starts from rest)
$t = T$	$t = T$
$x - x_0 = d$	$x - x_0 = d$

Use the equation $x - x_0 = v_0 t + \frac{1}{2}at^2$ for each object.

truck: $d = (10\,m\cdot s^{-1})T$

auto: $d = \frac{1}{2}(2\,m\cdot s^{-2})T^2$

$\Rightarrow \quad (10\,m\cdot s^{-1})T = (1\,m\cdot s^{-2})T^2$

$T = \frac{10\,m\cdot s^{-1}}{1\,m\cdot s^{-2}} = 10\,s$

Then $d = (10\,m\cdot s^{-1})T = (10\,m\cdot s^{-1})(10\,s) = \underline{100\,m}$

(or $d = \frac{1}{2}(2\,m\cdot s^{-2})T^2 = \frac{1}{2}(2\,m\cdot s^{-2})(10\,s)^2 = 100\,m$)

b) For the auto, $v = v_0 + at = 0 + (2\,m\cdot s^{-2})(10\,s) = \underline{20\,m\cdot s^{-1}}$

Note that at this point the auto will be traveling faster than the truck. The motion in this problem is nicely illustrated by plotting $x - x_0$ versus t for both objects, on the same graph.

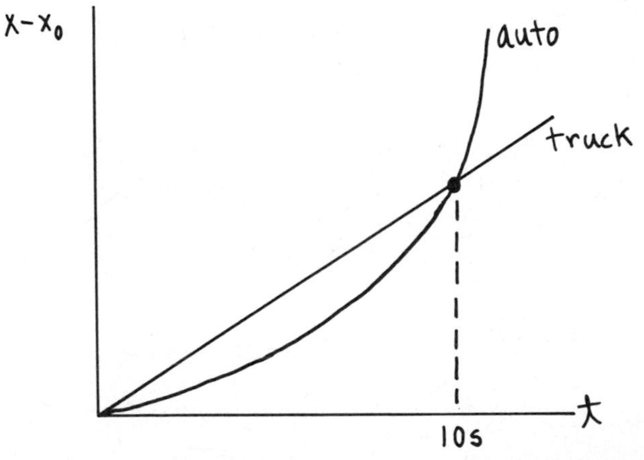

The graph shows that for $t < 10\,s$ the truck has traveled a greater distance than the auto, so is ahead.

At $t = 10\,s$ the displacements are the same, the two curves intersect, and the auto and truck are alongside each other.

The graph also shows that at $t = 10\,s$ the velocity of the auto (slope of the tangent to the curve) is greater than that of the truck, and that for $t > 10\,s$ the auto gets further and further ahead of the truck

2-39

We are given $v(t) = (4\,m\cdot s^{-2})t + (3\,m\cdot s^{-4})t^3$.

To calculate $a(t)$ we take the time derivative:

$$a = \frac{dv}{dt}$$

Since $\frac{d}{dt}(ct^n) = cnt^{n-1}$, $\underline{a = 4\,m\cdot s^{-2} + (9\,m\cdot s^{-4})t^2}$.

To calculate $x(t)$ we integrate:

$$x_2 - x_1 = \int_{t_1}^{t_2} v\,dt$$

Take $t_1 = 0$, $t_2 = t$. We are told that $x_0 = 0$.

$$\Rightarrow x = \int_0^t v\,dt = \int_0^t [(4\,m\cdot s^{-2})t + (3\,m\cdot s^{-4})t^3]\,dt$$

$$x = (4\,m\cdot s^{-2})\int_0^t t\,dt + (3\,m\cdot s^{-4})\int_0^t t^3\,dt = (4\,m\cdot s^{-2})\left(\tfrac{1}{2}t^2\right) + (3\,m\cdot s^{-4})\left(\tfrac{1}{4}t^4\right)$$

(We have used that $\int t^n\,dt = \frac{1}{n+1}t^{n+1}$.)

Thus $\underline{x = (2\,m\cdot s^{-2})t^2 + (\tfrac{3}{4}\,m\cdot s^{-4})t^4}$.

2-41

a) $x = A\sin\omega t$

$$v = \frac{dx}{dt} = \frac{d}{dt}(A\sin\omega t) = A\frac{d}{dt}(\sin\omega t) = A\omega\cos\omega t$$

b) $a = \frac{dv}{dt} = \frac{d}{dt}(A\omega\cos\omega t) = A\omega\frac{d}{dt}\cos\omega t = -A\omega^2\sin\omega t$

c) It is easier to relate $\sin^2\omega t$ and $\cos^2\omega t$ than to relate $\sin\omega t$ and $\cos\omega t$, so start with $x^2 = A^2\sin^2\omega t$ and $v^2 = A^2\omega^2\cos^2\omega t$

$$\sin^2\omega t + \cos^2\omega t = 1 \Rightarrow \cos^2\omega t = 1 - \sin^2\omega t = 1 - \frac{x^2}{A^2}$$

Thus $v^2 = A^2\omega^2\left(1 - \frac{x^2}{A^2}\right) = \omega^2(A^2 - x^2)$

$$\Rightarrow v = \pm\omega\sqrt{A^2 - x^2}\quad ; \quad |v| = \omega\sqrt{A^2 - x^2}$$

This is <u>largest</u> when <u>$x = 0$</u> and <u>smallest</u> when <u>$x = \pm A$</u>.

d) $a = -A\omega^2\sin\omega t = -\omega^2(A\sin\omega t)$

But the quantity in parenthesis is just x, so $a = -\omega^2 x$. $|a| = \omega^2|x|$

This is largest when $x = \pm A$ and smallest when $x = 0$.

2-41 (cont)

Note that a is max when $v=0$ (at $x=\pm A$) and that v is max when $a=0$ (at $x=0$).

e) $x = A\sin\omega t$. The maximum value of $\sin\omega t$ is $1 \Rightarrow x_{max} = A$.

f) $v = \pm\omega\sqrt{A^2 - x^2} \Rightarrow v_{max} = \pm\omega A$

g) $a = -\omega^2 x \Rightarrow a_{max} = \pm\omega^2 A$

h) $x(t) = A\sin\omega t$ $\qquad\qquad$ $v(t) = A\omega\cos\omega t$

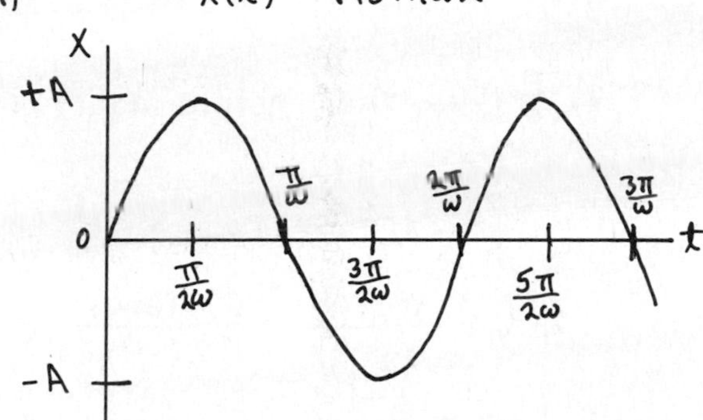

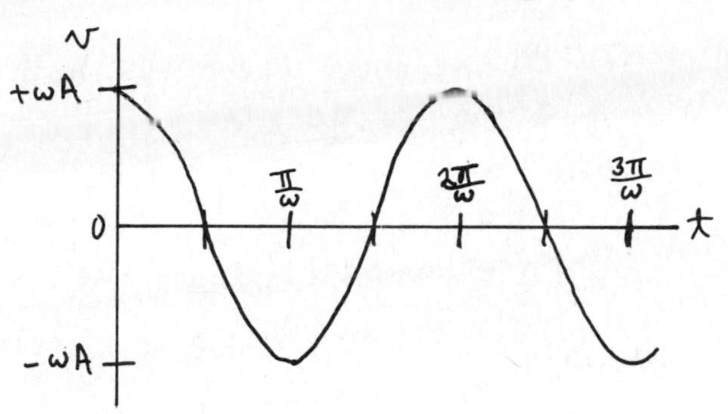

$a(t) = -\omega^2 A\sin\omega t$

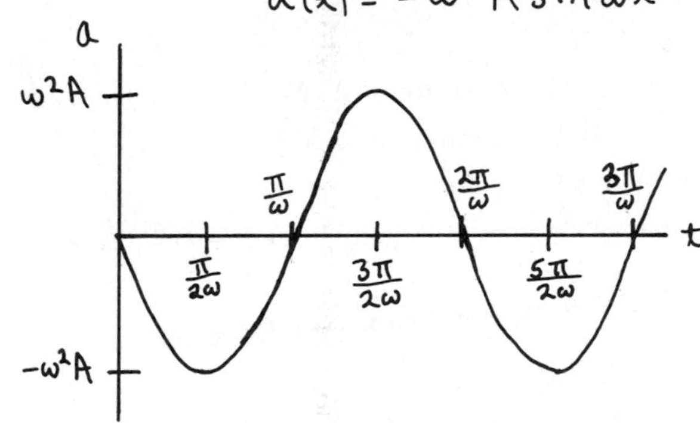

2-45

a) Use the information given about the motion of the first ball.

Let the direction down the incline be positive.

$v_0 = 0$ (released from rest)

$x - x_0 = 18\,m$

$t = 3\,s$

$a = ?$

$$x - x_0 = v_0 t^{\,\nearrow 0} + \tfrac{1}{2}at^2$$

$$a = \frac{2(x-x_0)}{t^2} = \frac{2(18\,m)}{(3\,s)^2} = \underline{4\,m\cdot s^{-2}}$$

b) Keep the same positive direction as in (a):

For the second ball:

2-45 (cont)

$a = 4\,m\cdot s^{-2}$
$t = 3\,s$
$x - x_0 = 0$
$v_0 = ?$

} Second ball is back at starting point after 3s.

$$\cancel{x - x_0}^{\,0} = v_0 t + \tfrac{1}{2}at^2 \Rightarrow v_0 = -\tfrac{1}{2}at = -\tfrac{1}{2}(4\,m\cdot s^{-2})(3s) = \underline{-6\,m\cdot s^{-1}}$$

(The minus sign means the direction of v_0 is up the incline.)

c) Let's now let up the incline be positive, since here all the motion is in that direction.

At the point where the ball has gone as far up the incline as it is going to go $v = 0$.
Thus:

$v = 0$
$a = -4\,m\cdot s^{-2}$ (down the incline)
$v_0 = +6\,m\cdot s^{-1}$ (up the incline)
$x - x_0 = ?$

$$v^2 = v_0^2 + 2a(x - x_0)$$

$$x - x_0 = \frac{v^2 - v_0^2}{2a} = \frac{0^2 - (6\,m\cdot s^{-1})^2}{2(-4\,m\cdot s^{-2})}$$

$$x - x_0 = \underline{4.5\,m}$$

2-47

Take downward to be positive, since all motion here is downward.
For the first marble take $y_0 = 0$, so for the second marble $y_0 = 5.0\,m$.
After time $t = T$ the first marble is at $y = d_1$.
After time $t = T - 1.0\,s$ the second marble is at $y = d_2$. We want $d_1 - d_2 = 15\,m$.

first marble

$v_0 = 0$
$y_0 = 0$
$y = d_1$
$a = +9.8\,m\cdot s^{-2}$ (downward)
$t = T$

$$y - y_0 = \cancel{v_0 t}^{\,0} + \tfrac{1}{2}at^2$$
$$d_1 = \tfrac{1}{2}(9.8\,m\cdot s^{-2})T^2$$

second marble

$v_0 = 0$
$y_0 = 5\,m$
$y = d_2$
$a = +9.8\,m\cdot s^{-2}$
$t = T - 1.0\,s$

$$y - y_0 = \cancel{v_0 t}^{\,0} + \tfrac{1}{2}at^2$$
$$d_2 - 5\,m = \tfrac{1}{2}(9.8\,m\cdot s^{-2})(T - 1s)^2$$
$$d_2 = 5\,m + \tfrac{1}{2}(9.8\,m\cdot s^{-2})(T - 1s)^2$$

Subtract the second equation from the first

$$\Rightarrow d_1 - d_2 = 4.9\,m\cdot s^{-2}\,T^2 - 5\,m - 4.9\,m\cdot s^{-2}\,(T^2 - (2s)T + 1s^2)$$
$$d_1 - d_2 = (9.8\,m\cdot s^{-1})T - 4.9\,m - 5\,m = (9.8\,m\cdot s^{-1})T - 9.9\,m$$

We want T such that $d_1 - d_2 = 15\,m$

$$15\,m = (9.8\,m\cdot s^{-1})T - 9.9\,m \quad \Rightarrow T = \frac{15\,m + 9.9\,m}{9.8\,m\cdot s^{-1}} = \underline{2.54\,s}$$ (after first marble is dropped)

2-49

a) It is very convenient to work in coordinates attached to the truck. Note that these coordinates move at constant velocity relative to the earth. In these coordinates the truck is at rest, and the initial velocity of the car is $v_0 = 0$. Also, the car's acceleration in these coordinates is the same as in coordinates fixed to the earth.

First, let's calculate how far the car must travel relative to the truck:

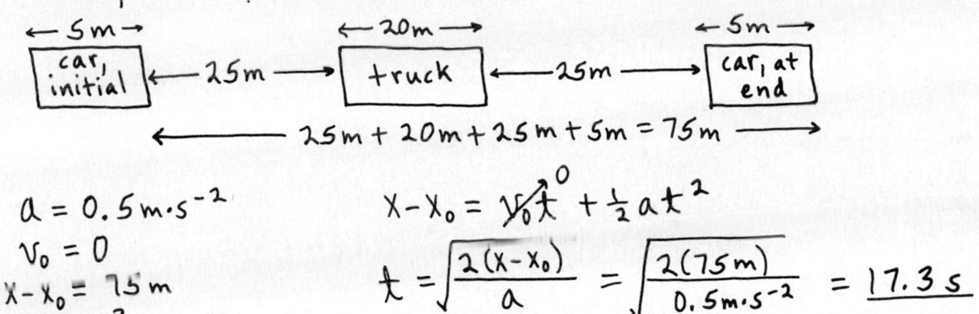

$a = 0.5\,m\cdot s^{-2}$

$v_0 = 0$

$x - x_0 = 75\,m$

$t = ?$

$$x - x_0 = \cancel{v_0 t}^{\,0} + \tfrac{1}{2}at^2$$

$$t = \sqrt{\frac{2(x-x_0)}{a}} = \sqrt{\frac{2(75\,m)}{0.5\,m\cdot s^{-2}}} = \underline{17.3\,s}$$

b) Need how far the car travels relative to the earth, so go now to coordinates fixed in the earth. In these coordinates $v_0 = 20\,m\cdot s^{-1}$!

$v_0 = 20\,m\cdot s^{-1}$

$a = 0.5\,m\cdot s^{-2}$

$t = 17.3\,s$

$x - x_0 = ?$

$$x - x_0 = v_0 t + \tfrac{1}{2}at^2$$

$$x - x_0 = (20\,m\cdot s^{-1})(17.3\,s) + \tfrac{1}{2}(0.5\,m\cdot s^{-2})(17.3\,s)^2$$

$$x - x_0 = 346\,m + 75\,m = \underline{421\,m}$$

c) Again, in coordinates fixed to the earth:

$v_0 = 20\,m\cdot s^{-1}$

$a = 0.5\,m\cdot s^{-2}$

$t = 17.3\,s$

$v = ?$

$$v = v_0 + at = 20\,m\cdot s^{-1} + (0.5\,m\cdot s^{-2})(17.3\,s)$$

$$v = \underline{28.6\,m\cdot s^{-1}}$$

CHAPTER 3

Exercises 1, 5, 9, 11, 15, 17, 19

Problems 23, 27, 29, 31, 33

Exercises

3-1

a) $x(t) = 2\,\text{m} - (4\,\text{m}\cdot\text{s}^{-1})\,t \quad ; \quad y(t) = (3\,\text{m}\cdot\text{s}^{-2})\,t^2$

$v_x = \dfrac{dx}{dt} = -4\,\text{m}\cdot\text{s}^{-1}$

$v_y = \dfrac{dy}{dt} = (6\,\text{m}\cdot\text{s}^{-2})\,t$

$\Rightarrow \vec{v}(t) = (-4\,\text{m}\cdot\text{s}^{-1})\vec{i} + ([6\,\text{m}\cdot\text{s}^{-2}]t)\vec{j}$

$a_x = \dfrac{dv_x}{dt} = 0$, since v_x is constant

$a_y = \dfrac{dv_y}{dt} = 6\,\text{m}\cdot\text{s}^{-2}$

$\Rightarrow \vec{a}(t) = (6\,\text{m}\cdot\text{s}^{-2})\vec{j}$

b) $t = 3\,\text{s} \Rightarrow v_x = -4\,\text{m}\cdot\text{s}^{-1}, \quad v_y = 18\,\text{m}\cdot\text{s}^{-1}$

$v = \sqrt{v_x^2 + v_y^2} = \sqrt{(-4\,\text{m}\cdot\text{s}^{-1})^2 + (18\,\text{m}\cdot\text{s}^{-1})^2} = \underline{18.4\,\text{m}\cdot\text{s}^{-1}}$

$\vec{v}$ Y 18 m·s⁻¹ φ θ X 4 m·s⁻¹

$\tan\phi = \dfrac{4\,\text{m}\cdot\text{s}^{-1}}{18\,\text{m}\cdot\text{s}^{-1}} = 0.222 \Rightarrow \phi = 12.5^\circ$

Thus $\theta = 90^\circ + \phi = \underline{102.5^\circ}$, measured counterclockwise from the +x-axis.

$t = 3\,\text{s} \Rightarrow a_x = 0, \quad a_y = 6\,\text{m}\cdot\text{s}^{-2}$

$a = \sqrt{a_x^2 + a_y^2} = \underline{6\,\text{m}\cdot\text{s}^{-2}}$

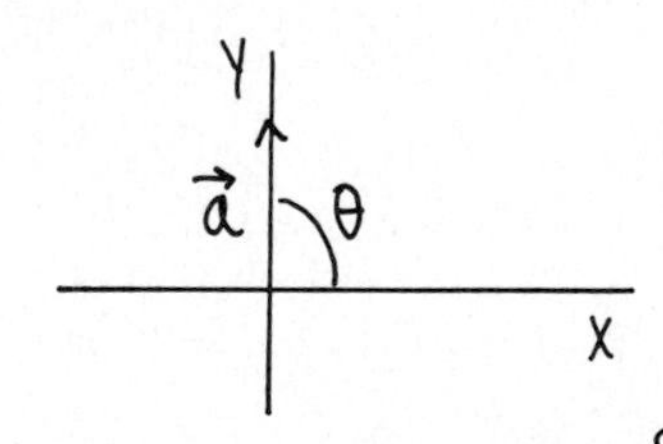

$\vec{a}$ is in the +y-direction; $\theta = 90^\circ$ measured counterclockwise from the +x-axis.

3-5

a) x y $\rightarrow v_0 = 4\,\text{m}\cdot\text{s}^{-1}$? ?

y-component	x-component
$a_y = 9.8\,\text{m}\cdot\text{s}^{-2}$	$a_x = 0$
$v_{0y} = 0$	$v_{0x} = 4\,\text{m}\cdot\text{s}^{-1}$
$t = 0.5\,\text{s}$	$t = 0.5\,\text{s}$
$y - y_0 = ?$	$x - x_0 = ?$

3-5 (cont)

y-comp $\Rightarrow y - y_0 = v_{0y} t + \frac{1}{2} a_y t^2$

$y - y_0 = 0 + \frac{1}{2}(9.8\,m \cdot s^{-2})(0.5\,s)^2 = \underline{1.22\,m}$ (height of table)

b) x-comp $\Rightarrow x - x_0 = v_{0x} t + \frac{1}{2} a_x t^2$ (with $a_x = 0$)

$x - x_0 = (4\,m \cdot s^{-1})(0.5\,s) = \underline{2.00\,m}$

c) x-comp $\Rightarrow v_x = v_{0x} + a_x t$ (with $a_x = 0$) $\Rightarrow v_x = \underline{4\,m \cdot s^{-1}}$

y-comp $\Rightarrow v_y = v_{0y} + a_y t = 0 + (9.8\,m \cdot s^{-2})(0.5\,s) = \underline{4.9\,m \cdot s^{-1}}$

3-9

y, x

$v_{0y} = 31.9\,m \cdot s^{-1}$

$v_0 = 40\,m \cdot s^{-1}$

$53°$

$v_{0x} = 24.1\,m \cdot s^{-1}$

a) At the maximum height $v_y = 0$.
Use the y-comp of the motion:

$v_y = 0$

$a_y = -9.8\,m \cdot s^{-2}$

$v_{0y} = +31.9\,m \cdot s^{-1}$

$y - y_0 = ?$

$$v_y^2 = v_{0y}^2 + 2a_y(y - y_0)$$

$$y - y_0 = \frac{v_y^2 - v_{0y}^2}{2a_y} = \frac{0^2 - (31.9\,m \cdot s^{-1})^2}{2(-9.8\,m \cdot s^{-2})} = \underline{51.9\,m}$$

b) $t = ?$

$$v_y = v_{0y} + a_y t \Rightarrow t = \frac{v_y - v_{0y}}{a_y} = \frac{0 - 31.9\,m \cdot s^{-1}}{-9.8\,m \cdot s^{-2}} = \underline{3.26\,s}$$

c) y-comp

$y - y_0 = 25\,m$

$a_y = -9.8\,m \cdot s^{-2}$

$v_{0y} = 31.9\,m \cdot s^{-1}$

$t = ?$

$$y - y_0 = v_{0y} t + \frac{1}{2} a_y t^2$$

$$25\,m = (31.9\,m \cdot s^{-1})t - (4.9\,m \cdot s^{-2})t^2$$

$$4.9t^2 - 31.9t + 25 = 0$$ [For convenience, don't write the units.]

The quadratic formula gives

$$t = \frac{1}{9.8}\left[31.9 \pm \sqrt{(31.9)^2 - 4(4.9)(25)}\right]$$

$t = 3.26\,s \pm 2.34\,s \Rightarrow t = \underline{0.92\,s}$ and $t = \underline{5.60\,s}$

25 m is less than the maximum height of 51.9 m. These two times correspond to the baseball passing through a height of 25 m on the way up and then again on the way down.

$t = 0$, $t = 0.92\,s$, $t = 3.26\,s$, $t = 5.60\,s$

51.9 m, 25 m

3-9 (cont)

d) $t = 0.92s$

$v_x = v_{ox}$ (since $a_x = 0$) $\Rightarrow v_x = +24.1\ m\cdot s^{-1}$

$v_y = v_{oy} + a_y t = 31.9\ m\cdot s^{-1} + (-9.8\ m\cdot s^{-2})(0.92s) = +22.9\ m\cdot s^{-1}$

$t = 5.60s$

$v_x = v_{ox} = +24.1\ m\cdot s^{-1}$

$v_y = v_{oy} + a_y t = 31.9\ m\cdot s^{-1} + (-9.8\ m\cdot s^{-2})(5.60s) = -23.0\ m\cdot s^{-1}$

e) $v_x = v_{ox} = 24.1\ m\cdot s^{-1}$ (for all t)

$y - y_0 = 0$ (when the baseball returns to the level from which it was thrown)

$v_y^2 = v_{oy}^2 + 2a_y(y - y_0) \Rightarrow v_y = \pm v_{oy} \Rightarrow v_y = -v_{oy} = -31.9\ m\cdot s^{-1}$

$v = \sqrt{v_x^2 + v_y^2} = \sqrt{(24.1\ m\cdot s^{-1})^2 + (-31.9\ m\cdot s^{-1})^2} = 40.0\ m\cdot s^{-1}$

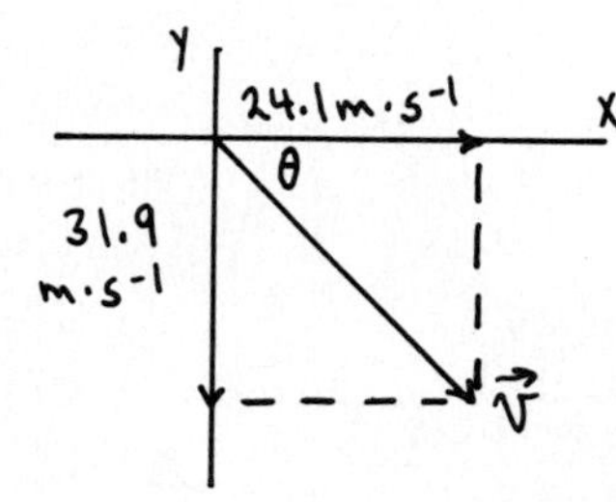

$\tan\theta = \frac{31.9\ m\cdot s^{-1}}{24.1\ m\cdot s^{-1}} = 1.324$

$\theta = 52.9°$, below the horizontal

One can see that the results from (d) and (e) make sense from a sketch of the baseball's path. Recall that the velocity is tangent to the path at each point.

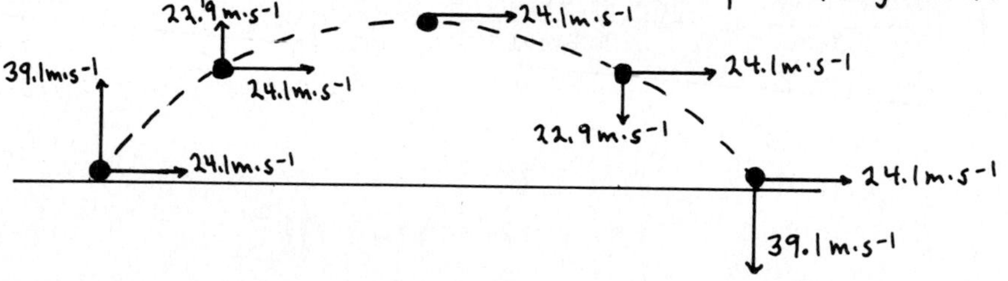

3-11

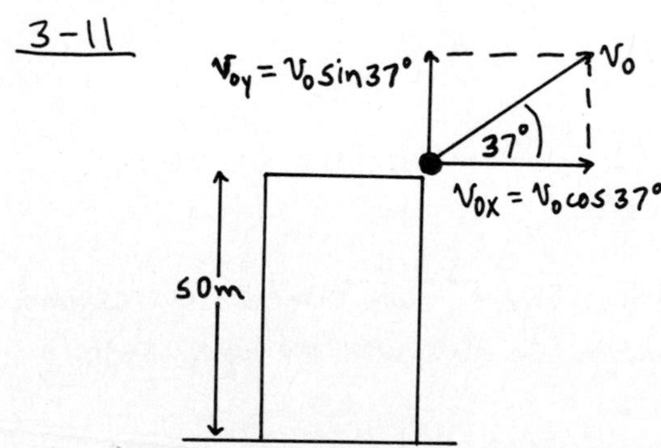

$v_{ox} = v_o \cos 37° = 47.9\ m\cdot s^{-1}$

$v_{oy} = v_o \sin 37° = 36.1\ m\cdot s^{-1}$

a) At the maximum height, $v_y = 0$.

Use the y-comp:

$a_y = -9.8\ m\cdot s^{-2}$

3-11 (Cont)

$v_{oy} = +36.1\ m\cdot s^{-1}$

$v_y = 0$

$y - y_0 = ?$

$$v_y^2 = v_{oy}^2 + 2a_y(y - y_0)$$

$$y - y_0 = \frac{v_y^2 - v_{oy}^2}{2a_y} = \frac{0^2 - (36.1\ m\cdot s^{-1})^2}{2(-9.8\ m\cdot s^{-2})} = \underline{66.5\ m}$$

b) $v_x = v_{ox}$ (since $a_x = 0$) $\Rightarrow v_x = 47.9\ m\cdot s^{-1}$

$v_y = ?$

$y - y_0 = -50\ m$ (at the ground the rock is 50m below its initial position)

$a_y = -9.8\ m\cdot s^{-2}$

$v_{oy} = +36.1\ m\cdot s^{-1}$

$$v_y^2 = v_{oy}^2 + 2a_y(y - y_0)$$

$$v_y = \pm\sqrt{v_{oy}^2 + 2a_y(y - y_0)}$$

$$v_y = \pm\sqrt{(36.1\ m\cdot s^{-1})^2 + 2(-9.8\ m\cdot s^{-2})(-50\ m)}$$

$v_y = -47.8\ m\cdot s^{-1}$ ← since is moving downward at this point

$$v = \sqrt{v_x^2 + v_y^2} = \sqrt{(47.9\ m\cdot s^{-1})^2 + (-47.8\ m\cdot s^{-1})^2} = \underline{67.7\ m\cdot s^{-1}}$$

c) x-comp

$x - x_0 = ?$

$a_x = 0$

$v_{ox} = 47.9\ m\cdot s^{-1}$

Need to know t (time rock is in the air) before can solve. Use the y-component to find t.

y-comp

$v_y = -47.8\ m\cdot s^{-1}$ (from (b))

$a_y = -9.8\ m\cdot s^{-2}$

$v_{oy} = +36.1\ m\cdot s^{-1}$

$t = ?$

$$v_y = v_{oy} + a_y t$$

$$t = \frac{v_y - v_{oy}}{a_y} = \frac{-47.8\ m\cdot s^{-1} - 36.1\ m\cdot s^{-1}}{-9.8\ m\cdot s^{-2}}$$

$$t = 8.56\ s$$

(Of course we also know that $y - y_0 = -50\ m$, so other eqs. could be used to find t, but the algebra and arithmetic in the above approach is particularly simple.)

Now can solve for $x - x_0$:

$$x - x_0 = v_{ox}t + \tfrac{1}{2}a_x t^2 = (47.9\ m\cdot s^{-1})(8.56\ s) = \underline{410\ m}$$

(with $a_x = 0$)

3-15

a)

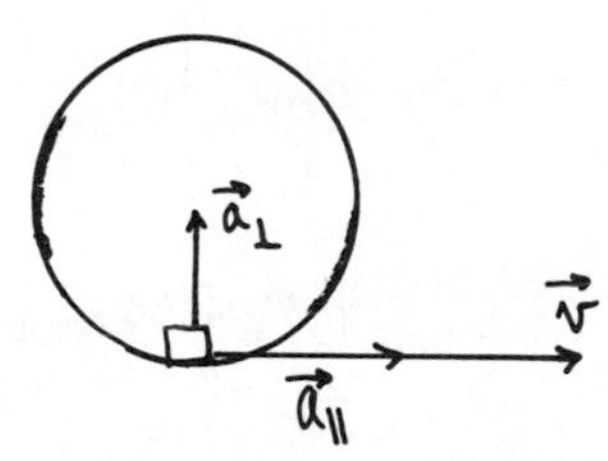

v constant $\Rightarrow a_\parallel = 0$

$$a_\perp = \frac{v^2}{R} = \frac{(9\ m\cdot s^{-1})^2}{12\ m} = 6.75\ m\cdot s^{-2}\text{; direction is upward}$$

$\Rightarrow$ Resultant acceleration is $\underline{6.75\ m\cdot s^{-2}}$, $\underline{\text{upward}}$

b) $v = \dfrac{2\pi R}{\tau}$

$$\tau = \frac{2\pi R}{v} = \frac{2\pi(12\ m)}{9\ m\cdot s^{-1}} = \underline{8.38\ s}$$

3-17

a) The velocity vectors involved are:

$\vec{v}_{PE}$ = velocity of the plane relative to the earth

$\vec{v}_{PA}$ = velocity of the plane relative to the air

$\vec{v}_{AE}$ = velocity of the air relative to the earth (the wind velocity)

The rule for relative velocities says that $\vec{v}_{PE} = \vec{v}_{PA} + \vec{v}_{AE}$.

The problem tells us that $v_{PA} = 290\ \text{km}\cdot\text{hr}^{-1}$ and $v_{AE} = 96\ \text{km}\cdot\text{hr}^{-1}$. It also says that $\vec{v}_{PE}$ is due north (the direction the pilot wishes to fly) and that $\vec{v}_{AE}$ is to the west. With these two directions and the fact that $\vec{v}_{PA}$ and $\vec{v}_{AE}$ must add to give $\vec{v}_{PE}$ we have the vector diagram

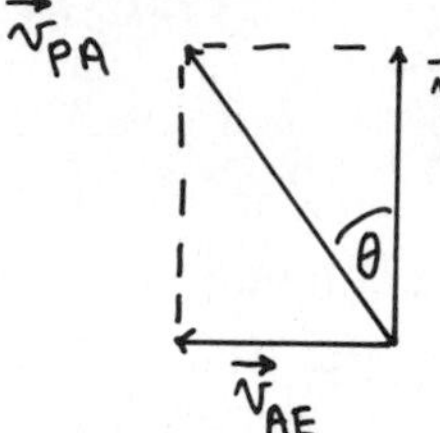

$$\sin\theta = \frac{v_{AE}}{v_{PA}} = \frac{96\ \text{km}\cdot\text{hr}^{-1}}{290\ \text{km}\cdot\text{hr}^{-1}} = 0.331 \Rightarrow \theta = 19.3^\circ$$

The pilot should head <u>19.3° W of N.</u>

b) From the vector diagram in (a),

$$v_{PA}^2 = v_{AE}^2 + v_{PE}^2 \Rightarrow v_{PE} = \sqrt{v_{PA}^2 - v_{AE}^2} = \underline{274\ \text{km}\cdot\text{hr}^{-1}}$$

3-19

The relevant velocity vectors are

$\vec{v}_{ME}$ = the velocity of the man relative to the earth

$\vec{v}_{MW}$ = the velocity of the man relative to the water

$\vec{v}_{WE}$ = the velocity of the water relative to the earth (the current)

The rule for relative velocities says $\vec{v}_{ME} = \vec{v}_{MW} + \vec{v}_{WE}$.

The problem tells us that $\vec{v}_{WE}$ is directed to the north and has a magnitude of $2\,\text{m}\cdot\text{s}^{-1}$, and that $\vec{v}_{MW}$ is directed to the east and has magnitude $3\,\text{m}\cdot\text{s}^{-1}$.

With these two directions the vector diagram showing the vector addition of the relative velocities is

a) $v_{ME} = \sqrt{v_{MW}^2 + v_{WE}^2} = \sqrt{(3\,\text{m}\cdot\text{s}^{-1})^2 + (2\,\text{m}\cdot\text{s}^{-1})^2} = \underline{3.61\,\text{m}\cdot\text{s}^{-1}}$

3-19 (cont)

$$\sin\theta = \frac{v_{WE}}{v_{ME}} = \frac{2\,m\cdot s^{-1}}{3.61\,m\cdot s^{-1}} = 0.554 \Rightarrow \theta = \underline{33.6^\circ \text{ N of E}}$$

b) This is a tricky question. To cross the river the man must travel 1000 m due east relative to the earth. The man's velocity relative to the earth is $\vec{v}_{ME}$.

But the eastward component of this velocity equals $v_{MW} = 3\,m\cdot s^{-1}$.

Thus $$t = \frac{x - x_0}{v_x} = \frac{1000\,m}{3\,m\cdot s^{-1}} = \underline{333\,s}$$

c) The northward component of $\vec{v}_{ME}$ equals $v_{ME} = 2\,m\cdot s^{-1}$. Therefore in 333 s the distance north the man travels relative to the earth is

$$y - y_0 = v_y t = (2\,m\cdot s^{-1})(333\,s) = \underline{666\,m}$$

Problems

3-23

a)

$$\vec{v} = (2\,m\cdot s^{-1} - 3\,m\cdot s^{-3}t^2)\,\vec{i} + (5\,m\cdot s^{-2}t)\,\vec{j}$$

$$v_x = 2\,m\cdot s^{-1} - (3\,m\cdot s^{-3})t^2 \;;\quad v_y = (5\,m\cdot s^{-2})t$$

$\vec{r}(t) = ?$

$$x_2 - x_1 = \int_{t_1}^{t_2} v_x\,dt$$

Take $t_1 = 0$ and $t_2 = t$. $x_1 = x_0 = 0 \Rightarrow x = \int_0^t v_x\,dt$

$$x = \int_0^t [2\,m\cdot s^{-1} - (3\,m\cdot s^{-3})t^2]\,dt = 2\,m\cdot s^{-1}\int_0^t dt - (3\,m\cdot s^{-3})\int_0^t t^2\,dt$$

$$x = (2\,m\cdot s^{-1})t - (3\,m\cdot s^{-3})(\tfrac{1}{3}t^3) = (2\,m\cdot s^{-1})t - (1\,m\cdot s^{-3})t^3$$

$$y_2 - y_1 = \int_{t_1}^{t_2} v_y\,dt$$

Take $t_1 = 0$ and $t_2 = t$. $y_1 - y_0 = 0 \Rightarrow y = \int_0^t v_y\,dt$

$$y = \int_0^t (5\,m\cdot s^{-2})t\,dt = (5\,m\cdot s^{-2})\int_0^t t\,dt = (5\,m\cdot s^{-2})(\tfrac{1}{2}t^2) = (2.5\,m\cdot s^{-2})t^2$$

$$\vec{r}(t) = x(t)\,\vec{i} + y(t)\,\vec{j}$$

$$\boxed{\vec{r}(t) = [(2\,m\cdot s^{-1})t - (1\,m\cdot s^{-3})t^3]\,\vec{i} + [(2.5\,m\cdot s^{-2})t^2]\,\vec{j}}$$

$\vec{a}(t) = ?$

$$a_x = \frac{dv_x}{dt} = \frac{d}{dt}(2\,m\cdot s^{-1} - (3\,m\cdot s^{-3})t^2) = -(6\,m\cdot s^{-3})t$$

$$a_y = \frac{dv_y}{dt} = \frac{d}{dt}(5\,m\cdot s^{-2}t) = 5\,m\cdot s^{-2}$$

3-23 (cont)

$$\vec{a}(t) = a_x(t)\,\vec{i} + a_y(t)\,\vec{j}$$

$$\boxed{\vec{a}(t) = [(-6\,m\cdot s^{-3})t]\,\vec{i} + [5\,m\cdot s^{-2}]\,\vec{j}}$$

b) Find the value of t that gives $x = 0$:

From (a), $x(t) = (2\,m\cdot s^{-1})t - (1\,m\cdot s^{-3})t^3$

$x = 0 \Rightarrow (2\,m\cdot s^{-1})t - (1\,m\cdot s^{-3})t^3 = 0$

$t\,[2\,m\cdot s^{-1} - (1\,m\cdot s^{-3})t^2] = 0$

$t = 0$, and $2\,m\cdot s^{-1} - (1\,m\cdot s^{-3})t^2 = 0 \Rightarrow t = \sqrt{\dfrac{2\,m\cdot s^{-1}}{1\,m\cdot s^{-3}}} = 1.41\,s$

Then calculate y at this t:

From (a), $y(t) = (2.5\,m\cdot s^{-2})t^2$

$t = 1.41\,s \Rightarrow y = (2.5\,m\cdot s^{-2})(1.41\,s)^2 = \underline{5.0\,m}$

3-27

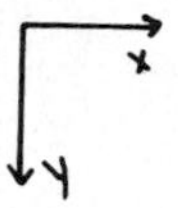

The initial velocity of the bag equals the velocity of the plane.

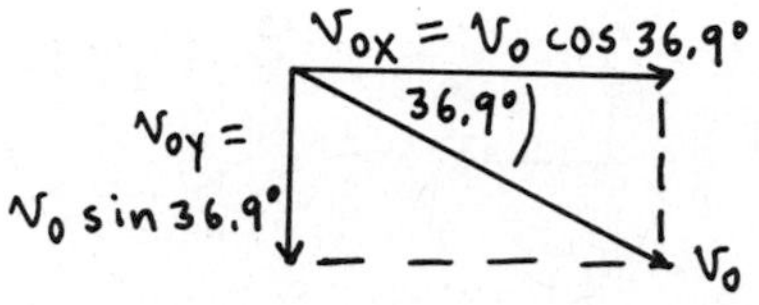

a) Use the y-component:

$a_y = +9.8\,m\cdot s^{-2}$

$v_{oy} = v_o \sin 36.9° = 0.6\,v_o = ?$

$t = 5\,s$

$y - y_o = +800\,m$

$$y - y_o = v_{oy}t + \tfrac{1}{2}a_y t^2$$

$$v_{oy} = \frac{y - y_o}{t} - \tfrac{1}{2}a_y t$$

$$v_{oy} = \frac{800\,m}{5\,s} - \tfrac{1}{2}(9.8\,m\cdot s^{-2})(5\,s) = 136\,m\cdot s^{-1}$$

Then $v_o = \dfrac{v_{oy}}{0.6} = \dfrac{136\,m\cdot s^{-1}}{0.6} = \underline{227\,m\cdot s^{-1}}$.

b) Use the x-component:

$a_x = 0$

$t = 5\,s$

$v_{ox} = v_o \cos 36.9° = (0.8)(227\,m\cdot s^{-1}) = 182\,m\cdot s^{-1}$

$x - x_o = ?$

$$x - x_o = v_{ox}t + \tfrac{1}{2}a_x t^2 \quad (a_x \to 0)$$

$$x - x_o = (182\,m\cdot s^{-1})(5\,s) = \underline{910\,m}$$

c) $v_x = v_{ox}$ (since $a_x = 0$) $= \underline{182\,m\cdot s^{-1}}$

$v_y = ?$

$t = 5\,s$

$a_y = 9.8\,m\cdot s^{-2}$

$v_{oy} = 136\,m\cdot s^{-1}$

$$v_y = v_{oy} + a_y t$$

$$v_y = 136\,m\cdot s^{-1} + (9.8\,m\cdot s^{-1})(5\,s) = \underline{185\,m\cdot s^{-1}} \text{ (downward)}$$

3-29

First we remind ourselves of the formula that relates the range R to the projection angle θ:

(sketch: axes x, y; velocity v_0 at angle θ, components $v_0 \sin\theta$ and $v_0 \cos\theta$)

$y - y_0 = 0$
$v_{oy} = v_0 \sin\theta$
$a_y = -g$
$y - y_0 = v_{oy}t + \frac{1}{2}a_y t^2$

$0 = v_0 \sin\theta\, t - \frac{1}{2}g t^2$

$\Rightarrow t = \frac{2 v_0 \sin\theta}{g}$

$a_x = 0 \Rightarrow R = v_{ox}t = (v_0 \cos\theta)\left(\frac{2 v_0 \sin\theta}{g}\right) = \left(\frac{2 v_0^2}{g}\right)\sin\theta\cos\theta$

Thus, $R_1 = \frac{2 v_0^2}{g}\sin\theta_0 \cos\theta_0$, for $\theta = \theta_0$

$R_2 = \frac{2 v_0^2}{g}\sin(90° - \theta_0)\cos(90° - 0)$, for $\theta = 90° - \theta_0$

Our task is to show that $R_1 = R_2$.

$\cos(90° - \theta_0) = \cos(\theta_0 - 90°) = \sin\theta_0$ (from Appendix B)

and $\sin(90° - \theta_0) = -\sin(\theta_0 - 90°) = -(-\cos\theta_0) = \cos\theta_0$

Thus $R_2 = \frac{2 v_0^2}{g}\cos\theta_0 \sin\theta_0 = R_1$, as desired.

3-31

a) $r = \sqrt{x^2 + y^2}$

$x = R\cos\omega t$
$y = R\sin\omega t$
$\Rightarrow r = \sqrt{R^2\cos^2\omega t + R^2\sin^2\omega t} = R\sqrt{\cos^2\omega t + \sin^2\omega t}$

But $\cos^2\omega t + \sin^2\omega t = 1 \Rightarrow r = R$, independent of t

b) $\vec{v}$ is perpendicular to $\vec{r}$ if $\vec{v}\cdot\vec{r} = 0$.

$\vec{r} = x\vec{i} + y\vec{j} = R\cos\omega t\,\vec{i} + R\sin\omega t\,\vec{j}$

$v_x = \frac{dx}{dt} = \frac{d}{dt}(R\cos\omega t) = -R\omega\sin\omega t$

$v_y = \frac{dy}{dt} = \frac{d}{dt}(R\sin\omega t) = R\omega\cos\omega t$

$\vec{v} = v_x\vec{i} + v_y\vec{j} = (-R\omega\sin\omega t)\vec{i} + (R\omega\cos\omega t)\vec{j}$

$\vec{v}\cdot\vec{r} = v_x x + v_y y = (-R\omega\sin\omega t)(R\cos\omega t) + (R\omega\cos\omega t)(R\sin\omega t)$
$= R^2\omega[-\sin\omega t\cos\omega t + \sin\omega t\cos\omega t] = 0$

c) $a_x = \frac{dv_x}{dt} = \frac{d}{dt}(-R\omega\sin\omega t) = -R\omega^2\cos\omega t$

$a_y = \frac{dv_y}{dt} = \frac{d}{dt}(R\omega\cos\omega t) = -R\omega^2\sin\omega t$

3-31 (cont)

$$\vec{a} = a_x\vec{i} + a_y\vec{j} = (-R\omega^2\cos\omega t)\vec{i} + (-R\omega^2\sin\omega t)\vec{j}$$

$$\vec{a} = -\omega^2[(R\cos\omega t)\vec{i} + (R\sin\omega t)\vec{j}] = -\omega^2\vec{r}$$

Thus $\vec{a}$ is in the direction opposite to $\vec{r}$.

The magnitude is

$$a = \sqrt{a_x^2 + a_y^2} = \sqrt{R^2\omega^4\cos^2\omega t + R^2\omega^4\sin^2\omega t} = R\omega^2\sqrt{\cos^2\omega t + \sin^2\omega t} = R\omega^2$$

d) $$v = \sqrt{v_x^2 + v_y^2} = \sqrt{R^2\omega^2\sin^2\omega t + R^2\omega^2\cos^2\omega t} = R\omega\sqrt{\sin^2\omega t + \cos^2\omega t} = R\omega$$

e) $v = R\omega \Rightarrow \omega = \frac{v}{R}$

Then $a = R\omega^2 = R\left(\frac{v}{R}\right)^2 = \frac{v^2}{R}$, as was to be shown.

3-33

N Y
W E x
S

The relevant vectors are:

$\vec{v}_{PE}$ = velocity of the plane relative to the earth.

$\vec{v}_{PA}$ = velocity of the plane relative to the air

$\vec{v}_{AE}$ = velocity of the air relative to the earth (the wind velocity)

The rule for combining relative velocities gives $\vec{v}_{PE} = \vec{v}_{PA} + \vec{v}_{AE}$.

a) We are told that $\vec{v}_{PE}$ has components $\frac{150\text{km}}{\frac{1}{2}\text{hr}} = 300\text{ km}\cdot\text{hr}^{-1}$ west, and $\frac{40\text{km}}{\frac{1}{2}\text{hr}} = 80\text{km}\cdot\text{hr}^{-1}$ south.

We are also told that $\vec{v}_{PA}$ is due west, with a magnitude of $240\text{ km}\cdot\text{hr}^{-1}$.

The vector diagram showing the relative velocity addition is hence

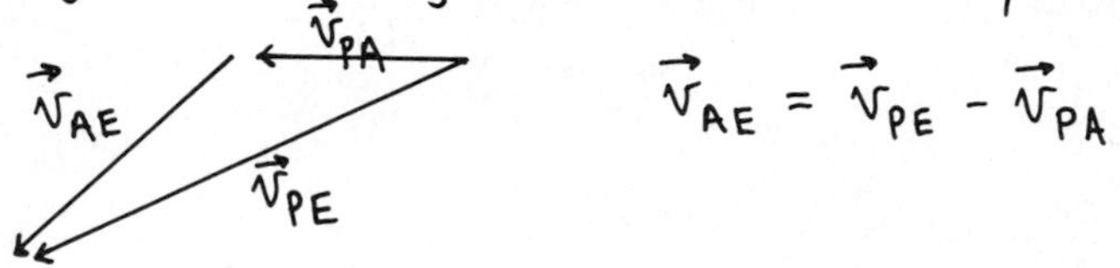

$$\vec{v}_{AE} = \vec{v}_{PE} - \vec{v}_{PA}$$

x-comp (E-W)

$v_{AE,x} = v_{PE,x} - v_{PA,x}$

$v_{PE,x} = -300\text{km}\cdot\text{hr}^{-1}$, $v_{PA,x} = -240\text{km}\cdot\text{hr}^{-1}$

$\Rightarrow v_{AE,x} = -300\text{km}\cdot\text{hr}^{-1} - (-240\text{km}\cdot\text{hr}^{-1})$

$v_{AE,x} = -60\text{ km}\cdot\text{hr}^{-1}$

y-comp (N-S)

$v_{AE,y} = v_{PE,y} - v_{PA,y}$

$v_{PE,y} = -80\text{km}\cdot\text{hr}^{-1}$, $v_{PA,y} = 0$

$\Rightarrow v_{AE,y} = -80\text{km}\cdot\text{hr}^{-1} - 0$

$v_{AE,y} = -80\text{km}\cdot\text{hr}^{-1}$

$$v_{AE} = \sqrt{v_{AE,x}^2 + v_{AE,y}^2} = \sqrt{(-60\text{km}\cdot\text{hr}^{-1})^2 + (-80\text{km}\cdot\text{hr}^{-1})^2} = \underline{100\text{km}\cdot\text{hr}^{-1}}$$

3-33 (cont)

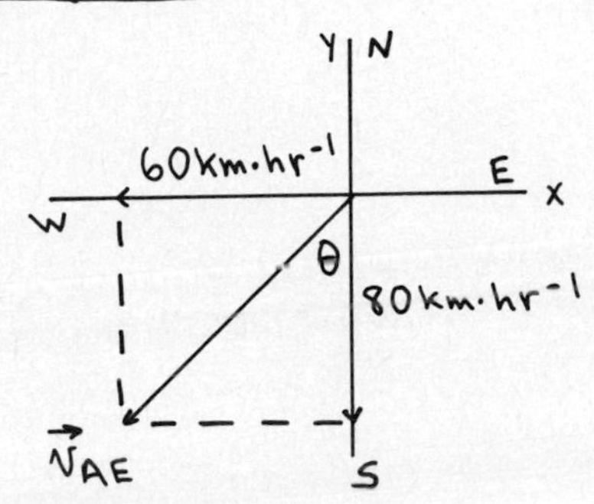

$$\tan\theta = \frac{60\,\text{km}\cdot\text{hr}^{-1}}{80\,\text{km}\cdot\text{hr}^{-1}} = 0.75$$

$\theta = \underline{36.9°, \text{ W of S}}$

b) $\vec{v}_{AE} = 120\ \text{km}\cdot\text{hr}^{-1}$ due south

$\vec{v}_{PE}$ due west

The relative velocity addition diagram corresponding to $\vec{v}_{PE} = \vec{v}_{PA} + \vec{v}_{AE}$ becomes

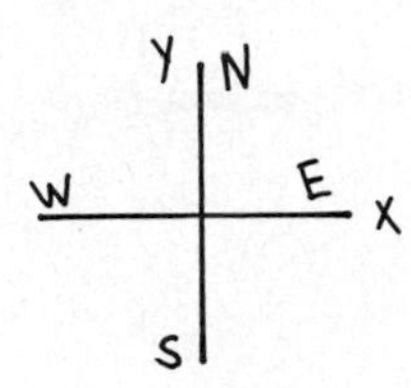

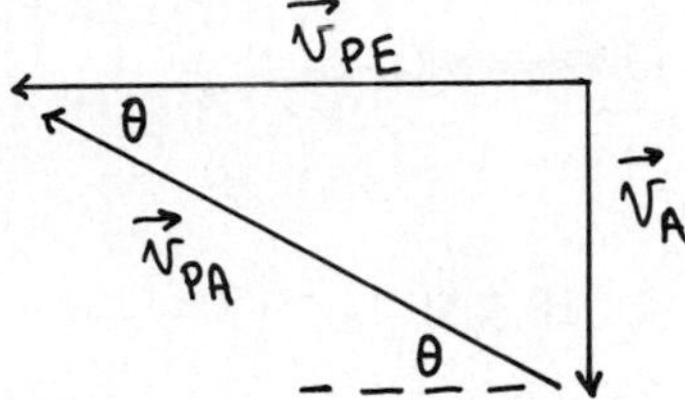

$$\sin\theta = \frac{v_{AE}}{v_{PA}} = \frac{120\,\text{km}\cdot\text{hr}^{-1}}{240\,\text{km}\cdot\text{hr}^{-1}}$$

$\sin\theta = 0.5 \Rightarrow \theta = 30°$

The direction of $\vec{v}_{PA}$ (the direction in which the pilot should set his course) is thus $\underline{30° \text{ N of W}}$.

CHAPTER 4

Exercises 3, 7, 9, 11, 15

Problems 17, 19, 21

Exercises

4-3

Let the x-axis be in the direction of $\vec{F}_A$, the force applied by man A.

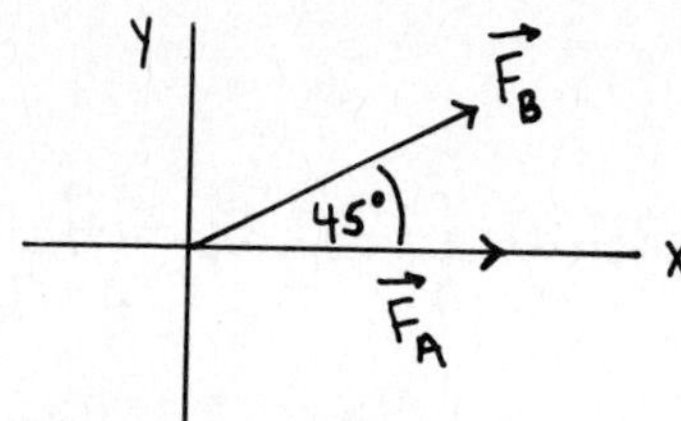

$F_A = 68$ lb

$F_B = 50$ lb

$F_{AX} = F_A = 68$ lb

$F_{AY} = 0$

$F_{Bx} = F_B \cos 45° = 35.4$ lb

$F_{By} = F_B \sin 45° = 35.4$ lb

$\vec{R} = \vec{F}_A + \vec{F}_B$

$R_x = F_{AX} + F_{BX} = 68 \text{ lb} + 35.4 \text{ lb} = 103.4 \text{ lb}$

$R_y = F_{Ay} + F_{By} = 0 + 35.4 \text{ lb} = 35.4 \text{ lb}$

Thus $R = \sqrt{R_x^2 + R_y^2} = \sqrt{(103.4 \text{ lb})^2 + (35.4 \text{ lb})^2} = \underline{109 \text{ lb}}$

Y, $\vec{R}$, 35.4 lb, θ, 103 lb, X

$\tan\theta = \dfrac{35.4 \text{ lb}}{103 \text{ lb}} = 0.344 \Rightarrow \theta = \underline{19.0°}$

The direction of $\vec{R}$ is 19.0° counterclockwise from the direction of $\vec{F}_A$.

4-7

$w = mg$

The mass of the watermelon is constant; its weight differs on earth and Mars.

Use the information about the watermelon's weight on earth to calculate its mass:

$w = mg \Rightarrow m = \dfrac{w}{g} = \dfrac{52 \text{N}}{9.8 \text{ m·s}^{-2}} = 5.31$ kg, on earth.

On Mars $m = 5.31$ kg, the same as on earth.

Then the weight on Mars is $w = mg = (5.31 \text{ kg})(3.7 \text{ m·s}^{-2}) = \underline{19.6 \text{ N}}$

4-9

a) The free-body force diagram for the bottle is

m

$\downarrow w = mg$

The only force on the bottle is the force of gravity.

b)

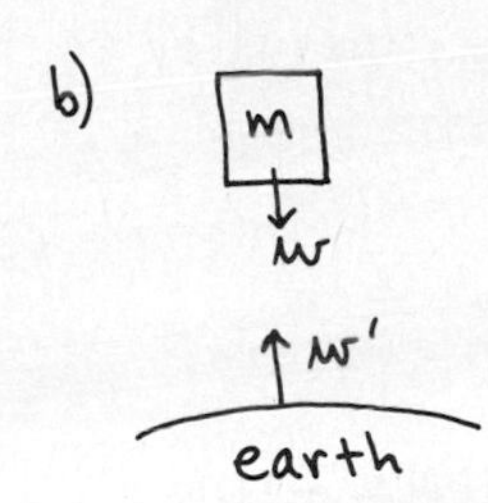

w is the force of gravity that the earth exerts on the bottle. The reaction to this force is w', the gravity force that the bottle exerts on the earth. Note that these two equal and opposite forces produce quite different accelerations because the bottle and the earth have quite different masses.

4-11

a) N, y, x, m, F = 40N, a, $w = mg$

Use the kinematic information to find the acceleration a of the block, and then $\Sigma F_x = ma_x$ to calculate m:

$v_0 = 0$
$t = 5\,s$
$x - x_0 = 100\,m$
$a = ?$

$$x - x_0 = \cancel{v_0 t}^{\,0} + \tfrac{1}{2}at^2$$

$$a = \frac{2(x-x_0)}{t^2} = \frac{2(100\,m)}{(5\,s)^2} = 8.0\,m\cdot s^{-2}$$

$$\Sigma F_x = ma_x \Rightarrow F = ma \;;\; m = \frac{F}{a} = \frac{40\,N}{8\,m\cdot s^{-2}} = \underline{5.0\,kg}$$

b) Find the velocity produced by the acceleration during the first 5 s:

$v = ?$
$v_0 = 0$
$t = 5\,s$
$a = 8\,m\cdot s^{-2}$

$$v = v_0 + at = 0 + (8\,m\cdot s^{-2})(5\,s) = 40\,m\cdot s^{-1}$$

The force ceases to act $\Rightarrow a = 0$. Thus in the next 5 s,

$v_0 = 40\,m\cdot s^{-1}$
$a = 0$
$t = 5\,s$
$x - x_0 = ?$

$$x - x_0 = v_0 t + \cancel{\tfrac{1}{2}at^2}^{\,0} = (40\,m\cdot s^{-1})(5\,s) = \underline{200\,m}$$

4-15

$$x(t) = At + Bt^3$$

Find $a(t)$, then $R = \Sigma F = ma$.

$$v(t) = \frac{dx}{dt} = A + 3Bt^2$$

$$a(t) = \frac{dv}{dt} = 6Bt \rightarrow \underline{R = 6Bmt}$$

Problems

4-17

$$\vec{F}_1 + \vec{F}_2 = \vec{R}$$

It is stated that $\vec{R}$ and $\vec{F}_1$ are perpendicular to each other. Select coordinates with the x-axis parallel to $\vec{F}_1$ and the y-axis parallel to $\vec{R}$.

$$F_{1x} = F_1 = 10N \qquad R_x = 0$$
$$F_{1y} = 0 \qquad R_y = 10N$$

$$\vec{F}_1 + \vec{F}_2 = \vec{R} \Rightarrow \vec{F}_2 = \vec{R} - \vec{F}_1 \; ; \; F_{2x} = R_x - F_{1x} = 0 - 10N = -10N$$
$$F_{2y} = R_y - F_{1y} = 10N - 0 = +10N$$

thus

$$F_2 = \sqrt{F_{2x}^2 + F_{2y}^2} = \sqrt{(-10N)^2 + (10N)^2} = \underline{14.1N}$$

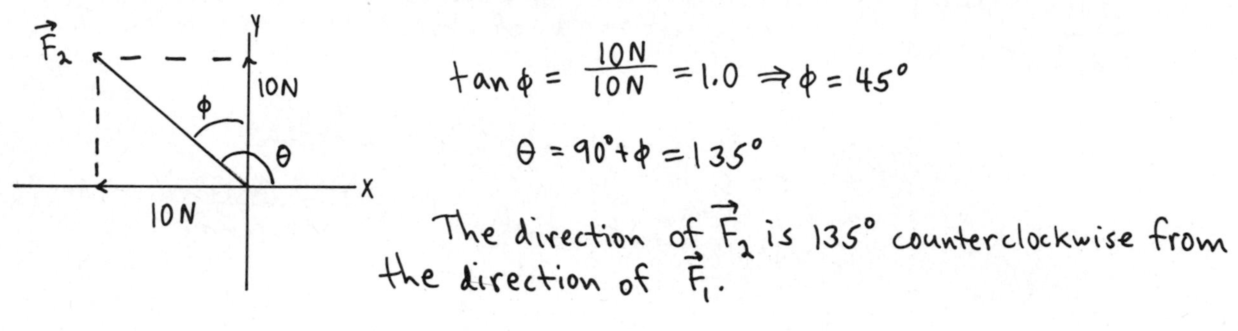

$$\tan\phi = \frac{10N}{10N} = 1.0 \Rightarrow \phi = 45°$$

$$\theta = 90° + \phi = 135°$$

The direction of $\vec{F}_2$ is 135° counterclockwise from the direction of $\vec{F}_1$.

4-19

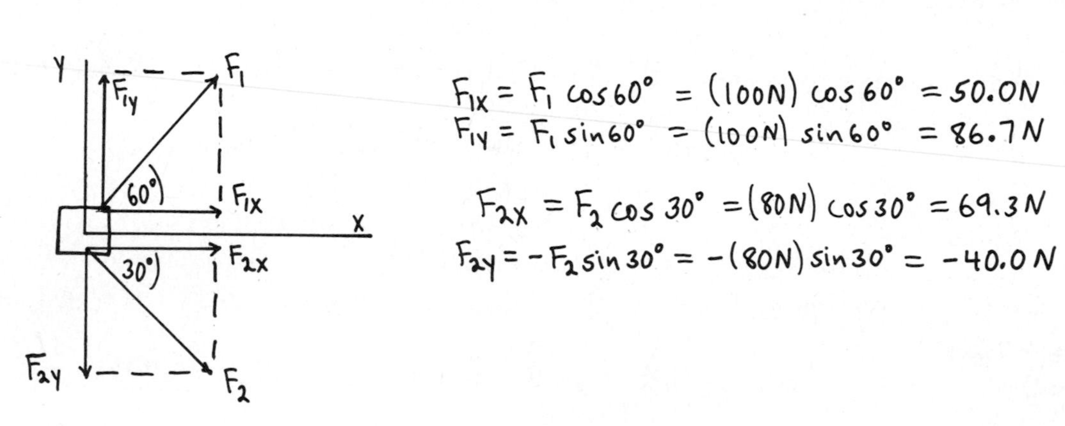

$$F_{1x} = F_1 \cos 60° = (100N)\cos 60° = 50.0N$$
$$F_{1y} = F_1 \sin 60° = (100N)\sin 60° = 86.7N$$

$$F_{2x} = F_2 \cos 30° = (80N)\cos 30° = 69.3N$$
$$F_{2y} = -F_2 \sin 30° = -(80N)\sin 30° = -40.0N$$

$\vec{F}_1 + \vec{F}_2 + \vec{F}_3 = \vec{R}$, where $\vec{F}_3$ is the force the boy exerts.
We require that $\vec{a}$ and hence $\vec{R}$ be in the +x-direction $\Rightarrow R_y = 0$.

$$R_y = F_{1y} + F_{2y} + F_{3y} = 0$$
$$F_{3y} = -(F_{1y} + F_{2y}) = -(86.7N - 40.0N) = -46.7N$$

$$R_x = F_{1x} + F_{2x} + F_{3x}$$

The boy **must** push so that $R_y = 0$, but he doesn't have to contribute to R_x.
His smallest force will be when $F_{3x} = 0 \Rightarrow F = \sqrt{F_{3x}^2 + F_{3y}^2} = |F_{3y}| = \underline{46.7N}$

4-19 (cont)

The direction of his force is in the $-y$-direction in our coordinates, and hence at 60° clockwise from $\vec{F}_2$.

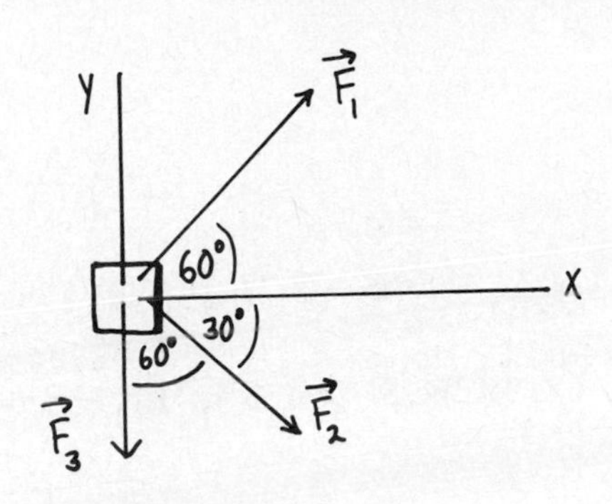

4-21

a)

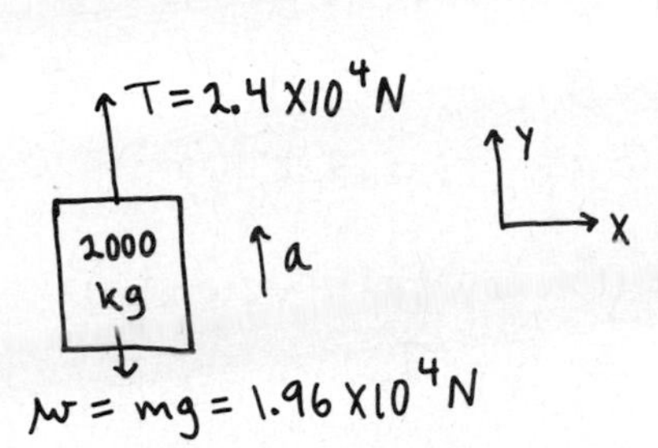

$w = mg = 1.96 \times 10^4\,N$

(We get the maximum upward acceleration with the maximum upward force, which is when the cables have the largest tension they can without breaking.)

$$\Sigma F_y = ma_y$$

$$T - w = ma \Rightarrow a = \frac{T-w}{m} = \frac{2.4\times10^4\,N - 1.96\times10^4\,N}{2000\,kg} = \underline{2.2\,m\cdot s^{-2}}$$

b) What changes here is the weight of the elevator,

$$w = mg = (2000\,kg)(1.67\,m\cdot s^{-2}) = 3.34\times10^3\,N$$

$$a = \frac{T-w}{m} = \frac{2.4\times10^4\,N - 3.34\times10^3\,N}{2000\,kg} = \underline{10.3\,m\cdot s^{-2}}$$

(Note that this is much larger than the answer in (a).)

CHAPTER 5

Exercises 3, 7, 11, 13, 17, 19, 21, 27, 29, 31, 33

Problems 37, 39, 41, 43, 49, 53, 55, 57, 61

Exercises

5-3

a)

n, y, x, f_k, 0.6 slugs, F, $\vec{v}$

$w = mg = 19.2 \text{ lb}$

$\Sigma F_y = ma_y = 0$

$n - w = 0 \Rightarrow n = 19.2 \text{ lb}$

$\Sigma F_x = ma_x = 0$ (constant velocity)

$F - f_k = 0 \Rightarrow F = f_k$

$f_k = \mu_k n = (0.20)(19.2 \text{ lb}) = 3.84 \text{ lb} \Rightarrow F = \underline{3.84 \text{ lb}}$

b)

n, y, x, f_k, 0.6 slugs, $\vec{v}$

$w = mg = 19.2 \text{ lb}$

$\Sigma F_y = ma_y = 0$

$n - w = 0 \Rightarrow n = 19.2 \text{ lb}$

$\Sigma F_x = ma_x$

$-f_k = ma \Rightarrow a = -\frac{f_k}{m} = -\frac{\mu_k n}{m} = -\frac{(0.2)(19.2 \text{ lb})}{0.6 \text{ slugs}}$

$a = -6.4 \text{ ft}\cdot\text{s}^{-2}$

$a = -6.4 \text{ ft}\cdot\text{s}^{-2}$
$v_0 = 15 \text{ ft}\cdot\text{s}^{-1}$
$v = 0$
$t = ?$

$v = v_0 + at$

$t = \frac{v - v_0}{a} = \frac{0 - 15 \text{ ft}\cdot\text{s}^{-1}}{-6.4 \text{ ft}\cdot\text{s}^{-2}} = \underline{2.34 \text{ s}}$

5-7

y, x, n, $\vec{v}$, F_{horiz}, $w = mg$

$\Sigma F_y = ma_y$

$n - mg = 0$

$n = mg$

$\Sigma F_x = ma_x$

$-F_{horiz} = ma$

$F_{horiz} = \mu_{tr} n = \mu_{tr} mg$

$\Rightarrow -\mu_{tr} \cancel{m} g = \cancel{m} a$

$a = -\mu_{tr} g$

Consider the kinematics of the motion:

$v^2 = v_0^2 + 2a(x - x_0)$ with $v = \frac{1}{2} v_0$

$\Rightarrow \frac{1}{4} v_0^2 = v_0^2 + 2a(x - x_0)$

$-\frac{3}{4} v_0^2 = 2a(x - x_0) = -2\mu_{tr} g(x - x_0) \Rightarrow \mu_{tr} = \frac{3 v_0^2}{8g(x - x_0)}$

5-7 (cont)

low pressure tire

$$X - X_0 = 45.6\,m$$

$$\mu_{tr} = \frac{3(5\,m\cdot s^{-1})^2}{8(9.8\,m\cdot s^{-2})(45.6\,m)} = \underline{0.0210}$$

high pressure tire

$$X - X_0 = 213\,m$$

$$\mu_{tr} = \frac{3(5\,m\cdot s^{-1})^2}{8(9.8\,m\cdot s^{-2})(213\,m)} = \underline{0.00449}$$

5-11

a)

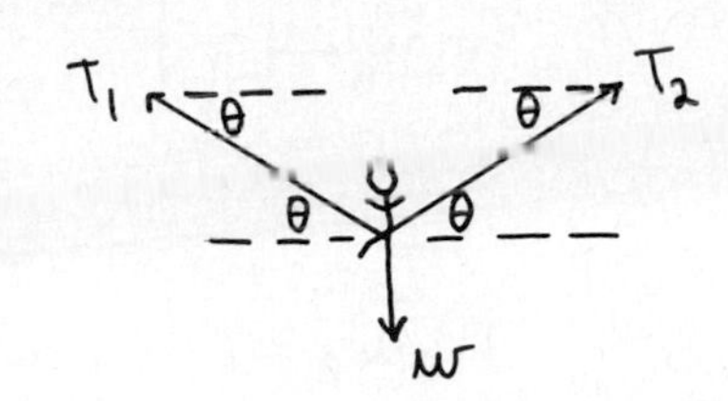

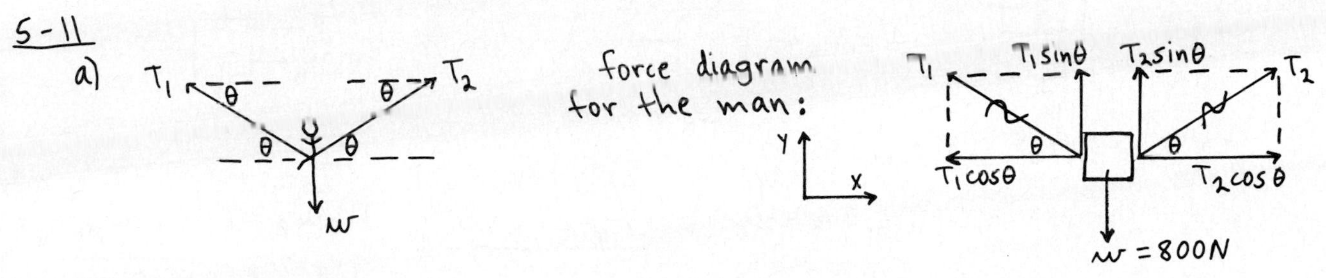

$$\Sigma F_x = ma_x$$
$$T_1\cos\theta - T_2\cos\theta = 0$$
$$\Rightarrow T_1 = T_2 = T$$

$$\Sigma F_y = ma_y$$
$$2T\sin\theta - w = 0$$
$$T = \frac{w}{2\sin\theta} = \frac{800\,N}{2(\sin 15^\circ)} = \underline{1545\,N}$$

b) $T = 20{,}000\,N \quad ; \quad \theta = ?$

$$2T\sin\theta - w = 0 \Rightarrow \sin\theta = \frac{w}{2T} = \frac{800N}{2(20{,}000N)} = 0.02 \quad ; \quad \theta = \underline{1.15^\circ}$$

5-13

a)

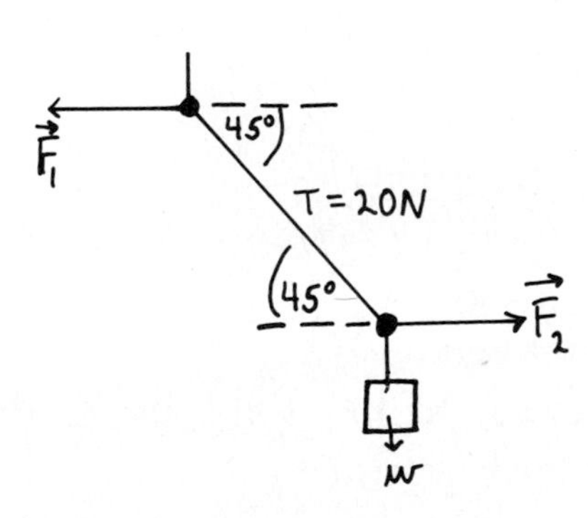

Force diagram for the upper knot:

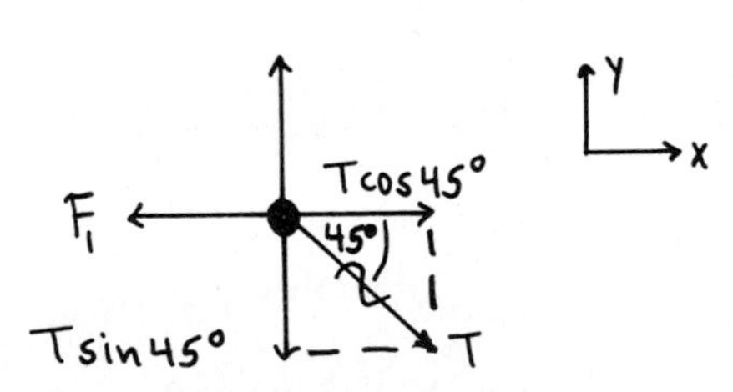

$$\Sigma F_x = ma_x$$
$$T\cos 45^\circ - F_1 = 0$$
$$F_1 = T\cos 45^\circ = (20N)(0.707) = \underline{14.1\,N}$$

Force diagram for the lower knot:

T, Tsin45°, Y, X, 45°, Tcos45°, F_2, T′

$$\Sigma F_x = ma_x$$
$$T\cos 45^\circ - F_2 = 0$$
$$F_2 = T\cos 45^\circ = \underline{14.1\,N}$$

5-13 (cont)

Forces on the block:

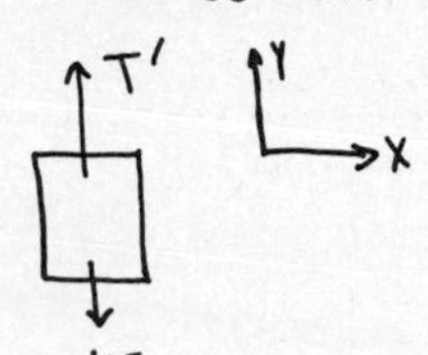

$\Sigma F_y = ma_y$

$T' - w = 0$

$T' = w$

Forces on the lower knot:

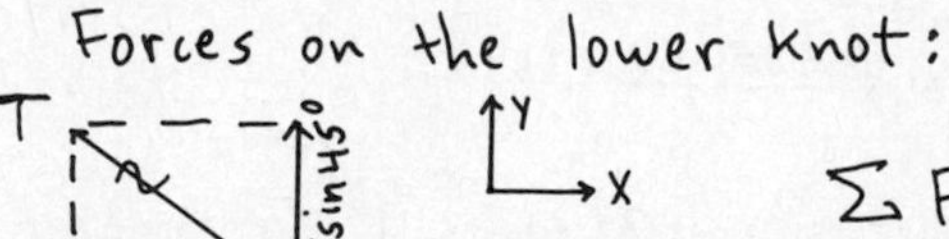
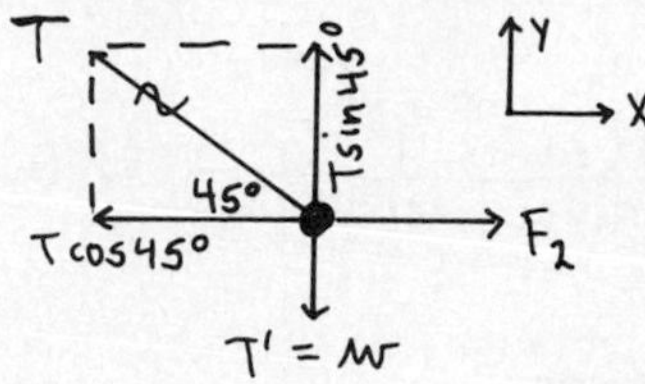

$\Sigma F_y = ma_y$

$T \sin 45° - w = 0$

$w = T \sin 45°$

$w = (20N)(0.707)$

$w = \underline{14.1\ N}$

5-17

a)

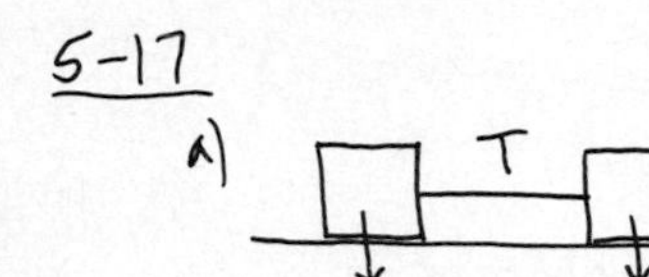

constant $v \Rightarrow a = 0$

n_A, $\mu_K n_A$, T, w_A, Y, X

n_B, T, $\mu_K n_B$, P, w_B, Y, X

$\Sigma F_y = ma_y$

$n_A - w_A = 0$

$n_A = w_A$

$\Sigma F_x = ma_x$

$T - \mu_K n_A = 0$

$\boxed{T = \mu_K w_A}$

$\Sigma F_y = ma_y$

$n_B - w_B = 0$

$n_B = w_B$

$\Sigma F_x = ma_x$

$P - T - \mu_K n_B = 0$

$\boxed{P = T + \mu_K w_B}$

Combine these eqs. $\Rightarrow P = \mu_K w_A + \mu_K w_B$

$$\boxed{P = \mu_K (w_A + w_B)}$$

b) $T = \mu_K w_A$

5-19

constant speed $\Rightarrow a = 0$

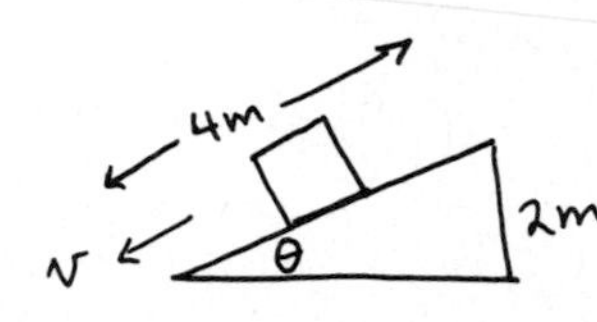

$\sin\theta = \frac{2m}{4m} = 0.5$; $\theta = 30°$

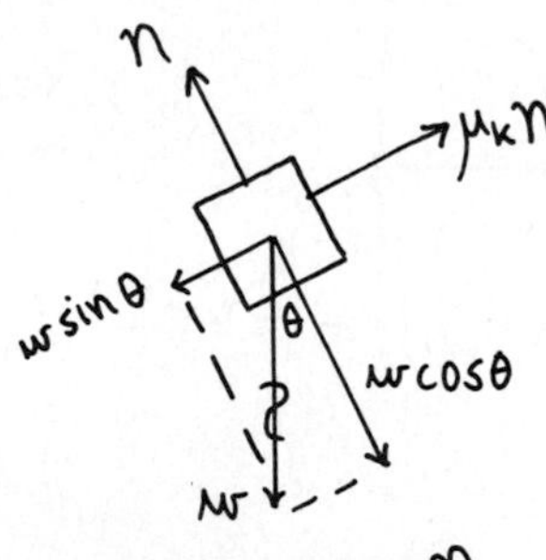

$w \sin\theta = (2000N) \sin 30° = 1000\ N$

$n = w \cos\theta = (2000N) \cos 30° = 1732N$

$\mu_K n = (0.30)(1732N) = 520\ N$

$w \sin\theta > \mu_K n \Rightarrow$ safe needs to be <u>held back</u> if it is to slide at constant velocity

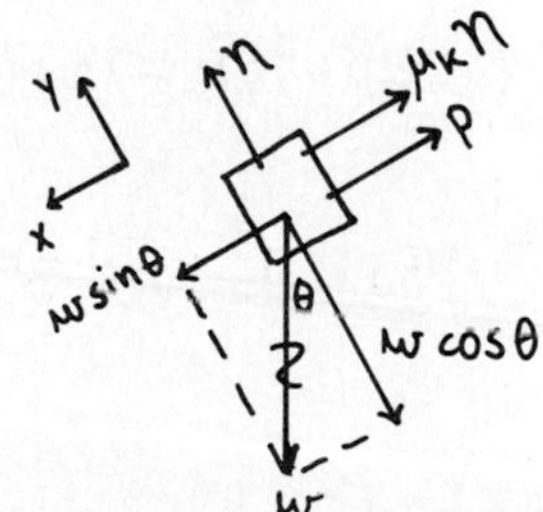

$\Sigma F_x = ma_x$

$w \sin\theta - \mu_K n - P = 0$

$P = w \sin\theta - \mu_K n = 1000N - 520\ N$

$P = \underline{480N}$

5-21

a)

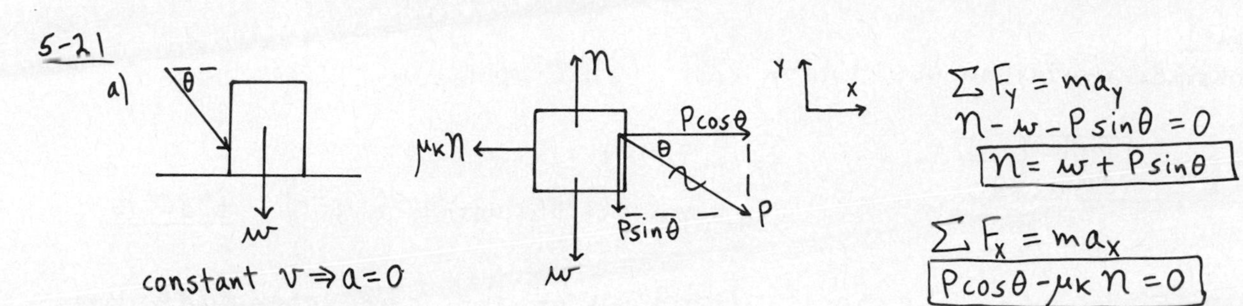

$\Sigma F_y = ma_y$

$n - w - P\sin\theta = 0$

$\boxed{n = w + P\sin\theta}$

$\Sigma F_x = ma_x$

$\boxed{P\cos\theta - \mu_K n = 0}$

constant $v \Rightarrow a = 0$

Combine these two equations:

$$P\cos\theta - \mu_K(w + P\sin\theta) = 0$$

$$\boxed{P = \frac{\mu_K w}{\cos\theta - \mu_K \sin\theta}}$$

b) "starting the crate to slide" ⇒ the force diagram is as in (a) except that $\mu_K \to \mu_s$

Thus $P = \dfrac{\mu_s w}{\cos\theta - \mu_s \sin\theta}$

$P \to \infty$ if $\cos\theta - \mu_s \sin\theta = 0 \Rightarrow \mu_s = \dfrac{\cos\theta}{\sin\theta} = \dfrac{1}{\tan\theta}$

5-27

2700 slugs	T_4	2700 slugs	T_3	2700 slugs	T_2	2700 slugs	T_1	engine	$\rightarrow a = 1.5\,\text{ft}\cdot\text{s}^{-2}$

a) Consider all four cars as a combined object:

Y, X — [4(2700) slugs] $\rightarrow T_1$, $\rightarrow a$

$\Sigma F_x = ma_x$

$T_1 = ma = 4(2700\,\text{slugs})(1.5\,\text{ft}\cdot\text{s}^{-2}) = \underline{16{,}200\,\text{lb}}$

b) Consider the last three cars as a combined object:

Y, X — [3(2700) slugs] $\rightarrow T_2$, $\rightarrow a$

$\Sigma F_x = ma_x$

$T_2 = ma = 3(2700\,\text{slugs})(1.5\,\text{ft}\cdot\text{s}^{-2}) = \underline{12{,}150\,\text{lb}}$

c) Consider the last two cars as a combined object:

Y, X — [2(2700) slugs] $\rightarrow T_3$, $\rightarrow a$

$\Sigma F_x = ma_x$

$T_3 = ma = 2(2700\,\text{slugs})(1.5\,\text{ft}\cdot\text{s}^{-2}) = \underline{8100\,\text{lb}}$

5-27 (cont)

d) Consider forces on the fourth car:

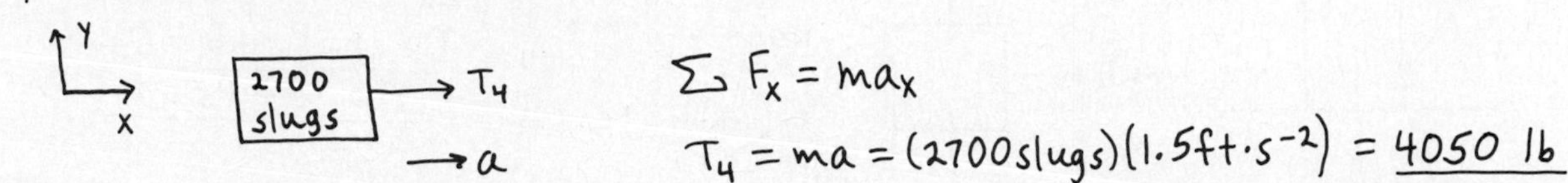

$\Sigma F_x = ma_x$

$T_4 = ma = (2700\,\text{slugs})(1.5\,\text{ft}\cdot\text{s}^{-2}) = \underline{4050\text{ lb}}$

e) ← $a = 1.5\,\text{ft}\cdot\text{s}^{-2}$

The forces would all be the same magnitude, but in the opposite direction.

5-29

a)

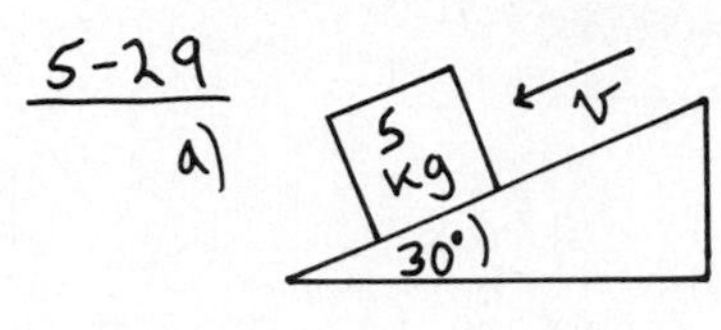

No Friction

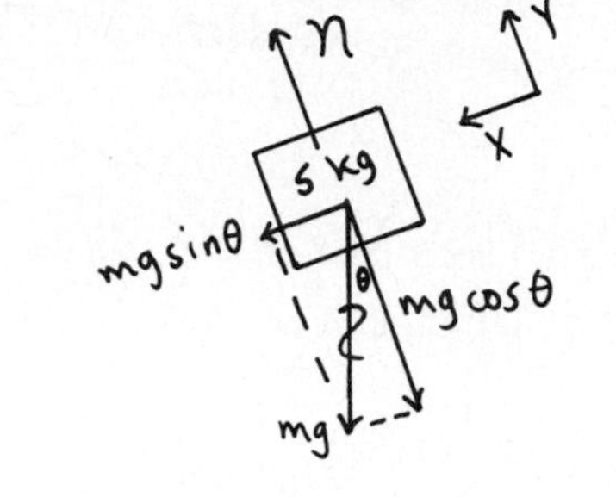

$\Sigma F_x = ma_x$

$mg\sin\theta = ma$

$a = g\sin\theta = (9.8\,\text{m}\cdot\text{s}^{-2})\sin 30°$

$a = \underline{4.9\,\text{m}\cdot\text{s}^{-2}}$

b) Now there is a kinetic friction force, directed up the incline:

n, $\mu_k n$, mg sinθ, θ, mg cosθ, mg, Y, X

$\Sigma F_y = ma_y$

$n - mg\cos\theta = 0$

$n = mg\cos\theta$

$\Sigma F_x = ma_x$

$mg\sin\theta - \mu_k n = ma$

$mg\sin\theta - \mu_k mg\cos\theta = ma$

$a = g(\sin\theta - \mu_k\cos\theta)$

$a = (9.8\,\text{m}\cdot\text{s}^{-2})(\sin 30° - 0.2\cos 30°)$

$a = \underline{3.20\,\text{m}\cdot\text{s}^{-2}}$

Note that the acceleration is less when friction is present, and that in both cases the acceleration is independent of the mass of the block.

5-31

a) Consider the forces on the man. The reading on the scale equals the upward normal force on the man.

Y, X, $n = 800\text{N}$, m, $mg = 600\text{N}$

$w = mg \Rightarrow m = \frac{w}{g} = \frac{600\text{N}}{9.8\,\text{m}\cdot\text{s}^{-2}} = 61.2\,\text{kg}$

$\Sigma F_y = ma_y$

$n - mg = ma \Rightarrow a = \frac{n - mg}{m} = \frac{800\text{N} - 600\text{N}}{61.2\,\text{kg}} = \underline{3.27\,\text{m}\cdot\text{s}^{-2}}$

a comes out to be positive, so a is <u>upward</u>.

5-31 (cont)

b) Free-body diagram: $n = 450\,N$ (up), $mg = 600\,N$ (down)

$$\Sigma F_y = ma_y$$

$$n - mg = ma \Rightarrow a = \frac{n - mg}{m}$$

$$a = \frac{450N - 600N}{61.2\,kg} = \underline{-2.45\,m\cdot s^{-2}}$$

a comes out to be negative $\Rightarrow$ a is <u>downward</u>

c) Free-body diagram: $n = 0$ (up), mg (down)

$$\Sigma F_y = ma_y$$

$$-mg = ma \Rightarrow \underline{a = -g} \quad \text{(free-fall!)}$$

5-33

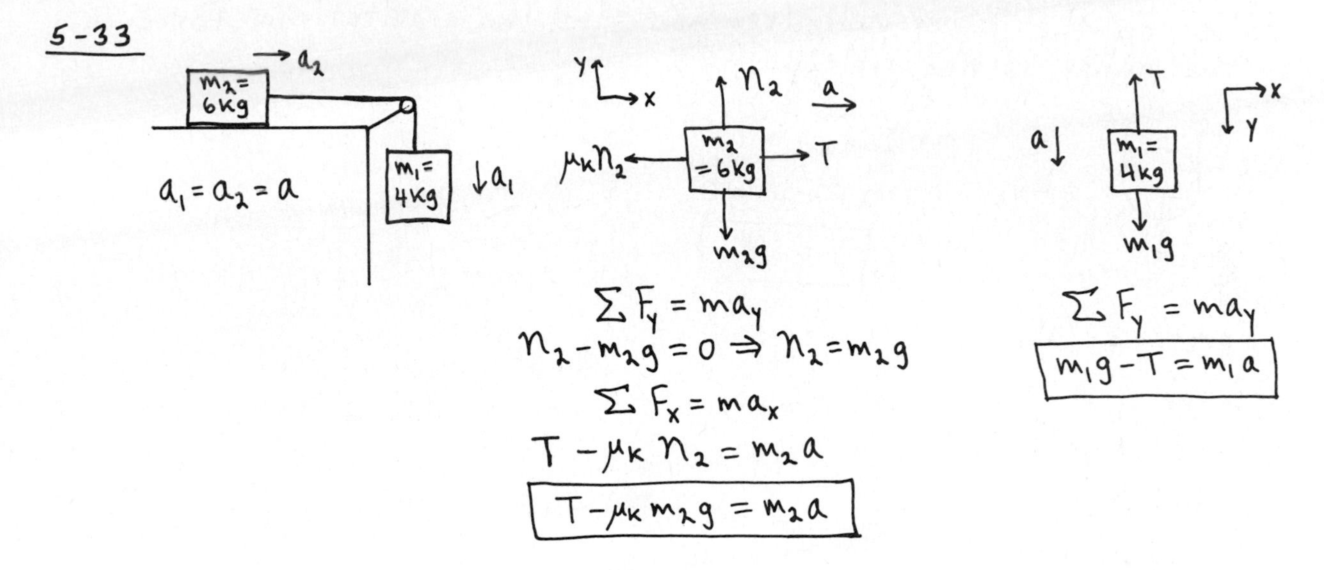

$$\Sigma F_y = ma_y$$

$$n_2 - m_2 g = 0 \Rightarrow n_2 = m_2 g$$

$$\Sigma F_x = ma_x$$

$$T - \mu_K n_2 = m_2 a$$

$$\boxed{T - \mu_K m_2 g = m_2 a}$$

$$\Sigma F_y = ma_y$$

$$\boxed{m_1 g - T = m_1 a}$$

a) Add these two equations:

$$T - \mu_K m_2 g = m_2 a$$
$$m_1 g - T = m_1 a$$
$$\overline{(m_1 - \mu_K m_2)\, g = (m_1 + m_2)\, a}$$

$$a = g\left(\frac{m_1 - \mu_K m_2}{m_1 + m_2}\right) = (9.8\,m\cdot s^{-2})\left(\frac{4kg - (0.5)(6kg)}{4kg + 6kg}\right) = \underline{0.98\,m\cdot s^{-2}}$$

b) $m_1 g - T = m_1 a \Rightarrow T = m_1(g - a) = 4kg\,(9.8\,m\cdot s^{-2} - 0.98\,m\cdot s^{-2}) = \underline{35.3\,N}$

or $T - \mu_K m_2 g = m_2 a \Rightarrow T = m_2(\mu_K g + a) = 6kg\,[0.5(9.8\,m\cdot s^{-2}) + (0.98\,m\cdot s^{-2})]$

$T = 35.3\,N$ ✓

Problems

5-37

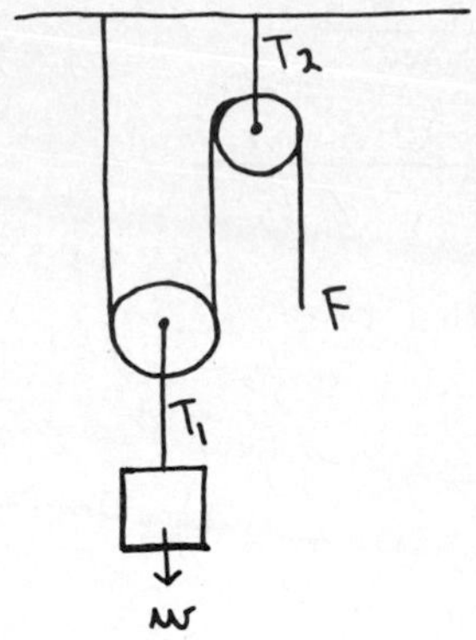

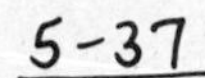

The tension in the rope is F.
T_1 and T_2 are the tensions in the chains.
Constant velocity $\Rightarrow a = 0$.

Forces on the weight:

a = 0

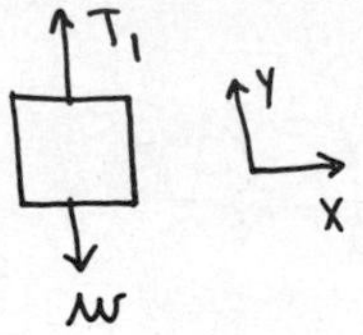

$$\sum F_y = ma_y$$
$$T_1 - w = 0$$
$$\boxed{T_1 = w}$$

Forces on the lower pulley (remember that the gravitational force on the pulley is negligible):

a = 0

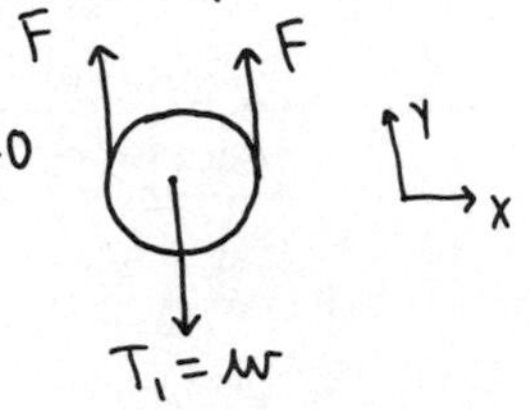

$$\sum F_y = ma_y$$
$$2F - w = 0$$
$$\boxed{F = \frac{w}{2}}$$

Forces on the upper pulley:

a = 0

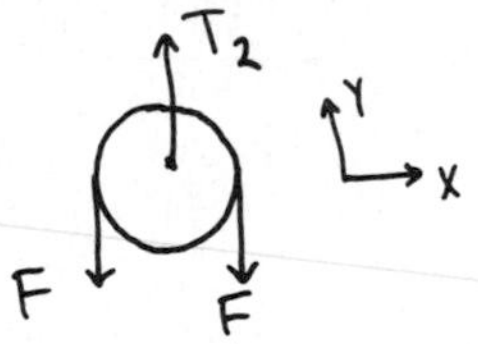

$$\sum F_y = ma_y$$
$$T_2 - 2F = 0$$
$$T_2 = 2F = 2\left(\frac{w}{2}\right) = w \quad ; \quad \boxed{T_2 = w}$$

5-39

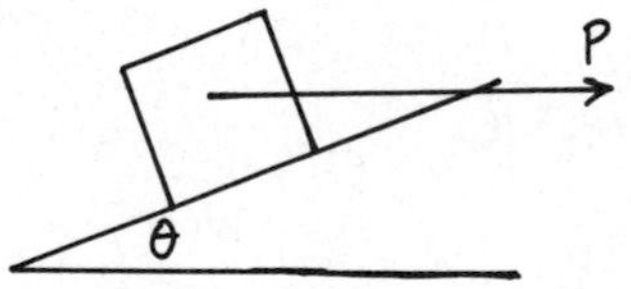

No friction.
constant $v \Rightarrow a = 0$.

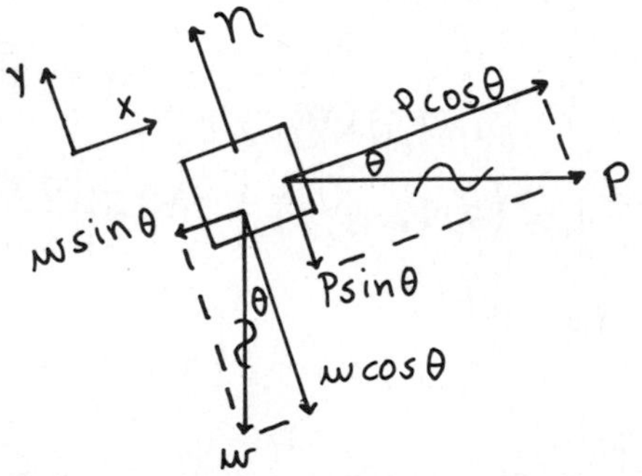

$$\sum F_x = 0$$
$$P\cos\theta - w\sin\theta = 0$$
$$\boxed{P = w\tan\theta}$$

or, could do this particular problem (a=0) with a different choice of axis:

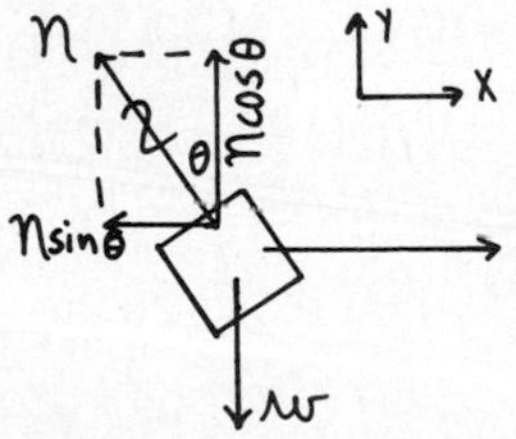

$$\sum F_y = ma_y$$
$$n\cos\theta - w = 0$$
$$n = \frac{w}{\cos\theta}$$

$$\sum F_x = ma_x$$
$$P - n\sin\theta = 0$$
$$P = \left(\frac{w}{\cos\theta}\right)\sin\theta$$
$$P = w\tan\theta, \text{ which checks}$$

5-41

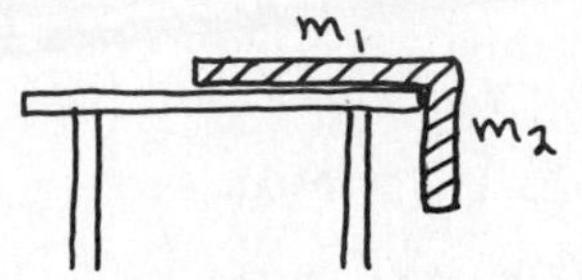

Let m_1 be the mass of that part of the rope that is on the table, and m_2 be the mass of that part of the rope that is hanging over the edge.
($m_1 + m_2 = m$, the total mass of the rope)

Forces on the hanging part of the rope:

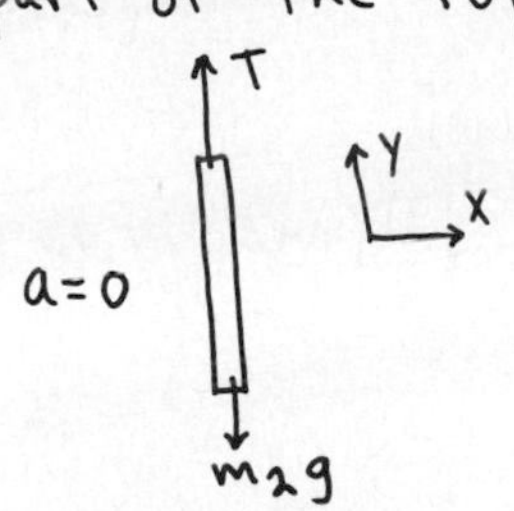

$$\Sigma F_y = ma_y$$
$$T - m_2 g = 0$$
$$\boxed{T = m_2 g}$$

Forces on the part of the rope on the table:

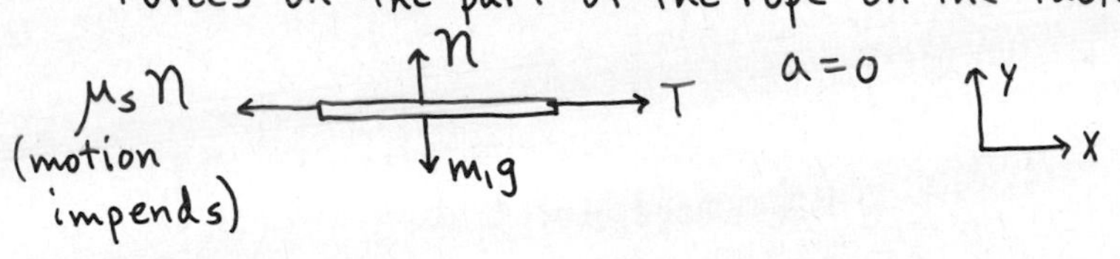

$$\Sigma F_y = ma_y$$
$$n - m_1 g = 0$$
$$n = m_1 g$$

$$\Sigma F_x = ma_x$$
$$T - \mu_s n = 0$$
$$T = \mu_s n$$
$$\boxed{T = \mu_s m_1 g}$$

Equate these two eqs. for T $\Rightarrow$ $m_2 g = \mu_s m_1 g$

$$m_2 = \mu_s m_1$$

The fraction that can hang over the edge is

$$\frac{m_2}{m} = \frac{m_2}{m_1 + m_2} = \frac{\mu_s m_1}{m_1 + \mu_s m_1} = \underline{\frac{\mu_s}{1 + \mu_s}}$$

5-43

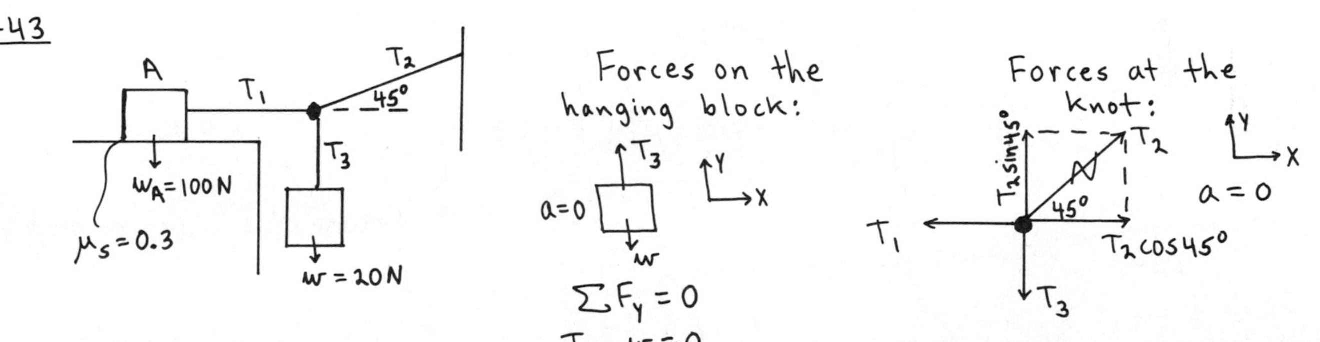
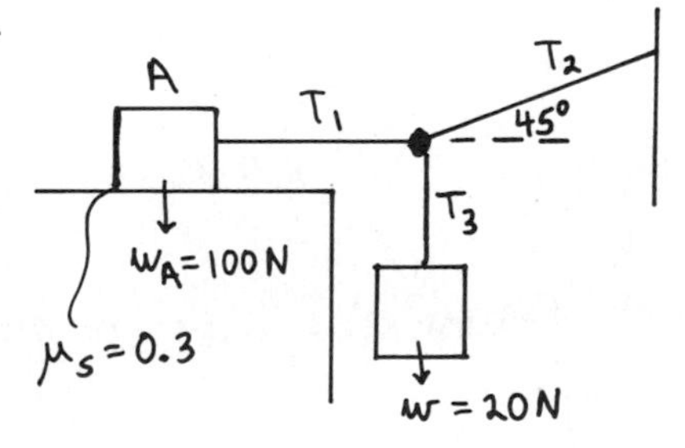

Forces on the hanging block:

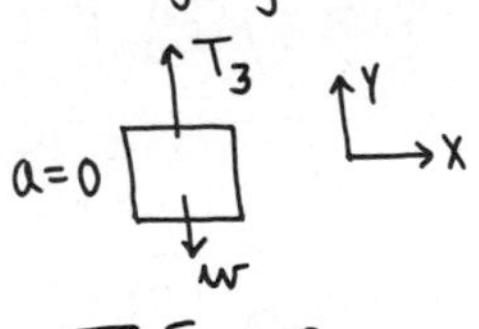

$$\Sigma F_y = 0$$
$$T_3 - w = 0$$
$$T_3 = w = 20\text{N}$$

Forces at the knot:

$$\Sigma F_y = ma_y$$
$$T_2 \sin 45° - T_3 = 0$$
$$T_2 \sin 45° = T_3$$
$$T_2 = \frac{20\text{ N}}{\sin 45°} = 28.3\text{N}$$

$$\Sigma F_x = ma_x$$
$$T_2 \cos 45° - T_1 = 0$$
$$T_1 = T_2 \cos 45°$$
$$T_1 = 20\text{N}$$

5-43 (cont)

Forces on block A:

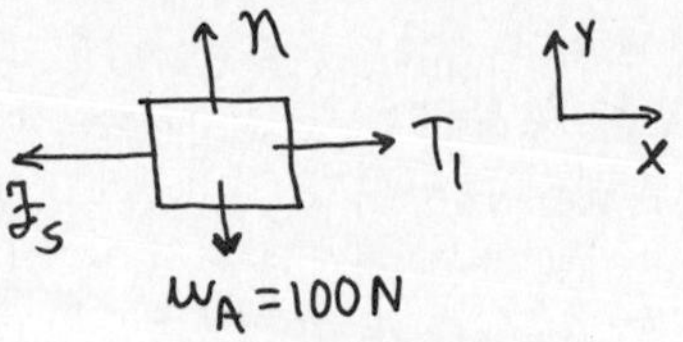

$\Sigma F_x = ma_x$

$T_1 - f_s = 0$

$f_s = \underline{20N}$

$\Sigma F_y = ma_y$

$n - w_A = 0$

$n = 100N$

$\mu_s n = (0.3)(100N) = 30N$

$f_s < \mu_s n$; motion is not impending

b) maximum $w \Rightarrow$ motion impends $\Rightarrow f_s = \mu_s n$

Use the above force diagrams and eqs., but with $f_s = \mu_s n$, and w as unknown.

Forces on block A: $f_s = \mu_s n = 30N$

$T_1 = f_s = 30N$

Forces at the knot: $T_2 \cos 45° = T_1 = 30N \Rightarrow T_2 = 42.4N$

$T_2 \sin 45° - T_3 = 0 \Rightarrow T_3 = T_2 \sin 45° = 30N$

Forces on the hanging block: $w = T_3 = \underline{30N}$

5-49

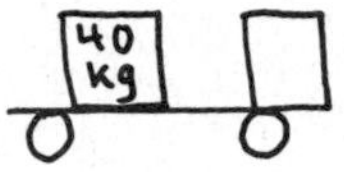

First calculate the maximum acceleration that the static friction force can give to the case:

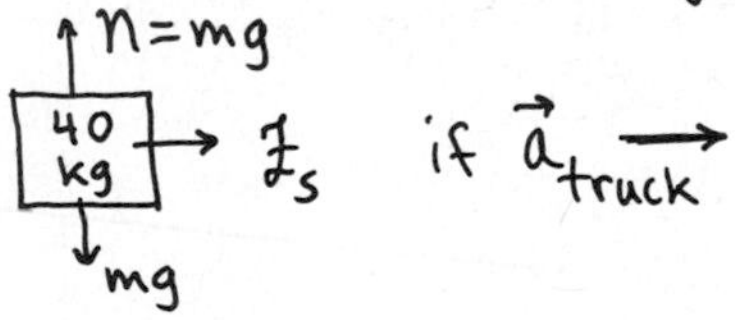

or

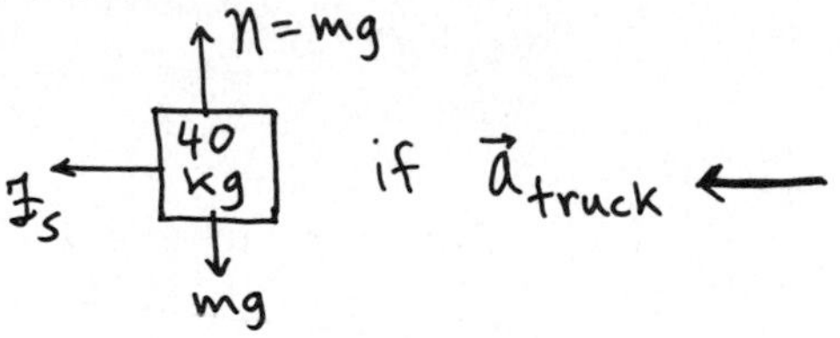

max $a \Rightarrow f_s = \mu_s n = \mu_s mg$

$f_s = ma$

$\mu_s \cancel{m} g = \cancel{m} a \Rightarrow a = \mu_s g = (0.3)(9.8 m\cdot s^{-2}) = 2.95 m\cdot s^{-2}$

If $a_{truck} > 2.95 m\cdot s^{-2}$ the case cannot follow the motion and slides relative to the truck. In this instance $f = f_k = \mu_k n$.

If $a_{truck} < 2.95 m\cdot s^{-2}$ the case will not slip. It will have the same acceleration as the truck.

In either situation the friction force on the case will give rise to an acceleration in the same direction as the acceleration of the truck; the friction tries to keep the case moving along with the truck.

5-49 (cont)

a) $a_{truck} = 2.5\ m\cdot s^{-2}$ $\qquad \longrightarrow \vec{a}_{truck}$

This is less than the maximum of $2.95\ m\cdot s^{-2}$ that can be given to the case by static friction $\Rightarrow a_{case} = 2.5\ m\cdot s^{-2}$.

$n = mg = 392\ N$; $a \rightarrow$; box: 40 kg, $\rightarrow f_s$; $mg \downarrow$; axes y, x

$$\Sigma F_x = ma_x$$

$$f_s = ma = (40\ kg)(2.5\ m\cdot s^{-2}) = \underline{100\ N};$$

toward the front of the truck

b) $a_{truck} = -3.0\ m\cdot s^{-2}$ $\qquad \longleftarrow \vec{a}_{truck}$

$|a_{truck}| > 2.95\ m\cdot s^{-2} \Rightarrow |a_{truck}| > |a_{case}|$, and the case slips

Thus $f = f_k = \mu_k n = (0.2)(392\ N) = \underline{78.4\ N}$

$\vec{a}_{truck}$ and $\vec{a}_{case}$ are in the same direction $\Rightarrow$ the friction force is directed toward the rear of the truck.

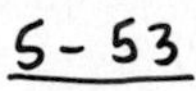

5-53

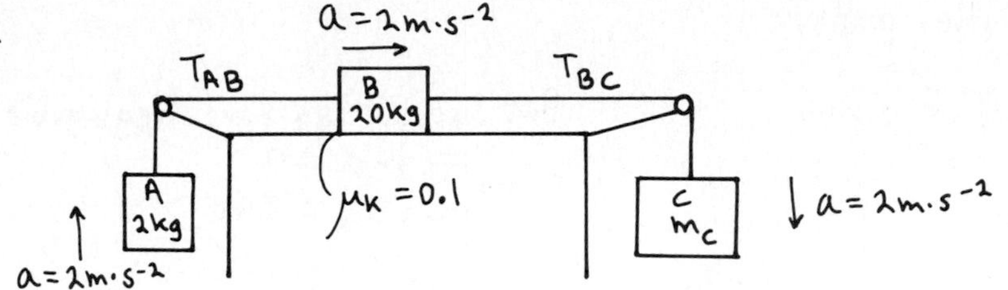

Parts (a) and (b) will be done together.

Consider the forces on each block. For each block let the direction of $\vec{a}$ be a positive coordinate direction.

Block A: $T_{AB}\uparrow$; $a = 2\ m\cdot s^{-2}\uparrow$; A, 2 kg; $m_A g = 19.6\ N \downarrow$; axes y, x

$$\Sigma F_y = ma_y$$

$$T_{AB} - m_A g = m_A a$$

$$T_{AB} = m_A(g + a) = 2\ kg\,(9.8\ m\cdot s^{-2} + 2.0\ m\cdot s^{-2}) = \underline{23.6\ N}$$

Block B: $n\uparrow$; $a = 2\ m\cdot s^{-2} \rightarrow$; $T_{AB}\leftarrow$; $\mu_k n \leftarrow$; B, 20 kg; $\rightarrow T_{BC}$; $m_B g = 196\ N\downarrow$

$$\Sigma F_y = ma_y$$

$$n - m_B g = 0$$

$$n = 196\ N$$

$$\mu_k n = 19.6\ N$$

$$\Sigma F_x = ma_x$$

$$T_{BC} - T_{AB} - \mu_k n = m_B a$$

$$T_{BC} = T_{AB} + \mu_k n + m_B a$$

$$T_{BC} = 23.6\ N + 19.6\ N + (20\ kg)(2\ m\cdot s^{-2})$$

$$T_{BC} = \underline{83.2\ N}$$

5-53 (cont)

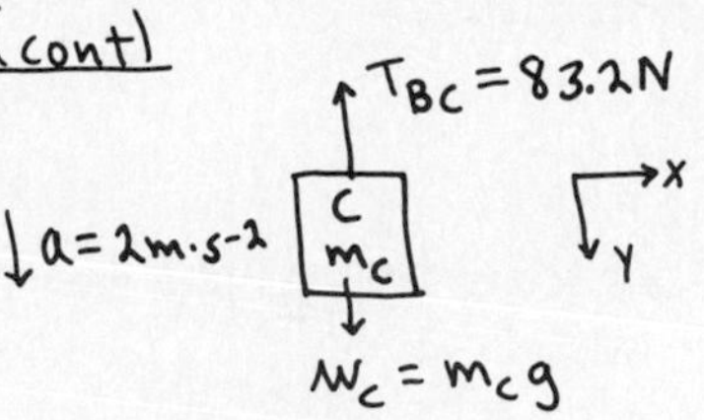

$$\sum F_y = ma_y$$

$$m_c g - T_{BC} = m_c a$$

$$m_c = \frac{T_{BC}}{g-a} = \frac{83.2N}{9.8 m \cdot s^{-2} - 2.0 m \cdot s^{-2}} = \underline{10.7 kg}$$

5-55

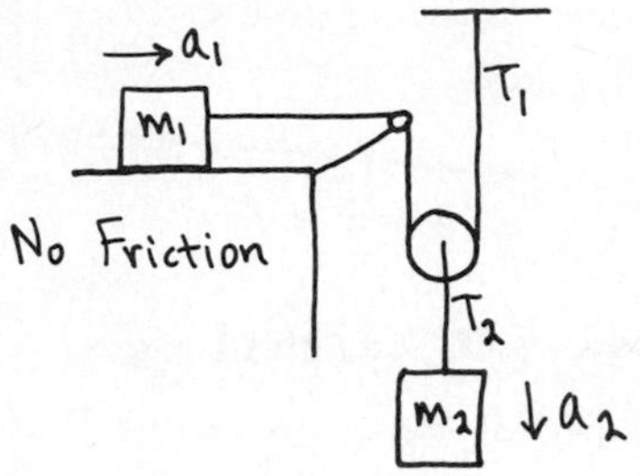

Forces on m_1:

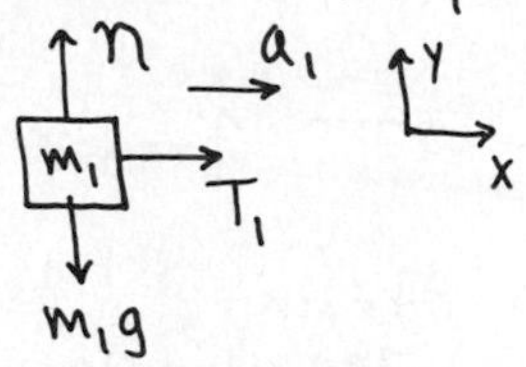

$$\sum F_x = ma_x$$

$$\boxed{T_1 = m_1 a_1}$$

Forces on m_2:

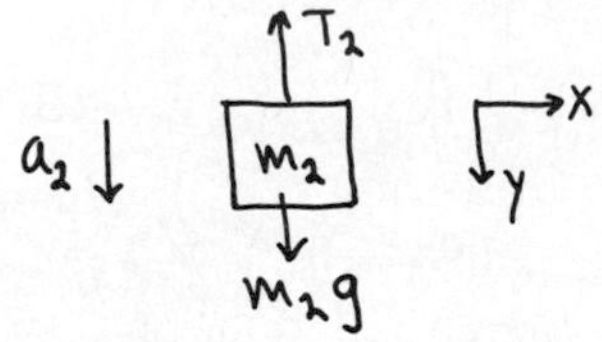

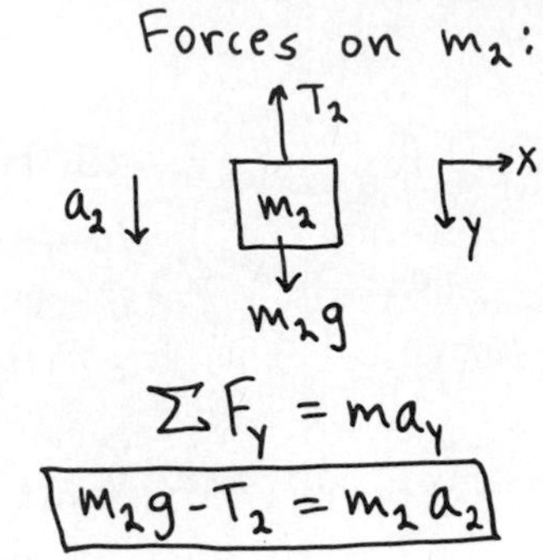

$$\sum F_y = ma_y$$

$$\boxed{m_2 g - T_2 = m_2 a_2}$$

This gives us two equations, but 4 unknowns (T_1, T_2, a_1 and a_2).

Forces on the movable pulley:

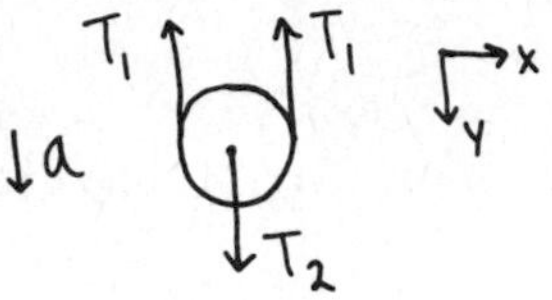

$$\sum F_y = ma_y$$

$$T_2 - 2T_1 = ma$$

But our pulleys have negligible mass

$\Rightarrow ma = 0$

$\Rightarrow T_2 = 2T_1$

Use this in the second of the above equations $\Rightarrow \boxed{m_2 g - 2T_1 = m_2 a_2}$

This reduces the unknowns to 3 (T_1, a_1 and a_2), but we still have only two equations.

But the accelerations a_1 and a_2 are related. In a given amount of time if m_1 moves to the right a distance d then in the same time m_2 moves downward a distance d/2. Thus,

since $v = \lim_{\Delta t \to 0} \frac{\Delta x}{\Delta t}$, $v_1 = 2v_2$. And furthermore, since $a = \lim_{\Delta t \to 0} \frac{\Delta v}{\Delta t}$,

$\boxed{a_1 = 2a_2}$. This is the third equation that we needed.

Use the above in the first eq:

$$T_1 = m_1 a_1 = 2m_1 a_2$$

Substitute this into $m_2 g - 2T_1 = m_2 a_2$

$$\Rightarrow m_2 g - 2(2m_1 a_2) = m_2 a_2$$

$$a_2(m_2 + 4m_1) = m_2 g \Rightarrow \boxed{a_2 = \frac{m_2 g}{4m_1 + m_2}}$$

5-49 (cont)

a) $a_{truck} = 2.5\ m\cdot s^{-2}$ $\longrightarrow \vec{a}_{truck}$

This is less than the maximum of $2.95\ m\cdot s^{-2}$ that can be given to the case by static friction $\Rightarrow a_{case} = 2.5\ m\cdot s^{-2}$.

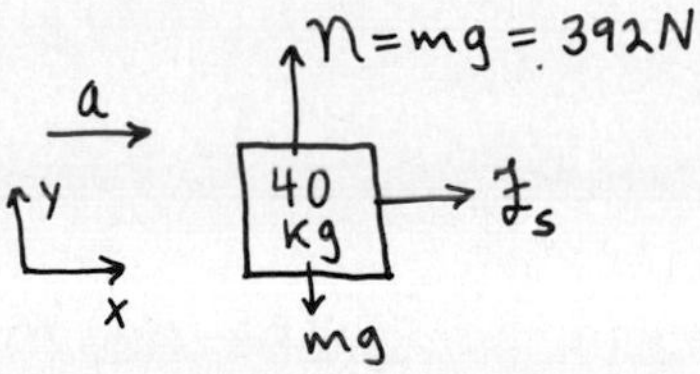

n=mg=392N
a
40 kg
f_s
mg
y
x

$\Sigma F_x = ma_x$

$f_s = ma = (40kg)(2.5m\cdot s^{-2}) = \underline{100\ N}$; toward the front of the truck

b) $a_{truck} = -3.0\ m\cdot s^{-2}$ $\longleftarrow \vec{a}_{truck}$

$|a_{truck}| > 2.95\ m\cdot s^{-2} \Rightarrow |a_{truck}| > |a_{case}|$, and the case slips

Thus $f = f_K = \mu_K n = (0.2)(392N) = \underline{78.4N}$

$\vec{a}_{truck}$ and $\vec{a}_{case}$ are in the same direction $\Rightarrow$ the friction force is directed toward the rear of the truck.

5-53

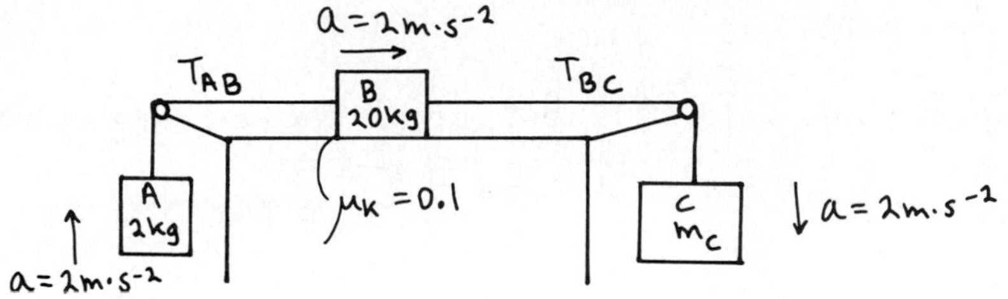

a=2m·s-2
T_AB
B 20kg
T_BC
A 2kg
μK = 0.1
C m_C
a=2m·s-2
a=2m·s-2

Parts (a) and (b) will be done together.

Consider the forces on each block. For each block let the direction of $\vec{a}$ be a positive coordinate direction.

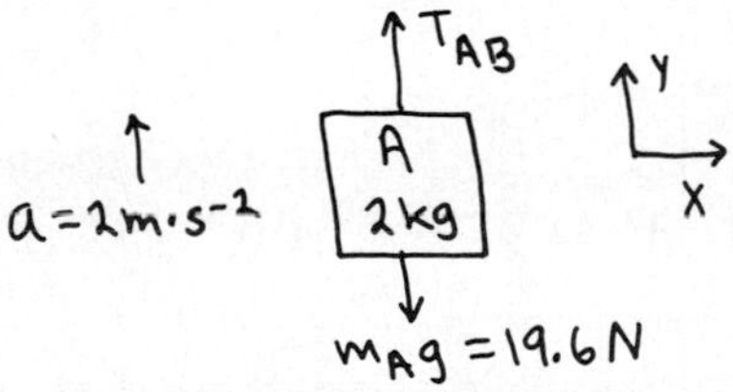

T_AB
a=2m·s-2
A 2kg
m_A g = 19.6N
y
x

$\Sigma F_y = ma_y$

$T_{AB} - m_A g = m_A a$

$T_{AB} = m_A(g+a) = 2kg(9.8m\cdot s^{-2} + 2.0m\cdot s^{-2}) = \underline{23.6N}$

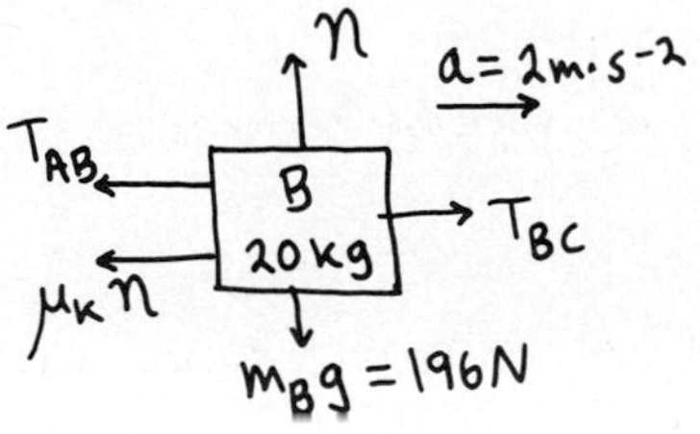

n
a=2m·s-2
T_AB
B 20kg
T_BC
μK n
m_B g = 196N

$\Sigma F_y = ma_y$

$n - m_B g = 0$

$n = 196N$

$\mu_K n = 19.6\ N$

$\Sigma F_x = ma_x$

$T_{BC} - T_{AB} - \mu_K n = m_B a$

$T_{BC} = T_{AB} + \mu_K n + m_B a$

$T_{BC} = 23.6N + 19.6N + (20kg)(2m\cdot s^{-2})$

$T_{BC} - \underline{83.2\ N}$

5-53 (cont)

$T_{BC} = 83.2\,N$

$a = 2\,m\cdot s^{-2}$ (down), block C, mass m_c, $w_c = m_c g$

$$\sum F_y = m a_y$$

$$m_c g - T_{BC} = m_c a$$

$$m_c = \frac{T_{BC}}{g-a} = \frac{83.2\,N}{9.8\,m\cdot s^{-2} - 2.0\,m\cdot s^{-2}} = \underline{10.7\,kg}$$

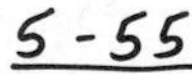

5-55

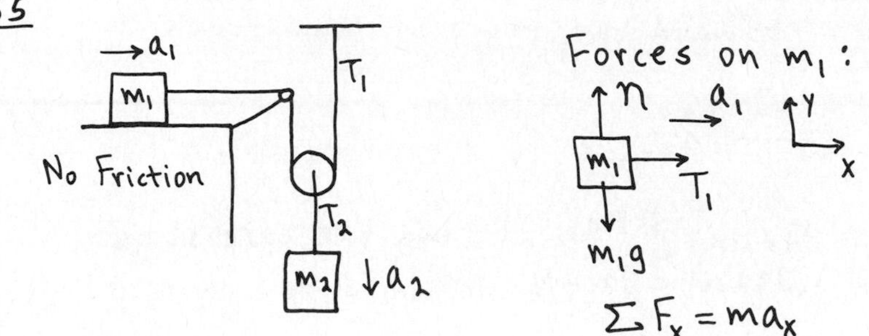

Forces on m_1:

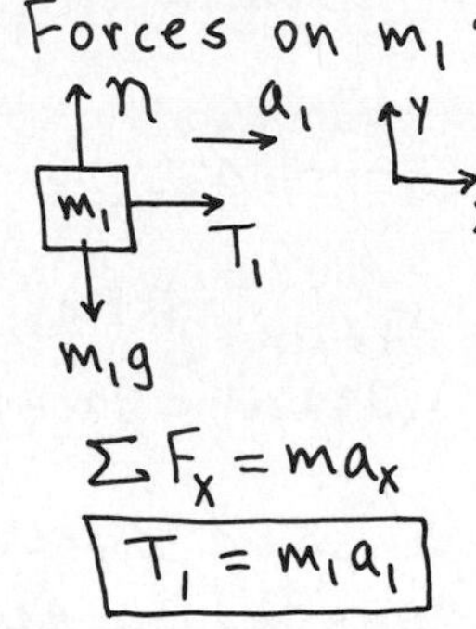

$$\sum F_x = m a_x$$

$$\boxed{T_1 = m_1 a_1}$$

Forces on m_2:

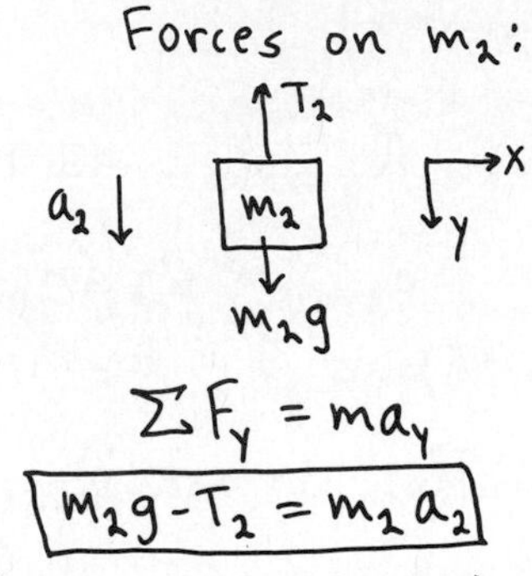

$$\sum F_y = m a_y$$

$$\boxed{m_2 g - T_2 = m_2 a_2}$$

This gives us two equations, but 4 unknowns (T_1, T_2, a_1 and a_2).

Forces on the movable pulley:

T_1 (up), T_1 (up), T_2 (down), a (down)

$$\sum F_y = m a_y$$

$$T_2 - 2T_1 = ma$$

But our pulleys have negligible mass

$$\Rightarrow ma = 0$$

$$\Rightarrow T_2 = 2T_1$$

Use this in the second of the above equations $\Rightarrow \boxed{m_2 g - 2T_1 = m_2 a_2}$

This reduces the unknowns to 3 (T_1, a_1 and a_2), but we still have only two equations.

But the accelerations a_1 and a_2 are related. In a given amount of time if m_1 moves to the right a distance d then in the same time m_2 moves downward a distance $d/2$. Thus,

since $v = \lim_{\Delta t \to 0} \frac{\Delta x}{\Delta t}$, $v_1 = 2v_2$. And furthermore, since $a = \lim_{\Delta t \to 0} \frac{\Delta v}{\Delta t}$,

$\boxed{a_1 = 2a_2}$. This is the third equation that we needed.

Use the above in the first eq:

$$T_1 = m_1 a_1 = 2m_1 a_2$$

Substitute this into $m_2 g - 2T_1 = m_2 a_2$

$$\Rightarrow m_2 g - 2(2m_1 a_2) = m_2 a_2$$

$$a_2(m_2 + 4m_1) = m_2 g \Rightarrow \boxed{a_2 = \frac{m_2 g}{4m_1 + m_2}}$$

5-55 (cont)

Then $a_1 = 2a_2 \Rightarrow$ $\boxed{a_1 = \dfrac{2m_2 g}{4m_1 + m_2}}$

5-57

F = 200N; 7 kg; 4kg; 5 kg

Note that the mass of the rope is given, and that it is not negligible compared to the other masses.

a) Treat the rope and two blocks all together as a single object, of mass 16kg:

F = 200N; a; 16 kg; Y; X

$mg = (16\text{kg})(9.8\text{m}\cdot\text{s}^{-2}) = 156.8\text{N}$

$$\Sigma F_y = ma_y$$

$$F - mg = ma$$

$$a = \frac{F - mg}{m} = \frac{200\text{N} - 156.8\text{N}}{16\text{kg}} = \underline{2.7\text{m}\cdot\text{s}^{-2}}$$

b) Consider the forces on the 7kg block, since the tension in the top of the rope (T_t) will be one of these forces.

F = 200N; $a = 2.7\ \text{m}\cdot\text{s}^{-2}$; 7 kg; mg; T_t; Y; X

$$\Sigma F_y = ma_y$$

$$F - mg - T_t = ma$$

$$T_t = F - mg - ma$$

$$T_t = 200\text{N} - (7\text{kg})(9.8\text{m}\cdot\text{s}^{-2}) - (7\text{kg})(2.7\text{m}\cdot\text{s}^{-2})$$

$$T_t = 200\text{N} - 68.6\text{N} - 18.9\text{N} = \underline{112.5\text{N}}$$

c) Consider the forces on the top half of the rope. Let T_m be the tension at the middle of the rope. The top half of the rope has mass $\frac{1}{2}(4\text{kg}) = 2\text{kg}$.

T_t; $a = 2.7\ \text{m}\cdot\text{s}^{-2}$; 2 kg; mg; T_m; Y; X

$$\Sigma F_y = ma_y$$

$$T_t - T_m - mg = ma$$

$$T_m = T_t - mg - ma$$

$$T_m = 112.5\text{N} - (2\text{kg})(9.8\text{m}\cdot\text{s}^{-2}) - (2\text{kg})(2.7\text{m}\cdot\text{s}^{-2})$$

$$T_m = 112.5\text{N} - 19.6\text{N} - 5.4\text{N} = \underline{87.5\text{N}}$$

5-61

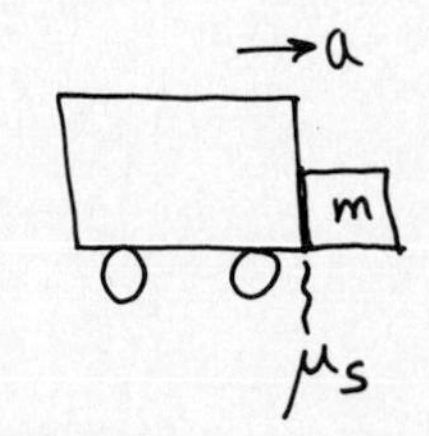

The cart and the block have the same acceleration. The normal force exerted by the cart on the block is perpendicular to the cart surface. The friction force on the block is directed so as to hold the block up against the downward pull of gravity.

Forces on the block:

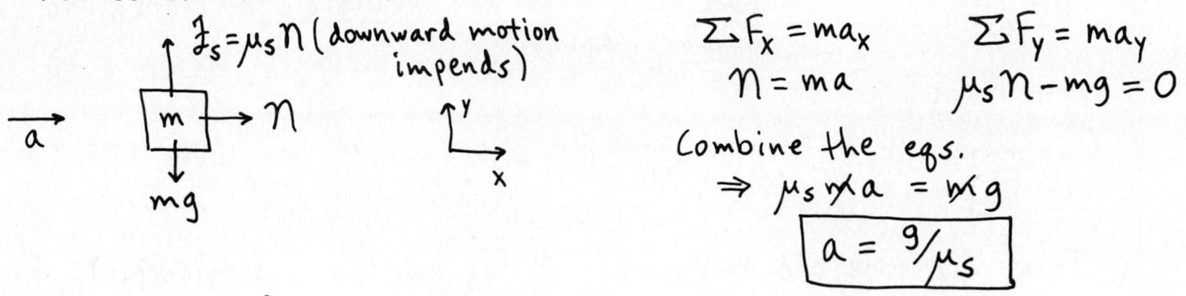

Note that $\Sigma \vec{F} = m\vec{a}$ does not apply in a coordinate frame attached to the cart; such an accelerated frame is non-inertial.

CHAPTER 6

Exercises 1, 7, 9, 13, 15, 17, 21, 23

Problems 27, 31, 33, 35, 39

Exercises

6-1

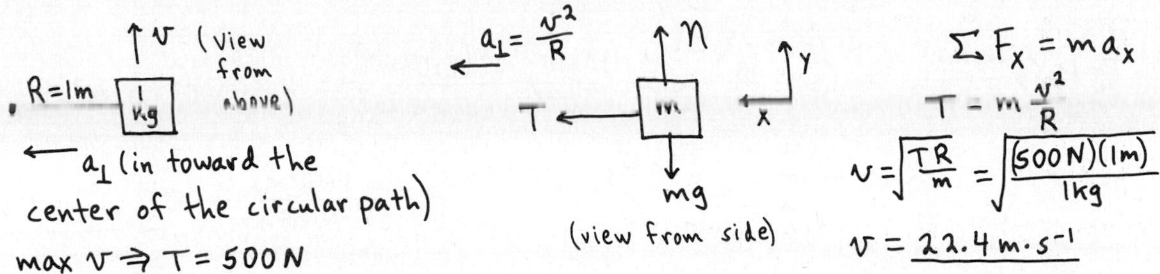

6-7

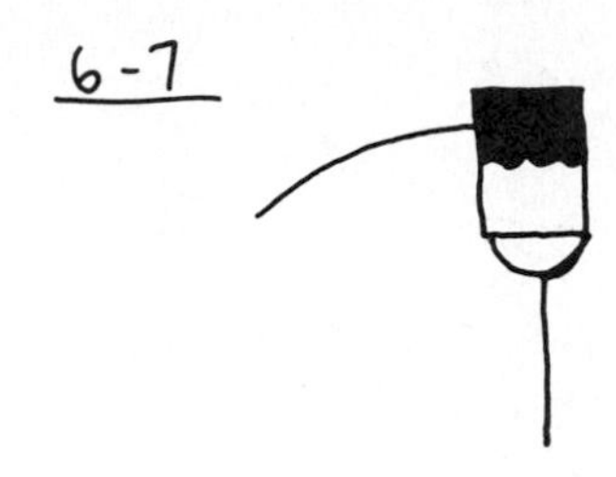

Consider the forces on the water:

v $\quad a_\perp = \frac{v^2}{R}$ $\quad mg \quad n \quad x \quad y$

n is the force exerted by the bottom of the bucket on the water.

$\vec{a}_\perp$ is downward at this point, since that is the direction in toward the center of the circular path.

$$\Sigma F_y = ma_y$$
$$mg + n = m\frac{v^2}{R}$$

At the critical velocity the water is just ready to lose contact with the bottom of the bucket, so at this velocity $n \to 0$. (Note that the force n cannot be upward.)

$$n \to 0 \Rightarrow \cancel{m}g = \cancel{m}\frac{v_c^2}{R} \Rightarrow v_c = \sqrt{gR} = \sqrt{(9.8\,\text{m·s}^{-2})(1\,\text{m})} = \underline{3.13\,\text{m·s}^{-1}}$$

6-9

a)

R

$v = 180\ \text{m·s}^{-1}$

$$a_\perp = \frac{v^2}{R}$$
$$a_\perp = 7g \Rightarrow 7g = \frac{v^2}{R}$$
$$R = \frac{v^2}{7g} = \frac{(180\,\text{m·s}^{-1})^2}{7(9.8\,\text{m·s}^{-2})} = \underline{472\,\text{m}}$$

6-9 (cont)

b) Consider the forces on the pilot:

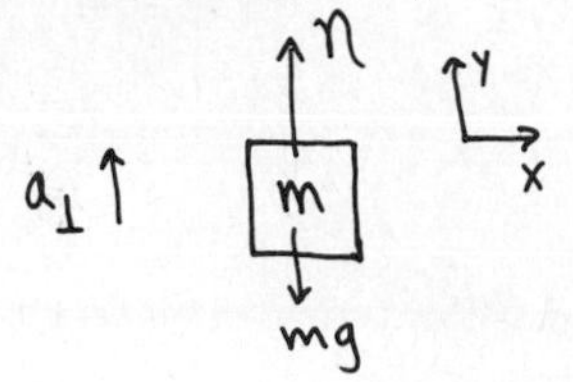

n is the normal force exerted on the pilot by his chair. This force is his apparent weight.

$$\Sigma F_y = ma_y$$
$$n - mg = ma_\perp$$
$$n = m(g + a_\perp)$$
$$a_\perp = 7g \Rightarrow n = 8mg$$

The pilot's weight is 8 times his true weight.

6-13

$$F_g = G\frac{m_1 m_2}{r^2} \Rightarrow G = \frac{F_g r^2}{m_1 m_2} = \frac{(13\times10^{-11}\,N)(0.04\,m)^2}{(0.8\,kg)(0.004\,kg)} = 6.50\times10^{-11}\,N\cdot m^2\cdot kg^{-2}$$

$$g = G\frac{m_E}{R_E^2} \text{ (eq. 6-15)} \Rightarrow m_E = \frac{gR_E^2}{G} = \frac{(9.8\,m\cdot s^{-2})(6.4\times10^6\,m)^2}{6.50\times10^{-11}\,N\cdot m^2\cdot kg^{-2}} = \underline{6.18\times10^{24}\,kg}$$

6-15

$$g = \frac{F_g}{m'} \;;\; F_g = G\frac{mm'}{r^2} \Rightarrow g = G\frac{m}{r^2}$$

$$g = (6.67\times10^{-11}\,N\cdot m^2\cdot kg^{-2})\frac{5\,kg}{(3\,m)^2} = \underline{3.71\times10^{-11}\,m\cdot s^{-2}}$$

6-17

$$\vec{g} = \frac{\vec{F}}{m} \Rightarrow g_x = \frac{F_x}{m},\quad g_y = \frac{F_y}{m}$$

$$F_x = -0.1\,N \Rightarrow g_x = \frac{-0.1\,N}{0.01\,kg} = \underline{-10\,m\cdot s^{-2}}$$

$$F_y = +0.4\,N \Rightarrow g_y = \frac{+0.4\,N}{0.01\,kg} = \underline{+40\,m\cdot s^{-2}}$$

6-21

a) $$T = \frac{2\pi r}{v} = \frac{2\pi(7\times10^6\,m)}{7550\,m\cdot s^{-1}} = 5825\,s = \underline{97.1\,min}$$

(Note: $v = \sqrt{\frac{Gm_E}{r}} = \sqrt{\frac{(6.67\times10^{-11}\,N\cdot m^2\cdot kg^{-2})(5.98\times10^{24}\,kg)}{7.0\times10^6\,m}} = 7550\,m\cdot s^{-1}$, so the data given is consistent with $\Sigma\vec{F} = m\vec{a}$.)

b) $$g = G\frac{m_E}{r^2} = (6.67\times10^{-11}\,N\cdot m^2\cdot kg^{-2})\left(\frac{5.98\times10^{24}\,kg}{7.0\times10^6\,m}\right) = \underline{8.14\,m\cdot s^{-2}}$$

6-23

a) $w = mg = (2\,kg)(25\,m\cdot s^{-2}) = \underline{50\,N}$; this is the true weight w_0 of the object.

6-23 (cont)

b)

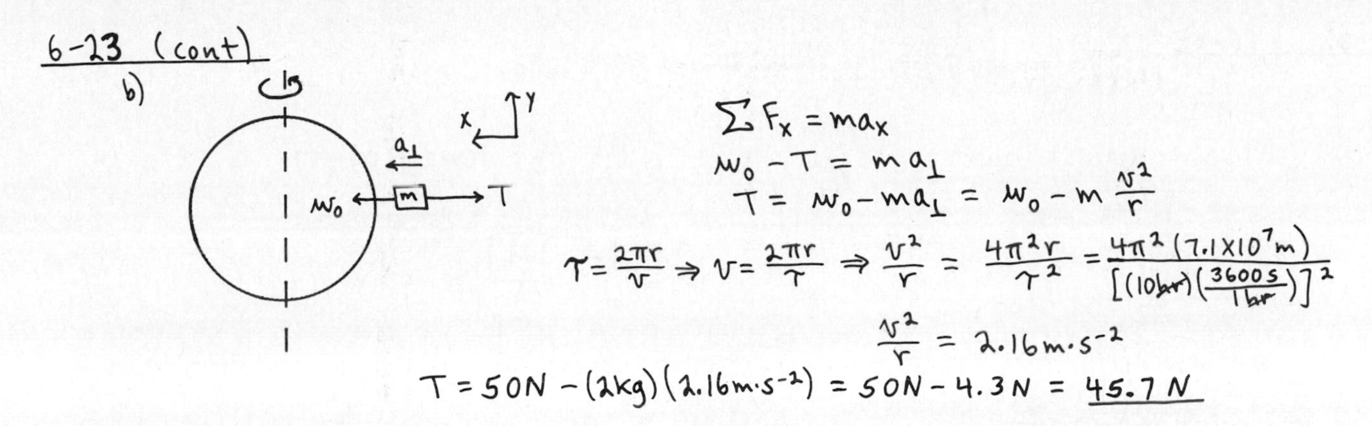

$$\Sigma F_x = ma_x$$
$$w_0 - T = ma_\perp$$
$$T = w_0 - ma_\perp = w_0 - m\frac{v^2}{r}$$

$$\mathcal{T} = \frac{2\pi r}{v} \Rightarrow v = \frac{2\pi r}{\mathcal{T}} \Rightarrow \frac{v^2}{r} = \frac{4\pi^2 r}{\mathcal{T}^2} = \frac{4\pi^2(7.1\times10^7\,\text{m})}{\left[(10\,\text{hr})\left(\frac{3600\,\text{s}}{1\,\text{hr}}\right)\right]^2}$$

$$\frac{v^2}{r} = 2.16\,\text{m}\cdot\text{s}^{-2}$$

$$T = 50\,\text{N} - (2\,\text{kg})(2.16\,\text{m}\cdot\text{s}^{-2}) = 50\,\text{N} - 4.3\,\text{N} = \underline{45.7\,\text{N}}$$

Problems

6-27

a) The person is held up against gravity by the static friction force exerted on him by the wall.

3m

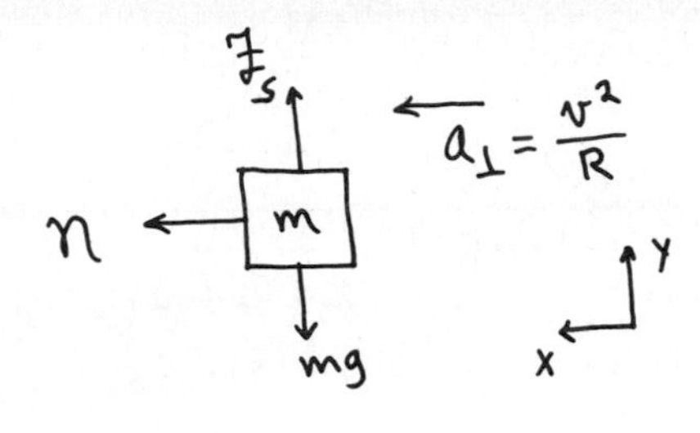

b) min $\mu_s \Rightarrow \mathcal{F}_s = \mu_s n$ (motion impends)

$$\Sigma F_y = ma_y \qquad \Sigma F_x = ma_x$$
$$\mathcal{F}_s - mg = 0 \qquad n = m\frac{v^2}{r}$$
$$\mu_s n = mg$$

Combine these two eqs.:

$$\mu_s m \frac{v^2}{r} = mg$$
$$\boxed{\mu_s = \frac{gr}{v^2}}$$

Calculate v: $v = \frac{\text{distance}}{\text{time}} = \frac{(0.5\,\text{rev})\left(\frac{2\pi r}{1\,\text{rev}}\right)}{1\,\text{s}} = \frac{\pi(3\,\text{m})}{1\,\text{s}} = 9.42\,\text{m}\cdot\text{s}^{-1}$

$$\mu_s = \frac{(9.8\,\text{m}\cdot\text{s}^{-2})(3\,\text{m})}{(9.42\,\text{m}\cdot\text{s}^{-1})^2} = \underline{0.331}$$

c) No, the mass of the person divides out of the equations.

6-31

a) n_t, $\downarrow a_\perp = \frac{v^2}{r}$, mg; n_b, $\uparrow a_\perp = \frac{v^2}{r}$, mg

The normal force exerted on the person by the chair in which he sits is the apparent weight. Note that at each point in the circular motion the acceleration $a_\perp$ is directed toward the center of the circle. When we apply $\Sigma\vec{F} = m\vec{a}$ we will in each case take the direction of $\vec{a}_\perp$ to be positive.

6-31 (cont)

Note: $v = \frac{2\pi r}{t} = \frac{2\pi(5m)}{10s} = 3.14 m \cdot s^{-1}$

At the highest point:

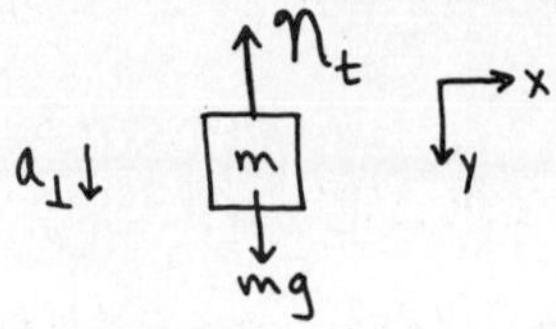

$\sum F_y = ma_y$

$mg - n_t = m\frac{v^2}{r}$

$\boxed{n_t = m(g - \frac{v^2}{r})}$

At the lowest point:

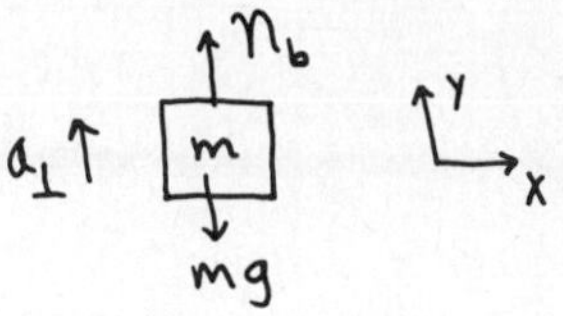

$\sum F_y = ma_y$

$n_b - mg = m\frac{v^2}{r}$

$\boxed{n_b = m(g + \frac{v^2}{r})}$

Subtract $\Rightarrow n_b - n_t = 2m\frac{v^2}{r}$

$$\frac{n_b - n_t}{mg} = \frac{2v^2}{gr} = \frac{2(3.14 m\cdot s^{-1})^2}{(9.8 m\cdot s^{-2})(5m)} = \underline{0.403}$$

That is, the difference in the person's apparent weight at these two points is 0.403 times his weight.

b) $n_t = m(g - \frac{v^2}{r})$

$n_t \to 0 \Rightarrow 0 = m(g - \frac{v^2}{r})$

$\Rightarrow g = \frac{v^2}{r}$; $v = \sqrt{gr} = \sqrt{(9.8 m\cdot s^{-2})(5m)} = 7.0 m\cdot s^{-1}$

$v = \frac{2\pi r}{t} \Rightarrow t = \frac{2\pi r}{v} = \frac{2\pi(5m)}{7 m\cdot s^{-1}} = \underline{4.49 s}$

c) $n_b = m(g + \frac{v^2}{r})$

From part (b), $\frac{v^2}{r} = g \Rightarrow n_b = m(g+g) = \underline{2mg}$

His apparent weight would be twice his true weight.

d)

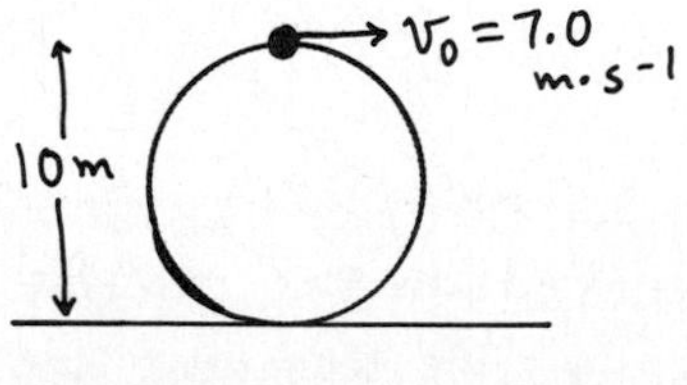

The person moves as a projectile, under the influence of gravity alone.

$a_x = 0$

$a_y = g = 9.8 m\cdot s^{-2}$

Calculate where this unfortunate person strikes the ground:

$\underline{y\text{-comp}}$; find the time in the air

$a_y = +9.8 m\cdot s^{-2}$

$y - y_0 = +10m$

$v_{0y} = 0$

$t = ?$

$y - y_0 = v_{0y}t + \frac{1}{2}a_y t^2$ (with $v_{0y}t \to 0$)

$t = \sqrt{\frac{2(y-y_0)}{a_y}} = \sqrt{\frac{2(10m)}{9.8 m\cdot s^{-2}}}$

$t = \underline{1.43 s}$

6-31 (cont)

(Note: The initial velocity is tangent to the circular path, and hence is horizontal at the top of the circle.)

X-comp

$x - x_0 = ?$

$a_x = 0$

$v_{0x} = 7.0\,m\cdot s^{-1}$

$t = 1.43\,s$

$$x - x_0 = v_{0x} t + \tfrac{1}{2} a_x t^2$$

(with $a_x = 0$)

$$x - x_0 = (7.0\,m\cdot s^{-1})(1.43\,s) = \underline{10.0\,m}$$

The person hits the ground 10m from the base of the ferris wheel.

6-33

Let F_S and F_E be the gravitational forces exerted by the sun and earth, respectively. The earth to sun distance is $r = 1.49 \times 10^{11}\,m$. In the sketch the point P is a distance x from the earth, and hence is a distance $r - x$ from the sun.

$$F_E = G\frac{m_E m}{x^2}\ ;\ F_S = G\frac{m_S m}{(r-x)^2}$$

$$F_E = F_S \Rightarrow G\frac{m_E m}{x^2} = G\frac{m_S m}{(r-x)^2} \Rightarrow \frac{m_E}{x^2} = \frac{m_S}{(r-x)^2}$$

$$(r-x)^2 = x^2\left(\frac{m_S}{m_E}\right)$$

$$r^2 - 2rx + x^2 = x^2(m_S/m_E) \Rightarrow r^2 - 2rx + \left(1 - \frac{m_S}{m_E}\right)x^2 = 0$$

$$\text{Let } \gamma = \frac{m_S}{m_E} - 1 \Rightarrow r^2 - 2rx - \gamma x^2 = 0 \Rightarrow \boxed{\gamma x^2 + 2rx - r^2 = 0}$$

Apply the quadratic formula:

$$x = \frac{1}{2\gamma}\left[-2r \pm \sqrt{4r^2 + 4\gamma r^2}\right]$$

$$\gamma = \frac{m_S}{m_E} - 1 = \frac{1.99 \times 10^{30}\,kg}{5.98 \times 10^{24}\,kg} - 1 = 3.33 \times 10^5$$

Thus

$$x = \frac{1}{2(3.33 \times 10^5)}\left[-2(1.49 \times 10^{11}\,m) \pm \sqrt{4(1.49 \times 10^{11}\,m)^2 + 4(3.33 \times 10^5)(1.49 \times 10^{11}\,m)^2}\right]$$

$$x = -4.474 \times 10^5\,m \pm 2.582 \times 10^8\,m$$

x must be positive $\Rightarrow x = 2.582 \times 10^8\,m - 4.474 \times 10^5\,m = \underline{2.58 \times 10^8\,m}$, from the earth

Note: An accurate approximate calculation can be made much more simply if one notes that the point must be much closer to the earth than to the sun, due to the earth's much smaller mass. Thus it is accurate to approximate $r - x$ by r.

6-33 (cont)

Then

$$\frac{m_E}{x^2} = \frac{m_S}{(r-x)^2} \Rightarrow \frac{m_E}{x^2} = \frac{m_S}{r^2} \Rightarrow x = r\sqrt{\frac{m_E}{m_S}}$$

$x = 1.49 \times 10^{11}\,m \sqrt{\frac{5.98 \times 10^{24}\,kg}{1.99 \times 10^{30}\,kg}} = \underline{2.58 \times 10^{8}\,m}$, which agrees with the above result where we didn't make this simplifying assumption.

6-35

a)

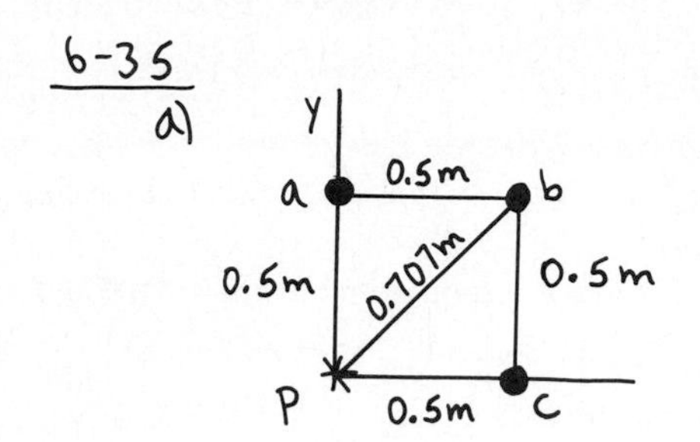

Label the masses a, b, and c. $m_a = 1\,kg$, $m_b = 2\,kg$, $m_c = 1\,kg$.

The gravitational field $\vec{g}$ at P is the sum of the fields due to the individual point masses. For a point mass, $g = G\frac{m}{r^2}$ and is directed toward the point mass.

Thus

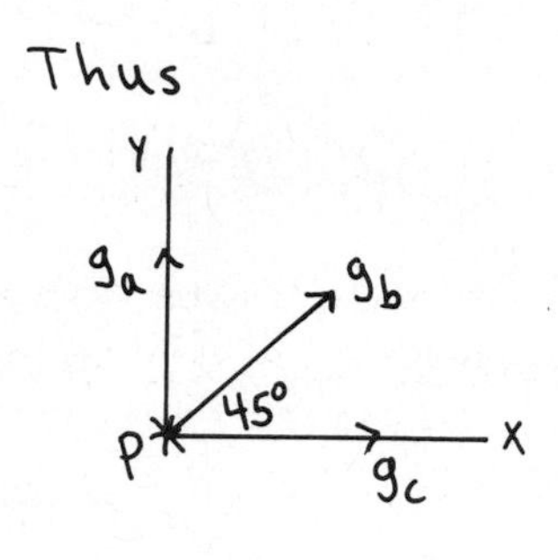

$$g_a = g_c = G\frac{1\,kg}{(0.5\,m)^2} = 2.67 \times 10^{-10}\,m \cdot s^{-2}$$

$$g_b = G\frac{2\,kg}{(0.707\,m)^2} = 2.67 \times 10^{-10}\,m \cdot s^{-2}$$

$$g_x = g_c + g_b \cos 45° = 2.67 \times 10^{-10}\,m \cdot s^{-2}(1 + \cos 45°)$$

$$g_x = \underline{4.56 \times 10^{-10}\,m \cdot s^{-2}}$$

$$g_y = g_a + g_b \sin 45° = 2.67 \times 10^{-10}\,m \cdot s^{-2}(1 + \sin 45°) = \underline{4.56 \times 10^{-10}\,m \cdot s^{-2}}$$

b) $\vec{F} = m\vec{g} \Rightarrow F_x = m g_x = (0.01\,kg)(4.56 \times 10^{-10}\,m \cdot s^{-2}) = 4.56 \times 10^{-12}\,N$

$F_y = m g_y = (0.01\,kg)(4.56 \times 10^{-10}\,m \cdot s^{-2}) = 4.56 \times 10^{-12}\,N$

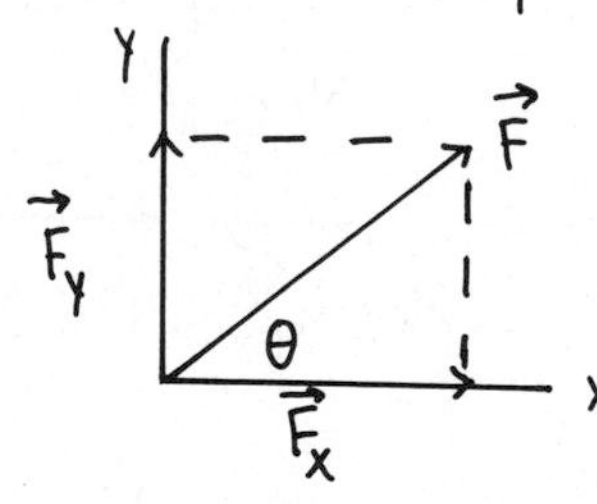

$$F = \sqrt{F_x^2 + F_y^2} = \underline{6.44 \times 10^{-12}\,N}$$

$\tan\theta = \frac{F_y}{F_x} = 1.0 \Rightarrow \theta = \underline{45°}$; the force is directed at 45° above the x-axis.

6-39

At the equator $g = g_0 - \frac{v^2}{R}$ (eq. 6-27)

$g = 0 \Rightarrow g_0 = \frac{v^2}{R}$, where $g_0 = G\frac{m_E}{R^2} = 9.8\,m \cdot s^{-2}$

$$v = \sqrt{g_0 R} = \sqrt{(9.8\,m \cdot s^{-2})(6.38 \times 10^{6}\,m)} = 7.91 \times 10^{3}\,m \cdot s^{-1}$$

$$t = \frac{2\pi R}{v} = \frac{2\pi\,(6.38 \times 10^{6}\,m)}{7.91 \times 10^{3}\,m \cdot s^{-1}} = 5068\,s = \underline{1.41\,hr}$$

(The earth would be a quite different place on which to live!)

CHAPTER 7

Exercises 3, 5, 7, 11, 15, 21, 23, 25, 27, 29, 35, 39, 41, 43, 45

Problems 47, 49, 51, 53, 59, 63, 65, 67

Exercises

7-3

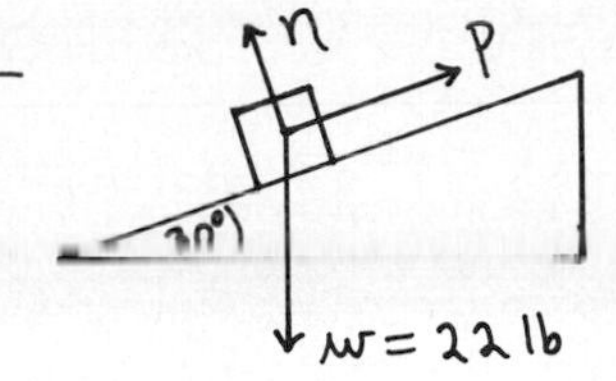

a) $W_P = Ps$ (both P and $s = 15$ ft are parallel to the incline)

$W_P = (18\ \text{lb})(15\ \text{ft}) = \underline{270\ \text{ft}\cdot\text{lb}}$

b) The component of the weight in the direction of the displacement is $-w\sin\theta$.

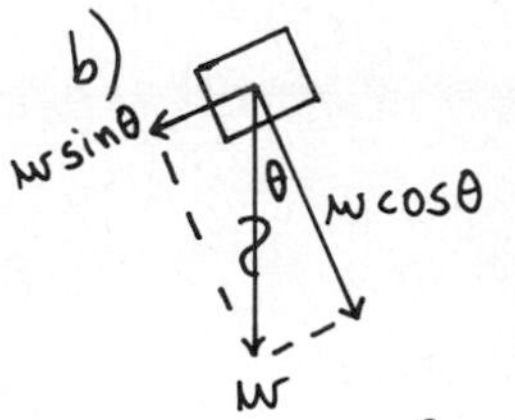

Hence $W_w = -(w\sin\theta)\,s = -(22\ \text{lb}\sin 30°)(15\ \text{ft}) = \underline{-165\ \text{ft}\cdot\text{lb}}$

(The object has an upward component of displacement, the gravity force is downward, so the work done by gravity is negative.)

c) The normal force is perpendicular to the displacement $\Rightarrow W_n = 0$.

d) $W_{total} = W_P + W_w + W_n = 270\ \text{ft}\cdot\text{lb} - 165\ \text{ft}\cdot\text{lb} + 0 = \underline{+105\ \text{ft}\cdot\text{lb}}$

7-5

Use the information given to calculate the force constant k of the spring:

$$F = kx \Rightarrow k = \frac{F}{x} = \frac{100\ \text{N}}{0.4\ \text{m}} = 250\ \text{N}\cdot\text{m}^{-1}$$

a) $F = kx$

$x = +0.1\ \text{m} \Rightarrow F = (250\ \text{N}\cdot\text{m}^{-1})(0.1\ \text{m}) = \underline{25.0\ \text{N}}$

$x = -0.2\ \text{m} \Rightarrow F = (250\ \text{N}\cdot\text{m}^{-1})(-0.2\ \text{m}) = \underline{-50.0\ \text{N}}$

b) $$W = \int_{x_1}^{x_2} F\,dx = \int_{x_1}^{x_2} kx\,dx = \tfrac{1}{2}kx^2\Big|_{x_1}^{x_2} = \tfrac{1}{2}k(x_2^2 - x_1^2)$$

$x_1 = 0,\ x_2 = +0.1\ \text{m} \Rightarrow W = \tfrac{1}{2}(250\ \text{N}\cdot\text{m}^{-1})\left((0.1\ \text{m})^2 - 0\right) = \underline{1.25\ \text{J}}$

$x_1 = 0,\ x_2 = -0.2\ \text{m} \Rightarrow W = \tfrac{1}{2}(250\ \text{N}\cdot\text{m}^{-1})\left([-0.2\ \text{m}]^2 - 0\right) = \underline{5.00\ \text{J}}$

7-7

a) $F = -(6\ \text{N}\cdot\text{m}^{-3})\,x^3$; the negative sign shows that the force is directed toward the origin (in the negative x-direction).

$x_a = 1\ \text{m} \Rightarrow F_a = -(6\ \text{N}\cdot\text{m}^{-3})(1\ \text{m})^3 = \underline{-6\ \text{N}}$

b) $x_b = 2\ \text{m} \Rightarrow F_b = -(6\ \text{N}\cdot\text{m}^{-3})(2\ \text{m})^3 = \underline{-48\ \text{N}}$

7-7 (cont)

c) $W = \int_{x_a}^{x_b} F(x)\,dx = \int_{x_a}^{x_b} (-6\,N\cdot m^{-3})\,x^3\,dx = (-6\,N\cdot m^{-3})\int_{x_a}^{x_b} x^3\,dx = (-6\,N\cdot m^{-3})\left(\tfrac{1}{4}x^4\Big/_{x_a}^{x_b}\right)$

$W = -1.5\,N\cdot m^{-3}\,(x_b^4 - x_a^4) = -1.5\,N\cdot m^{-3}\left((2m)^4 - (1m)^4\right) = \underline{-22.5\,J}$

This work is negative, as given by the above calculation. We know this is correct. The force is toward the origin, but the object moves away from the origin.

7-11

$W = K_2 - K_1$

$W = Fs = (25N)(4m) = 100\,J$

$K_1 = \tfrac{1}{2}mv_1^2 = \tfrac{1}{2}(2kg)(10\,m\cdot s^{-1})^2 = 100\,J$

$K_2 = W + K_1 = 100\,J + 100\,J = 200\,J$

$K_2 = \tfrac{1}{2}mv_2^2 \Rightarrow v_2 = \sqrt{\dfrac{2K_2}{m}} = \sqrt{\dfrac{2(200J)}{2kg}} = \underline{14.1\,m\cdot s^{-1}}$

7-15

a) $W = K_2 - K_1$

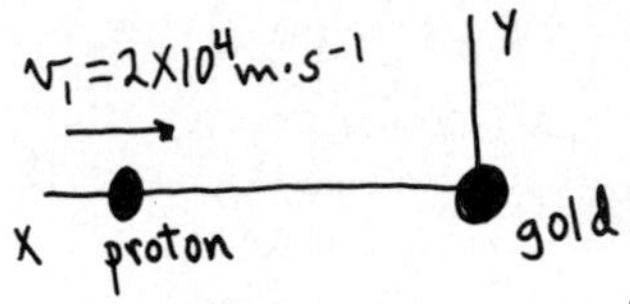

$x_1 = 4m$

$x_2 = 1.0\times10^{-7}m$

$v_2 = ?$

$K_1 = \tfrac{1}{2}mv_1^2 = \tfrac{1}{2}(1.67\times10^{-27}kg)(2\times10^4 m\cdot s^{-1})^2$

$K_1 = 3.34\times10^{-19}\,J$

$W = \int_{x_1}^{x_2} F(x)\,dx = \int_{x_1}^{x_2} \dfrac{1.82\times10^{-26}\,N\cdot m^2}{x^2}\,dx$ (Note: proton is repelled $\Rightarrow$ F is in our +X direction.)

$W = (1.82\times10^{-26}\,N\cdot m^2)\int_{x_1}^{x_2}\dfrac{dx}{x^2} = (1.82\times10^{-26}\,N\cdot m^2)\left(-\tfrac{1}{x}\Big/_{x_1}^{x_2}\right) = -(1.82\times10^{-26}\,N\cdot m^2)\left(\tfrac{1}{x_2} - \tfrac{1}{x_1}\right)$

$W = -(1.82\times10^{-26}\,N\cdot m^2)\left(\dfrac{1}{1\times10^{-7}m} - \dfrac{1}{4m}\right) = -1.82\times10^{-19}\,J$

(Note: In this problem 4m is a very large distance, essentially infinity. Also, W is negative as it must be. In our sketch the force is to the left and the displacement is to the right.)

$K_2 = W + K_1 = -1.82\times10^{-19}\,J + 3.34\times10^{-19}\,J = 1.52\times10^{-19}\,J$

$K_2 = \tfrac{1}{2}mv_2^2 \Rightarrow v_2 = \sqrt{\dfrac{2K_2}{m}} = \sqrt{\dfrac{2(1.52\times10^{-19}J)}{1.67\times10^{-27}kg}} = \underline{1.35\times10^4\,m\cdot s^{-1}}$

b) $W = K_2 - K_1$

Stops momentarily at distance of closest approach $\Rightarrow K_2 = 0$.

$K_1 = 3.34\times10^{-19}\,J$, as in (a)

$W = -(1.82\times10^{-26}\,N\cdot m^2)\left(\tfrac{1}{x_2} - \tfrac{1}{x_1}\right)$ (from (a))

$K_2 = 0 \Rightarrow W = -K_1$

7-15 (cont)

$$-(1.82\times10^{-26}\,N\cdot m^2)\left(\frac{1}{x_2}-\frac{1}{x_1}\right) = -3.34\times10^{-19}\,J$$

$$\frac{1}{x_2} - \underbrace{\frac{1}{4m}}_{\text{neglect}} = 1.84\times10^{7}\,m^{-1} \Rightarrow x_2 = \underline{5.45\times10^{-8}\,m}$$

(Note: The radius of a nucleus is on the order of $1\,fm = 1\times10^{-15}\,m$, so the proton is still well outside of the gold nucleus.)

7-21

a)

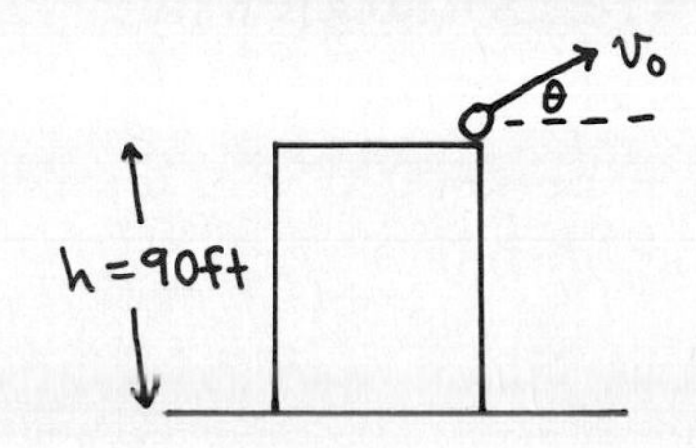

$W_{other} = 0$ (The only force on the ball while it is in the air is gravity.)

$\Rightarrow K_1 + U_1 = K_2 + U_2$

Let $U=0$ at the ground.

$K_1 = \frac{1}{2}mv_0^2$ (independent of θ!)

$U_1 = mgh$

$K_2 = \frac{1}{2}mv_2^2$

$U_2 = 0$

$$K_1 + U_1 = K_2 + U_2 \Rightarrow \frac{1}{2}mv_0^2 + mgh = \frac{1}{2}mv_2^2$$

$$v_2 = \sqrt{v_0^2 + 2gh} = \sqrt{(60\,ft\cdot s^{-1})^2 + 2(32\,ft\cdot s^{-2})(90\,ft)} = \underline{96.7\,ft\cdot s^{-1}}$$

b) The expression for v_2 obtained in (a) was found to be independent of θ, so the velocity would again be $\underline{96.7\,ft\cdot s^{-1}}$.

7-23

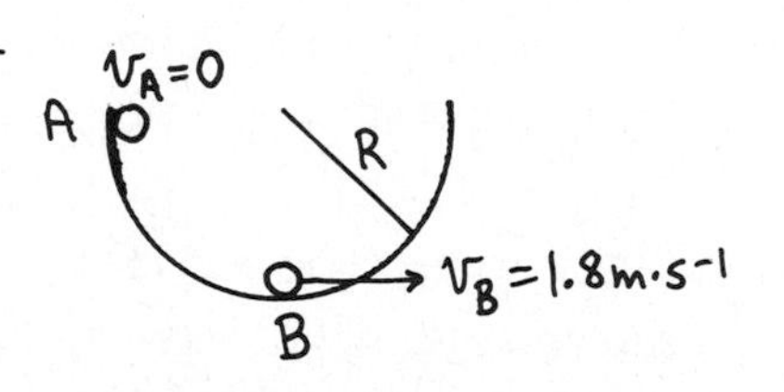

The forces on the object are gravity, the normal force, and friction.

The normal force is at all points in the motion perpendicular to the displacement $\Rightarrow W_n = 0$.

Hence $W_{other} = W_f$, the work done by friction.

$$W_{other} = \Delta K + \Delta U = (K_B - K_A) + (U_B - U_A)$$

Let $U=0$ at point B.

$K_B = \frac{1}{2}mv_B^2 = \frac{1}{2}(0.1\,kg)(1.8\,m\cdot s^{-1})^2 = 0.162\,J$

$K_A = 0$

$U_B = 0$

$U_A = mgy_A = mgR = (0.1\,kg)(9.8\,m\cdot s^{-2})(0.4\,m) = 0.392\,J$

Thus

$$W_f = (0.162\,J - 0) + (0 - 0.392\,J) = -0.230\,J$$

(The friction work is negative, as expected.)

7-25

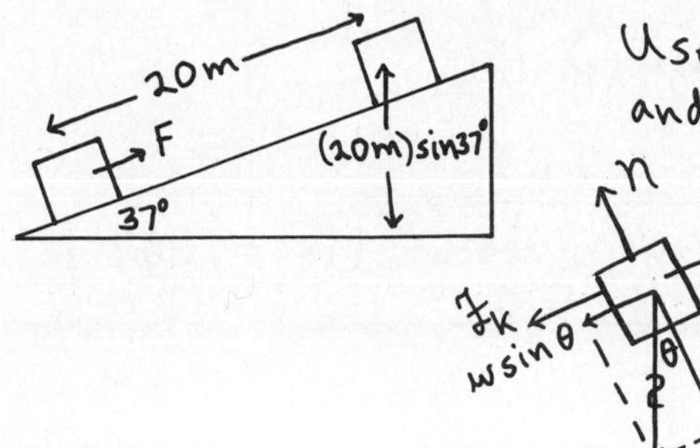

Use $\Sigma \vec{F} = m\vec{a}$ to calculate the normal force, and from that the friction force:

$$\Sigma F_y = ma_y$$
$$n - w\cos\theta = 0$$
$$n = w\cos\theta$$

$$\mathcal{f}_k = \mu_k n = (0.25)(12\,kg)(9.8\,m\cdot s^{-2})\cos 37°$$
$$\mathcal{f}_k = 23.5\,N$$

a) $W_F = Fs = (120\,N)(20\,m) = \underline{2400\,J}$

b) $W_{\mathcal{f}_k} = -\mathcal{f}_k s = -(23.5\,N)(20\,m) = \underline{-470\,J}$

c) $\Delta U = U_2 - U_1 = mgy_2 - mgy_1$

Let $U = 0$ at the bottom of the plane $\Rightarrow y_1 = 0$, $y_2 = (20\,m)\sin 37°$

$$\Delta U = (12\,kg)(9.8\,m\cdot s^{-2})(20\,m \sin 37°) = \underline{1415\,J}$$

d) $W_{other} = W_F + W_{\mathcal{f}_k} = 2400\,J - 470\,J = 1930\,J$

$W_{other} = \Delta K + \Delta U \Rightarrow \Delta K = W_{other} - \Delta U = 1930\,J - 1415\,J = \underline{515\,J}$

e) $\Sigma F_x = ma_x$

$$F - \mathcal{f}_k - w\sin 37° = ma$$

$$a = \frac{F - \mathcal{f}_k - mg\sin 37°}{m} = \frac{120\,N - 23.5\,N - (12\,kg)(9.8\,m\cdot s^{-2})\sin 37°}{12\,kg} = 2.15\,m\cdot s^{-2}$$

$a = 2.15\,m\cdot s^{-2}$
$v_0 = 0$
$x - x_0 = 20\,m$
$v = ?$

$$v_2^2 = v_0^2 + 2a(x - x_0)$$
$$v_2 = \sqrt{v_0^2 + 2a(x - x_0)}$$
$$v_2 = \sqrt{0^2 + 2(2.15\,m\cdot s^{-2})(20\,m)}$$
$$v_2 = 9.27\,m\cdot s^{-1}$$

$$K_2 = \tfrac{1}{2}mv_2^2 = \tfrac{1}{2}(12\,kg)(9.27\,m\cdot s^{-1})^2$$
$$K_2 = 515\,J$$

$K_1 = 0 \Rightarrow \Delta K = 515\,J$, in agreement with the answer we got in (d).

7-27

Use eq. (7-23), with the mass m_T of Toro in place of the mass of the earth:

$$\tfrac{1}{2}mv_1^2 - G\frac{mm_T}{r_1} = \tfrac{1}{2}mv_2^2 - G\frac{mm_T}{r_2}$$

For the escape velocity (v_1) calculation, let $r_1 = R_T$ (the radius of Toro), $r_2 \to \infty$ and $v_2 = 0$.

$$\Rightarrow \tfrac{1}{2}mv_1^2 - G\frac{mm_T}{R_T} = 0$$

The mass m of the escaping object divides out and we have

7-27 (cont)

$$v_1 = \sqrt{\frac{2Gm_T}{R_T}} = \sqrt{\frac{2(6.67\times10^{-11}\,N\cdot m^2\cdot kg^{-2})(2\times10^{15}\,kg)}{5.0\times10^{3}\,m}} = \underline{7.30\,m\cdot s^{-1}}$$

Can a person run at this speed? Well, running 100m in 10s requires world class sprinting, so $7.30\,m\cdot s^{-1}$ is just barely possible for an average person.

7-29

Use the information given to calculate k for the spring.

$$F = kx \Rightarrow k = \frac{F}{x} = \frac{1200\,N}{0.1\,m} = 1.2\times10^{4}\,N\cdot m^{-1}$$

a) $U = \frac{1}{2}kx^2$

$x = 0.1\,m \Rightarrow U = \frac{1}{2}(1.2\times10^{4}\,N\cdot m^{-1})(0.1\,m)^2 = \underline{60\,J}$

$x = -0.05\,m \Rightarrow U = \frac{1}{2}(1.2\times10^{4}\,N\cdot m^{-1})(-0.05\,m)^2 = \underline{15\,J}$

b) First calculate the amount the hanging mass stretches the spring. Forces on the hanging mass:

$a = 0$

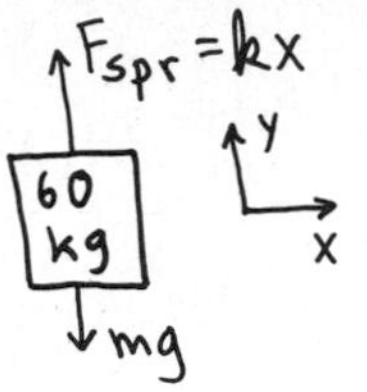

$$\Sigma F_y = ma_y$$
$$kx - mg = 0$$
$$x = \frac{mg}{k} = \frac{(60\,kg)(9.8\,m\cdot s^{-2})}{1.2\times10^{4}\,N\cdot m^{-1}} = 0.049\,m$$

Then $U = \frac{1}{2}kx^2 = \frac{1}{2}(1.2\times10^{4}\,N\cdot m^{-1})(0.049\,m)^2 = \underline{14.4\,J}$

7-35

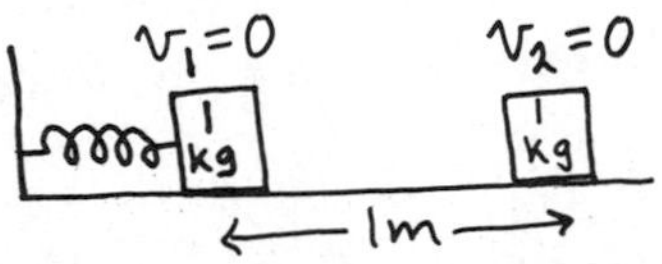

Work is done on the block by the spring force and by friction $\Rightarrow W_{other} = W_{f_k}$

$$W_{other} = \Delta K + \Delta U = (K_2 - K_1) + (U_2 - U_1)$$

$W_{other} = W_{f_k} = -\mu_k mg x_2$

$K_1 = K_2 = 0$

$U_1 = \frac{1}{2}kx_1^2 = \frac{1}{2}(100\,N\cdot m^{-1})(0.2\,m)^2 = 2.0\,J$

$U_2 = 0$

$$-\mu_k mg x_2 = (0-0) + (0 - 0.5\,J)$$

$$\mu_k = \frac{2.0\,J}{mg x_2} = \frac{2.0\,J}{(1\,kg)(9.8\,m\cdot s^{-2})(1.0\,m)} = \underline{0.204}$$

7-39

a) $\Delta U = mg\Delta y = (70\,kg)(9.8\,m\cdot s^{-2})(12\,m) = \underline{8.23\times10^{3}\,J}$

7-39 (cont)

b) $P(\text{average}) = \frac{W}{t}$

The work W he performs equals his increase in potential energy.

$\Rightarrow P = \frac{8.23\times10^3\,J}{20\,s} = \underline{412\text{ Watts}}$

7-41

2000 megawatts $\Rightarrow P = 2.0\times10^{19}\,W$

Thus, in 1 second the work done by gravity on the water must be $2.0\times10^9\,J$.

$W_{grav} = -\Delta U = -mg(y_2 - y_1) = +mgh$, where h is the height of the dam.
(The water falls from $y_1 = h$ to $y_2 = 0$.)

Use this to calculate the mass of water that flows over the dam each second:

$W = mgh \Rightarrow m = \frac{W}{gh} = \frac{2.0\times10^9\,W}{(9.8\,m\cdot s^{-2})(170\,m)} = 1.2\times10^6\,kg$ of water

$\rho = \frac{m}{V} \Rightarrow V = \frac{m}{\rho} = \frac{1.2\times10^6\,kg}{1.0\times10^3\,kg\cdot m^{-3}} = 1200\,m^3$

Thus $1200\,m^3\cdot s^{-1}$ of water over the dam is required.

7-43

$P = Fv \Rightarrow F = \frac{P}{v} = \frac{30\times10^3\,W}{10\,m\cdot s^{-1}} = \underline{3.0\times10^3\,N}$

(This is the resisting force against which the motor drives the boat, so is the tension in the towrope if that is how the boat is being powered.)

7-45

a) $F_{air} = \frac{1}{2}CA\rho v^2$

$C = 1.0$
$A = 0.463\,m^2$
$\rho = 1.2\,kg\cdot m^{-3}$ (density of air)
$v = 14\,m\cdot s^{-1}$

$F_{air} = \frac{1}{2}(1)(0.463\,m^2)(1.2\,kg\cdot m^{-3})(14\,m\cdot s^{-1})^2$
$F_{air} = 54.4\,N$

$F_{roll} = \mu_r n = 0.0045(712\,N + 111\,N) = 3.7\,N$

$F_{tot} = F_{air} + F_{roll} = 54.4\,N + 3.7\,N = 58.1\,N$

$P = F_{tot}v = (58.1\,N)(14\,m\cdot s^{-1}) = \underline{813\,W}$ (1.09 hp)

b) $F_{air} = \frac{1}{2}(0.88)(0.366\,m^2)(1.2\,kg\cdot m^{-3})(14\,m\cdot s^{-1})^2 = 37.9\,N$

$F_{roll} = 0.003(712\,N + 111\,N) = 2.47\,N$

7-45 (cont)

$$F_{tot} = F_{air} + F_{roll} = 37.9\,N + 2.5\,N = 40.4\,N$$
$$P = F_{tot}\,v = (40.4\,N)(14\,m\cdot s^{-1}) = \underline{566\,W} \quad (0.758\,hp)$$

c) $F_{air} = \frac{1}{2}(0.88)(0.366\,m^2)(1.2\,kg\cdot m^{-3})(7\,m\cdot s^{-1})^2 = 9.47\,N$

$F_{roll} = 2.47\,N$ (independent of v)

$F_{tot} = F_{air} + F_{roll} = 9.47\,N + 2.47\,N = 11.9\,N$

$P = F_{tot}\,v = (11.9\,N)(7\,m\cdot s^{-1}) = \underline{83.3\,W}$ (0.112 hp)

Problems

7-47

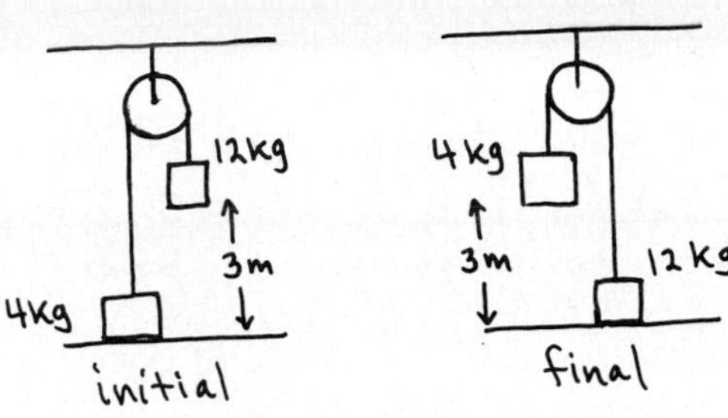

Let $U = 0$ at the floor.
For the system of two blocks, $W_{other} = 0$.

(The tension does positive work on the 4kg block and an equal amount of negative work on the 12kg block, so the net work done by T is zero.)

$K_1 + U_1 = K_2 + U_2$

$U_1 = (12\,kg)(9.8\,m\cdot s^{-2})(3\,m)$ [4kg block has $U = 0$] $\Rightarrow U_1 = 353\,J$

$U_2 = (4\,kg)(9.8\,m\cdot s^{-2})(3\,m)$ [12kg block has $U = 0$] $\Rightarrow U_2 = 118\,J$

$K_1 = 0$ (system is released from rest)

$K_2 = \frac{1}{2}(4\,kg)\,v^2 + \frac{1}{2}(12\,kg)\,v^2 = (8\,kg)\,v^2$

(The blocks have equal speeds, since they are connected by the rope.)

$\Rightarrow 0 + 353\,J = (8\,kg)\,v^2 + 118\,J$

$$v = \sqrt{\frac{353\,J - 118\,J}{8\,kg}} = \underline{5.42\,m\cdot s^{-1}}$$

7-49

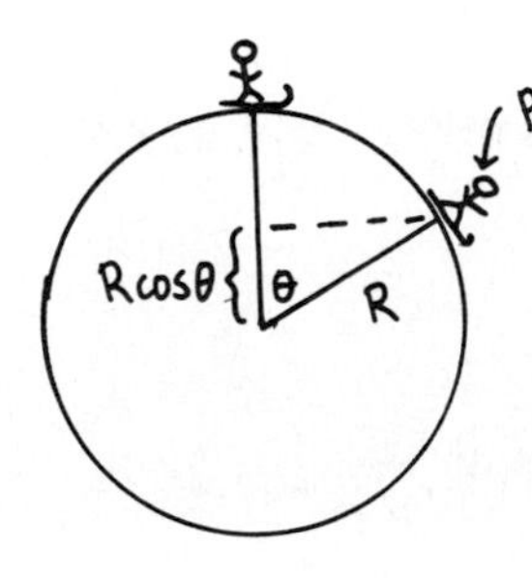

Forces on the skier:

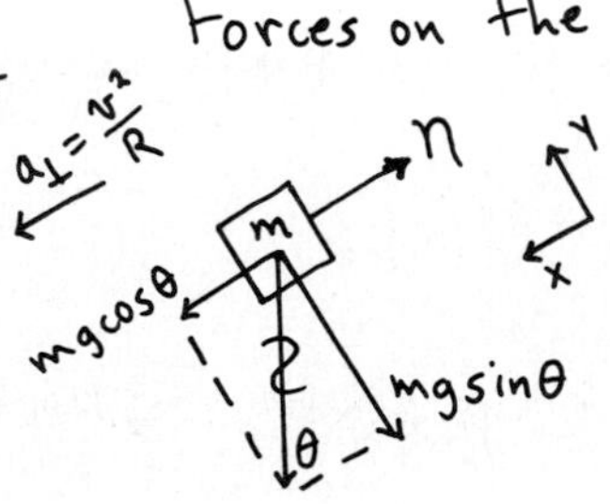

$\Sigma F_x = ma_x$

$mg\cos\theta - n = m\frac{v^2}{R}$

loses contact $\Rightarrow n \to 0$

$\cancel{m}g\cos\theta = \cancel{m}\frac{v^2}{R}$

$\boxed{v^2 = gR\cos\theta}$

v as a function of θ is determined by energy conservation:

$W_n = 0 \Rightarrow W_{other} = 0$

$K_1 + U_1 = K_2 + U_2$

7-49 (cont)

Take $U=0$ at the height of the center of the snowball.

$U_1 = mgR \qquad K_1 = 0$

$U_2 = mgR\cos\theta \qquad K_2 = \frac{1}{2}mv^2$

$$0 + mgR = \frac{1}{2}mv^2 + mgR\cos\theta$$

$$\boxed{v^2 = 2gR(1-\cos\theta)}$$

Combine this result with the equation we obtained by applying $\Sigma\vec{F} = m\vec{a}$.

$$\Rightarrow \quad 2gR(1-\cos\theta) = gR\cos\theta$$

$$2 - 2\cos\theta = \cos\theta \quad \Rightarrow \cos\theta = \tfrac{2}{3} \Rightarrow \theta = \underline{48.2^\circ}$$

7-51

a) 0.5 kg, $v_1 = 0$, 1 m, 1 m, v_2

The tension in the string is at all points perpendicular to the motion, so $W_T = 0$. Thus the only work is that done by gravity, hence $W_{other} = 0$.

$$W_{other} = 0 \Rightarrow K_1 + U_1 = K_2 + U_2$$

Let $U=0$ at the lowest point.

$K_1 = 0 \qquad U_1 = mgh, \quad h = 1\text{m}$

$K_2 = \frac{1}{2}mv_2^2 \qquad U_2 = 0$

$$0 + mgh = \tfrac{1}{2}mv_2^2 + 0 \Rightarrow v_2 = \sqrt{2gh} = \sqrt{2(9.8\,\text{m}\cdot\text{s}^{-2})(1\text{m})} = \underline{4.43\,\text{m}\cdot\text{s}^{-1}}$$

b) Consider forces on the ball, and apply $\Sigma\vec{F} = m\vec{a}$:

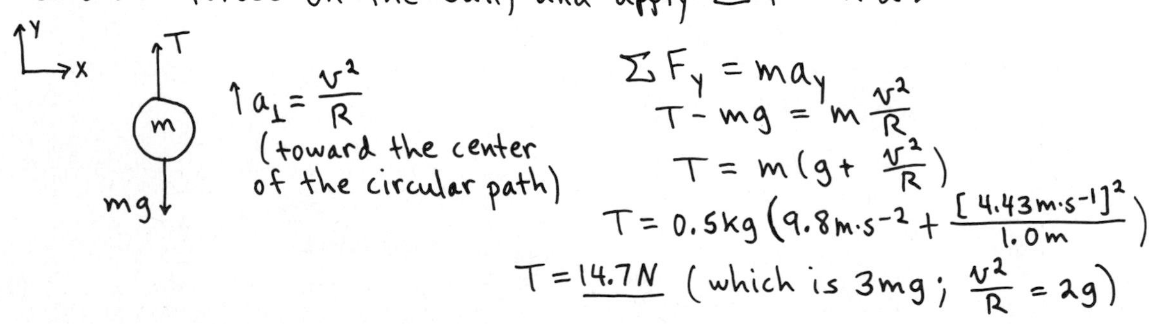

$\uparrow a_\perp = \frac{v^2}{R}$ (toward the center of the circular path)

$$\Sigma F_y = ma_y$$

$$T - mg = m\frac{v^2}{R}$$

$$T = m\left(g + \frac{v^2}{R}\right)$$

$$T = 0.5\,\text{kg}\left(9.8\,\text{m}\cdot\text{s}^{-2} + \frac{[4.43\,\text{m}\cdot\text{s}^{-1}]^2}{1.0\,\text{m}}\right)$$

$T = \underline{14.7\,N}$ (which is $3mg$; $\frac{v^2}{R} = 2g$)

7-53

a) A, B, C, 1 m

Consider the motion from B to C:

$$W_{other} = \Delta K + \Delta U$$

$$W_{other} = W_{f_K} = -\mu_K mgs$$

$$W_{f_K} = (K_C - K_B) + (U_C - U_B)$$

Let $U=0$ at the level of points B and C.

$U_B = U_C = 0$

7-53 (cont)

$K_B = \frac{1}{2}mv_B^2 \qquad \Rightarrow \qquad -\mu_k m g s = (0 - \frac{1}{2}m v_B^2) + (0-0)$

$K_C = 0$

$\mu_k = \frac{v_B^2}{2gs} = \frac{(4\,m\cdot s^{-1})^2}{2(9.8\,m\cdot s^{-2})(3m)} = \underline{0.272}$

b) Consider the motion from A to B:

$W_{other} = W_{f_k}$

$\Rightarrow W_{f_k} = (K_B - K_A) + (U_B - U_A)$

$K_A = 0$

$K_B = \frac{1}{2}mv_B^2$

$U_A = mgh \quad (h = 1m)$

$U_B = 0$

$\Rightarrow$

$W_{f_k} = \frac{1}{2}mv_B^2 - mgh$

$W_{f_k} = \frac{1}{2}(2kg)(4\,m\cdot s^{-1})^2 - (2kg)(9.8\,m\cdot s^{-2})(1m)$

$W_{f_k} = 16J - 19.6J = \underline{-3.6J}$

7-59

$\vec{F} = 3xy\,\vec{i}$

a) $y = 4m$, $x = 2m$

$W = \int \vec{F}\cdot d\vec{l}$

$d\vec{l} = dx\,\vec{i} \Rightarrow \vec{F}\cdot d\vec{l} = (3xy\,\vec{i})\cdot dx\,\vec{i} = 3xy\,dx$

$W = \int_{x_1=0}^{x_2=2m} 3xy\,dx$; $y = 4m$ along the path

$W = 3(4)\int_{x_1}^{x_2} x\,dx = 12\left(\frac{1}{2}x^2\Big|_{x_1}^{x_2}\right) = 6(x_2^2 - x_1^2) = 6(2^2 - 0) = \underline{24J}$

b)

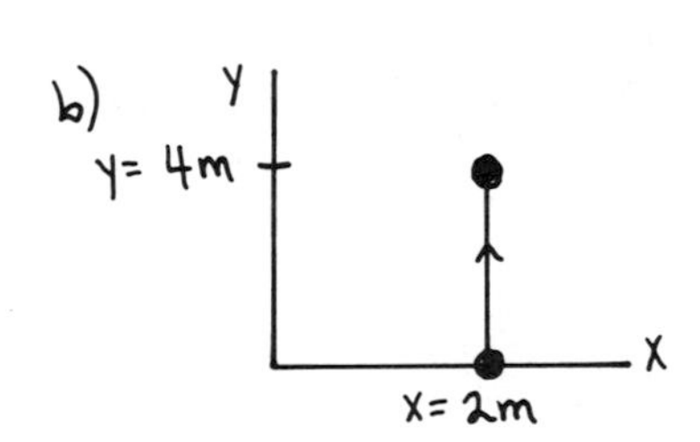

$d\vec{l} = dy\,\vec{j}$

$\vec{F}\cdot d\vec{l} = (3xy\,\vec{i})\cdot dy\,\vec{j} = 0$, since $\vec{i}\cdot\vec{j} = 0$

$W = \int \vec{F}\cdot d\vec{l} = \underline{0}$

c)

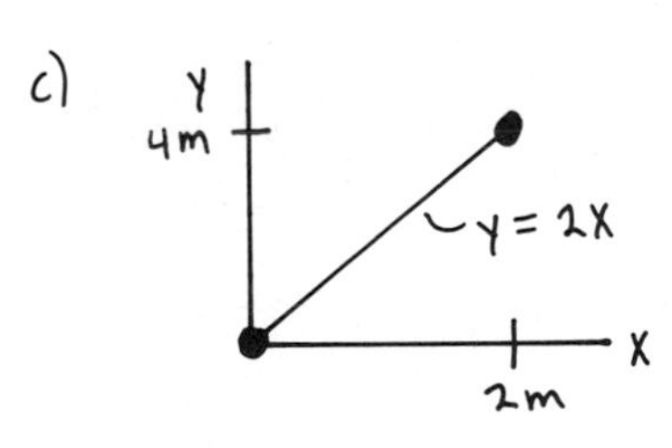

$d\vec{l} = dx\,\vec{i} + dy\,\vec{j}$

$\vec{F}\cdot d\vec{l} = (3xy\,\vec{i})\cdot(dx\,\vec{i} + dy\,\vec{j}) = 3xy\,dx$

On the path $y = 2x \Rightarrow \vec{F}\cdot d\vec{l} = 6x^2\,dx$

$W = \int \vec{F}\cdot d\vec{l} = \int_{x_1=0}^{x_2=2m} 6x^2\,dx = 6\int_{x_1}^{x_2} x^2\,dx = 6\left(\frac{1}{3}x^3\Big|_{x_1}^{x_2}\right)$

$W = 2(x_2^3 - x_1^3) = 2((2)^3 - 0) = \underline{16J}$

7-63

$\frac{1}{2}mv_1^2 - G\frac{mm_E}{r_1} = \frac{1}{2}mv_2^2 - G\frac{mm_E}{r_2}$ (eq. 7-23)

$v_1 = 0$

$v_2 = ?$

$r_1 = (R_E + h)$

$r_2 = R_E$

$\Rightarrow -G\frac{m m_E}{R_E + h} = \frac{1}{2}m v_2^2 - G\frac{m m_E}{R_E}$

7-63 (cont)

$$v^2 = 2Gm_E\left(\frac{1}{R_E} - \frac{1}{R_E+h}\right) = \frac{2Gm_E}{R_E}\left(1 - \frac{R_E}{R_E+h}\right)$$

Write in terms of g: $g = G\frac{m_E}{R_E^2} \Rightarrow Gm_E = gR_E^2$

$$\Rightarrow v = \sqrt{2gR_E\left(1 - \frac{R_E}{R_E+h}\right)}$$

7-65

a) $F(x) = -80x - 15x^2$

$W_{spr} = -\Delta U = -(U_2 - U_1)$

$$W_{spr} = \int_{x_1}^{x_2} F(x)\,dx = \int_{x_1}^{x_2}(-80x - 15x^2)\,dx = -80\int_{x_1}^{x_2} x\,dx - 15\int_{x_1}^{x_2} x^2\,dx$$

$$W_{spr} = -40(x_2^2 - x_1^2) - 5(x_2^3 - x_1^3)$$

Let $x_1 = 0$ and x_2 be some arbitrary point $x \Rightarrow U_1 = 0$ and $U_2 = U(x)$, and

$W_{spr} = -40x^2 - 5x^3$

$W_{spr} = -U(x)$ $\Rightarrow U(x) = 40x^2 + 5x^3$ (x in m $\Rightarrow U$ in J)

b) $W_{other} = 0 \Rightarrow K_1 + U_1 = K_2 + U_2$

$K_1 = 0$ $\quad U_1 = U(x_1)$; $x_1 = 1.0\text{ m} \Rightarrow U_1 = 40(1)^2 + 5(1)^3 = 45\text{ J}$

$K_2 = \frac{1}{2}mv_2^2$ $\quad U_2 = U(x_2)$; $x_2 = 0.5\text{ m} \Rightarrow U_2 = 40(0.5)^2 + 5(0.5)^3 = 10.6\text{ J}$

$$0 + 45\text{ J} = \tfrac{1}{2}mv_2^2 + 10.6\text{ J}$$

$$v_2 = \sqrt{\frac{2(45\text{ J} - 10.6\text{ J})}{2\text{ kg}}} = \underline{5.87\text{ m}\cdot\text{s}^{-1}}$$

7-67

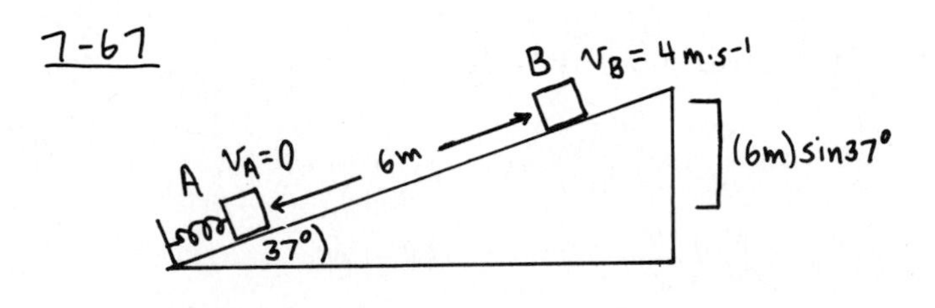

Force diagram for the block:

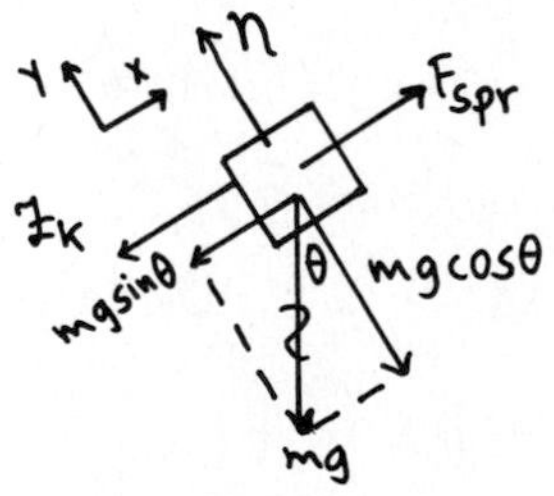

$\sum F_y = ma_y$

$n = mg\cos 37°$

$f_k = \mu_k mg\cos 37°$

$W_{other} = W_{f_k} = -(\mu_k mg\cos 37°)s$

$W_{other} = (K_B - K_A) + (U_B - U_A)$

Let $U_{grav} = 0$ at point A.

$K_A = 0$ $\quad U_A = U_{spr}$ ($U_{grav} = 0$ at A)

$K_B = \frac{1}{2}mv_B^2$ $\quad U_B = mg(6\text{ m}\sin 37°)$ ($U_{spr} = 0$ at B)

$$\Rightarrow -(\mu_k mg\cos 37°)s = \tfrac{1}{2}mv_B^2 - 0 + mg(6\text{ m}\sin 37°) - U_{spr}$$

$$U_{spr} = mg(6\text{ m}\sin 37°) + \mu_k mgs\cos 37° + \tfrac{1}{2}mv_B^2$$

$$U_{spr} = (5\text{ kg})(9.8\text{ m}\cdot\text{s}^{-2})(6\text{ m}\sin 37°) + (0.5)(5\text{ kg})(9.8\text{ m}\cdot\text{s}^{-2})(6\text{ m})\cos 37° + \tfrac{1}{2}(5\text{ kg})(4\text{ m}\cdot\text{s}^{-1})^2$$

$$U_{spr} = 177\text{ J} + 117\text{ J} + 40\text{ J} = \underline{334\text{ J}}$$

CHAPTER 8

Exercises 3, 7, 9, 11, 13, 15, 21, 23, 27, 31

Problems 37, 39, 41, 45, 49, 51, 53

Exercises

8-3

a) (axes: y up, x right) $p_1 = mv_1 = (2\,kg)(+6\,m\cdot s^{-1}) = 12\,kg\cdot m\cdot s^{-1}$

The force is in the positive x-direction, so the impulse of the force is

$$J = F(t_2 - t_1) = (5\,N)(5\,s) = +25\,N\cdot s = +25\,kg\cdot m\cdot s^{-1}$$

$J = \Delta p = p_2 - p_1$

$p_2 = p_1 + J = 12\,kg\cdot m\cdot s^{-1} + 25\,kg\cdot m\cdot s^{-1} = 37\,kg\cdot m\cdot s^{-1}$

$$p_2 = mv_2 \Rightarrow v_2 = \frac{p_2}{m} = \frac{37\,kg\cdot m\cdot s^{-1}}{2\,kg} = 18.5\,m\cdot s^{-1} \text{ (to the right)}$$

b) Now the force is in the negative x-direction.

$J = F(t_2 - t_1) = (-7\,N)(5\,s) = -35\,N\cdot s = -35\,kg\cdot m\cdot s^{-1}$

$p_2 = p_1 + J = +12\,kg\cdot m\cdot s^{-1} - 35\,kg\cdot m\cdot s^{-1} = -23\,kg\cdot m\cdot s^{-1}$

$$v_2 = \frac{p_2}{m} = \frac{-23\,kg\cdot m\cdot s^{-1}}{2\,kg} = -11.5\,m\cdot s^{-1} \text{ (to the left)}$$

8-7

a) (axes: y up, x right) $$J = \int_{t_1}^{t_2} F(t)\,dt = \int_0^{t_2} (A + Bt^2)\,dt = At_2 + \tfrac{1}{3}Bt_2^3$$

b) $J = p_2 - p_1 \Rightarrow p_2 = p_1 + J$

Initially at rest $\Rightarrow p_1 = 0 \Rightarrow p_2 = At_2 + \frac{1}{3}Bt_2^3$

$$v_2 = \frac{p_2}{m} = \frac{t_2}{m}\left(A + \tfrac{1}{3}Bt_2^2\right)$$

8-9

a) initial: $v_{A1} = 13\,m\cdot s^{-1}$ (A moving right), $v_{B1} = 4\,m\cdot s^{-1}$ (B moving left); final: $v_{A2} = 8\,m\cdot s^{-1}$ (A moving right), v_{B2}; axes: y up, x right

$$m_A = \frac{w_A}{g} = \frac{756\,N}{9.8\,m\cdot s^{-1}} = 77.1\,kg \;;\; m_B = \frac{w_B}{g} = \frac{900\,N}{9.8\,m\cdot s^{-1}} = 91.8\,kg$$

$p_1 = m_A v_{A1} + m_B v_{B1} = (77.1\,kg)(13\,m\cdot s^{-1}) + (91.8\,kg)(-4\,m\cdot s^{-1}) = 1002\,kg\cdot m\cdot s^{-1} - 367\,kg\cdot m\cdot s^{-1}$

$p_1 = 635\,kg\cdot m\cdot s^{-1}$

8-9 (cont)

$$p_2 = m_A v_{A2} + m_B v_{B2} = (77.1\,\text{kg})(8\,\text{m}\cdot\text{s}^{-1}) + (91.8\,\text{kg})v_{B2} = 617\,\text{kg}\cdot\text{m}\cdot\text{s}^{-1} + (91.8\,\text{kg})v_{B2}$$

Conservation of linear momentum in the x-direction $\Rightarrow p_1 = p_2$

$$635\,\text{kg}\cdot\text{m}\cdot\text{s}^{-1} = 617\,\text{kg}\cdot\text{m}\cdot\text{s}^{-1} + (91.8\,\text{kg})v_{B2}$$

$$v_{B2} = \underline{0.196\,\text{m}\cdot\text{s}^{-1}}$$, in the same direction as Gretzky

b) $K_1 = \frac{1}{2}m_A v_{A1}^2 + \frac{1}{2}m_B v_{B1}^2 = \frac{1}{2}(77.1\,\text{kg})(13\,\text{m}\cdot\text{s}^{-1})^2 + \frac{1}{2}(91.8\,\text{kg})(-4\,\text{m}\cdot\text{s}^{-1})^2$

$K_1 = 6515\,\text{J} + 734\,\text{J} = 7249\,\text{J}$

$K_2 = \frac{1}{2}m_A v_{A2}^2 + \frac{1}{2}m_B v_{B2}^2 = \frac{1}{2}(77.1\,\text{kg})(8\,\text{m}\cdot\text{s}^{-1})^2 + \frac{1}{2}(91.8\,\text{kg})(0.196\,\text{m}\cdot\text{s}^{-1})^2$

$K_2 = 2467\,\text{J} + 2\,\text{J} = 2469\,\text{J}$

$\Delta K = K_2 - K_1 = 2469\,\text{J} - 7249\,\text{J} = \underline{-4780\,\text{J}}$ (decrease)

8-11

a)

$v_{A1} = 30\,\text{m}\cdot\text{s}^{-1}$ (m) A; $v_{B1} = 0$ (m) B — initial

$v_{A2}\sin 30°$, v_{A2}, 30°, A (m) $v_{A2}\cos 30°$; B (m) $v_{B2}\cos 45°$, 45°, $v_{B2}\sin 45°$ — final

Y, X

conservation of linear momentum $\Rightarrow \vec{P}_1 = \vec{P}_2$ ($P_{1x} = P_{2x}$ and $P_{1y} = P_{2y}$)

$$P_{1x} = P_{2x} \Rightarrow m_A v_{A1x} + m_B v_{B1x} = m_A v_{A2x} + m_B v_{B2x}$$

$$\cancel{m}(30\,\text{m}\cdot\text{s}^{-1}) + 0 = \cancel{m}(v_{A2}\cos 30°) + \cancel{m}(v_{B2}\cos 45°)$$

$$\boxed{30\,\text{m}\cdot\text{s}^{-1} = v_{A2}\cos 30° + v_{B2}\cos 45°}$$

$$P_{1y} = P_{2y} \Rightarrow 0 + 0 = \cancel{m}v_{A2}\sin 30° + \cancel{m}(-v_{B2}\sin 45°)$$

$$\boxed{0 = v_{A2}\sin 30° - v_{B2}\sin 45°}$$

Add these two eqs., and use that $\cos 45° = \sin 45°$

$$\Rightarrow 30\,\text{m}\cdot\text{s}^{-1} = v_{A2}(\cos 30° + \sin 30°)$$

$$v_{A2} = \frac{30\,\text{m}\cdot\text{s}^{-1}}{\cos 30° + \sin 30°} = \underline{22.0\,\text{m}\cdot\text{s}^{-1}}$$

Then $0 = v_{A2}\sin 30° - v_{B2}\sin 45° \Rightarrow v_{B2} = v_{A2}\dfrac{\sin 30°}{\sin 45°} = (22.0\,\text{m}\cdot\text{s}^{-1})\left(\dfrac{\sin 30°}{\sin 45°}\right) = \underline{15.5\,\text{m}\cdot\text{s}^{-1}}$

b) $K_1 = \frac{1}{2}m v_{A1}^2 + \frac{1}{2}m v_{B1}^2 = \frac{1}{2}m(30\,\text{m}\cdot\text{s}^{-1})^2 = (450\,\text{m}^2\cdot\text{s}^{-2})m$

$K_2 = \frac{1}{2}m v_{A2}^2 + \frac{1}{2}m v_{B2}^2 = \frac{1}{2}m(22.0\,\text{m}\cdot\text{s}^{-1})^2 + \frac{1}{2}m(15.5\,\text{m}\cdot\text{s}^{-1})^2 = (362\,\text{m}^2\cdot\text{s}^{-2})m$

Fraction "lost" is $\dfrac{\Delta K}{K_1} = \dfrac{K_2 - K_1}{K_1} = \dfrac{(362\,\text{m}^2\cdot\text{s}^{-2})\cancel{m} - (450\,\text{m}^2\cdot\text{s}^{-2})\cancel{m}}{(450\,\text{m}^2\cdot\text{s}^{-2})\cancel{m}} = \underline{-0.196}$

(19.6 %)

8-13

a)

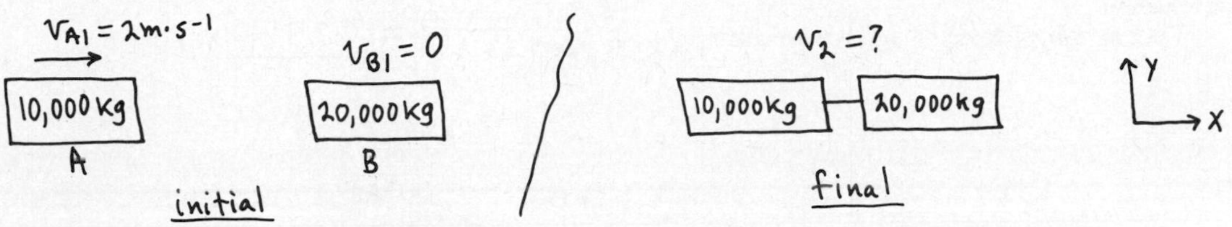

Conservation of momentum in the x-direction $\Rightarrow p_1 = p_2$.

$$p_1 = m_A v_{A1} + m_B v_{B1} = (10{,}000\,kg)(2\,m\cdot s^{-1}) = 20{,}000\,kg\cdot m\cdot s^{-1}$$

$$p_2 = (m_A + m_B)\, v_2 = (30{,}000\,kg)\, v_2$$

$$p_1 = p_2 \Rightarrow 20{,}000\,kg\cdot m\cdot s^{-1} = (30{,}000\,kg)\, v_2 \Rightarrow v_2 = \underline{0.667\,m\cdot s^{-1}}$$

b) $K_1 = \frac{1}{2} m_A v_{A1}^2 + \frac{1}{2} m_B v_{B1}^2 = \frac{1}{2}(10{,}000\,kg)(2\,m\cdot s^{-1})^2 + 0 = 20{,}000\,J$

$K_2 = \frac{1}{2}(m_A + m_B)\, v^2 = \frac{1}{2}(30{,}000\,kg)(0.667\,m\cdot s^{-1})^2 = 6673\,J$

$\Delta K = K_2 - K_1 = -\underline{1.33\times 10^4\,J}$ (decrease)

c) Again $p_1 = p_2$.

$v_2 = 0 \Rightarrow p_2 = 0 \Rightarrow p_1 = 0$

$$p_1 = m_A v_{A1} + m_B v_{B1} = (10{,}000\,kg)(2\,m\cdot s^{-1}) + (20{,}000\,kg)\, v_{B1}$$

$$p_1 = 0 \Rightarrow v_{B1} = -\frac{(10{,}000\,kg)(2\,m\cdot s^{-1})}{20{,}000\,kg} = \underline{-1.0\,m\cdot s^{-1}}$$

8-15

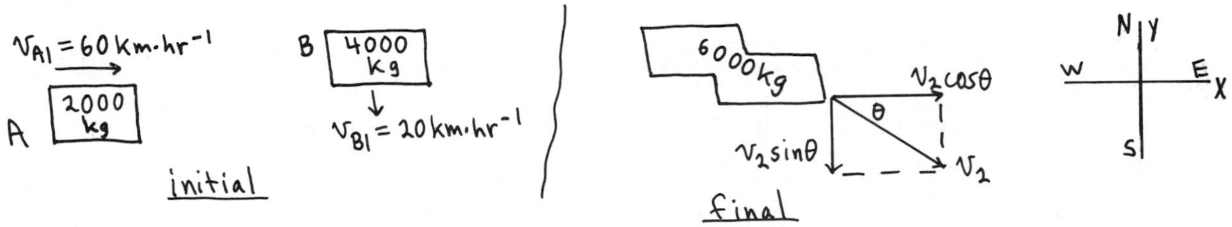

Conservation of linear momentum $\Rightarrow \vec{p}_1 = \vec{p}_2 \Rightarrow p_{1x} = p_{2x}$ and $p_{1y} = p_{2y}$

$p_{1x} = p_{2x} \Rightarrow m_A v_{A1x} + m_B v_{B1x} = (m_A + m_B)\, v_{2x}$

$$(2000\,kg)(60\,km\cdot hr^{-1}) + 0 = (6000\,kg)(v_2 \cos\theta)$$

$$\boxed{v_2 \cos\theta = 20.0\,km\cdot hr^{-1}}$$

$p_{1y} = p_{2y} \Rightarrow m_A v_{A1y} + m_B v_{B1y} = (m_A + m_B)\, v_{2y}$

$$0 + (4000\,kg)(-20\,km\cdot hr^{-1}) = (6000\,kg)(-v_2 \sin\theta)$$

$$\boxed{v_2 \sin\theta = 13.3\,km\cdot hr^{-1}}$$

Divide one equation by the other

$$\Rightarrow \frac{v_2 \sin\theta}{v_2 \cos\theta} = \frac{13.3\,km\cdot hr^{-1}}{20.0\,km\cdot hr^{-1}} \Rightarrow \tan\theta = 0.665 \Rightarrow \theta = \underline{33.6^\circ}$$

(S of E, as shown in the sketch)

8-15 (cont)

Then $v_2 \cos\theta = 20\,km\cdot hr^{-1} \Rightarrow v_2 = \frac{20\,km\cdot hr^{-1}}{\cos 33.6^\circ} = \underline{23.9\,km\cdot hr^{-1}}$

8-21

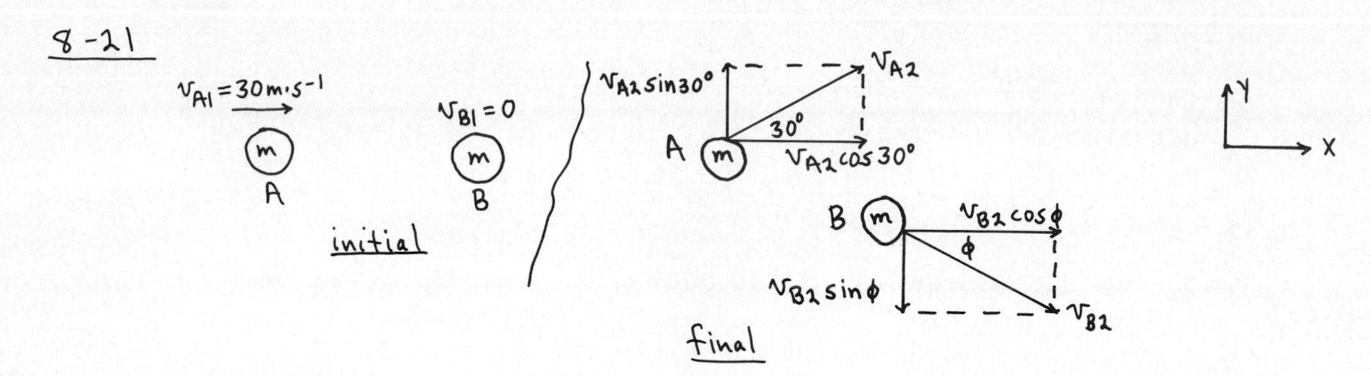

Conservation of linear momentum $\Rightarrow \vec{p}_1 = \vec{p}_2$ ($p_{1x} = p_{2x}$ and $p_{1y} = p_{2y}$)

$$p_{1x} = p_{2x} \Rightarrow m_A v_{A1x} + m_B v_{B1x} = m_A v_{A2x} + m_B v_{B2x}$$
$$m v_{A1} + 0 = m v_{A2} \cos 30^\circ + m v_{B2} \cos\phi$$
$$\boxed{30\,m\cdot s^{-1} = v_{A2} \cos 30^\circ + v_{B2} \cos\phi}$$

$$p_{1y} = p_{2y} \Rightarrow m_A v_{A1y} + m_B v_{B1y} = m_A v_{A2y} + m_B v_{B2y}$$
$$0 + 0 = m v_{A2} \sin 30^\circ - m v_{B2} \sin\phi$$
$$\boxed{0 = v_{A2} \sin 30^\circ - v_{B2} \sin\phi}$$

elastic collision $\Rightarrow K_1 = K_2$

$$\tfrac{1}{2} m_A v_{A1}^2 + \tfrac{1}{2} m_B v_{B1}^2 = \tfrac{1}{2} m_A v_{A2}^2 + \tfrac{1}{2} m_B v_{B2}^2$$
$$\tfrac{1}{2} m (30\,m\cdot s^{-1})^2 + 0 = \tfrac{1}{2} m v_{A2}^2 + \tfrac{1}{2} m v_{B2}^2$$
$$\boxed{(30\,m\cdot s^{-1})^2 = v_{A2}^2 + v_{B2}^2}$$

Three equations in three unknowns: v_{A2}, v_{B2} and ϕ.
Square the momentum conservation eqs. and add them:

$$(30\,m\cdot s^{-1})^2 = v_{A2}^2 \cos^2 30^\circ + v_{B2}^2 \cos^2\phi + 2 v_{A2} v_{B2} \cos 30^\circ \cos\phi$$
$$0 = v_{A2}^2 \sin^2 30^\circ + v_{B2}^2 \sin^2\phi - 2 v_{A2} v_{B2} \sin 30^\circ \sin\phi$$
$$(30\,m\cdot s^{-1})^2 = v_{A2}^2 (\sin^2 30^\circ + \cos^2 30^\circ) + v_{B2}^2 (\sin^2\phi + \cos^2\phi) + 2 v_{A2} v_{B2} (\cos 30^\circ \cos\phi - \sin 30^\circ \sin\phi)$$

But $\sin^2\theta + \cos^2\theta = 1$
and $\cos\theta_1 \cos\theta_2 - \sin\theta_1 \sin\theta_2 = \cos(\theta_1 + \theta_2)$
Thus
$$(30\,m\cdot s^{-1})^2 = v_{A2}^2 + v_{B2}^2 + 2 v_{A2} v_{B2} \cos(30^\circ + \phi)$$

Comparison of this result with the conservation of kinetic energy equation
$\Rightarrow 2 v_{A2} v_{B2} \cos(30^\circ + \phi) = 0$

$\cos(30^\circ + \phi) = 0 \Rightarrow 30^\circ + \phi = 90^\circ \Rightarrow \underline{\phi = 60^\circ}$ (ϕ is defined in the sketch)

8-21 (cont)

Now that ϕ is known, solve for v_{A2} and v_{B2}:

$$0 = v_{A2}\sin 30^\circ - v_{B2}\sin\phi$$

$$\Rightarrow v_{B2} = v_{A2}\left(\frac{\sin 30^\circ}{\sin 60^\circ}\right) = 0.577\,v_{A2}$$

Substitute this into the other eq., $30\,\mathrm{m\cdot s^{-1}} = v_{A2}\cos 30^\circ + v_{B2}\cos\phi$

$$\Rightarrow 30\,\mathrm{m\cdot s^{-1}} = (0.866)\,v_{A2} + (0.500)(0.577\,v_{A2})$$

$v_{A2} = \underline{26.0\,\mathrm{m\cdot s^{-1}}}$; $v_{B2} = 0.577\,v_{A2} = \underline{15.0\,\mathrm{m\cdot s^{-1}}}$

8-23

a) $v_{A1} = 0$ $v_{B1} = 0$ v_{A2} $v_{B2} = 0.5\,\mathrm{m\cdot s^{-1}}$

A B initial A B final

Conservation of linear momentum $\Rightarrow P_1 = P_2$

$P_1 = 0$; $P_2 = -m_A v_{A2} + m_B v_{B2} \Rightarrow 0 = -m_A v_{A2} + m_B v_{B2}$

$$v_{A2} = \left(\frac{m_B}{m_A}\right)v_{B2} = \left(\frac{2\,\mathrm{kg}}{1\,\mathrm{kg}}\right)(0.5\,\mathrm{m\cdot s^{-1}}) = \underline{1.0\,\mathrm{m\cdot s^{-1}}}$$

b) Use conservation of energy:

The spring force is the only force that does work on the blocks $\Rightarrow W_{other} = 0$.

$$K_1 + U_1 = K_2 + U_2$$

$K_1 = 0$ $K_2 = \frac{1}{2}m_A v_{A2}^2 + \frac{1}{2}m_B v_{B2}^2 = \frac{1}{2}(1\,\mathrm{kg})(1.0\,\mathrm{m\cdot s^{-1}})^2 + \frac{1}{2}(2\,\mathrm{kg})(0.5\,\mathrm{m\cdot s^{-1}})^2 = 0.75\,\mathrm{J}$

$U_1 = U_{spr}$ $U_2 = 0$ (No potential energy is left stored in the spring.)

$$0 + U_{spr} = 0.75\,\mathrm{J} + 0 \Rightarrow U_{spr} = \underline{0.75\,\mathrm{J}}$$

8-27

$$X = \frac{m_E X_E + m_m X_m}{m_E + m_m}$$

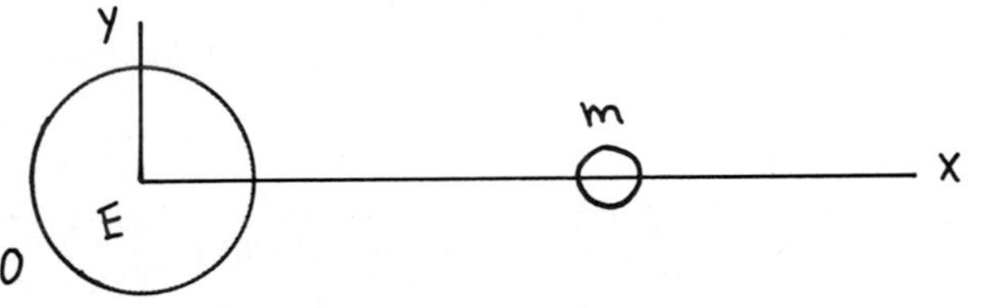

Use coordinates with origin at the center of the earth $\Rightarrow X_E = 0$

$$X = \frac{0 + (7.36\times10^{22}\,\mathrm{kg})(0.38\times10^{9}\,\mathrm{m})}{5.98\times10^{24}\,\mathrm{kg} + 7.36\times10^{22}\,\mathrm{kg}} = \underline{4.62\times10^{6}\,\mathrm{m}}$$

, from the center of the earth and on the line connecting the centers of the moon and earth.

8-31

a) Eq. (8-35) $\Rightarrow v = v_0 + v_r \ln\left(\frac{m_0}{m}\right) - gt$

No gravitational force $\Rightarrow g = 0$.

8-31 (cont)

$v_0 = 0$

$v_r = 2400\,m \cdot s^{-1}$

$v = 0.001c = 3.0 \times 10^5\,m \cdot s^{-1}$

$3 \times 10^5\,m \cdot s^{-1} = (2400\,m \cdot s^{-1}) \ln\left(\frac{m_0}{m}\right)$

$\ln\left(\frac{m_0}{m}\right) = 125$

$\Rightarrow \frac{m_0}{m} = e^{125} = 1.94 \times 10^{54}$

m is the final mass of the rocket, after all the fuel has been expended.
m_0 is the initial velocity of the rocket, with its fuel.

The problem asks for $\frac{m}{m_0}$; $\frac{m}{m_0} = e^{-125} = \underline{5.2 \times 10^{-55}}$.

(This is obviously <u>not</u> feasible, for so little of the initial mass to be <u>not</u> fuel.)

b) $v = 3000\,m \cdot s^{-1}$

$3000\,m \cdot s^{-1} = (2400\,m \cdot s^{-1}) \ln\left(\frac{m_0}{m}\right)$

$\ln\left(\frac{m_0}{m}\right) = 1.25 \Rightarrow \frac{m_0}{m} = e^{1.25} = 3.49 \Rightarrow \frac{m}{m_0} = \underline{0.287}$

(28.7% of initial mass not fuel, or 71.3% fuel; this is much more feasible!)

Problems

8-37

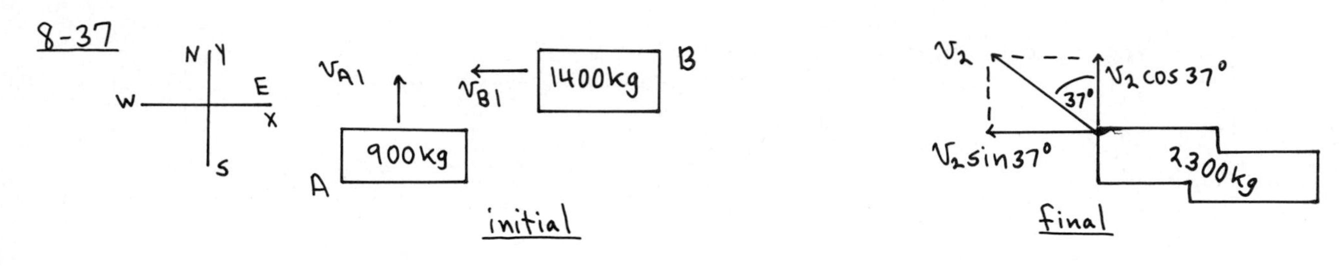

Conservation of linear momentum $\Rightarrow \vec{P}_1 = \vec{P}_2$ ($P_{1x} = P_{2x}$ and $P_{1y} = P_{2y}$)

$P_{1x} = P_{2x} \Rightarrow -(1400\,kg)\,v_{B1} = (2300\,kg)(-v_2 \sin 37°)$

$v_{B1} = \left(\frac{2300\,kg}{1400\,kg}\right)(9\,m \cdot s^{-1}) \sin 37° = \underline{8.90\,m \cdot s^{-1}}$ (station wagon)

$P_{1y} = P_{2y} \Rightarrow (900\,kg)\,v_{A1} = (2300\,kg)(v_2 \cos 37°)$

$v_{A1} = \left(\frac{2300\,kg}{900\,kg}\right)(9\,m \cdot s^{-1}) \cos 37° = \underline{18.4\,m \cdot s^{-1}}$ (pickup truck)

8-39

a) Apply conservation of momentum to the collision

$v_{A1} = 500\,m \cdot s^{-1}$

$m_A = 2 \times 10^{-3}\,kg$

$v_{B1} = 0$

$m_B = 1\,kg$

initial

$v_{B2} = ?$

m_B

$v_{A2} = 100\,m \cdot s^{-1}$

Y

X

final

8-39 (cont)

$$P_{1x} = P_{2x} \Rightarrow m_A v_{A1} + m_B v_{B1} = m_A v_{A2} + m_B v_{B2}$$

$$(2\times10^{-3}\,kg)(500\,m\cdot s^{-1}) + 0 = (2\times10^{-3}\,kg)(100\,m\cdot s^{-1}) + (1\,kg)\,v_{B2}$$

$$v_{B2} = \frac{1.00\,kg\cdot m\cdot s^{-1} - 0.20\,kg\cdot m\cdot s^{-1}}{1\,kg} = \underline{0.80\,m\cdot s^{-1}}$$

Apply conservation of energy to the subsequent motion of the block <u>after</u> the collision:

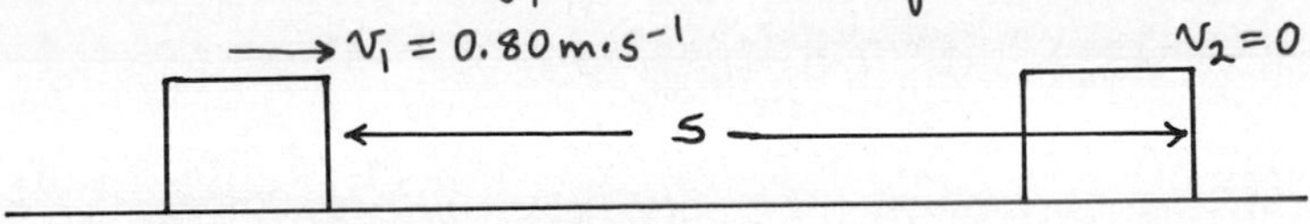

$$W_{other} = W_f = -\mu_k m g s$$

$$W_{other} = (K_2 - K_1) + (U_2 - U_1)$$

Let $U_{grav} = 0$ at the surface $\Rightarrow U_1 = U_2 = 0$

$$K_1 = \tfrac{1}{2} m v_1^2 \quad , \quad K_2 = 0$$

$$\Rightarrow -\mu_k m g s = (0 - \tfrac{1}{2} m v_1^2) + (0 - 0)$$

$$\mu_k \cancel{m} g s = \tfrac{1}{2} \cancel{m} v_1^2 \Rightarrow \mu_k = \frac{v_1^2}{2gs} = \frac{(0.8\,m\cdot s^{-1})^2}{2(9.8\,m\cdot s^{-2})(0.20\,m)} = \underline{0.163}$$

b) For the <u>bullet</u>:

$$K_{A1} = \tfrac{1}{2} m_A v_{A1}^2 = \tfrac{1}{2}(2\times10^{-3}\,kg)(500\,m\cdot s^{-1})^2 = 250\,J$$

$$K_{A2} = \tfrac{1}{2} m_A v_{A2}^2 = \tfrac{1}{2}(2\times10^{-3}\,kg)(100\,m\cdot s^{-1})^2 = 10\,J$$

$$\Delta K_{bullet} = K_{A2} - K_{A1} = 10\,J - 250\,J = -\underline{240\,J} \text{ (decrease)}$$

c) For the <u>block</u>,

$$K = \tfrac{1}{2} m_B v_{B2}^2 = \tfrac{1}{2}(1\,kg)(0.8\,m\cdot s^{-1})^2 = \underline{0.32\,J}$$

(Note that the collision is highly inelastic. The bullet loses 240 J of kinetic energy, but only 0.32 J is gained by the block. But <u>momentum was conserved</u> in the collision. All of the momentum lost by the bullet was gained by the block.)

8-41

Find the force constant of the spring:

$F_{spr} = kx$; $a = 0$; 0.20 kg ; mg ; y, x

$$\Sigma F_y = m a_y$$

$$kx - mg = 0$$

$$k = \frac{mg}{x} = \frac{(0.20\,kg)(9.8\,m\cdot s^{-2})}{(0.10\,m)} = 19.6\,N\cdot m^{-1}$$

Next find the velocity of the putty just before it strikes the frame:

8-41 (cont)

$v_0 = 0$
$(y - y_0) = 0.30\,m$
$a = 9.8\,m \cdot s^{-2}$
$v = ?$

$$v^2 = v_0^2 + 2a(y - y_0)$$

$$v = \sqrt{2a(y-y_0)} = \sqrt{2(9.8\,m\cdot s^{-2})(0.30\,m)} = 2.42\,m\cdot s^{-1}$$

Now consider the collision between the frame and the putty:

$m_A = 0.20\,kg$ $\downarrow v_{A1} = 2.42\,m\cdot s^{-1}$

$m_B = 0.20\,kg$ $v_{B1} = 0$

before

after $\downarrow v_2$

$$P_{1y} = P_{2y}$$

$$m_A v_{A1} + m_B v_{B1} = (m_A + m_B) v_2$$

$$(0.20\,kg)(2.42\,m\cdot s^{-1}) = (0.40\,kg)\, v_2 \Rightarrow v_2 = 1.21\,m\cdot s^{-1}$$

Finally, apply energy conservation to the motion of the frame + putty after the collision:

$\downarrow v_1 = 1.21\,m\cdot s^{-1}$ $h = ?$ $v_2 = 0$

The spring force and gravity do work $\Rightarrow W_{other} = 0$

$$K_1 + U_1 = K_2 + U_2$$

Let $U_{grav} = 0$ at the final position of the frame.

$$K_1 = \tfrac{1}{2}(0.40\,kg)(1.21\,m\cdot s^{-1})^2 = 0.293\,J$$

$$K_2 = 0$$

$$U_1 = U_{1,spr} + U_{1,grav} = \tfrac{1}{2}k(0.10\,m)^2 + mgh = \tfrac{1}{2}(19.6\,N\cdot m^{-1})(0.10\,m)^2 + (0.4\,kg)(9.8\,m\cdot s^{-2})h$$

$$U_1 = 0.098\,J + (3.92\,kg\cdot m\cdot s^{-2})h$$

(Note: The spring starts off stretched 0.10m, due to the weight of the frame.)

$$U_2 = U_{2,spr} + U_{2,grav} = \tfrac{1}{2}k(0.1\,m + h)^2 = (9.8\,N\cdot m^{-1})(0.1\,m + h)^2$$

$$K_1 + U_1 = K_2 + U_2 \Rightarrow 0.293\,J + 0.098\,J + (3.92\,kg\cdot m\cdot s^{-2})h = 0 + (9.8\,N\cdot m^{-1})(0.1\,m + h)^2$$

$$0.391 + 3.92h = 9.8(0.01 + 0.2h + h^2)$$

$$0.391 + 3.92h = 0.098 + 1.96h + 9.8h^2$$

$$\boxed{9.8h^2 - 1.96h - 0.293 = 0}$$

Use the quadratic formula $\Rightarrow h = \frac{1}{19.6}\left[1.96 \pm \sqrt{(1.96)^2 + 4(9.8)(0.293)}\right]$

$h = 0.10\,m \pm 0.20\,m \Rightarrow h = \underline{0.30\,m}$ (whew!)

8-45

a) $\overrightarrow{v_{A1}} = 400\,m\cdot s^{-1}$ $\quad m_A = 2.5 \times 10^{-3}\,kg$ $\quad v_{B1} = 0$ $\quad m_B$ $0.10\,kg$

initial

$v_{B2}\sin\theta$ $\quad v_{B2}$ $\quad m_B$ $\quad \theta$ $\quad v_{B2}\cos\theta$

final $\quad \downarrow v_{A2} = 300\,m\cdot s^{-1}$

8-45 (cont)

Conservation of linear momentum $\Rightarrow \vec{P}_1 = \vec{P}_2$ ($P_{1x} = P_{2x}$ and $P_{1y} = P_{2y}$)

$$P_{1x} = P_{2x} \Rightarrow m_A v_{A1x} + m_B v_{B1x} = m_A v_{A2x} + m_B v_{B2x}$$
$$(2.5\times10^{-3}\,\text{kg})(400\,\text{m}\cdot\text{s}^{-1}) + 0 = 0 + (0.10\,\text{kg})\, v_{B2}\cos\theta$$
$$\boxed{v_{B2}\cos\theta = 10\,\text{m}\cdot\text{s}^{-1}}$$

$$P_{1y} = P_{2y} \Rightarrow m_A v_{A1y} + m_B v_{B1y} = m_A v_{A2y} + m_B v_{B2y}$$
$$0 + 0 = (0.10\,\text{kg})\, v_{B2}\sin\theta - (2.5\times10^{-3}\,\text{kg})(300\,\text{m}\cdot\text{s}^{-1})$$
$$\boxed{v_{B2}\sin\theta = 7.5\,\text{m}\cdot\text{s}^{-1}}$$

Divide one equation by the other $\Rightarrow \dfrac{v_{B2}\sin\theta}{v_{B2}\cos\theta} = \dfrac{7.5\,\text{m}\cdot\text{s}^{-1}}{10\,\text{m}\cdot\text{s}^{-1}}$

$\tan\theta = 0.75 \Rightarrow \theta = \underline{36.9^\circ}$, defined as in the sketch

Then $v_{B2}\cos\theta = 10\,\text{m}\cdot\text{s}^{-1} \Rightarrow v_{B2} = \dfrac{10\,\text{m}\cdot\text{s}^{-1}}{\cos 36.9^\circ} = \underline{12.5\,\text{m}\cdot\text{s}^{-1}}$

b) To answer this question compare K_1 and K_2 for the system:

$$K_1 = \tfrac{1}{2} m_A v_{A1}^2 + \tfrac{1}{2} m_B v_{B1}^2 = \tfrac{1}{2}(2.5\times10^{-3}\,\text{kg})(400\,\text{m}\cdot\text{s}^{-1})^2 + 0 = 200\,\text{J}$$
$$K_2 = \tfrac{1}{2} m_A v_{A2}^2 + \tfrac{1}{2} m_B v_{B2}^2 = \tfrac{1}{2}(2.5\times10^{-3}\,\text{kg})(300\,\text{m}\cdot\text{s}^{-1})^2 + \tfrac{1}{2}(0.10\,\text{kg})(12.5\,\text{m}\cdot\text{s}^{-1})^2 = 112\,\text{J} + 7.8\,\text{J} = 120\,\text{J}$$
$$\Delta K = K_2 - K_1 = 120\,\text{J} - 200\,\text{J} = -80\,\text{J}$$

The kinetic energy decreases by 80 J as a result of the collision; the collision is not elastic.

8-49

Apply momentum conservation in the xy-plane (see Fig. 8-18):

$$P_{1x} = P_{2x} \Rightarrow m v_0 = m \frac{v_0}{2}\cos 10^\circ + m v_1 \cos 45^\circ + m v_2 \cos 30^\circ$$
$$3.0\times10^6\,\text{m}\cdot\text{s}^{-1} = (1.5\times10^6\,\text{m}\cdot\text{s}^{-1})\cos 10^\circ + v_1\cos 45^\circ + v_2 \cos 30^\circ$$
$$\boxed{1.52\times10^6\,\text{m}\cdot\text{s}^{-1} = v_1\cos 45^\circ + v_2\cos 30^\circ}$$

$$P_{1y} = P_{2y} \Rightarrow 0 = m \frac{v_0}{2}\sin 10^\circ + m v_1 \sin 45^\circ - m v_2 \sin 30^\circ$$
$$0 = (1.5\times10^6\,\text{m}\cdot\text{s}^{-1})\sin 10^\circ + v_1 \sin 45^\circ - v_2\sin 30^\circ$$
$$\boxed{2.60\times10^5\,\text{m}\cdot\text{s}^{-1} = -v_1\sin 45^\circ + v_2 \sin 30^\circ}$$

Add these two equations, and use that $\sin 45^\circ = \cos 45^\circ$

$$1.78\times10^6\,\text{m}\cdot\text{s}^{-1} = v_2(\sin 30^\circ + \cos 30^\circ) \Rightarrow v_2 = \underline{1.30\times10^6\,\text{m}\cdot\text{s}^{-1}}$$

The first equation then implies $v_1 = \dfrac{1.52\times10^6\,\text{m}\cdot\text{s}^{-1} - (1.30\times10^6\,\text{m}\cdot\text{s}^{-1})\cos 30^\circ}{\cos 45^\circ} = \underline{5.54\times10^5\,\text{m}\cdot\text{s}^{-1}}$

Apply momentum conservation in the z-plane:

$$P_{1z} = P_{2z} \Rightarrow 0 = m_{Ba} v_{Ba} - m_{Kr} v_{Kr}$$

8-49 (cont)

$$v_{Kr} = \left(\frac{m_{Ba}}{m_{Kr}}\right) v_{Ba} = \left(\frac{2.3\times10^{-25}\,\text{kg}}{1.5\times10^{-25}\,\text{kg}}\right) v_{Ba} = \underline{1.53\, v_{Ba}}$$

(We can't say what these velocities are, but they must satisfy this relation. What v_{Kr} and v_{Ba} individually will be depends on energy considerations.)

8-51

a) $K = \frac{1}{2} m_A v_A'^2 + \frac{1}{2} m_B v_B'^2$

$M = m_A + m_B \quad ; \quad \vec{V} = \frac{m_A \vec{v}_A + m_B \vec{v}_B}{M}$

Let's write out the terms in the expression we want to prove:

$$\tfrac{1}{2} m_A v_A'^2 = \tfrac{1}{2} m_A (\vec{v}_A - \vec{V}) \cdot (\vec{v}_A - \vec{V}) = \tfrac{1}{2} m_A (v_A^2 + V^2 - 2\vec{v}_A \cdot \vec{V})$$

(We have used that for a vector $\vec{A}$, $A^2 = \vec{A}\cdot\vec{A}$.)

Similarly,

$$\tfrac{1}{2} m_B v_B'^2 = \tfrac{1}{2} m_B (\vec{v}_B - \vec{V}) \cdot (\vec{v}_B - \vec{V}) = \tfrac{1}{2} m_B (v_B^2 + V^2 - 2\vec{v}_B \cdot \vec{V})$$

$$\Rightarrow \tfrac{1}{2} m_A v_A'^2 + \tfrac{1}{2} m_B v_B'^2 = \tfrac{1}{2} m_A v_A^2 + \tfrac{1}{2} m_B v_B^2 + \tfrac{1}{2}(m_A + m_B) V^2 - \vec{V}\cdot(m_A \vec{v}_A + m_B \vec{v}_B)$$

But $m_A + m_B = M$ and $m_A \vec{v}_A + m_B \vec{v}_B = M\vec{V}$, so

$$\tfrac{1}{2} m_A v_A'^2 + \tfrac{1}{2} m_B v_B'^2 = \tfrac{1}{2} m_A v_A^2 + \tfrac{1}{2} m_B v_B^2 + \tfrac{1}{2} M V^2 - M V^2$$

And

$$\tfrac{1}{2} m_A v_A'^2 + \tfrac{1}{2} m_B v_B'^2 + \tfrac{1}{2} M V^2 = \tfrac{1}{2} m_A v_A^2 + \tfrac{1}{2} m_B v_B^2 + \cancel{\tfrac{1}{2} M V^2} - \cancel{M V^2} + \cancel{\tfrac{1}{2} M V^2},$$

so the expression given is equal to K.

b) In the collision $M\vec{V}$ is constant $\Rightarrow \frac{1}{2} M V^2$ is unchanged. The asteroids can lose all their relative energy, but $\frac{1}{2} M V^2$ must remain.

8-53

a) Eq (8-35) $\quad v = v_0 + v_r \ln\left(\frac{m_0}{m}\right) - gt$

Neglect gravity $\Rightarrow$ set $g = 0$.

$$v_0 = 0 \Rightarrow v = v_r \ln\left(\frac{m_0}{m}\right) = v_r \ln\left(\frac{13{,}000\,\text{kg}}{3000\,\text{kg} + 250\,\text{kg}}\right) = \underline{1.39\, v_r}$$

(Note: m is the mass that is not fuel; this is 12,000 kg − 9000 kg in the first stage and 1000 kg − 750 kg = 250 kg in the second stage.)

b) $$v = \cancelto{0}{v_0} + v_r \ln\left(\frac{m_0}{m}\right) = v_r \ln\left(\frac{13{,}000\,\text{kg}}{4000\,\text{kg}}\right) = \underline{1.18\, v_r}$$

(Note: m = 4000 kg; 9000 kg of first stage fuel has been burned.)

8-53 (cont)

c) $v = v_0 + v_r \ln\left(\frac{m_0}{m}\right) = 1.18\, v_r + v_r \ln\left(\frac{1000\,\text{kg}}{250\,\text{kg}}\right) = 1.18\, v_r + 1.39\, v_r = \underline{2.57\, v_r}$

(After the first stage separates the second stage is left, with an initial velocity of $1.18\, v_r$, an initial mass of 1000 kg, and a mass of 250 kg after all the fuel has burned.)

d) $v = 8\ \text{km}\cdot\text{s}^{-1}$

$v = 2.57\, v_r \Rightarrow v_r = \frac{v}{2.57} = \frac{8\ \text{km}\cdot\text{s}^{-1}}{2.57} = \underline{3.12\ \text{km}\cdot\text{s}^{-1}}$

CHAPTER 9

Exercises 3, 5, 7, 11, 15, 17, 19, 23, 27, 29, 31, 37, 39, 43, 45, 47, 51, 53

Problems 55, 57, 61, 65, 67, 69, 71, 73

Exercises

9-3

a) $\theta(t) = (2\,\text{rad}\cdot s^{-1})t + (0.05\,\text{rad}\cdot s^{-3})t^3$

$\omega = \frac{d\theta}{dt} = \underline{2\,\text{rad}\cdot s^{-1} + (0.15\,\text{rad}\cdot s^{-3})t^2}$

b) $t = 0 \Rightarrow \omega = \omega_0 = 2\,\text{rad}\cdot s^{-1}$

c) $t = 5s \Rightarrow \omega = 2\,\text{rad}\cdot s^{-1} + (0.15\,\text{rad}\cdot s^{-3})(5s)^2 = 2\,\text{rad}\cdot s^{-1} + 3.75\,\text{rad}\cdot s^{-1} = \underline{5.75\,\text{rad}\cdot s^{-1}}$

$\omega_{av} = \frac{\Delta\theta}{\Delta t}$

$t = 0 \Rightarrow \theta = 0$

$t = 5s \Rightarrow \theta = (2\,\text{rad}\cdot s^{-1})(5s) + (0.05\,\text{rad}\cdot s^{-3})(5s)^3 = 10\,\text{rad} + 6.25\,\text{rad} = 16.2\,\text{rad}$

$\omega_{av} = \frac{16.2\,\text{rad} - 0}{5s - 0} = \underline{3.24\,\text{rad}\cdot s^{-1}}$

9-5

$\theta(t) = a + bt^2 + ct^3$

$\omega = \frac{d\theta}{dt} = 2bt + 3ct^2$

$\alpha = \frac{d\omega}{dt} = 2b + 6ct$

9-7

a) $\omega_0 = 1000\,\text{rev}\cdot\text{min}^{-1}(1\,\text{min}/60s) = 16.7\,\text{rev}\cdot s^{-1}$

$\omega = 400\,\text{rev}\cdot\text{min}^{-1}(1\,\text{min}/60s) = 6.67\,\text{rev}\cdot s^{-1}$

$t = 5s$

$\alpha = ?$

$\omega = \omega_0 + \alpha t$

$\alpha = \frac{\omega - \omega_0}{t} = \frac{6.67\,\text{rev}\cdot s^{-1} - 16.7\,\text{rev}\cdot s^{-1}}{5s}$

$\alpha = \underline{-2.00\,\text{rev}\cdot s^{-2}}$

$\theta - \theta_0 = ?$

$\theta - \theta_0 = \omega_0 t + \frac{1}{2}\alpha t^2$

$\theta - \theta_0 = (16.7\,\text{rev}\cdot s^{-1})(5s) + \frac{1}{2}(-2.00\,\text{rev}\cdot s^{-2})(5s)^2 = 83.5\,\text{rev} - 25.0\,\text{rev}$

$\theta - \theta_0 = \underline{58.5\,\text{rev}}$

9-7 (cont)

b) $\omega = 0$

$\omega_0 = 400\,\text{rev}\cdot\text{min}^{-1} = 6.67\,\text{rev}\cdot\text{s}^{-1}$

$\alpha = -2.00\,\text{rev}\cdot\text{s}^{-2}$

$t = ?$

$$\omega = \omega_0 + \alpha t$$

$$t = \frac{\omega - \omega_0}{\alpha} = \frac{0 - 6.67\,\text{rev}\cdot\text{s}^{-1}}{-2.00\,\text{rev}\cdot\text{s}^{-2}} = \underline{+3.34\,\text{s}}$$

(Note that consistent units were used, so that the answers would come out in the units indicated.)

9-11

a) $v = r\omega$, but in this equation ω must be in $\text{rad}\cdot\text{s}^{-1}$.

$$\omega = (750\,\text{rev}\cdot\text{min}^{-1})\left(\frac{2\pi\,\text{rad}}{1\,\text{rev}}\right)\left(\frac{1\,\text{min}}{60\,\text{s}}\right) = 78.5\,\text{rad}\cdot\text{s}^{-1}$$

$$v = r\omega = (0.075\,\text{m})(78.5\,\text{rad}\cdot\text{s}^{-1}) = 5.89\,\text{m}\cdot\text{s}^{-1}$$

b) $$v = r\omega \Rightarrow \omega = \frac{v}{r} = \frac{0.60\,\text{m}\cdot\text{s}^{-1}}{0.025\,\text{m}} = 24.0\,\text{rad}\cdot\text{s}^{-1}$$

(In $v = r\omega$ the units of ω are $\text{rad}\cdot\text{s}^{-1}$.)

Convert the units of ω to $\text{rev}\cdot\text{min}^{-1} \Rightarrow \omega = (24\,\text{rad}\cdot\text{s}^{-1})\left(\frac{1\,\text{rev}}{2\pi\,\text{rad}}\right)\left(\frac{60\,\text{s}}{1\,\text{min}}\right) = \underline{229\,\text{rev}\cdot\text{min}^{-1}}$

9-15

a) At the start:

flywheel starts from rest $\Rightarrow \omega = \omega_0 = 0$

$a_\perp = r\omega^2 = \underline{0}$

$a_\parallel = r\alpha = (0.30\,\text{m})(0.50\,\text{rad}\cdot\text{s}^{-2}) = \underline{0.15\,\text{m}\cdot\text{s}^{-2}}$

$a = \sqrt{a_\perp^2 + a_\parallel^2} = \underline{0.15\,\text{m}\cdot\text{s}^{-2}}$

b) $\theta - \theta_0 = 120°$

$$\theta - \theta_0 = 120°\left(\frac{2\pi\,\text{rad}}{360°}\right) = 2.09\,\text{rad}$$

Calculate ω, because need it to calculate $a_\perp$:

$\theta - \theta_0 = 2.09\,\text{rad}$

$\omega_0 = 0$

$\alpha = 0.50\,\text{rad}\cdot\text{s}^{-2}$

$\omega = ?$

$$\omega^2 = \omega_0^2 + 2\alpha(\theta - \theta_0) \quad (\omega_0^2 \to 0)$$

$$\omega = \sqrt{2(0.50\,\text{rad}\cdot\text{s}^{-2})(2.09\,\text{rad})}$$

$$\omega = 1.45\,\text{rad}\cdot\text{s}^{-1}$$

$a_\perp = r\omega^2 = (0.30\,\text{m})(1.45\,\text{rad}\cdot\text{s}^{-1})^2 = \underline{0.631\,\text{m}\cdot\text{s}^{-2}}$

$a_\parallel = r\alpha = 0.15\,\text{m}\cdot\text{s}^{-2}$, as in (a)

$a = \sqrt{a_\perp^2 + a_\parallel^2} = \sqrt{(0.631\,\text{m}\cdot\text{s}^{-2})^2 + (0.15\,\text{m}\cdot\text{s}^{-2})^2} = \underline{0.649\,\text{m}\cdot\text{s}^{-2}}$

9-15 (cont)

c) $\theta - \theta_0 = 240°$

$\theta - \theta_0 = 240° \left(\frac{2\pi \text{ rad}}{360°}\right) = 4.19 \text{ rad}$

$\omega_0 = 0$

$\alpha = 0.50 \text{ rad} \cdot s^{-2}$

$\omega = ?$

$\omega^2 = \omega_0^2 + 2\alpha(\theta - \theta_0)$ (with $\omega_0^2 = 0$)

$\omega = \sqrt{2(0.50 \text{ rad} \cdot s^{-2})(4.19 \text{ rad})} = 2.05 \text{ rad} \cdot s^{-1}$

$a_\perp = r\omega^2 = (0.30 m)(2.05 \text{ rad} \cdot s^{-1})^2 = \underline{1.26 m \cdot s^{-2}}$

$a_\parallel = r\alpha = 0.15 m \cdot s^{-2}$ (as in (a))

$a = \sqrt{a_\perp^2 + a_\parallel^2} = \sqrt{(1.26 m \cdot s^{-2})^2 + (0.15 m \cdot s^{-2})^2} = \underline{1.27 m \cdot s^{-2}}$

9-17

The diagonal of the square has length $d = \sqrt{(0.5 m)^2 + (0.5 m)^2} = 0.707 m$, so the distance from the center to each corner is $\frac{d}{2} = 0.354 m$.

a) Each mass is 0.354 m from this axis, so

$I = \Sigma mr^2 = 4(3 kg)(0.354 m)^2 = \underline{1.50 kg \cdot m^2}$

b) Each mass is $\frac{0.50 m}{2} = 0.25 m$ from this axis, so

$I = \Sigma mr^2 = 4(3 kg)(0.25 m)^2 = \underline{0.75 kg \cdot m^2}$

9-19

$I = \Sigma mr^2 \Rightarrow I = I_{rim} + I_{spokes}$

$I_{rim} = mR^2 = (1.0 kg)(0.3 m)^2 = 0.090 kg \cdot m^2$ (The rim is a thin-walled hollow cylinder.)

$I_{spokes} = 4\left(\frac{1}{12} ml^2\right)$ (Each spoke is a rod of length $l = 2R = 0.6 m$, and the axis is at the center of each spoke.)

$I_{spokes} = 4\left(\frac{1}{12}\right)(0.40 kg)(0.6 m)^2 = 0.048 kg \cdot m^2$

Thus $I = 0.090 kg \cdot m^2 + 0.048 kg \cdot m^2 = \underline{0.138 kg \cdot m^2}$

9-23

a)

0.4m, 0.3m, axis, m = 24 kg

The mass distribution about the axis is the same as for a rod with the axis through its center.

$$\Rightarrow I = \frac{1}{12} m l^2 = \frac{1}{12}(24\,\text{kg})(0.3\,\text{m})^2 = \underline{0.18\,\text{kg}\cdot\text{m}^2}$$

b)

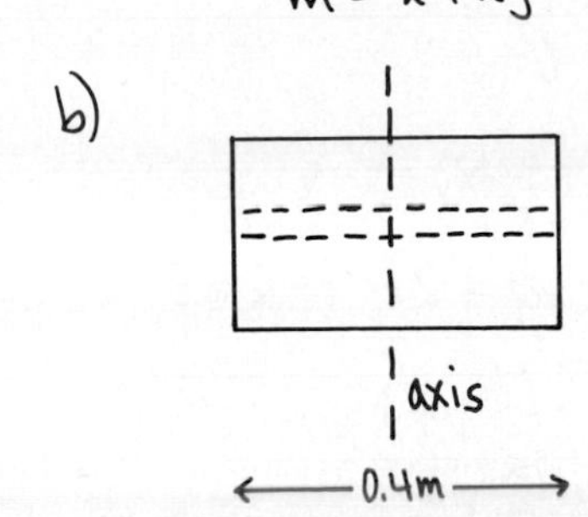

The formula for a rod again applies, now with $l = 0.4\,\text{m}$.

$$I = \frac{1}{12} m l^2 = \frac{1}{12}(24\,\text{kg})(0.4\,\text{m})^2 = \underline{0.32\,\text{kg}\cdot\text{m}^2}$$

c)

0.4m, 0.3m

From Fig. 9-7 (b), $I = \frac{1}{12} m (a^2 + b^2)$

$$I = \frac{1}{12}(24\,\text{kg})\left[(0.4\,\text{m})^2 + (0.3\,\text{m})^2\right] = \underline{0.50\,\text{kg}\cdot\text{m}^2}$$

9-27

a)

4ft, axis, F = 20 lb

$$\Gamma = Fl = (20\,\text{lb})(4\,\text{ft}) = \underline{80\,\text{ft}\cdot\text{lb}}$$

↺ counterclockwise

b)

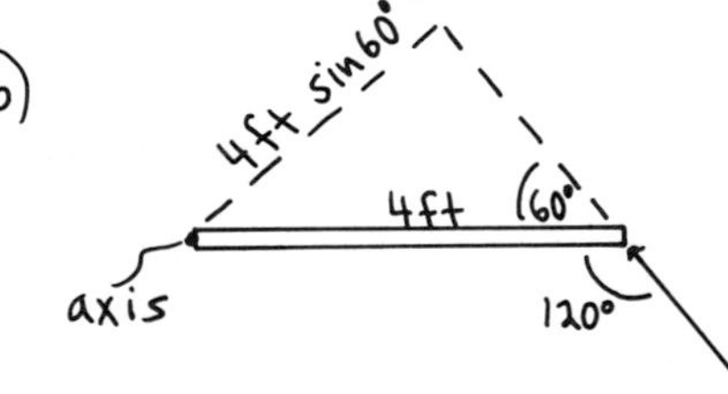

$$l = (4\,\text{ft}) \sin 60° = 3.46\,\text{ft}$$

$$\Gamma = Fl = (20\,\text{lb})(3.46\,\text{ft}) = \underline{69.2\,\text{ft}\cdot\text{lb}}$$

↺ counterclockwise

c)

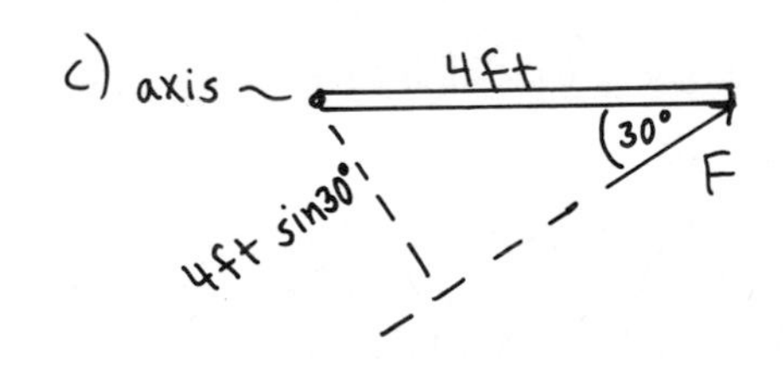

$$l = (4\,\text{ft}) \sin 30° = 2\,\text{ft}$$

$$\Gamma = Fl = (20\,\text{lb})(2\,\text{ft}) = \underline{40.0\,\text{ft}\cdot\text{lb}}$$

↺ counterclockwise

d)

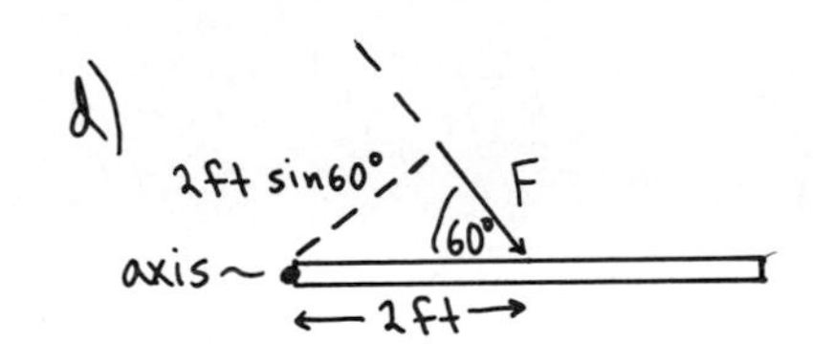

$$l = (2\,\text{ft}) \sin 60° = 1.73\,\text{ft}$$

$$\Gamma = Fl = (20\,\text{lb})(1.73\,\text{ft}) = \underline{34.6\,\text{ft}\cdot\text{lb}}$$

↻ clockwise

9-27 (cont)

e)

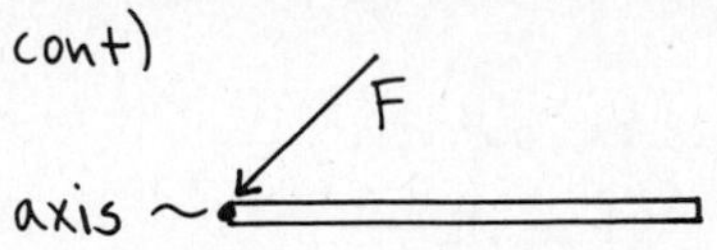

$\ell = 0 \Rightarrow \Gamma = F\ell = \underline{0}$

f)

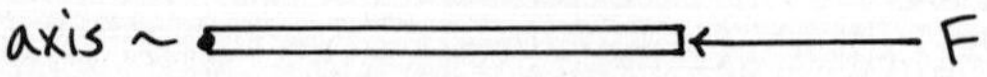

$\ell = 0 \Rightarrow \Gamma = F\ell = \underline{0}$

9-29

$\vec{F} = (2N)\vec{i} - (3N)\vec{j}$

$\vec{r} = (-0.3m)\vec{i} + (0.5m)\vec{j}$

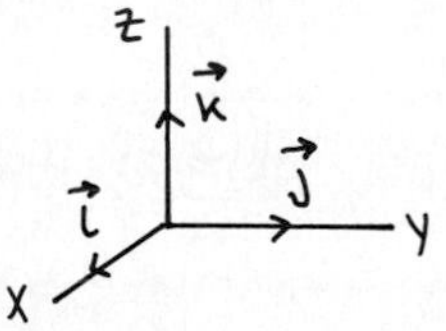

recall that

$\vec{i}\times\vec{i} = \vec{j}\times\vec{j} = \vec{k}\times\vec{k} = 0$

$\vec{i}\times\vec{j} = \vec{k}, \ \vec{j}\times\vec{k} = \vec{i}, \ \vec{k}\times\vec{i} = \vec{j}$

$\vec{\Gamma} = \vec{r}\times\vec{F} = [(-0.3m)\vec{i} + (0.5m)\vec{j}]\times[(2N)\vec{i} + (-3N)\vec{j}]$

$\vec{\Gamma} = +0.9\,N\cdot m(\vec{i}\times\vec{j}) + 1.0\,N\cdot m(\vec{j}\times\vec{i}) = (0.9\,N\cdot m)\vec{k} + (1.0\,N\cdot m)(-\vec{k}) = \underline{-(0.1\,N\cdot m)\vec{k}}$

9-31

Use the Kinematic information to solve for the angular acceleration of the grindstone:

$\omega_0 = 900\,rev\cdot min^{-1}(2\pi rad/1 rev)\left(\frac{1 min}{60s}\right) = 94.2\,rad\cdot s^{-1}$

$t = 10s$

$\omega = 0$

$\alpha = ?$

$\omega = \omega_0 + \alpha t$

$\alpha = \frac{\omega - \omega_0}{t} = \frac{0 - 94.2\,rad\cdot s^{-1}}{10s}$

$\alpha = -9.42\,rad\cdot s^{-2}$

Now consider the torques on the grindstone, and apply $\Sigma\Gamma = I\alpha$:

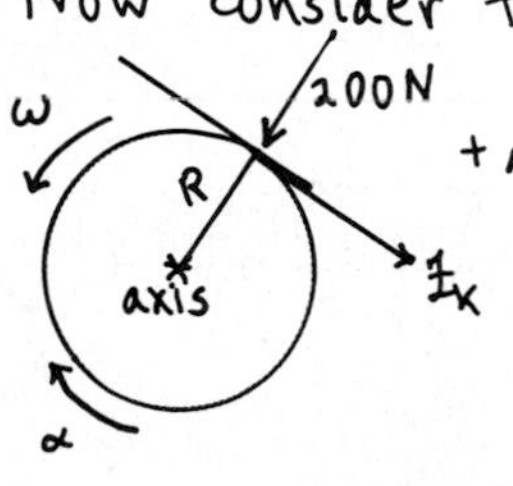

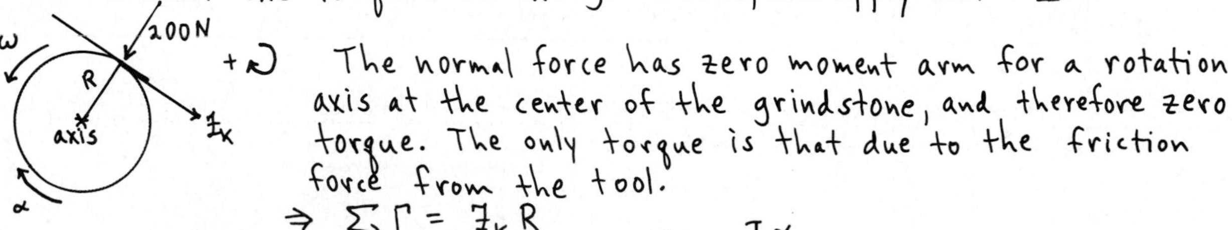

$+\circlearrowleft$ The normal force has zero moment arm for a rotation axis at the center of the grindstone, and therefore zero torque. The only torque is that due to the friction force from the tool.

$\Rightarrow \Sigma\Gamma = f_k R$

$\Sigma\Gamma = I\alpha \Rightarrow f_k R = I\alpha \Rightarrow f_k = \frac{I\alpha}{R}$

$I = \frac{1}{2}MR^2 = \frac{1}{2}(50kg)(0.5m)^2 = 6.25\,kg\cdot m^2$

$f_k = \frac{(6.25\,kg\cdot m^2)(9.42\,rad\cdot s^{-2})}{(0.5m)} = 118\,N$

$\mu_k = \frac{f_k}{n} = \frac{118N}{200N} = \underline{0.590}$

9-37

a) $P = \Gamma\omega \Rightarrow \Gamma = \frac{P}{\omega}$, but ω must be in $rad\cdot s^{-1}$

$\omega = (2400\,rev\cdot min^{-1})(2\pi rad/1 rev)(1 min/60s) = 251\,rad\cdot s^{-1}$

$\Gamma = \frac{P}{\omega} = \frac{1.5\times10^6 W}{251\,rad\cdot s^{-1}} = \underline{5.98\times10^3\,N\cdot m}$

9-37 (cont)

b) For the drum $\omega = \text{constant} \Rightarrow \alpha = 0$

$\Sigma\Gamma = 0 \Rightarrow \Gamma_{motor} = \Gamma_T = TR$

$T = \frac{\Gamma}{R} = \frac{5.98\times10^3\ N\cdot m}{0.25\ m} = \underline{2.39\times10^4\ N}$

$\uparrow v$ constant $\Rightarrow a = 0 \Rightarrow T = mg$

c) $v = R\omega = (0.25\ m)(251\ rad\cdot s^{-1}) = \underline{62.8\ m\cdot s^{-1}}$

9-39

a) $\Sigma\vec{F} = m\vec{a}$, for the motion of the center of mass: $\Sigma F_y = ma_y$

$\boxed{mg - T = ma}$

$+\circlearrowright\ \Sigma\Gamma = I\alpha$, for rotation about the center of mass (which is at the center of the hoop).

$\Sigma\Gamma = TR$ (mg has zero torque)

$I = mR^2$

$\Sigma\Gamma = I\alpha \Rightarrow TR = mR^2\alpha$

$T = mR\alpha$

But α and a are related by $\alpha R = a$

$\Rightarrow \boxed{T = ma}$

Substitute the second eq. into the first $\Rightarrow mg - ma = ma \Rightarrow a = g/2 = 4.9\ m\cdot s^{-2}$

$T = ma = (1.2\ kg)(4.9\ m\cdot s^{-2}) = \underline{5.88\ N}$

b) Use results from (a) for the motion of the center of mass:

$a = 4.9\ m\cdot s^{-2}$

$y - y_0 = 0.5\ m$

$v_0 = 0$ (released from rest)

$t = ?$

$y - y_0 = v_0 t + \frac{1}{2}at^2$ (with $v_0 t = 0$)

$t = \sqrt{\frac{2(y-y_0)}{a}} = \sqrt{\frac{2(0.5\ m)}{4.9\ m\cdot s^{-2}}} = \underline{0.452\ s}$

c) Calculate v of the cm, and obtain ω by $\omega = \frac{v}{R}$:

$v = v_0 + at = 0 + (4.9\ m\cdot s^{-2})(0.452\ s) = 2.21\ m\cdot s^{-1}$

$\omega = \frac{v}{R} = \frac{2.21\ m\cdot s^{-1}}{0.08\ m} = \underline{27.6\ rad\cdot s^{-1}}$

9-43

$v = 12\ m\cdot s^{-1}$

$r = 8\ m \sin 37° = 4.81\ m$

$L = mvr$ (eq. 9-38)

$L = (2\ kg)(12\ m\cdot s^{-1})(4.81\ m) = \underline{115\ kg\cdot m^2\cdot s^{-1}}$

9-45

a) $I_i = I_{man} + I_{dumbbells} = 5\,kg\cdot m^2 + 2mr^2 = 5\,kg\cdot m^2 + 2(5\,kg)(0.6\,m)^2 = 8.6\,kg\cdot m^2$

$L_i = I_i\omega_i = (8.6\,kg\cdot m^2)(5\,rad\cdot s^{-1}) = \underline{43\,kg\cdot m^2\cdot s^{-1}}$

b) $I_f = I_{man} + I_{dumbbells} = 5\,kg\cdot m^2 + 2(5\,kg)(0.2\,m)^2 = 5.4\,kg\cdot m^2$

$L_i = L_f$

$I_i\omega_i = I_f\omega_f \Rightarrow \omega_f = \left(\frac{I_i}{I_f}\right)\omega_i = \left(\frac{8.6\,kg\cdot m^2}{5.4\,kg\cdot m^2}\right)(5\,rad\cdot s^{-1}) = \underline{7.96\,rad\cdot s^{-1}}$

c) $K_i = \frac{1}{2}I_i\omega_i^2 = \frac{1}{2}(8.6\,kg\cdot m^2)(5.0\,rad\cdot s^{-1})^2 = 108\,J$

$K_f = \frac{1}{2}I_f\omega_f^2 = \frac{1}{2}(5.4\,kg\cdot m^2)(7.96\,rad\cdot s^{-1})^2 = 171\,J$

$\Delta K = K_f - K_i = 171\,J - 108\,J = 63\,J$

The increase in kinetic energy comes from the work done by the man.

9-47

a) $L_i = L_f \Rightarrow I_i\omega_i = I_f\omega_f$

$I = mr^2 \Rightarrow I_i = mr_i^2\ ,\quad I_f = mr_f^2$

$\omega_f = \left(\frac{I_i}{I_f}\right)\omega_i = \left(\frac{\cancel{m}r_i^2}{\cancel{m}r_f^2}\right)\omega_i = \left(\frac{0.2\,m}{0.1\,m}\right)^2(3\,rad\cdot s^{-1}) = \underline{12\,rad\cdot s^{-1}}$

b) $K_i = \frac{1}{2}I_i\omega_i^2 = \frac{1}{2}mr_i^2\omega_i^2 = \frac{1}{2}(0.05\,kg)(0.2\,m)^2(3\,rad\cdot s^{-1})^2 = 0.0090\,J$

$K_f = \frac{1}{2}I_f\omega_f^2 = \frac{1}{2}mr_f^2\omega_f^2 = \frac{1}{2}(0.05\,kg)(0.1\,m)^2(12\,rad\cdot s^{-1})^2 = 0.0360\,J$

$\Delta K = K_f - K_i = 0.0360\,J - 0.0090\,J = \underline{+0.0270\,J}$ (increase)

c) The work-energy theorem says $W = \Delta K$.
The only work done on the puck is that done by the person

$\Rightarrow W_{person} = \underline{+0.027\,J}$

9-51

$L_i = L_f$

L_i is due to the linear motion of the mud. (The door is initially at rest.)

$L_i = mvr = (0.5\,kg)(10\,m\cdot s^{-1})(0.5\,m) = 2.5\,kg\cdot m^2\cdot s^{-1}$

L_f is due to the door + mud rotating about the hinges with angular velocity ω.

$L_f = I_{tot}\,\omega_f$

$I_{tot} = I_{door} + I_{mud}$

$I_{door} = \frac{1}{3}ml^2 = \frac{1}{3}(50\,kg)(1.0\,m)^2 = 16.7\,kg\cdot m^2$

$I_{mud} = mr^2 = (0.5\,kg)(0.5\,m)^2 = 0.125\,kg\cdot m^2$

$\Rightarrow I_{tot} = 16.8\,kg\cdot m^2$

9-51 (cont)

$$L_i = L_f \Rightarrow \omega_f = \frac{L_i}{I_{tot}} = \frac{2.5\ kg\cdot m^2\cdot s^{-1}}{16.8\ kg\cdot m^2} = \underline{0.149\ rad\cdot s^{-1}}$$

(The moment of inertia of the mud is $\frac{0.125}{16.8} \times 100\% = 0.7\%$ of the total.)

9-53

a) Find the amount of kinetic energy stored in the gyroscope when it is up to speed:

$$K = \tfrac{1}{2} I \omega^2, \quad \omega \text{ in } rad\cdot s^{-1}$$

$$\omega = (900\ rev\cdot min^{-1})(2\pi\ rad/1\ rev)(1\ min/60\ s) = 94.2\ rad\cdot s^{-1}$$

$$I = \tfrac{1}{2} m R^2 \text{ (disk)} \Rightarrow I = \tfrac{1}{2}(50{,}000\ kg)(2\ m)^2 = 1.0\times10^5\ kg\cdot m^2$$

$$K = \tfrac{1}{2} I \omega^2 = \tfrac{1}{2}(1.0\times10^5\ kg\cdot m^2)(94.2\ rad\cdot s^{-1})^2 = 4.44\times10^8\ J$$

$$P = \frac{W}{t} = \frac{\Delta K}{t} \quad \text{(by the work-energy relation)}$$

$$\Rightarrow t = \frac{\Delta K}{P} = \frac{4.44\times10^8\ J}{7.46\times10^4\ W} = 5.95\times10^3\ s = \underline{99.1\ min}$$

b) $\Omega = \frac{\Gamma}{L}$ (eq. 9-50) $\Rightarrow \Gamma = \Omega L$

$$L = I\omega = (1\times10^5\ kg\cdot m^2)(94.2\ rad\cdot s^{-1}) = 9.42\times10^6\ kg\cdot m^2\cdot s^{-1}$$

$$\Omega = 1^\circ\cdot s^{-1}\left(\frac{2\pi\ rad}{360^\circ}\right) = 0.0175\ rad\cdot s^{-1}$$

$$\Gamma = \Omega L = (0.0175\ rad\cdot s^{-1})(9.42\times10^6\ kg\cdot m^2\cdot s^{-1}) = \underline{1.65\times10^5\ N\cdot m}$$

Problems

9-55

The core radius is the radius of the earth minus the thickness of the mantle $\Rightarrow 6.38\times10^6\ m - 3.60\times10^6\ m = 2.78\times10^6\ m$.

The model represents the earth as the superposition of two spheres, one sphere with the earth's radius and a density of $5.0\times10^3\ kg\cdot m^{-3}$ and the other with the core radius and a density of $11\times10^3\ kg\cdot m^{-3} - 5\times10^3\ kg\cdot m^{-3} = 6\times10^3\ kg\cdot m^{-3}$.

The volume of a sphere is $\frac{4}{3}\pi R^3$, so the mass of a sphere is $m = \frac{4}{3}\pi R^3 \rho$. The moment of inertia of the earth in this model is the sum of $\frac{2}{5} m R^2$ for each of the two spheres.

$$\Rightarrow I = \tfrac{2}{5}\left(\tfrac{4}{3}\pi(6.38\times10^6 m)^3\right)(5.0\times10^3 kg\cdot m^{-3})(6.38\times10^6 m)^2$$

$$+ \tfrac{2}{5}\left(\tfrac{4}{3}\pi(2.78\times10^6 m)^3\right)(6.0\times10^3 kg\cdot m^{-3})(2.78\times10^6 m)^2$$

$$I = \frac{8\pi}{15}\left[(5.0\times10^3 kg\cdot m^{-3})(6.38\times10^6 m)^5 + (6.0\times10^3 kg\cdot m^{-3})(2.78\times10^6 m)^5\right]$$

$$I = 9.02\times10^{37}\ kg\cdot m^2$$

9-55 (cont)

Now express this in terms of MR^2 of the earth:

$$MR^2 = (5.98\times10^{24}\,kg)(6.38\times10^{6}\,m)^2 = 2.43\times10^{38}\,kg\cdot m^2$$

$$\Rightarrow I = \frac{9.02\times10^{37}}{2.43\times10^{38}} MR^2 = \underline{0.371\,MR^2}$$

The satellite data gives $I = 0.3308\,MR^2$, so this model is rather crude.

Note: The model has been chosen such that it gives the correct value of M, the mass of the earth:

$$M = \tfrac{4}{3}\pi(6.38\times10^{6}\,m)^3(5.0\times10^{3}\,kg\cdot m^{-3}) + \tfrac{4}{3}\pi(2.78\times10^{6}\,m)^3(6.0\times10^{3}\,kg\cdot m^{-3})$$

$M = 5.98\times10^{24}\,kg$, which is the correct value.

9-57

a) $W = \Delta K$

$W = -4000\,J$ (The flywheel gives up energy.)

$$K_i = \tfrac{1}{2} I \omega_i^2$$

$$\omega_i = 300\,\text{rev}\cdot\text{min}^{-1}(2\pi\,rad/1\,\text{rev})(1\,\text{min}/60\,s) = 31.4\,rad\cdot s^{-1}$$

$$K_i = \tfrac{1}{2} I\omega_i^2 = \tfrac{1}{2}(25\,kg\cdot m^2)(31.4\,rad\cdot s^{-1})^2 = 1.23\times10^4\,J$$

$$W = K_f - K_i \Rightarrow K_f = W + K_i = -4000\,J + 1.23\times10^4\,J = 8.3\times10^3\,J$$

$$K_f = \tfrac{1}{2}I\omega_f^2 \Rightarrow \omega_f = \sqrt{\frac{2K_f}{I}} = \sqrt{\frac{2(8.3\times10^3\,J)}{25\,kg\cdot m^2}} = 25.8\,rad\cdot s^{-1}$$

$$\omega_f = (25.8\,rad\cdot s^{-1})\left(\frac{1\,rev}{2\pi\,rad}\right)\left(\frac{60\,s}{1\,min}\right) = \underline{246\,rev\cdot min^{-1}}$$

b) $P = \frac{W}{t} = \frac{\Delta K}{t} = \frac{4000\,J}{5\,s} = \underline{800\,W}$

9-61

$n = m_A g$, $\mu_k n$, v, m_A, $m_A g$, I, m_B, v, d, m_B

$$W_{other} = W_{f_k} = -\mu_k n d = -\mu_k m_A g d$$

$$W_{other} = (K_2 - K_1) + (U_2 - U_1)$$

$$K_1 = 0$$

$$K_2 = \tfrac{1}{2}m_A v^2 + \tfrac{1}{2}m_B v^2 + \tfrac{1}{2}I\omega^2$$

$\omega = \frac{v}{R}$ for the pulley, where v is the velocity of each block

$$\Rightarrow K_2 = \tfrac{1}{2}(m_A + m_B + I/R^2)v^2$$

Let $U = 0$ at the final position of block B.

$\Rightarrow U_1 = m_B g d$, $U_2 = 0$

Thus $-\mu_k m_A g d = \left(\tfrac{1}{2}(m_A + m_B + I/R^2)v^2 - 0\right) + (0 - m_B g d)$

9-61 (cont)

$$\tfrac{1}{2}(m_A + m_B + I/R^2)\,v^2 = m_B g d - \mu_K m_A g d$$

$$v = \sqrt{\frac{2gd\,(m_B - \mu_K m_A)}{m_A + m_B + I/R^2}}$$

9-65

Use $\sum\Gamma = I\alpha$ to find α, and then use the constant α kinematic eqs. to solve for t:

F=180N   1m   hinge (axis)

$$\sum\Gamma = (180\,N)(1m) = 180\,N\cdot m$$

$$I = \tfrac{1}{3}ml^2 = \tfrac{1}{3}\left(\frac{600N}{9.8m\cdot s^{-2}}\right)(1m)^2 = 20.4\,kg\cdot m^2$$

$$\sum\Gamma = I\alpha \Rightarrow \alpha = \frac{180\,N\cdot m}{20.4\,kg\cdot m^2} = 8.82\,rad\cdot s^{-2}$$

$\theta - \theta_0 = 90° = \frac{\pi}{2}$ rad

$\omega_0 = 0$

$\alpha = 8.82\,rad\cdot s^{-2}$

$t = ?$

$$\theta - \theta_0 = \omega_0 t^{\,0} + \tfrac{1}{2}\alpha t^2$$

$$t = \sqrt{\frac{2(\theta-\theta_0)}{\alpha}} = \sqrt{\frac{2(\frac{\pi}{2}\,rad)}{8.82\,rad\cdot s^{-2}}} = \underline{0.597\,s}$$

9-67

a   5 kg T_1 R=0.1m T_2 No Friction   5 kg   a

Use kinematics to find a for each block:

$t = 2s$

$x - x_0 = 4m$

$v_0 = 0$

$a = ?$

$$x - x_0 = v_0 t^{\,0} + \tfrac{1}{2}at^2$$

$$a = \frac{2(x-x_0)}{t^2} = \frac{2(4m)}{(2s)^2} = 2.0\,m\cdot s^{-2}$$

$$a = R\alpha \Rightarrow \alpha = \frac{a}{R} = \frac{2.0\,m\cdot s^{-2}}{0.1m} = 20\,rad\cdot s^{-2}$$

a) Consider the forces on the block on the table:

y   x   n $a = 2.0\,m\cdot s^{-2}$ 5kg T_1 mg

$$\sum F_x = ma_x$$

$$T_1 = ma = (5kg)(2.0\,m\cdot s^{-1}) = \underline{10N}$$

Consider the forces on the hanging block:

x   y T_2 5kg $a = 2.0\,m\cdot s^{-2}$ mg = 49N

$$\sum F_y = ma_y$$

$$mg - T_2 = ma$$

$$T_2 = m(g-a) = 5kg\,(9.8\,m\cdot s^{-2} - 2.0\,m\cdot s^{-2}) = \underline{39N}$$

b) Apply $\sum\Gamma = I\alpha$ to the pulley:

The normal force n exerted by the axle and the weight Mg of the pulley both act at the axis and therefore give no torque.

T_1 n   + α Mg T_2

9-67 (cont)

Thus $\Sigma\Gamma = T_2 R - T_1 R = (T_2 - T_1)R$

$$\Sigma\Gamma = I\alpha \Rightarrow I = \frac{\Sigma\Gamma}{\alpha} = \frac{(T_2-T_1)R}{\alpha} = \frac{(39N - 10N)(0.1m)}{20\ rad\cdot s^{-2}} = \underline{0.145\ kg\cdot m^2}$$

9-69

a)

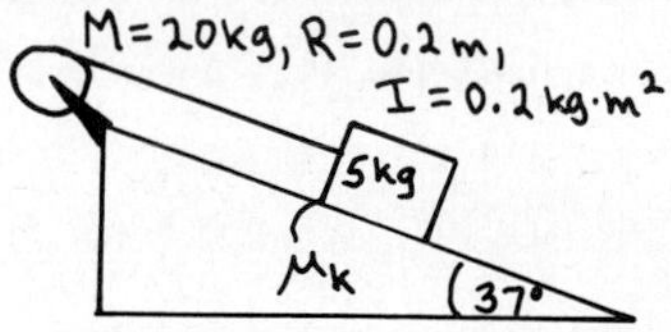

Apply $\Sigma\Gamma = I\alpha$ to the flywheel:

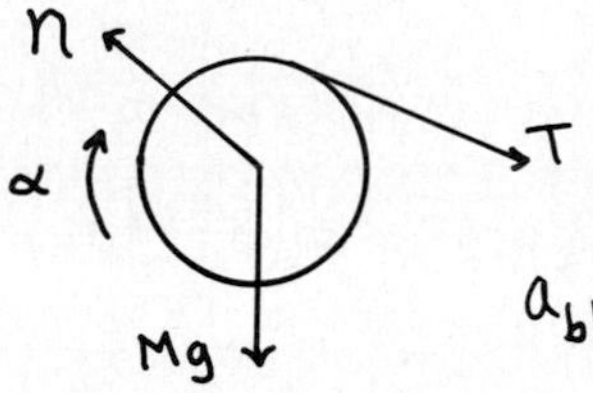

$+\circlearrowleft\ \Sigma\Gamma = I\alpha$

$\Gamma_n = \Gamma_{Mg} = 0 \Rightarrow \Sigma\Gamma = TR$

$TR = I\alpha$

$a_{block} = R\alpha_{wheel} \Rightarrow \alpha = \frac{a}{R}$

$\Rightarrow \boxed{T = \left(\frac{I}{R^2}\right)a}$

Apply $\Sigma\vec{F} = m\vec{a}$ to the block:

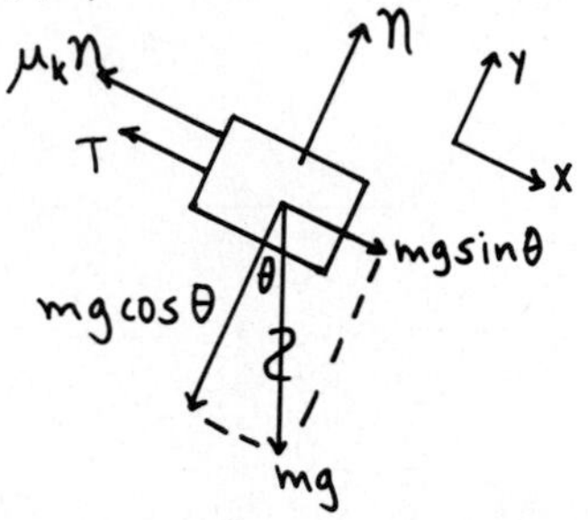

$\Sigma F_y = ma_y$

$n - mg\cos\theta = 0$

$n = mg\cos\theta$

$\Sigma F_x = ma_x$

$mg\sin\theta - \mu_K n - T = ma$

$\boxed{mg\sin\theta - \mu_K mg\cos\theta - T = ma}$

Use the above expression for T

$\Rightarrow mg(\sin\theta - \mu_K\cos\theta) - \left(\frac{I}{R^2}\right)a = ma$

$$a = \frac{mg(\sin\theta - \mu_K\cos\theta)}{(m + I/R^2)} = \frac{(5kg)(9.8m\cdot s^{-2})(\sin 37° - 0.25\cos 37°)}{5kg + \frac{0.2kg\cdot m^2}{(0.2m)^2}} = \underline{1.97m\cdot s^{-2}}$$

b) $T = \left(\frac{I}{R^2}\right)a = \frac{0.2kg\cdot m^2}{(0.2m)^2}(1.97m\cdot s^{-2}) = \underline{9.85N}$

9-71

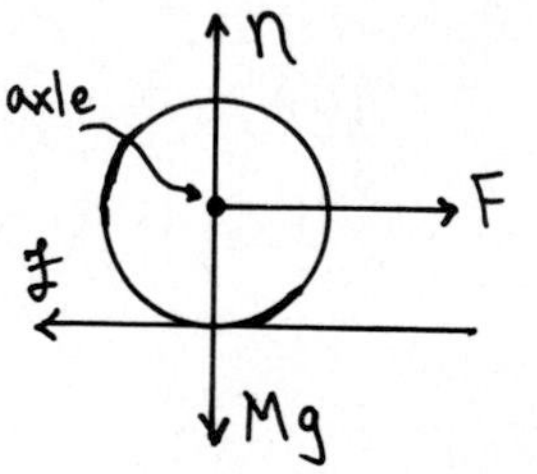

Apply $\Sigma\vec{F} = m\vec{a}$ to the linear motion of the center of mass:

$\Sigma F_x = ma_x$

$\boxed{F - f = Ma}$

Apply $\Sigma\Gamma = I\alpha$ to the rotation about the center of mass:

$+\circlearrowleft\ \Sigma\Gamma = fR$; all other forces have zero moment arms for an axis at the axle

$\Sigma\Gamma = I\alpha \Rightarrow fR = MR^2\alpha$, since $I = MR^2$

$f = MR\alpha$

But $R\alpha = a \Rightarrow \boxed{f = Ma}$

Combine these two equations $\Rightarrow F - Ma = Ma \Rightarrow \underline{a = \frac{F}{2M}}$

Then $f = Ma = \underline{\frac{F}{2}}$ is the friction force.

9-73

$$\sum F_x = ma_x$$

$$a_\perp = \frac{v^2}{R} \qquad T = ma_\perp = m\frac{v^2}{R}$$

So the velocity and radius when the cord breaks satisfy the equation

$$T = m\frac{v_f^2}{R_f}$$

Apply conservation of angular momentum, to calculate how v changes with R:

$$L_i = L_f$$

$$I_i\omega_i = I_f\omega_f$$

$$I = mR^2 \Rightarrow mR_i^2\omega_i = mR_f^2\omega_f$$

$$\text{But } \omega = \frac{v}{R} \Rightarrow R_i^2\left(\frac{v_i}{R_i}\right) = R_f^2\left(\frac{v_f}{R_f}\right)$$

$$R_i v_i = R_f v_f \Rightarrow v_f = \frac{R_i v_i}{R_f}$$

Use this in the equation for T:

$$T = m\frac{1}{R_f}\left(\frac{R_i v_i}{R_f}\right)^2 = \frac{mR_i^2 v_i^2}{R_f^3} \Rightarrow R_f = \left(\frac{mv_i^2 R_i^2}{T}\right)^{1/3}$$

$$R_f = \left[\frac{4\,\text{kg}\,(4\,\text{m}\cdot\text{s}^{-1})^2(0.5\,\text{m})^2}{600\,\text{N}}\right]^{1/3} = \underline{0.299\,\text{m}}$$

CHAPTER 10

Exercises 1, 5, 9, 13

Problems 17, 21, 23, 25, 27

Exercises

10-1

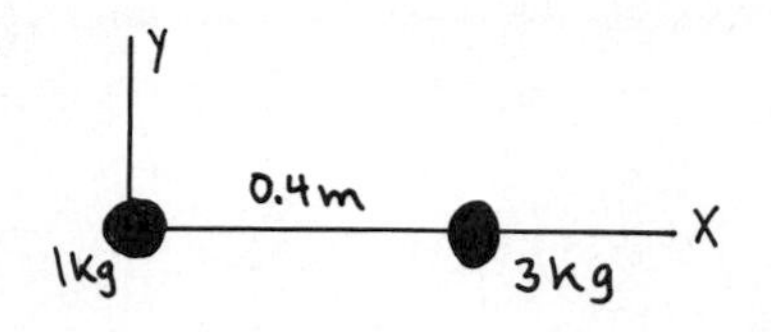

$$X = \frac{\Sigma wx}{\Sigma w}$$

With the coordinates indicated, the 1kg mass is at $x = 0$ and the 3kg mass is at $x = 0.4\,m$

$$\Rightarrow X = \frac{0 + (3\,kg)(9.8\,m \cdot s^{-2})(0.4\,m)}{(1\,kg + 3\,kg)(9.8\,m \cdot s^{-2})} = 0.3\,m$$

The center of gravity is between the balls, 0.3 m to the right of the 1kg ball.

10-5

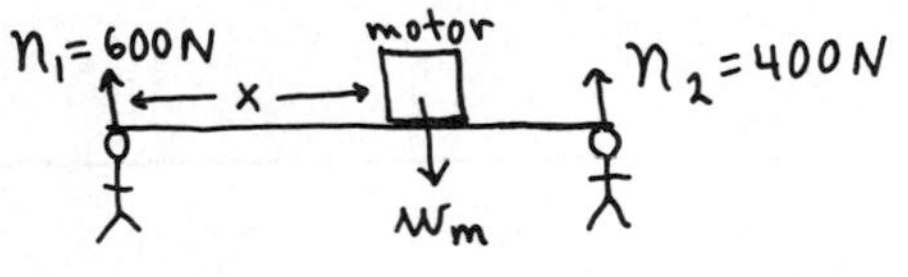

Consider the forces on the board. (The weight of the board is negligible.)

$$\Sigma F_y = ma_y$$

$$n_1 + n_2 - w_m = 0$$

$$w_m = n_1 + n_2 = 600\,N + 400\,N = \underline{1000\,N}$$

To find where the motor sits on the board use $\Sigma \Gamma = 0$, with the axis at the left-hand end:

$$\circlearrowleft^{+} \; \Sigma \Gamma = 0 \Rightarrow n_2 (2m) - w_m X = 0$$

$$X = \frac{n_2}{w_m}(2m) = \left(\frac{400\,N}{1000\,N}\right)(2m) = \underline{0.8\,m}$$

(The motor is 0.8m from the end where the 600N force is applied.)

10-9

a) Forces on the strut:

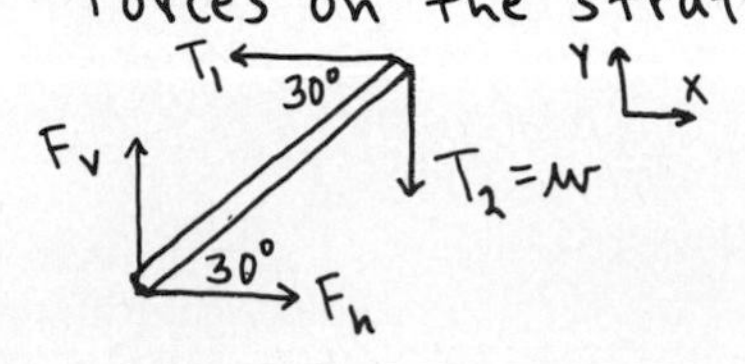

F_h and F_v are the horizontal and vertical components of the force exerted on the strut by the pivot.

10-9 (cont)

$$\Sigma F_x = ma_x \Rightarrow F_h = T_1$$
$$\Sigma F_y = ma_y \Rightarrow F_v = w$$

Use $\Sigma \Gamma = 0$, with ↺+ and the pivot at the lower end of the strut:
$\Gamma_{F_v} = \Gamma_{F_h} = 0$. Let L be the length of the strut.

$$\Sigma \Gamma = 0 \Rightarrow T_1 L \sin 30° - w L \cos 30° = 0$$
$$T_1 = w\left(\frac{\cos 30°}{\sin 30°}\right) = \frac{w}{\tan 30°} = 1.73w \;;\; F_h = 1.73w$$

Then the resultant force at the pivot is
$$F = \sqrt{F_h^2 + F_v^2} = \sqrt{(1.73w)^2 + w^2} = 2.0w$$

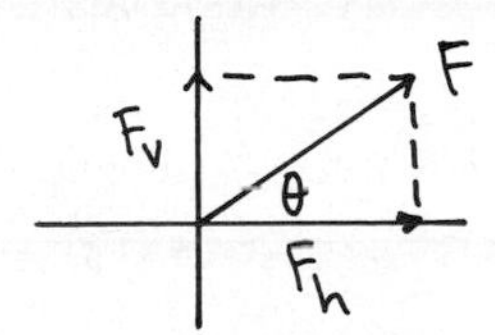

$$\tan\theta = \frac{F_v}{F_h} = \frac{w}{1.73w} = 0.578 \Rightarrow \theta = 30°$$

The resultant force exerted by the pivot is directed along the strut.

We could have known this direction also as follows:
Take $\Sigma\Gamma = 0$ with the pivot at the upper end of the strut
$\Rightarrow \Gamma_{T_1} = \Gamma_{T_2} = 0$. But then Γ_F is the only torque left, and hence must also be zero. But for this to happen the line of action of $\vec{F}$ must lie along the strut, so the moment arm will be zero.

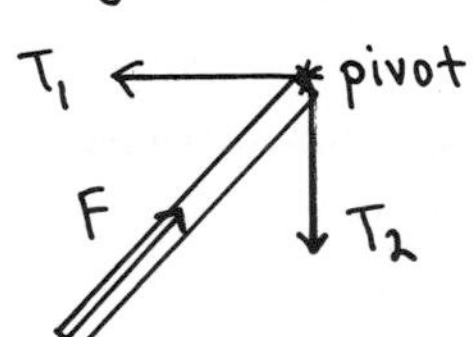

b)

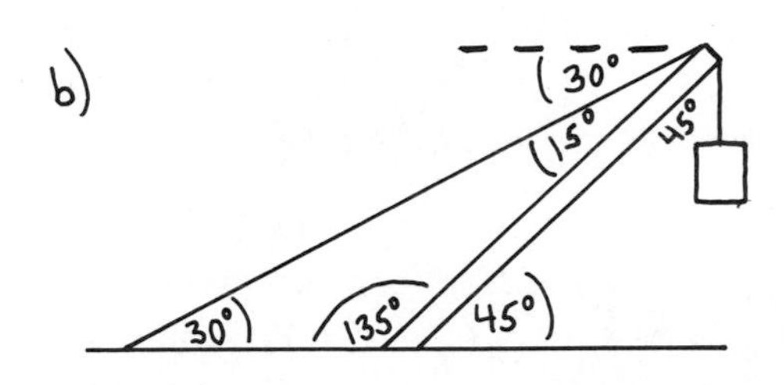

Forces on the strut:

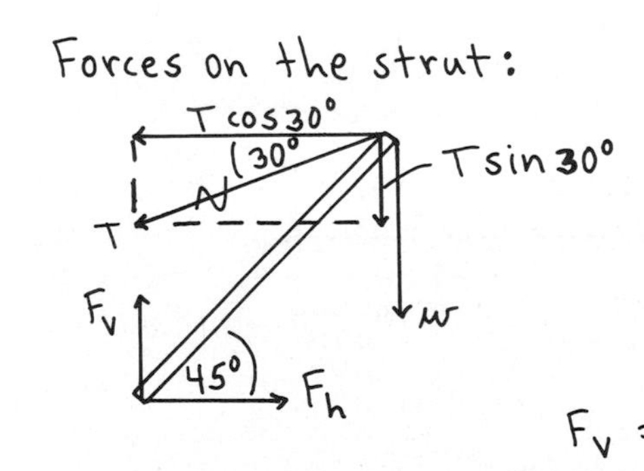

y, x

$$\Sigma F_x = ma_x$$
$$F_h = T\cos 30°$$
$$\Sigma F_y = ma_y$$
$$F_v = w + T\sin 30°$$

$\Sigma \Gamma = 0$, axis at lower end of strut, ↺+
$\Gamma_{F_v} = \Gamma_{F_h} = 0$. Let L be the length of the strut.
The moment arm for $T\cos 30°$ is $L\sin 45°$. For w and for $T\sin 30°$ the moment arm is $L\cos 45°$.

10-9 (cont)

$\Sigma\Gamma = 0 \Rightarrow (T\cos 30°)(L\sin 45°) - (T\sin 30°)(L\cos 45°) - w L\cos 45° = 0$

But $\sin 45° = \cos 45° \Rightarrow T\cos 30° - T\sin 30° - w = 0$

$$T = \frac{w}{\cos 30° - \sin 30°} = 2.73w$$

$F_h = T\cos 30° = (2.73w)\cos 30° = 2.37w$

$F_v = w + T\sin 30° = w + (2.73w)\sin 30° = 2.37w$

$F = \sqrt{F_h^2 + F_v^2} = \sqrt{(2.37w)^2 + (2.37w)^2} = \underline{3.35w}$

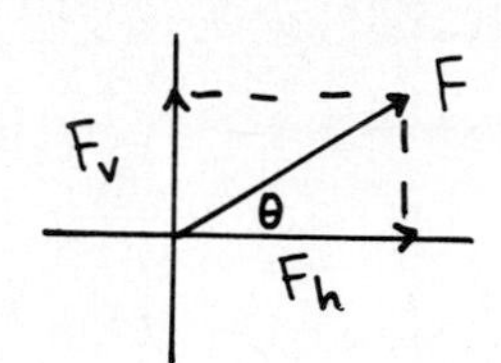

$\tan\theta = \frac{F_v}{F_h} = \frac{2.37w}{2.37w} = 1.0 \Rightarrow \theta = 45°$

The resultant force exerted at the pivot is directed along the strut.

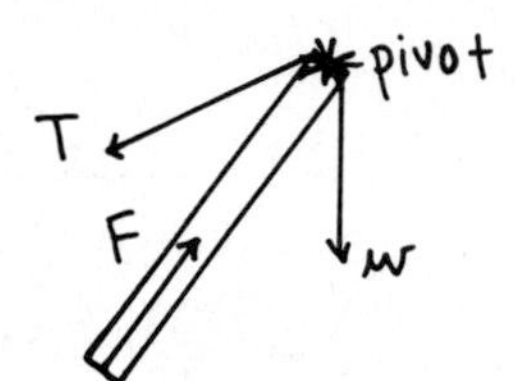

As in part (a) we can see that this force must have this direction, by considering $\Sigma\Gamma = 0$ with the axis at the upper end of the strut.

10-13

a)

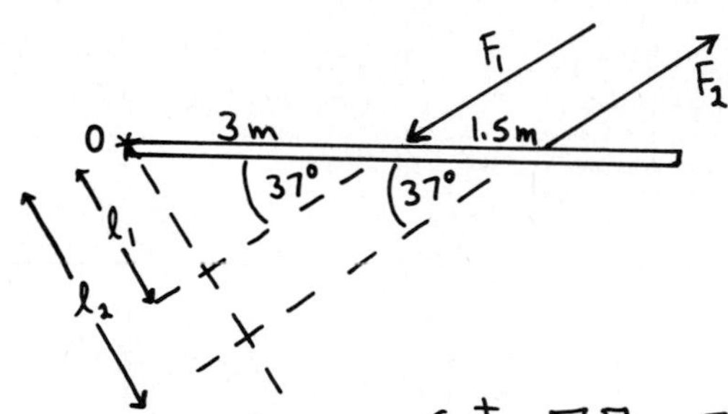

$l_1 = (3m)\sin 37° = 1.805\ m$ (moment arm for $\vec{F}_1$)

$l_2 = (4.5m)\sin 37° = 2.708m$ (moment arm for $\vec{F}_2$)

$\circlearrowleft^{+}\ \Sigma\Gamma_0 = F_2 l_2 - F_1 l_1 = (3N)(2.708m) - (3N)(1.805m) = \underline{2.71\ N\cdot m}$

b)

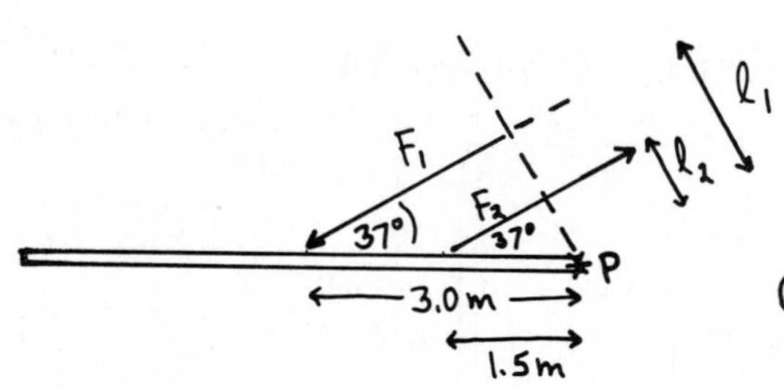

$l_1 = (3.0m)\sin 37° = 1.805m$

$l_2 = (1.5m)\sin 37° = 0.903\ m$

$\circlearrowleft^{+}\ \Sigma\Gamma_P = F_1 l_1 - F_2 l_2 = (3N)(1.805\ m) - (3N)(0.903m)$

$\Sigma\Gamma_P = \underline{2.71\ N\cdot m}$

c) Eq. (10-8) $\Rightarrow \Sigma\Gamma = lF$ for any axis.

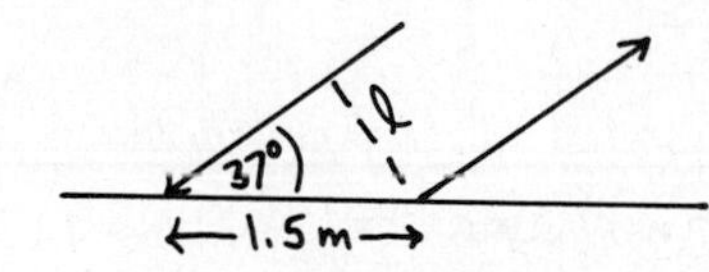

$l = (1.5m)\sin 37° = 0.903m$

$lF = (0.903m)(3N) = \underline{2.71\ N\cdot m}$

The results in (a) and (b) are equal, and agree with eq. (10-8).

Problems

10-17

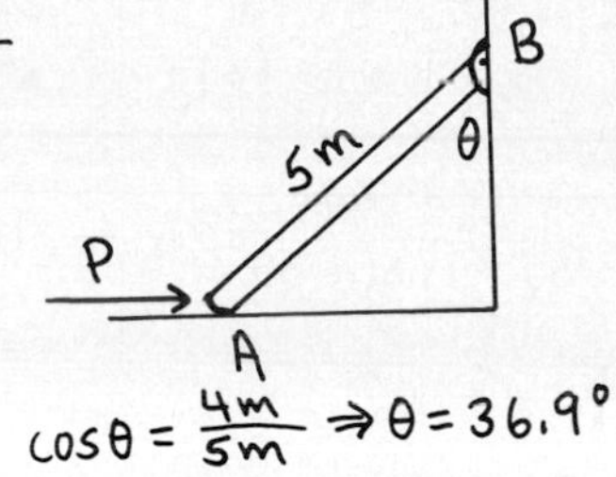

Forces on the bar:

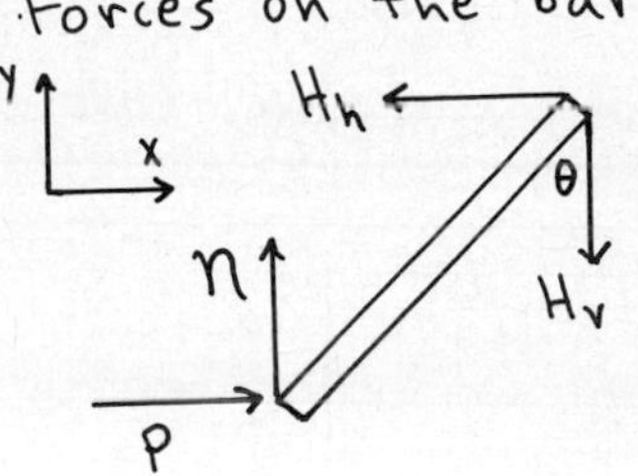

H_h and H_v are the components of the force exerted by the hinge on the bar. The components of the force of the bar on the hinge will be equal and opposite.
Note that there is no friction at A.

$\cos\theta = \frac{4m}{5m} \Rightarrow \theta = 36.9°$

$\Sigma F_x = ma_x \Rightarrow H_h = P = 60N$

$\Sigma F_y = ma_y \Rightarrow H_v = n$

$\Sigma\Gamma = 0$, pivot at B. ↺+ Let the length of the bar be L.

$PL\cos\theta - nL\sin\theta = 0$

$$n = P\left(\frac{\cos\theta}{\sin\theta}\right) = \frac{P}{\tan\theta} = \frac{60N}{\tan 36.9°} = 79.9N \Rightarrow H_v = \underline{79.9N}$$

Note:

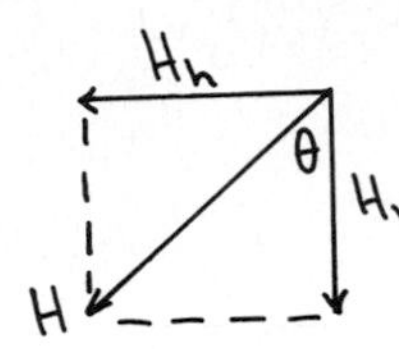

$$\tan\theta = \frac{H_h}{H_v} = \frac{60N}{79.9N} = 0.751 \Rightarrow \theta = 36.9°,$$

so $\vec{H}$ lies along the bar.

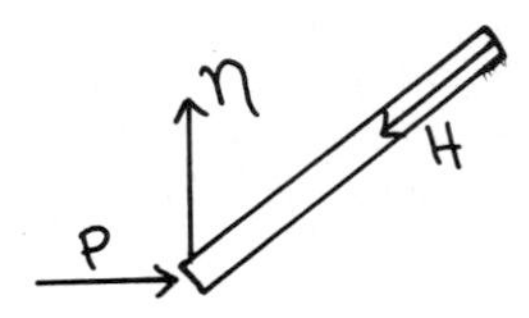

$\Sigma\Gamma = 0$ at A $\Rightarrow \vec{H}$ <u>must</u> lie along bar.

10-21

a) Consider the forces on one end of the swing frame:

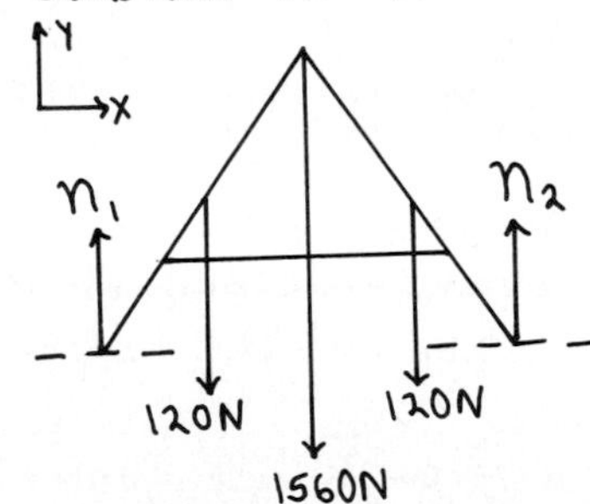

n_1 and n_2 are the normal forces exerted by the ground on each side piece. By symmetry $n_1 = n_2$.

$\Sigma F_y = ma_y$

$n_1 + n_2 - 1560N - 120N - 120N = 0$

$n_1 + n_2 = 1800N \Rightarrow n_1 = n_2 = \underline{900N}$

b) Consider the forces (and torques) on one of the side pieces:

H 5m 30° 1560N 900N 0.5m 30° 1.0m T 120N

$\vec{H}$ is the force exerted at the hinge by the other side piece.
T is the tension in the horizontal rod.

$\Sigma\Gamma = 0$, ↺+, pivot at hinge

$T(2m\cos 30°) + 120N(1.5m\sin 30°) - 900N(3.0m\sin 30°) = 0$

$$T = \frac{(900N)(3.0m\sin 30°) - (120N)(1.5m\sin 30°)}{2m\cos 30°} = \underline{727N}$$

10-23

a) Consider the force diagram for the gate. Let $\vec{A}$ and $\vec{B}$ be the forces exerted at the hinges. The problem specifies that $\vec{A}$ has no horizontal component.

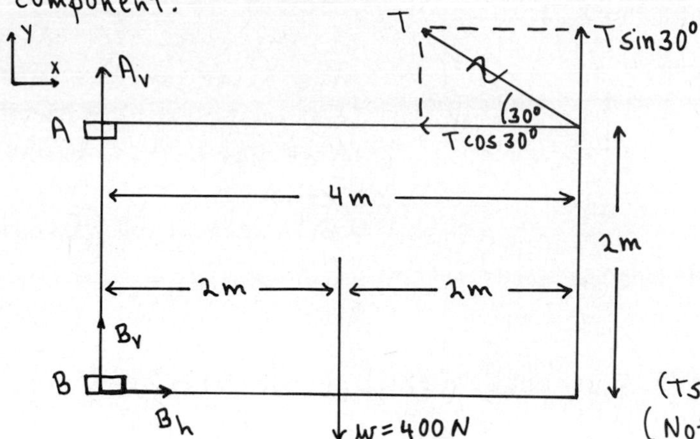

$$\Sigma F_y = ma_y$$
$$A_v + B_v + T\sin 30° - 400\,N = 0$$

$$\Sigma F_x = ma_x$$
$$B_h - T\cos 30° = 0$$

$\circlearrowleft^{+}$ $\Sigma \Gamma = 0$, pivot at hinge B

A_v, B_v, and B_h have zero torque

$$(T\sin 30°)(4m) + (T\cos 30°)(2m) - w\,(2m) = 0$$

(Note: It is simpler to calculate Γ for each component of $\vec{T}$ than for $\vec{T}$ itself.)

$$w = 400\,N \Rightarrow T = \frac{(400N)(2m)}{4m\sin 30° + 2m\cos 30°} = \underline{214\,N}$$

b) $B_h = T\cos 30° = (214N)\cos 30° = \underline{186\,N}$

c) $A_v + B_v = 400N - T\sin 30° = 400N - (214N)\sin 30° = \underline{293\,N}$

10-25

a) Find the location of the resultant normal force for $\theta = 15°$:

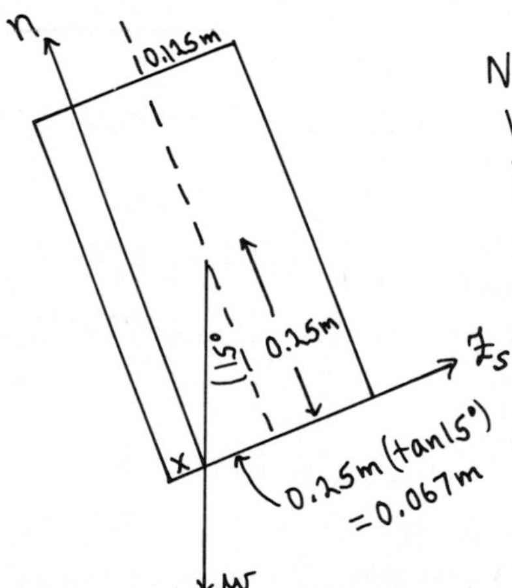

Note: Consider $\Sigma\Gamma$ about an axis where the $\vec{w}$ force line of action crosses the bottom of the block. $\Gamma_w = 0$ and $\Gamma_{f_s} = 0$ for this axis, since the lines of action for these forces pass through the axis. But then $\Sigma\Gamma = 0$ implies $\Gamma_n = 0$ also, so the line of action of $\vec{n}$ must pass through this point also, as shown in the sketch. Let x be the distance from the left-hand edge of the block to the line of action of the normal force.

The distance x can be found from simple geometry, as shown in the sketch. Thus $x = 0.125m - 0.067m = \underline{0.058\,m}$.

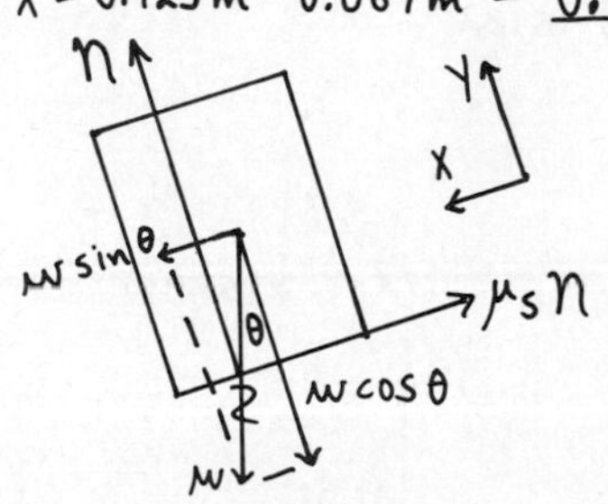

b) Find the angle when starts to slide:

Starts to slide $\Rightarrow f_s = \mu_s n$

$$\Sigma F_y = ma_y$$
$$n - w\cos\theta = 0$$
$$n = w\cos\theta$$

$$\Sigma F_x = ma_x$$
$$w\sin\theta - \mu_s n = 0$$
$$\not{w}\sin\theta - \mu_s \not{w}\cos\theta = 0$$

10-25 (cont)

$\tan\theta = \mu_s = 0.40 \Rightarrow \theta = 21.8°$ (for sliding)

Now find the angle where the block starts to tip:
Starts to tip ⇒ the line of action of $\vec{n}$ passes through the left-hand edge of the block. Consider $\Sigma\Gamma = 0$ with the pivot at the lower left-hand corner of the block. $\Gamma_n = 0$ and $\Gamma_{f_s} = 0 \Rightarrow \Gamma_w = 0$ also, and the line of action of $\vec{w}$ must pass through this point also.

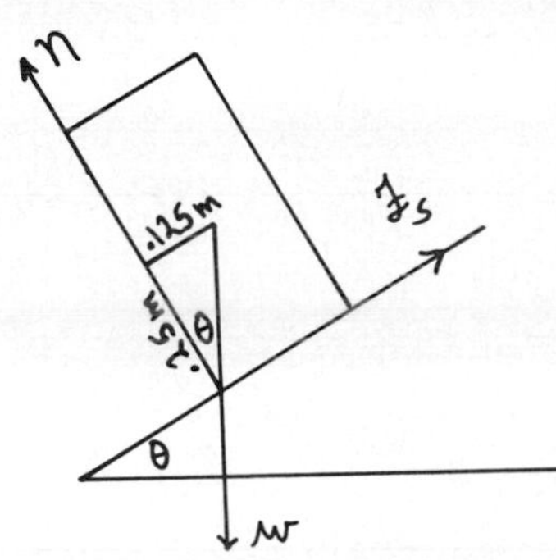

$$\tan\theta = \frac{0.125\text{ m}}{0.250\text{ m}} = 0.5 \Rightarrow \theta = 26.6°$$
(angle where tips)

$\theta = 21.8°$ to slide, $26.6°$ to tip over
⇒ slides

c) $\mu_s = 0.60$
to slide: $\tan\theta = \mu_s = 0.60 \Rightarrow \theta = 31.0°$
to tip: $\theta = 26.6°$, independent of μ_s
⇒ now will tip over

10-27

a) Force diagram for the door:

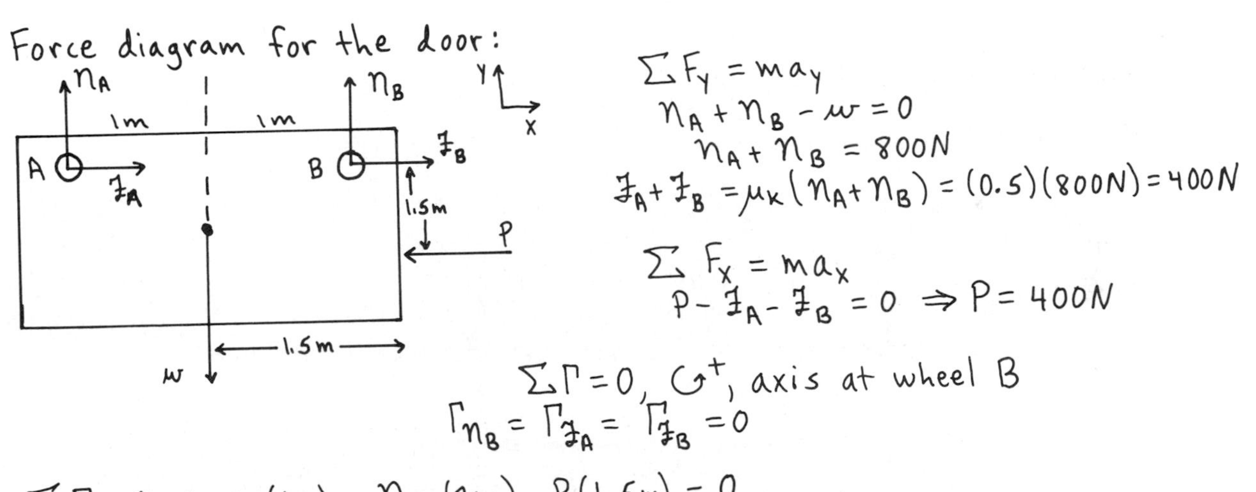

$\Sigma F_y = ma_y$
$n_A + n_B - w = 0$
$n_A + n_B = 800N$
$f_A + f_B = \mu_k(n_A + n_B) = (0.5)(800N) = 400N$

$\Sigma F_x = ma_x$
$P - f_A - f_B = 0 \Rightarrow P = 400N$

$\Sigma\Gamma = 0$, ↺+, axis at wheel B
$\Gamma_{n_B} = \Gamma_{f_A} = \Gamma_{f_B} = 0$

$\Sigma\Gamma = 0 \Rightarrow w(1m) - n_A(2m) - P(1.5m) = 0$

$$n_A = \frac{w(1m) - P(1.5m)}{2m} = \frac{(800N)(1m) - (400N)(1.5m)}{2m} = \underline{100N}$$

$n_A + n_B = 800N \Rightarrow n_B = \underline{700N}$

b) The torque due to $\vec{P}$ causes wheel A to tend to leave track.

Now the distance of $\vec{P}$ below B is the unknown h. Wheel A just leaves track ⇒ $n_A \to 0$.

10-27 (cont)

$\Sigma \Gamma = 0$, ↺+, axis at wheel B

$\Rightarrow w(1\text{m}) - Ph = 0$ (since $n_A \to 0$)

$h = \frac{w}{P}(1\text{m}) = \left(\frac{800\text{N}}{400\text{N}}\right)(1\text{m}) = \underline{2.0\text{m}}$

CHAPTER 11

Exercises 5, 7, 9, 11, 15, 17, 19, 23, 25

Problems 29, 31, 33, 37, 41

<u>Exercises</u>

<u>11-5</u>

$$f = 3\,Hz \Rightarrow \tau = \frac{1}{f} = \frac{1}{3}\,s = \underline{0.333\,s}$$

$$\omega = 2\pi f = 2\pi(3\,Hz) = \underline{18.8\,rad\cdot s^{-1}}$$

$$\omega = \sqrt{\frac{k}{m}} \Rightarrow m = \frac{k}{\omega^2} = \frac{200\,N\cdot m^{-1}}{(18.8\,rad\cdot s^{-1})^2} = \underline{0.566\,kg}$$

<u>11-7</u>

a) $\frac{1}{2}mv^2 + \frac{1}{2}kx^2 = \frac{1}{2}kA^2 \Rightarrow v = \pm\sqrt{\frac{k}{m}(A^2-x^2)}$

$v = v_{max}$ when $x=0 \Rightarrow v_{max} = A\sqrt{\frac{k}{m}}$

Use f to calculate $\sqrt{\frac{k}{m}}$: $f = \frac{1}{2\pi}\sqrt{\frac{k}{m}} \Rightarrow \sqrt{\frac{k}{m}} = 2\pi f$

Then $v_{max} = 2\pi f A = 2\pi(4\,s^{-1})(0.15\,m) = \underline{3.77\,m\cdot s^{-1}}$

$-kx = ma \Rightarrow a = -\frac{k}{m}x$

$a = a_{max}$ when $x = \pm A \Rightarrow a_{max} = \frac{k}{m}A$

But $\sqrt{\frac{k}{m}} = 2\pi f \Rightarrow \frac{k}{m} = 4\pi^2 f^2 \Rightarrow a_{max} = 4\pi^2 f^2 A = 4\pi^2(4\,s^{-1})^2(0.15\,m) = \underline{94.7\,m\cdot s^{-2}}$

b) $-kx = ma \Rightarrow a = -\frac{kx}{m} = -(2\pi f)^2 x$

$$a = -[(2\pi)4\,Hz]^2(0.09\,m) = \underline{-56.8\,m\cdot s^{-2}}$$

$\frac{1}{2}mv^2 + \frac{1}{2}kx^2 = \frac{1}{2}kA^2$

$v = \pm\sqrt{\frac{k}{m}(A^2-x^2)} = \pm 2\pi f\sqrt{A^2-x^2} = \pm 2\pi(4\,Hz)\sqrt{(0.15\,m)^2-(0.09\,m)^2} = \underline{\pm 3.02\,m\cdot s^{-1}}$

c) $x = A\sin\omega t$, $t = 0 \Rightarrow x = 0$

$$12\,cm = 15\,cm\,\sin(2\pi f t)$$

$$2\pi f t = \sin^{-1}\left(\frac{12\,cm}{15\,cm}\right) = 0.927$$

$$t = \frac{0.927}{2\pi(4\,Hz)} = \underline{0.0369\,s}$$

11-9

$v_0 = -6\,m\cdot s^{-1}$, $x_0 = \pm 0.2\,m$

$A = ?$

conservation of energy $\Rightarrow \frac{1}{2}mv_0^2 + \frac{1}{2}kx_0^2 = \frac{1}{2}kA^2$

$$A = \sqrt{\left(\frac{m}{k}\right)v_0^2 + x_0^2} = \sqrt{\left(\frac{4kg}{100N\cdot m^{-1}}\right)(6m\cdot s^{-1})^2 + (0.2m)^2} = \underline{1.22m}$$

$\theta_0 = ?$

$$\omega = \sqrt{\frac{k}{m}} = \sqrt{\frac{100N\cdot m^{-1}}{4kg}} = 5.0\,rad\cdot s^{-1}$$

$$\theta_0 = \tan^{-1}\left(-\frac{v_0}{\omega x_0}\right) = \tan^{-1}\left(-\frac{-6m\cdot s^{-1}}{(5s^{-1})(0.2m)}\right) = \tan^{-1}(6) = \underline{80.5°} = \underline{1.41\,rad}$$

$E = ?$

$$E = \frac{1}{2}mv_0^2 + \frac{1}{2}kx_0^2 = \frac{1}{2}kA^2 = \frac{1}{2}(100N\cdot m^{-1})(1.22m)^2 = \underline{74.4J}$$

$x(t) = ?$

$$x(t) = A\cos(\omega t + \theta_0) = (1.22m)\cos\left[(5s^{-1})t + 1.41\right]$$

11-11

a)

kd $(d = 0.2m)$

$a = 0$ [m] mg

$$\Sigma F_y = ma_y$$

$$kd - mg = 0$$

$$k = \frac{mg}{d} = \frac{(2kg)(9.8\,m\cdot s^{-2})}{(0.20m)} = \underline{98\,N\cdot m^{-1}}$$

b) $$T = 2\pi\sqrt{\frac{m}{k}} = 2\pi\sqrt{\frac{2kg}{98N\cdot m^{-1}}} = \underline{0.898\,s}$$

c) $$T = 2\pi\sqrt{\frac{4kg}{98N\cdot m^{-1}}} = \underline{1.27\,s}$$ (longer by a factor of $\sqrt{\frac{4}{2}}$)

11-15

From the statement of the problem, at equilibrium the spring is stretched a distance $d = 0.5m$.

kd

$a = 0$ [m] mg

$$\Sigma F_y = ma_y$$

$$kd - mg = 0$$

$$k = \frac{mg}{d} = \frac{(4kg)(9.8m\cdot s^{-2})}{0.5m} = 78.4N\cdot m^{-1}$$

a) $U_{grav} = \underline{0}$ (by the choice we are told to make for the zero of gravitational potential energy)

11-15 (cont)

$U_{spr} = \frac{1}{2}ky^2$; at this point the mass is $A = 0.5m$ below equilibrium, so the spring is stretched $0.5m + 0.5m = 1.0m$

$U_{spr} = \frac{1}{2}(78.4\,N\cdot m^{-1})(1.0m)^2 = \underline{39.2J}$

$K = 0$, since $v = 0$ when $y = \pm A$

$E = U_{grav} + U_{spr} + K = \underline{39.2J}$

b) The mass is $A = 0.5m$ above its lowest point

$\Rightarrow U_{grav} = mgh = (4kg)(9.8\,m\cdot s^{-2})(0.5m) = \underline{19.6J}$

The spring is stretched 0.5m

$\Rightarrow U_{spr} = \frac{1}{2}ky^2 = \frac{1}{2}(78.4\,N\cdot m^{-1})(0.5m)^2 = \underline{9.8J}$

$K = \frac{1}{2}mv^2$; $v = v_m = \sqrt{\frac{k}{m}}A$ at this point

$\Rightarrow K = \frac{1}{2}m\left(\frac{k}{m}A^2\right) = \frac{1}{2}kA^2 = \frac{1}{2}(78.4\,N\cdot m^{-1})(0.5m)^2 = \underline{9.8J}$

$E = U_{grav} + U_{spr} + K = 19.6J + 9.8J + 9.8J = \underline{39.2J}$

c) The mass is $2A = 1.0m$ above its lowest point

$\Rightarrow U_{grav} = mgh = (4kg)(9.8\,m\cdot s^{-2})(1.0m) = \underline{39.2J}$

The spring is unstretched (the problem tells us this)

$\Rightarrow U_{spr} = \underline{0}$.

At $x = \pm A$, $v = 0 \Rightarrow \underline{K = 0}$.

$E = U_{grav} + U_{spr} + K = \underline{39.2J}$

Note: At each point in the motion the total energy E is the same, as required by conservation of energy, but how it is divided up among U_{grav}, U_{spr}, and K changes.

11-17

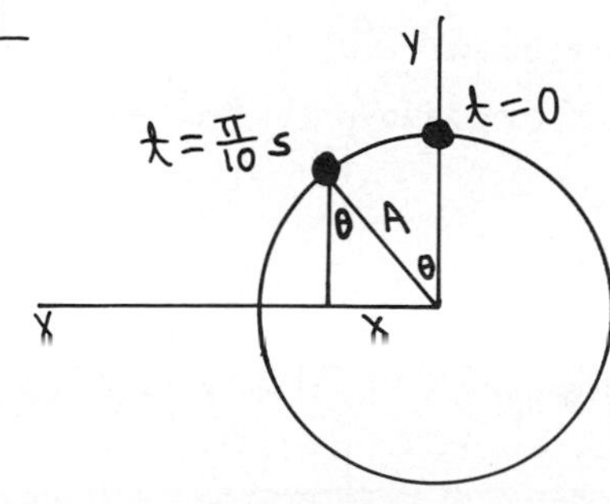

$x = A\sin\theta$

$\theta = \left(\frac{\frac{\pi}{10}s}{\frac{\pi}{2}s}\right)(360°) = 0.2(360°) = 72°$

$x = (0.2m)\sin 72° = \underline{0.190m}$

11-19

ticks four times each second, each tick is half a period $\Rightarrow \tau = 0.5\,s$

a) thin rim $\Rightarrow I = mR^2 = (0.8\times10^{-3}\,kg)(0.015\,m)^2 = \underline{1.8\times10^{-7}\,kg\cdot m^2}$

b) $\tau = 2\pi\sqrt{\frac{I}{k'}}$

$$\frac{I}{k'} = \left(\frac{\tau}{2\pi}\right)^2 \Rightarrow k' = I\left(\frac{2\pi}{\tau}\right)^2 = (1.8\times10^{-7}\,kg\cdot m^2)\left(\frac{2\pi}{0.5\,s}\right)^2 = \underline{2.84\times10^{-5}\,N\cdot m}$$

11-23

a)

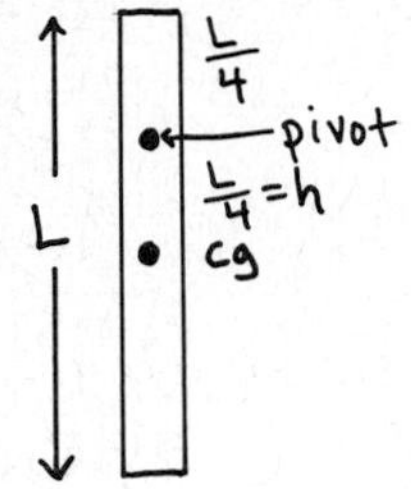

$$I_{cm} = \frac{1}{12}mL^2$$

parallel-axis theorem (eq. 9-22) $\Rightarrow I = I_{cm} + Mh^2$

$$I = \frac{1}{12}ML^2 + M\left(\frac{L}{4}\right)^2 = ML^2\left(\frac{1}{12} + \frac{1}{16}\right) = \frac{1}{4}mL^2\left(\frac{1}{3} + \frac{1}{4}\right)$$

$$\underline{I = \frac{7}{48}mL^2}$$

b) $$\tau = 2\pi\sqrt{\frac{I}{mgh}} = 2\pi\sqrt{\frac{\frac{7}{48}mL^2}{mg(L/4)}} = 2\pi\sqrt{\frac{L}{g}}\sqrt{\frac{7}{12}} = 2\pi\sqrt{\frac{7L}{12g}}$$

11-25

a) $$\omega' = \sqrt{\frac{k}{m} - \frac{b^2}{4m^2}} = \sqrt{\frac{300\,N\cdot m^{-1}}{0.4\,kg} - \frac{(5\,kg\cdot s^{-1})^2}{4(0.4\,kg)^2}} = 26.7\,rad\cdot s^{-1}$$

$$f' = \frac{\omega'}{2\pi} = \frac{26.7\,rad\cdot s^{-1}}{2\pi} = \underline{4.24\,Hz}$$

b) critical damping $\Rightarrow \frac{k}{m} - \frac{b^2}{4m^2} = 0 \quad (\omega' = 0)$

$$\Rightarrow b^2 = 4km \Rightarrow b = 2\sqrt{km} = 2\sqrt{(300\,N\cdot m^{-1})(0.4\,kg)} = \underline{21.9\,kg\cdot s^{-1}}$$

Problems

11-29

a) We are given information about v at a particular x. The expression relating these two quantities comes from conservation of energy:

$$\frac{1}{2}mv^2 + \frac{1}{2}kx^2 = \frac{1}{2}kA^2$$

We are asked to calculate the period, which we know to be given by $\tau = 2\pi\sqrt{\frac{m}{k}}$. We see that we can obtain $\sqrt{\frac{m}{k}}$ from the conservation of energy equation, and then use it to calculate the period.

11-29 (cont)

$$\tfrac{1}{2}mv^2 + \tfrac{1}{2}kx^2 = \tfrac{1}{2}kA^2 \Rightarrow \sqrt{\frac{m}{k}} = \sqrt{\frac{A^2-x^2}{v^2}}$$

$$T = 2\pi\sqrt{\frac{m}{k}} = 2\pi\sqrt{\frac{A^2-x^2}{v^2}} = 2\pi\sqrt{\frac{(0.10\,m)^2-(0.06\,m)^2}{(0.24\,m\cdot s^{-1})^2}} = \underline{2.09\,s}$$

b) Need to relate v to x; use $\tfrac{1}{2}mv^2 + \tfrac{1}{2}kx^2 = \tfrac{1}{2}kA^2$

$$\Rightarrow x = \sqrt{A^2 - \frac{mv^2}{k}}$$

Need $\frac{m}{k}$. From (a) $\frac{m}{k} = \frac{A^2-x^2}{v^2}$. Use that $v = 0.24\,m\cdot s^{-1}$ when $x = 0.06\,m$

$$\Rightarrow \frac{m}{k} = \frac{(0.10\,m)^2-(0.06\,m)^2}{(0.24\,m\cdot s^{-1})^2} = 0.111\,s^2$$

Then $x = \sqrt{(0.10\,m)^2 - (0.111\,s^2)(0.12\,m\cdot s^{-1})^2} = \underline{0.0916\,m}$

c) "very small object" $\Rightarrow$ doesn't alter the motion of the block.

For the block, $-kx = ma \Rightarrow a = -\frac{k}{m}x$

$$\Rightarrow a_{max} = \frac{k}{m}A = \frac{0.10\,m}{(0.111\,s^2)} \quad (\text{since } \tfrac{m}{k} = 0.111\,s^2 \text{ from (b)})$$

$$a_{max} = 0.90\,m\cdot s^{-2}$$

If the very small object doesn't slip it must be able to be given this much acceleration by the static friction force.
Forces on the very small object:

$n = mg$, $\rightarrow a_{max}$, $\mu_s n = \mu_s mg$, mg; axes y, x

$$\Sigma F_x = ma_x$$

$$\mu_s mg = m a_{max}$$

$$\mu_s = \frac{a_{max}}{g} = \frac{0.90\,m\cdot s^{-2}}{9.8\,m\cdot s^{-2}} = \underline{0.0918}$$

11-31

dock, raft, $x = +A$, $x = 0$, $x = -A$

We need to calculate the time it takes the raft to move from $x = A$ to $x = A - 1.0\,ft$.
Use $x = A\cos\omega t \Rightarrow x = A$ at $t = 0$.
What then is t when $x = A - 1\,ft$:

$$\omega = 2\pi f = \frac{2\pi}{T} \Rightarrow x = A\cos\left(\frac{2\pi}{T}t\right)$$

$$x = A - 1.0\,ft = 2.0\,ft - 1.0\,ft = 1.0\,ft$$

$$1.0\,ft = (2.0\,ft)\cos\left(\frac{2\pi}{5.0\,s}t\right) \Rightarrow \cos\left(\frac{2\pi}{5.0\,s}t\right) = \frac{1}{2}$$

$$\frac{2\pi t}{5.0\,s} = \frac{\pi}{3} \Rightarrow t = \frac{\pi}{3}\left(\frac{5.0\,s}{2\pi}\right) = 0.833\,s$$

11-31 (cont)

This is the time for $x = A$ to $x = A - 1.0\,ft$.

But people can also get off while the raft is moving from $x = A - 1.0\,ft$ to $x = A$; this motion also takes $0.833\,s$.

Thus the people have $2(0.833\,s) = \underline{1.67\,s}$.

11-33

a) $\frac{1}{2}mv^2 + \frac{1}{2}ky^2 = \frac{1}{2}kA^2$, where y is measured from the equilibrium position of the hanging mass.

equilibrium $\Rightarrow y = 0 \Rightarrow \frac{1}{2}mv^2 = \frac{1}{2}kA^2$

$$v = \sqrt{\frac{k}{m}}\,A$$

Use the known period to calculate $\sqrt{\frac{k}{m}}$; $\tau = 2\pi\sqrt{\frac{m}{k}} \Rightarrow \sqrt{\frac{k}{m}} = \frac{2\pi}{\tau}$

Thus $v = \frac{2\pi}{\tau}A = \frac{2\pi\,(0.10\,m)}{2.0\,s} = \underline{0.314\,m \cdot s^{-1}}$

b) (axes: x to the right, y downward) $ma = -ky \Rightarrow a = -\frac{k}{m}y$

$y = -0.05\,m \Rightarrow a = -\left(\frac{2\pi}{\tau}\right)^2(-0.05\,m) = +\left(\frac{2\pi}{2.0\,s}\right)^2(0.05\,m) = +\underline{0.493\,m \cdot s^{-2}}$, $\underline{\text{downward}}$

c)

t_2	☐	$y = -0.05\,m$
	☐	$y = 0$
t_1	☐	$y = +0.05\,m$
$t = 0$	☐	$y = +0.10\,m = A$

(axes: x to the right, y downward)

We want to calculate $t_2 - t_1$.

$y = +A$ at $t = 0 \Rightarrow y(t) = A\cos(\omega t)$; $\omega = \frac{2\pi}{\tau}$

$y = +0.05\,m \Rightarrow 0.05\,m = (0.10\,m)\cos\left(\frac{2\pi}{\tau}t_1\right)$

$\cos\left(\frac{2\pi}{2.0\,s}t_1\right) = \frac{1}{2} \Rightarrow \frac{2\pi}{2.0\,s}t_1 = \frac{\pi}{3}\,(60°)$

$t_1 = \frac{2.0\,s}{6} = 0.333\,s$

$y = -0.05\,m \Rightarrow -0.05\,m = (0.10\,m)\cos\left(\frac{2\pi}{\tau}t_2\right)$

$\cos\left(\frac{2\pi}{2.0\,s}t_2\right) = -\frac{1}{2} \Rightarrow \frac{2\pi}{2.0\,s}t_2 = \frac{2\pi}{3}\,(120°)$

$t_2 = \frac{2.0\,s}{3} = 0.667\,s$

The time is $t_2 - t_1 = 0.667\,s - 0.333\,s = \underline{0.334\,s}$

d) Calculate the distance d that the spring is stretched by the weight of the hanging mass. This then will be the amount the spring will shorten when the object is removed.

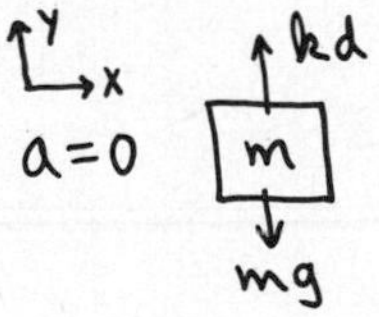

(free-body diagram: mass m, kd upward, mg downward; $a = 0$; axes y up, x right)

$\Sigma F_y = ma_y$

$kd - mg \Rightarrow d = \frac{mg}{k}$; $\sqrt{\frac{k}{m}} = \frac{2\pi}{\tau} \Rightarrow \frac{m}{k} = \left(\frac{\tau}{2\pi}\right)^2$ (from part (a))

$d = \left(\frac{\tau}{2\pi}\right)^2 g = \left(\frac{2.0\,s}{2\pi}\right)^2(9.8\,m \cdot s^{-2}) = \underline{0.993\,m}$

11-37

a)

$$\Sigma F_y = ma_y$$
$$n - mg = ma$$
$$n = mg + ma$$
$$a = -\frac{kx}{m} = -\omega^2 x = -(2\pi f)^2 x$$

Lens is moving in simple harmonic motion, and the ball bearing moves along with it $\Rightarrow x(t) = A\cos(\omega t + \theta_0) = A\cos(2\pi f t + \theta_0)$

Thus $n = m[g - (2\pi f)^2 A\cos(2\pi f t + \theta_0)]$

b) bounce $\Rightarrow$ ball bearing loses contact with the lens $\Rightarrow n \to 0$

$$n = 0 \Rightarrow g - (2\pi f)^2 A\cos(2\pi f t + \theta_0) = 0$$

The smallest frequency f_b where this happens is when $g - (2\pi f_b)^2 A = 0$, so that $n \to 0$ right at the point in the motion when $\cos(2\pi f t + \theta_0) = 1$ (which is when $x = A$).

$$\Rightarrow \quad g = (2\pi f_b)^2 A$$

11-41

a) $T = 2\pi\sqrt{\frac{L}{g}} \Rightarrow L = \left(\frac{T}{2\pi}\right)^2 g = \left(\frac{4.0\,s}{2\pi}\right)^2 (9.8\,m\cdot s^{-2}) = \underline{3.97\,m}$

b) Use a stick of length $L = 0.5\,m$. Place the pivot a distance h above the center. The stick oscillates as a physical pendulum. We will require that h be such that $T = 4.0\,s$.

Note: $T = 2\pi\sqrt{\frac{I}{mgh}}$

$h = \frac{L}{2}$ (pivot at top of stick) and $I = \frac{1}{3}mL^2$ for axis at one end

$$\Rightarrow T = 2\pi\sqrt{\frac{\frac{1}{3}mL^2}{mg(L/2)}} = 2\pi\sqrt{\frac{L}{g}}\sqrt{\frac{2}{3}} = 2\pi\sqrt{\frac{0.5\,m}{9.8\,m\cdot s^{-2}}}\sqrt{\frac{2}{3}} = 1.16\,s < 4.0\,s,$$

and $h \to 0 \Rightarrow T \to \infty$.

So there must be a value of h between 0 and $L/2$ that will give $T = 4.0\,s$.

Let's calculate it:

$$I_{pivot} = I_{cm} + mh^2 = \frac{1}{12}mL^2 + mh^2$$

$$T = 2\pi\sqrt{\frac{\frac{1}{12}mL^2 + mh^2}{mgh}} = 2\pi\sqrt{\frac{\frac{1}{12}L^2 + h^2}{gh}}$$

$$\left(\frac{T}{2\pi}\right)^2 = \frac{\frac{1}{12}L^2 + h^2}{gh}$$

11-41 (cont)

$$h^2 - g\left(\frac{T}{2\pi}\right)^2 h + \frac{1}{12}L^2 = 0$$

$T = 4.0\,s \Rightarrow h^2 - (3.972\,m)h + 0.02083\,m^2 = 0$

quadratic formula $\Rightarrow h = \frac{1}{2}\left[3.972\,m \pm \sqrt{(-3.972\,m)^2 - 4(0.02083\,m^2)}\right]$

$h = 1.986\,m \pm 1.981\,m$

But since the stick is only 0.5m long, must have $h < 0.25\,m$
$\Rightarrow$ take the minus sign in the quadratic formula

Thus $h = 1.986\,m - 1.981\,m = 5\times10^{-3}\,m = 0.5\,cm$

Note: There is a lot of numerical cancellation in the above expression. We have found that $h \approx 5\times10^{-3}\,m \Rightarrow h^2 \approx 2.5\times10^{-5}\,m$, which is much smaller than the $0.02083\,m^2$ term in the expression. It is therefore a good approximation to neglect the h^2 term in the quadratic equation, giving the simpler eq.

$$-(3.972\,m)h + 0.02083\,m^2 = 0$$

$$h = \frac{0.02083\,m^2}{3.972\,m} = 5.24\times10^{-3}\,m = \underline{0.524\,cm}$$

CHAPTER 12

Exercises 3, 5, 7, 9, 11

Problems 15, 17, 23

Exercises

12-3

$$\text{stress} = \frac{F}{A} = \frac{400\,N}{(0.2\times10^{-2}\,m)^2} = \underline{1.0\times10^{8}\,Pa}$$

$$Y = \frac{\text{stress}}{\text{strain}} \Rightarrow \text{strain} = \frac{\text{stress}}{Y} = \frac{1.0\times10^{8}\,Pa}{2\times10^{11}\,Pa} = \underline{5.0\times10^{-4}}$$

$$\text{strain} = \frac{\Delta l}{l_0} \Rightarrow \Delta l = l_0(\text{strain}) = (5m)(5\times10^{-4}) = \underline{2.5\times10^{-3}\,m}$$

$$\frac{\Delta w}{w_0} = -\sigma\frac{\Delta l}{l_0} = -\sigma(\text{strain}) = -(0.19)(5.0\times10^{-4}) = \underline{-9.5\times10^{-5}}$$

12-5

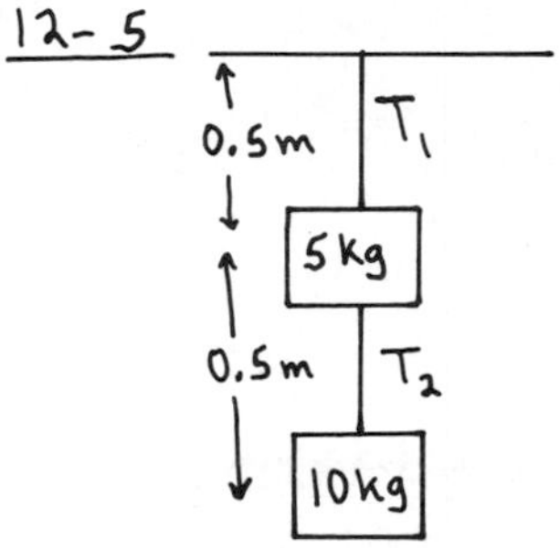

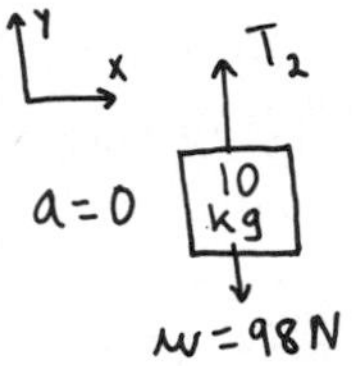

$\Sigma F_y = ma_y$

$T_2 - w = 0$

$T_2 = 98\,N$

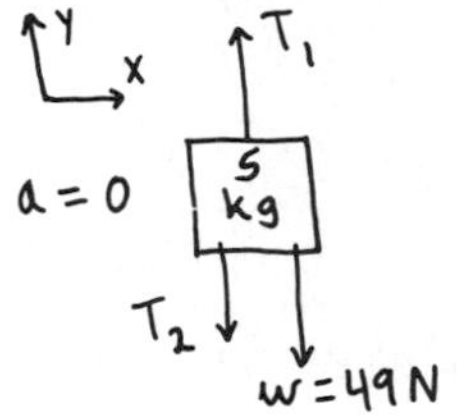

$\Sigma F_y = ma_y$

$T_1 - w - T_2 = 0$

$T_1 = w + T_2$

$T_1 = 49N + 98N$

$T_1 = 147\,N$

a) $Y = \frac{\text{stress}}{\text{strain}} \Rightarrow \text{strain} = \frac{\text{stress}}{Y} = \frac{F}{AY}$

upper wire: $\text{strain} = \frac{T_1}{AY} = \frac{147\,N}{(0.004\times10^{-4}\,m^2)(2\times10^{11}\,Pa)} = \underline{1.84\times10^{-3}}$

lower wire: $\text{strain} = \frac{T_2}{AY} = \frac{98\,N}{(0.004\times10^{-4}\,m^2)(2\times10^{11}\,Pa)} = \underline{1.23\times10^{-3}}$

b) $\text{strain} = \frac{\Delta l}{l_0} \Rightarrow \Delta l = l_0(\text{strain})$

upper wire: $\Delta l = (0.5m)(1.84\times10^{-3}) = \underline{9.20\times10^{-4}\,m}$

lower wire: $\Delta l = (0.5m)(1.23\times10^{-3}) = \underline{6.15\times10^{-4}\,m}$

12-7

$$Y = \frac{F/A}{\Delta l/l_0} \Rightarrow A = \frac{l_0 F}{Y\Delta l} = \frac{(2m)(300\,N)}{(2\times10^{11}\,Pa)(0.2\times10^{-2}\,m)} = 1.5\times10^{-6}\,m^2$$

$$A = \pi r^2 \Rightarrow r = \sqrt{\frac{A}{\pi}} = \sqrt{\frac{1.5\times10^{-6}\,m^2}{\pi}} = 6.91\times10^{-4}\,m$$

$$d = 2r = 1.38\times10^{-3}\,m = \underline{1.38\,mm}$$

12-9

$$B = -\frac{\Delta P}{\Delta V/V_0} = -\frac{12\times10^5\ \text{Pa}}{(-0.3\times10^{-6}\text{m}^3)/(1.0\times10^{-3}\text{m}^3)} = \underline{+4.0\times10^9\ \text{Pa}}$$

$$k = \frac{1}{B} = \underline{2.5\times10^{-10}\ \text{Pa}^{-1}}$$

12-11

shear stress $= \frac{F}{A} \Rightarrow F = A\times(\text{stress}) = \pi r^2 \times (\text{stress})$

For one rivet, $F = \pi(0.25\times10^{-2}\text{m})^2(6\times10^8\text{Pa}) = 1.18\times10^4\ \text{N}$

Each rivet carries $\frac{1}{4}$ the load, so the total force that can be applied is $4(1.18\times10^4\ \text{N}) = \underline{4.72\times10^4\ \text{N}}$

Problems

12-15

$a_\perp = r\omega^2$, T, 15kg, mg, y, x

First calculate the tension in the wire at this point:

$$\omega = (2\ \text{rev}\cdot\text{s}^{-1})(2\pi\ \text{rad}/1\ \text{rev}) = 4\pi\ \text{rad}\cdot\text{s}^{-1}$$

$$\Sigma F_y = ma_y$$

$$T - mg = ma_\perp$$

$$T = m(g + r\omega^2) = 15\text{kg}\left[9.8\ \text{m}\cdot\text{s}^{-2} + (0.5\text{m})(4\pi\ \text{rad}\cdot\text{s}^{-1})^2\right] = 1331\ \text{N}$$

Now calculate the Δl that results from this tension:

$$Y = \frac{F/A}{\Delta l/l_0} \Rightarrow \Delta l = \frac{l_0 F}{YA} = \frac{(0.5\text{m})(1331\text{N})}{(2\times10^{11}\text{Pa})(0.02\times10^{-4}\text{m}^2)} = 1.66\times10^{-3}\text{m} = \underline{1.66\text{mm}}$$

12-17

a) T_A, T_B, x, w

equal stresses $\Rightarrow$ same $\frac{T}{A}$ in each wire

$$\frac{T_A}{1\text{mm}^2} = \frac{T_B}{2\text{mm}^2} \Rightarrow T_B = 2T_A$$

Consider forces (and torques) on the bar:

T_A, T_B, x, w, y, x

$$\Sigma F_y = ma_y$$

$$T_A + T_B - w = 0$$

$$T_B = 2T_A \Rightarrow T_A + 2T_A - w = 0$$

$$T_A = \frac{w}{3},\quad T_B = \frac{2w}{3}$$

$\Sigma\Gamma = 0$, $\circlearrowleft^+$, pivot at wire A:

$$T_B(1.05\text{m}) - wx = 0$$

12-17 (cont)

$$x = 1.05\,m\left(\frac{T_B}{w}\right) = 1.05\,m\left(\frac{2w/3}{w}\right) = \frac{2}{3}(1.05\,m) = 0.70\,m$$

The weight should be hung 0.70m to the right of A.

b) equal strains:

$$Y = \frac{\text{stress}}{\text{strain}} = \frac{F/A}{\text{strain}} \Rightarrow \text{strain} = \frac{F}{AY}$$

$$\Rightarrow \frac{T_A}{1\,mm^2(2.4\times10^{11}\,Pa)} = \frac{T_B}{2\,mm^2(1.6\times10^{11}\,Pa)} \Rightarrow T_B = \left(\frac{2}{1}\right)\left(\frac{1.6}{2.4}\right)T_A = 1.33\,T_A$$

As in (a), $T_A + T_B = w$.

Thus $T_A + 1.33\,T_A = w \Rightarrow T_A = 0.429\,w,\; T_B = 0.571\,w$

Then, again as is (a), $x = 1.05\,m\left(\frac{T_B}{w}\right) = 1.05\,m\left(\frac{0.571\,w}{w}\right) = 0.600\,m$

The weight should be hung 0.60m to the right of A.

12-23

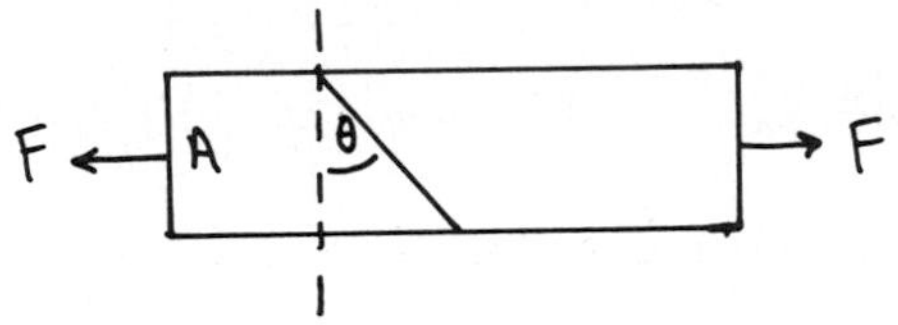

a)

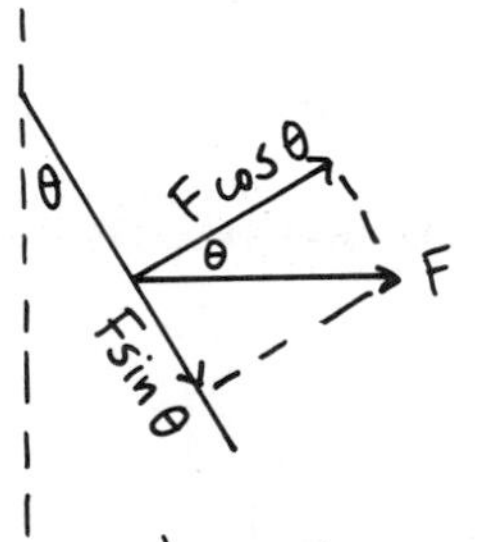

$$\text{tensile stress} = \frac{F_\perp}{A'} = \frac{F\cos\theta}{\frac{A}{\cos\theta}} = \frac{F\cos^2\theta}{A}$$

b) $$\text{shear stress} = \frac{F_\parallel}{A'} = \frac{F\sin\theta}{\frac{A}{\cos\theta}} = \frac{F\sin\theta\cos\theta}{A} = \frac{F(\sin 2\theta)}{2A}$$

c) From (a) the tensile stress is a maximum for $\cos\theta = 1 \Rightarrow \theta = 0°$

d) From (b) the shear stress is a maximum for $\sin 2\theta = 1$
$\Rightarrow 2\theta = 90° \Rightarrow \theta = 45°$.

CHAPTER 13

Exercises 3, 7, 9, 11, 13, 15, 17, 19, 23, 29, 31, 33

Problems 35, 39, 45, 49, 51, 53, 55, 57, 59, 61

Exercises

13-3

a) gauge pressure $p - p_a = \rho g h = (1.03\times10^3\,kg\cdot m^{-3})(9.8\,m\cdot s^{-2})(600\,m) = \underline{6.06\times10^6\,Pa}$

b)

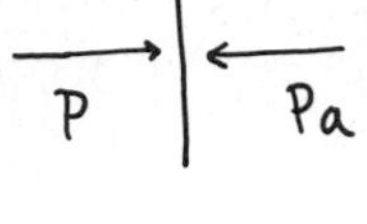

$F = pA - p_aA = (p - p_a)A = (6.06\times10^6\,Pa)[\pi(0.075\,m)^2]$

$F = \underline{1.07\times10^5\,N}$

13-7

$p_a = 970\text{ millibars} = 0.97\times10^5\,Pa$

a) $p = p_a + \rho g y_2 = 0.97\times10^5\,Pa + (13.6\times10^3\,kg\cdot m^{-3})(9.8\,m\cdot s^{-2})(0.08\,m) = \underline{1.08\times10^5\,Pa}$

b) $p = p_a + \rho g(0.05\,m) = 0.97\times10^5\,Pa + (13.6\times10^3\,kg\cdot m^{-3})(9.8\,m\cdot s^{-2})(0.05\,m) = \underline{1.04\times10^5\,Pa}$

c) The pressure in the tank equals that calculated in (b), $\underline{1.04\times10^5\,Pa}$

d) $p - p_a = \rho g(0.05\,m) = (13.6\times10^3\,kg\cdot m^{-3})(9.8\,m\cdot s^{-2})(0.05\,m) = \underline{6.66\times10^3\,Pa}$

e) The height $y_2 - y_1$ of the Hg column $\Rightarrow$ $\underline{5\,cm\text{ of Hg}}$.

f) The question is asking what $h = y_2 - y_1$ would be if the liquid in the tube was water ($\rho = 1.0\times10^3\,kg\cdot m^{-3}$):

$p - p_a = \rho g h = 6.66\times10^3\,Pa$ (the gauge pressure of the gas in the tank)

$$\Rightarrow h = \frac{6.66\times10^3\,Pa}{\rho g} = \frac{6.66\times10^3\,Pa}{(1.0\times10^3\,kg\cdot m^{-3})(9.8\,m\cdot s^{-2})} = 0.680\,m = \underline{68.0\,cm}$$

$$\left(\text{or, } h = \frac{\rho_{Hg}}{\rho_{H_2O}}(5\,cm) = \frac{13.6}{1}(5\,cm) = 68\,cm\right)$$

13-9

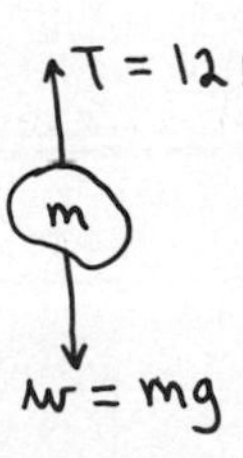

$w = 12\,N$

$$\Rightarrow m = \frac{12\,N}{9.8\,m\cdot s^{-2}} = 1.22\,kg$$

13-9 (cont)

$T = 10\,N$, B, $w = mg = 12\,N$

$$\Sigma F_y = 0$$
$$T + B - w = 0$$
$$B = w - T = 12N - 10N = 2\,N$$

Archimedes' principle $\Rightarrow B = \rho_{water} V_{object}\, g$

$$V_{obj} = \frac{B}{\rho_{water}\, g} = \frac{2N}{(1.0\times10^3\,kg\cdot m^{-3})(9.8\,m\cdot s^{-2})}$$

$$V_{obj} = \underline{2.04\times10^{-4}\,m^3}$$

$$\rho_{obj} = \frac{m_{obj}}{V_{obj}} = \frac{1.22\,kg}{2.04\times10^{-4}\,m^3} = \underline{5.98\times10^3\,kg\cdot m^{-3}}$$ (about 6 times the density of water)

13-11

The floating object is the slab of ice plus the man; the buoyant force must support both.

$\uparrow B = \rho_{water} V_{ice}\, g$

$a = 0$ — 80 kg + m_{ice}

$\downarrow (80\,kg + m_{ice})\,g$

$$\Sigma F_y = m a_y$$
$$B - (80\,kg + m_{ice})\,g = 0$$
$$\rho_{water} V_{ice}\, g - (80\,kg + m_{ice})\,g = 0$$

But $\rho = \frac{m}{V} \Rightarrow m_{ice} = \rho_{ice} V_{ice}$

$$\Rightarrow \rho_{water} V_{ice} - 80\,kg - \rho_{ice} V_{ice} = 0$$

$$V_{ice} = \frac{80\,kg}{\rho_{water} - \rho_{ice}} = \frac{80\,kg}{1000\,kg\cdot m^{-3} - 920\,kg\cdot m^{-3}} = \underline{1.00\,m^3}$$

13-13

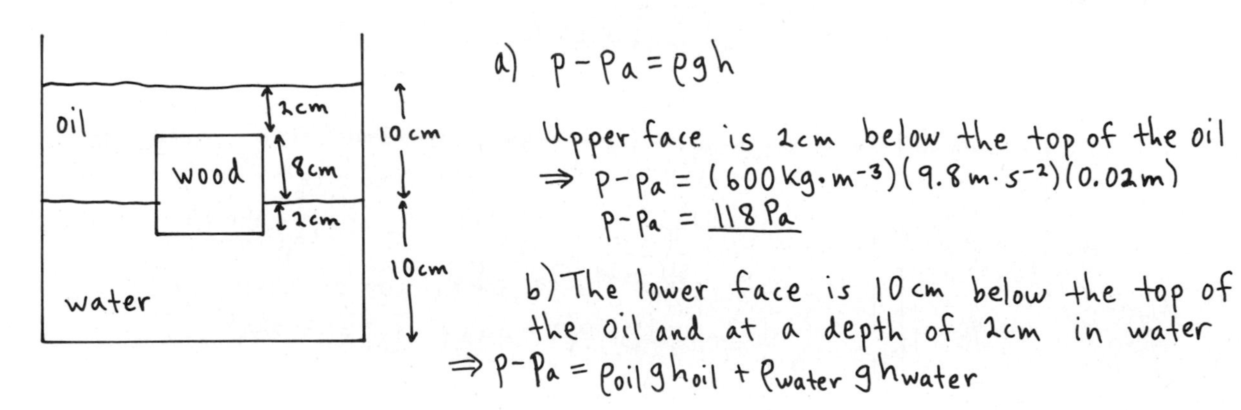

a) $p - p_a = \rho g h$

Upper face is 2cm below the top of the oil

$$\Rightarrow p - p_a = (600\,kg\cdot m^{-3})(9.8\,m\cdot s^{-2})(0.02\,m)$$
$$p - p_a = \underline{118\,Pa}$$

b) The lower face is 10 cm below the top of the oil and at a depth of 2cm in water

$$\Rightarrow p - p_a = \rho_{oil}\, g\, h_{oil} + \rho_{water}\, g\, h_{water}$$

$$p - p_a = (600\,kg\cdot m^{-3})(9.8\,m\cdot s^{-2})(0.10\,m) + (1000\,kg\cdot m^{-3})(9.8\,m\cdot s^{-2})(0.02\,m)$$

$$p - p_a = 588\,Pa + 196\,Pa = \underline{784\,Pa}$$

13-13 (cont)

c) Consider the vertical forces on the block:

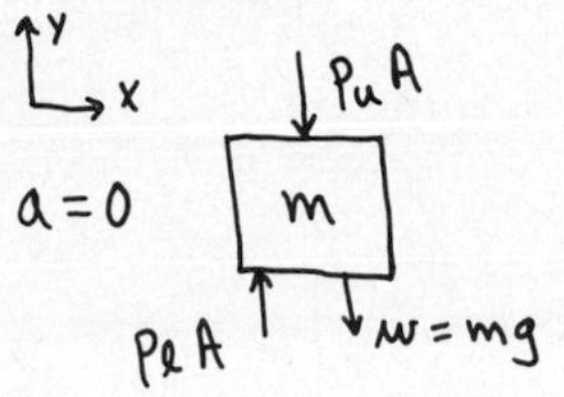

p_u and p_ℓ are the absolute pressures at the upper and lower faces of the block.

$$\Sigma F_y = ma_y \Rightarrow p_\ell A - p_u A - mg = 0$$
$$(p_\ell - p_u)A = mg$$

We note that $p_\ell - p_u = [p_\ell - p_a] - [p_u - p_a]$, where p_a is air pressure. (The <u>difference</u> in absolute pressures equals the difference in gauge pressures.)

$$p_\ell - p_u = 784\ \text{Pa} - 118\ \text{Pa} = 666\ \text{Pa}$$
$$A = (0.10\,\text{m})^2 = 0.01\,\text{m}^2$$

$$m = \frac{(p_\ell - p_u)A}{g} = \frac{(666\,\text{Pa})(0.01\,\text{m}^2)}{9.8\,\text{m}\cdot\text{s}^{-2}} = \underline{0.680\,\text{kg}}$$

13-15

$$\text{Eq}(13\text{-}10) \Rightarrow p - p_a = \frac{4\gamma}{R} = \frac{4\,(25\times10^{-3}\,\text{N}\cdot\text{m}^{-1})}{(0.25\,\text{m})} = \underline{4.00\,\text{Pa}}$$

13-17

a) volume flow rate $= vA = v\pi r^2$

$$v = \frac{\text{volume flow rate}}{\pi r^2} = \frac{0.8\,\text{m}^3\cdot\text{s}^{-1}}{\pi\,(0.2\,\text{m})^2} = \underline{6.37\,\text{m}\cdot\text{s}^{-1}}$$

b) vA = volume flow rate, which is constant. Thus at this second point the volume flow rate is also $0.8\,\text{m}^3\cdot\text{s}^{-1}$.

$$v\pi r^2 = 0.8\,\text{m}^3\cdot\text{s}^{-1} \Rightarrow r = \sqrt{\frac{0.8\,\text{m}^3\cdot\text{s}^{-1}}{\pi v}} = \frac{0.8\,\text{m}^3\cdot\text{s}^{-1}}{\pi\,(3.8\,\text{m}\cdot\text{s}^{-1})} = \underline{0.259\,\text{m}}$$

13-19

a)

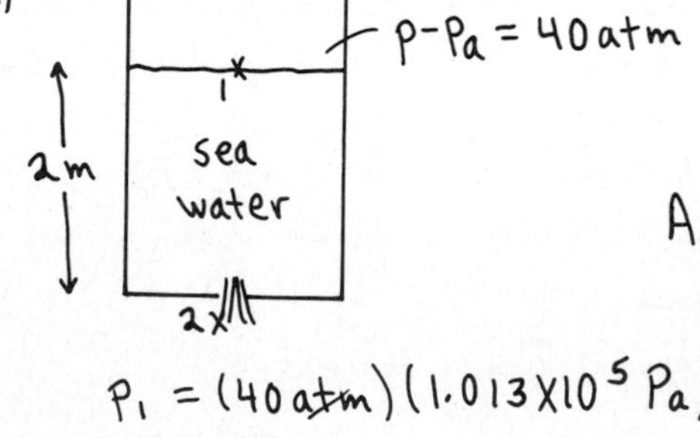

Apply Bernoulli's eq., with points 1 and 2 chosen as in the sketch:

$$p_1 + \rho g y_1 + \tfrac{1}{2}\rho v_1^2 = p_2 + \rho g y_2 + \tfrac{1}{2}\rho v_2^2 \quad ;\ v_2 = ?$$

$$A_1 \gg A_2 \Rightarrow \tfrac{1}{2}\rho v_1^2 \text{ can be neglected}$$

$$(p_1 - p_2) + \rho g (y_1 - y_2) = \tfrac{1}{2}\rho v_2^2$$

$p_1 = (40\,\text{atm})(1.013\times10^5\,\text{Pa}/1\,\text{atm}) + p_a$

$p_2 = p_a$

$\Rightarrow p_1 - p_2 = 4.05\times10^6\,\text{Pa}$

$y_1 - y_2 = 2\,\text{m}$

sea water $\Rightarrow \rho = 1.03\times10^3\,\text{kg}\cdot\text{m}^{-3}$

13-19 (cont)

$$\Rightarrow v_2 = \sqrt{\frac{2(p_1-p_2)}{\rho} + 2g(y_1-y_2)} = \sqrt{\frac{2(4.05\times10^6\,Pa)}{1.03\times10^3\,kg\cdot m^{-3}} + 2(9.8\,m\cdot s^{-2})(2m)} = \underline{88.9\,m\cdot s^{-1}}$$

b) $F = \frac{\Delta p}{\Delta t} = \frac{\Delta(mv)}{\Delta t}$

In time Δt the volume of water flowing out of the hole is $Av\Delta t$. The mass flowing out is $\rho Av\Delta t$. The momentum of this mass as it leaves the hole is $(\rho Av\Delta t)v = \rho Av^2\Delta t$. Its initial momentum (while inside the tank) is approximately zero (v_1 small) $\Rightarrow \Delta(mv) = \rho Av^2\Delta t$.

Thus $F = \frac{\Delta(mv)}{\Delta t} = \frac{\rho Av^2\Delta t}{\Delta t} = \rho Av^2$

$$F = (1.03\times10^3\,kg\cdot m^{-3})(10\times10^{-4}\,m^2)(88.9\,m\cdot s^{-1})^2 = \underline{8140N}$$

13-23

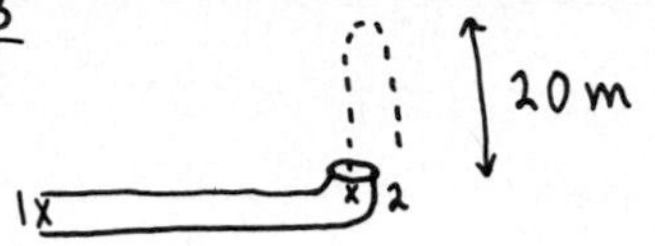

First apply conservation of energy to the motion of a drop of water, from when it leaves the end of the hose until it reaches its maximum height.

$$K_1 + U_1 = K_2 + U_2$$

($U_1 = 0$, $K_2 = 0$)

$$\tfrac{1}{2}mv^2 = mgh \Rightarrow v^2 = 2gh$$

Now apply Bernoulli's eq. to the flow of water through the fire hose. Let point 1 be in the water mains, and point 2 be at the discharge end of the hose.

$$p_1 + \rho g y_1 + \tfrac{1}{2}\rho v_1^2 = p_2 + \rho g y_2 + \tfrac{1}{2}\rho v_2^2$$

$A_1 \gg A_2 \Rightarrow$ neglect the $\frac{1}{2}\rho v_1^2$ term

$p_2 = p_{air}$

$y_1 = y_2$

$$\Rightarrow \quad p_1 - p_{air} = \tfrac{1}{2}\rho v_2^2$$

But from the conservation of energy analysis, $v_2^2 = 2gh$

$$\Rightarrow p_1 - p_{air} = \tfrac{1}{2}\rho(2gh) = \rho g h$$

$$p_1 - p_{air} = (1.0\times10^3\,kg\cdot m^{-3})(9.8\,m\cdot s^{-2})(20m) = \underline{1.96\times10^5\,Pa}$$

13-29

a) water at 20°C $\Rightarrow \eta = 1.005\times10^{-2}$ poise $= 1.005\times10^{-3}\,N\cdot s\cdot m^{-2}$

Volume flow rate $= \frac{dV}{dt}$

Eq. (13-28) $\Rightarrow \frac{dV}{dt} = \frac{\pi}{8}\,\frac{R^4}{\eta}\,\frac{(p_1-p_2)}{l}$

13-29 (cont)

P_1 = absolute pressure at pump
P_2 = pressure at open end of pipe = P_{air}
$\Rightarrow P_1 - P_2$ = gauge pressure at pump = 800 Pa

$$\frac{dV}{dt} = \frac{\pi}{8} \frac{(0.04\,m)^4}{1.005\times10^{-3}\,N\cdot s\cdot m^{-2}} \left(\frac{800\,Pa}{15\,m}\right) = \underline{0.0533\,m^3\cdot s^{-1}}$$

b) $R_2 = 0.02\,m$, $R_1 = 0.04\,m$; $\Delta P_1 = 800\,Pa$, $\Delta P_2 = ?$
(ΔP_1 and ΔP_2 are the pump gauge pressures for R_1 and R_2, respectively)

Same volume flow rate $\Rightarrow R^4 \Delta p$ must stay constant

$$R_1^4 \Delta P_1 = R_2^4 \Delta P_2$$

$$\Delta P_2 = \left(\frac{R_1}{R_2}\right)^4 \Delta P_1 = \left(\frac{0.04\,m}{0.02\,m}\right)^4 (800\,Pa) = \underline{1.28\times10^4\,Pa}$$

c) $T \rightarrow 60^\circ C \Rightarrow \eta = 0.469\times10^{-3}\,N\cdot s\cdot m^{-2}$

$$\frac{dV}{dt}\eta = \frac{\pi}{8} R^4 \frac{P_1 - P_2}{L} = \text{constant} \Rightarrow \left(\frac{dV}{dt}\right)_1 \eta_1 = \left(\frac{dV}{dt}\right)_2 \eta_2$$

(The subscript 1 refers to 20°C and 2 to 60°C.)

$$\left(\frac{dV}{dt}\right)_2 = \left(\frac{dV}{dt}\right)_1 \left(\frac{\eta_1}{\eta_2}\right) = 0.0533\,m^3\cdot s^{-1} \left(\frac{1.005\times10^{-3}\,N\cdot s\cdot m^{-2}}{0.469\times10^{-3}\,N\cdot s\cdot m^{-2}}\right) = \underline{0.114\,m^3\cdot s^{-1}}$$

13-31

$$w = mg$$

$$F = 6\pi\eta r v \text{ (viscous drag force)}$$

$$F = \frac{1}{4} w \Rightarrow 6\pi\eta r v = \frac{1}{4} mg$$

Write the mass in terms of the density: $m = \rho V = \rho \frac{4}{3}\pi r^3$

$$\Rightarrow 6\pi\eta r v = \frac{1}{4}\left(\rho \frac{4}{3}\pi r^3\right) g$$

$$v = \frac{\rho g r^2}{18\eta} = \frac{(19.3\times10^3\,kg\cdot m^{-3})(9.8\,m\cdot s^{-2})(4\times10^{-3}\,m)^2}{18\,(1.005\times10^{-3}\,N\cdot s\cdot m^{-2})} = \underline{167\,m\cdot s^{-1}}$$

13-33

a)
$$N_R = \frac{\rho v D}{\eta} = \frac{(1.0\times10^3\,kg\cdot m^{-3})(2\,m\cdot s^{-1})(4\times10^{-3}\,m)}{0.357\times10^{-3}\,N\cdot s\cdot m^{-2}} = \underline{22{,}400}$$

b) N_R is greater than 3000, so the flow is turbulent.

Problems

13-35

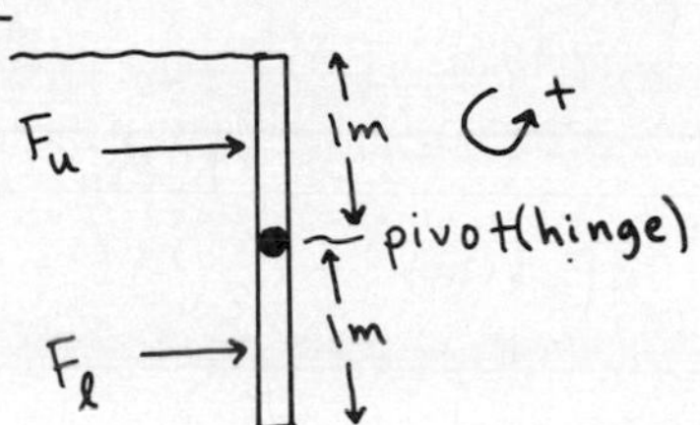

↻ Γ_u, the torque due to the force on the upper half of the gate.

↺ Γ_ℓ, the torque due to the force on the lower half of the gate.

Calculation of Γ_u (negative):

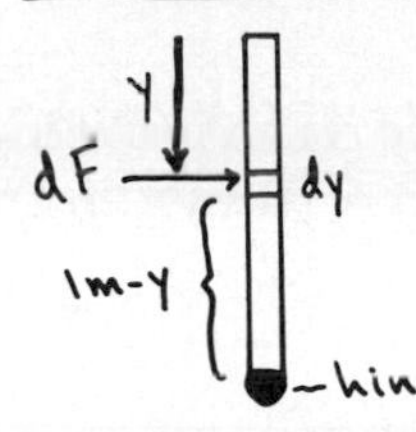

The strip of width dy and area L dy (L = 3m is the width of the gate) is a distance y below the water surface. Hence the pressure at the strip is $p_a + \rho g y$. The force on the strip is

$$dF = (p_a + \rho g y)\,L\,dy$$

The torque about the hinge due to this force is $d\Gamma_u = -dF\,\ell$, where $\ell = (1m - y)$ is the moment arm.

$$\Rightarrow d\Gamma_u = -(p_a + \rho g y)\,L\,dy\,(1m - y)$$

The total torque Γ_u on the upper half of the gate is obtained by integrating $d\Gamma_u$ over all the strips into which the upper half can be divided:

$$\Gamma_u = \int_0^{1m} d\Gamma_u = -\int_0^{1m} L(p_a + \rho g y)(1m - y)\,dy$$

$$\Gamma_u = -L\int_0^{1m} [(1m)p_a + (1m)\rho g y - p_a y - \rho g y^2]\,dy$$

$$\Gamma_u = -L\left[(1m)p_a y\Big|_0^{1m} + \tfrac{1}{2}(1m)\rho g y^2\Big|_0^{1m} - \tfrac{1}{2}p_a y^2\Big|_0^{1m} - \tfrac{1}{3}\rho g y^3\Big|_0^{1m}\right]$$

$$\Gamma_u = -L\left[p_a(1m)^2 + \tfrac{1}{2}\rho g(1m)^3 - \tfrac{1}{2}p_a(1m)^2 - \tfrac{1}{3}\rho g(1m)^3\right]$$

$$\Gamma_u = -L\left[\tfrac{1}{2}p_a(1m)^2 + \tfrac{1}{6}\rho g(1m)^3\right]$$

Calculation of Γ_ℓ (positive):

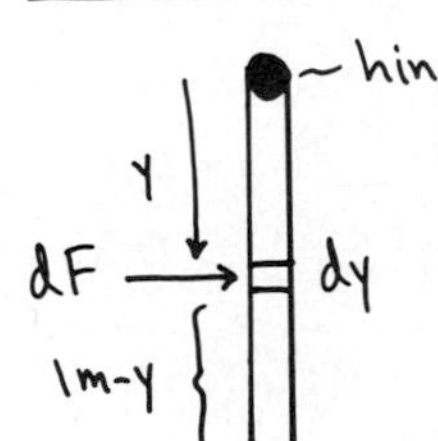

The small strip of width dy is a distance y below the hinge, and hence a distance (1m+y) below the water surface. Hence the pressure at the strip is $p_a + \rho g(1m + y)$. The force on the strip is

$$dF = [p_a + \rho g(1m + y)]\,L\,dy.$$

The torque about the hinge due to this force is $d\Gamma_\ell = +dF\,\ell$, where $\ell = y$ is the moment arm.

$$\Rightarrow d\Gamma_\ell = [p_a + \rho g(1m + y)]\,L\,y\,dy$$

The total torque Γ_ℓ on the lower half of the gate is obtained by integrating

13-35 (cont)

$d\Gamma_\ell$ over all strips into which the lower half can be divided:

$$\Rightarrow \Gamma_\ell = \int_0^{1m} d\Gamma_\ell = \int_0^{1m} L[p_a + \rho g(1m + y)]y\,dy = Lp_a \int_0^{1m} y\,dy + L\rho g(1m)\int_0^{1m} y\,dy + L\rho g \int_0^{1m} y^2\,dy$$

$$\Gamma_\ell = \tfrac{1}{2}Lp_a(1m)^2 + \tfrac{1}{2}L\rho g(1m)^3 + \tfrac{1}{3}L\rho g(1m)^3 = L[\tfrac{1}{2}p_a(1m)^2 + \tfrac{5}{6}\rho g(1m)^3]$$

The total torque is $\Gamma = \Gamma_u + \Gamma_\ell = \frac{2}{3}L\rho g(1m)^3$

$$\Gamma = \tfrac{2}{3}(3m)(1.0\times10^3\,kg\cdot m^{-3})(9.8\,m\cdot s^{-2})(1m)^3 = \underline{1.96\times10^4\,N\cdot m}$$

Note: We knew from the start that the constant p_a contribution to the pressure at each depth would have no net contribution. (Didn't we?)

13-39

a) Consider the forces on the dirigible: (Note that the total mass of the dirigible is 10,000 kg plus the mass of the gas in it.)

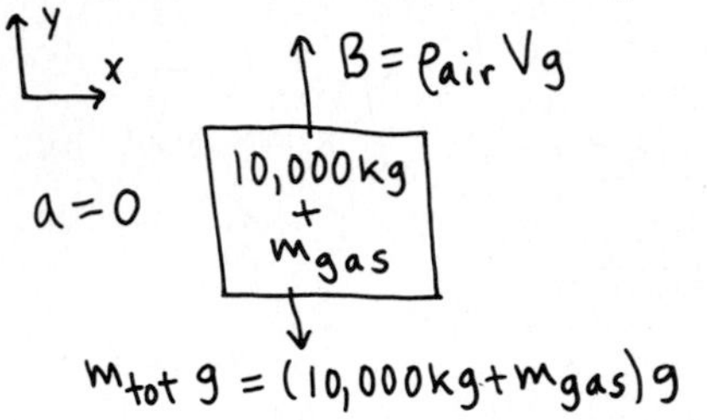

$$\sum F_y = 0$$

$B - m_{tot}\,g = 0$ (the dirigible floats in air)

$$\rho_{air} V g - (10{,}000\,kg + m_{gas})g = 0$$

$$\rho_{air} V - 10{,}000\,kg - m_{gas} = 0$$

But m_{gas} depends on V, so write $m_{gas} = \rho_{gas} V$

$$\Rightarrow \rho_{air} V - 10{,}000\,kg - \rho_{gas} V = 0$$

$$V = \frac{10{,}000\,kg}{\rho_{air} - \rho_{gas}} = \frac{10{,}000\,kg}{1.29\,kg\cdot m^{-3} - 0.0899\,kg\cdot m^{-3}} = \underline{8.33\times10^3\,m^3}$$

b) Instead of a lift of 10,000 kg, let the lift be m_{lift}, that we want to solve for. The volume is now known to be the $V = 8.33\times10^3\,m^3$ that we found in (a).

$$B - m_{tot}\,g \Rightarrow \rho_{air} V g - (m_{lift} + m_{gas})g = 0$$

$$m_{gas} = \rho_{gas} V \Rightarrow m_{lift} = (\rho_{air} - \rho_{gas})V$$

$$m_{lift} = (1.29\,kg\cdot m^{-3} - 0.178\,kg\cdot m^{-3})(8.33\times10^3\,m^3) = \underline{9260\,kg}$$

(Hydrogen gas is highly explosive. Helium is somewhat expensive but is chemically inert.)

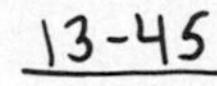

13-45

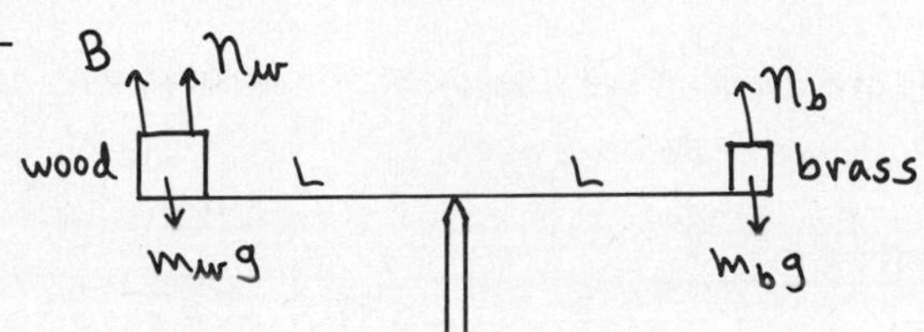

Consider forces and torques on the balance arm:

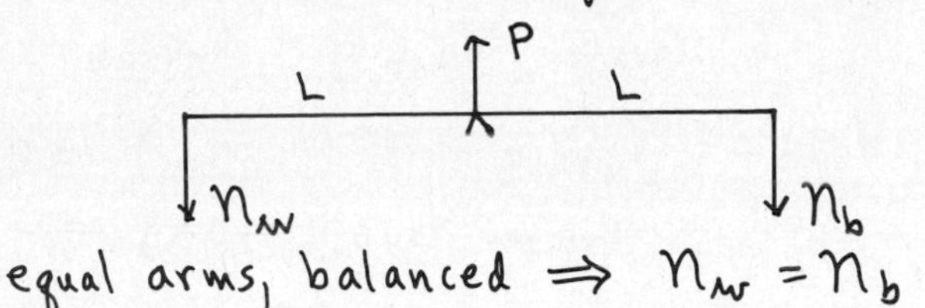

equal arms, balanced $\Rightarrow n_w = n_b$

Forces on the wood:

B, n_w; $a = 0$; m_w; $m_w g$; y, x

$$\Sigma F_y = ma_y$$
$$n_w + B - m_w g = 0$$
$$n_w = m_w g - B$$

Forces on the brass:

n_b; $a = 0$; m_b; $m_b g$; y, x

$$\Sigma F_y = ma_y$$
$$n_b - m_b g = 0$$
$$n_b = m_b g$$

Then $n_w = n_b \Rightarrow m_w g - B = m_b g$

And $B = \rho_{air} V_w g = \rho_{air}\left(\frac{m_w}{\rho_w}\right) g \Rightarrow m_w g\left(1 - \frac{\rho_{air}}{\rho_w}\right) = m_b g$

$$m_w = \frac{m_b}{\left(1 - \frac{\rho_{air}}{\rho_w}\right)}$$

(Note: In the limit of $\rho_{air} \ll \rho_{object}$, $m_{object} = m_b$.)

Specific gravity of wood $= 0.15 \Rightarrow \rho_w = 150\ \mathrm{kg \cdot m^{-3}}$

Thus $m_w = \dfrac{0.100\ \mathrm{kg}}{1 - \dfrac{1.29\ \mathrm{kg \cdot m^{-3}}}{150\ \mathrm{kg \cdot m^{-3}}}} = \underline{0.1009\ \mathrm{kg}}$

13-49

The resultant buoyant force acts at the geometrical center of the submerged portionof the object. The weight of the object acts at the center of gravity of the object. These two points are displaced from each other in Fig. 13-33 (b), and this gives rise to a restoring torque about the geometrical center of the block.

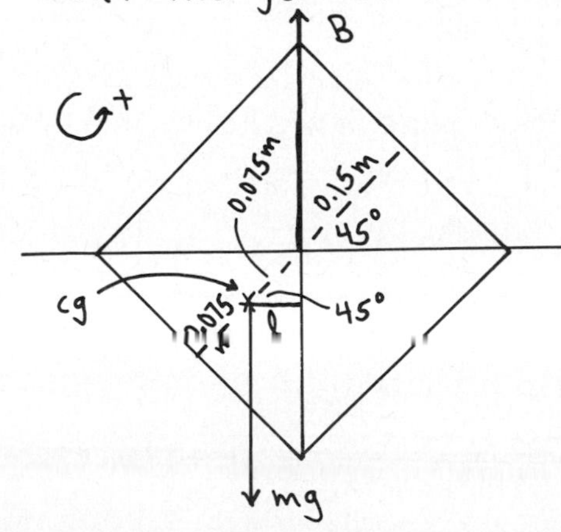

The block is floating $\Rightarrow B = mg$

Compute the resultant torque about the center of the block. $\vec{B}$ acts through the center $\Rightarrow \Gamma_B = 0$.

For this axis the mg force has moment arm

$$l = (0.075\ \mathrm{m}) \sin 45° = 0.0530\ \mathrm{m}$$

$$\Rightarrow \Gamma = mgl$$

13-49 (cont)

Use $B = mg$ to calculate the mass m of the block:

$B = \rho_{water} V_{sub}\, g \quad (V_{sub} = \text{volume submerged})$

$V_{sub} = \frac{1}{2} V_{block} = \frac{1}{2}(0.30\,m)^3 = 0.0135\,m^3$

$B = mg \Rightarrow \rho_{water} V_{sub}\, g = mg \Rightarrow m = (1.0 \times 10^3\,kg \cdot m^{-3})(0.0135\,m^3) = 135\,kg$

Thus $\Gamma = mg\,l = (135\,kg)(9.8\,m \cdot s^{-2})(0.0530\,m) = \underline{7.01\,N \cdot m}$

13-51

$vA = 1.40 \times 10^{-4}\,m^3 \cdot s^{-1}$

h, 1, 2

The water level in the vessel will rise until the volume flow rate into the vessel from the tube equals the volume flow rate out the hole in the bottom. Let points 1 and 2 be chosen as in the sketch.

$vA = v_2 A_2 \Rightarrow v_2 = \frac{1.4 \times 10^{-4}\,m^3 \cdot s^{-1}}{A_2} = \frac{1.4 \times 10^{-4}\,m^3 \cdot s^{-1}}{1 \times 10^{-4}\,m^2} = 1.4\,m \cdot s^{-1}$

Bernoulli's eq. $\Rightarrow p_1 + \rho g y_1 + \frac{1}{2}\rho v_1^2 = p_2 + \rho g y_2 + \frac{1}{2}\rho v_2^2$

$A_1 \gg A_2 \Rightarrow \frac{1}{2}\rho v_1^2 \ll \frac{1}{2}\rho v_2^2$; neglect the $\frac{1}{2}\rho v_1^2$ term

$p_1 = p_a = p_{air}$

Thus Bernoulli's eq. becomes $\rho g (y_1 - y_2) = \frac{1}{2}\rho v_2^2$

$y_1 - y_2 = h \Rightarrow h = \frac{v_2^2}{2g} = \frac{(1.4\,m \cdot s^{-1})^2}{2(9.8\,m \cdot s^{-2})} = \underline{0.10\,m}$

13-53

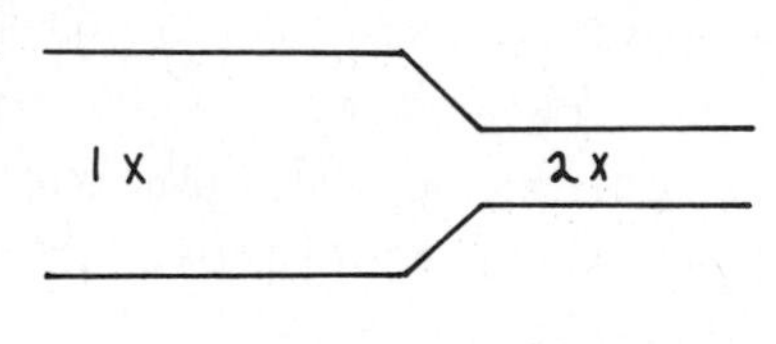

$A_1 = 10\,cm^2, \quad A_2 = 5\,cm^2$

$v_1 A_1 = v_2 A_2$

$v_2 = v_1 \left(\frac{A_1}{A_2}\right) = v_1 \left(\frac{10\,cm^2}{5\,cm^2}\right) = 2 v_1$

$v_2 > v_1 \Rightarrow p_1 > p_2 \; ; \; p_1 - p_2 = 300\,Pa$

To find the volume flow rate out of the pipe calculate $v_1 A_1$, which equals $v_2 A_2$:

$p_1 + \frac{1}{2}\rho v_1^2 + \rho g y_1 = p_2 + \frac{1}{2}\rho v_2^2 + \rho g y_2$

$y_1 = y_2 \Rightarrow \frac{1}{2}\rho (v_2^2 - v_1^2) = p_1 - p_2$

$v_2 = 2 v_1 \Rightarrow v_2^2 - v_1^2 = 4 v_1^2 - v_1^2 = 3 v_1^2$

Thus $\frac{3}{2}\rho v_1^2 = p_1 - p_2 \Rightarrow v_1 = \sqrt{\frac{2(p_1 - p_2)}{3\rho}} = \sqrt{\frac{2(300\,Pa)}{3(1.0 \times 10^3\,kg \cdot m^{-3})}} = 0.447\,m \cdot s^{-1}$

13-53 (cont)

The volume flow rate through the pipe is thus

$v_1 A_1 = (0.447\,\text{m}\cdot\text{s}^{-1})(10\times10^{-4}\,\text{m}^2) = 4.47\times10^{-4}\,\text{m}^3\cdot\text{s}^{-1} = \underline{0.0268\,\text{m}^3\cdot\text{min}^{-1}}$

13-55

x1

x2

$$\text{lift} = 1000\,\text{N}\cdot\text{m}^{-2} \Rightarrow p_2 - p_1 = 1000\,\text{N}\cdot\text{m}^{-2}$$

$$p_1 + \rho g y_1 + \tfrac{1}{2}\rho v_1^2 = p_2 + \rho g y_2 + \tfrac{1}{2}\rho v_2^2$$

$$v_1 = \sqrt{v_2^2 - 2g(y_1 - y_2) + \frac{2(p_2 - p_1)}{\rho}}$$

Note: $\dfrac{2(p_2-p_1)}{\rho} = \dfrac{2(1000\,\text{N}\cdot\text{m}^{-2})}{1.29\,\text{kg}\cdot\text{m}^{-3}} = 1550\,\text{m}^2\cdot\text{s}^{-2}$.

We don't know $y_1 - y_2$, but it is surely no more than a meter or so. Then $2g(y_1 - y_2) = 2(9.8\,\text{m}\cdot\text{s}^{-2})(1\,\text{m}) = 19.6\,\text{m}^2\cdot\text{s}^{-2}$.

So if $y_1 - y_2 \approx 1\,\text{m}$ this term is negligible.

Thus $v_1 = \sqrt{v_2^2 + \dfrac{2(p_2-p_1)}{\rho}} = \sqrt{(100\,\text{m}\cdot\text{s}^{-1})^2 + \dfrac{2(1000\,\text{N}\cdot\text{m}^{-2})}{1.29\,\text{kg}\cdot\text{m}^{-3}}} = \underline{107\,\text{m}\cdot\text{s}^{-1}}$

13-57

$$A_C = \tfrac{1}{2} A_D$$

$$v_C A_C = v_D A_D \Rightarrow v_D = v_C \frac{A_D}{A_C} = v_C \frac{\tfrac{1}{2}A_D}{A_D} = \tfrac{1}{2} v_C$$

1, 2, h_1

Use Bernoulli's eq. for points 1 and 2 to relate v_2 to h_1:

$$p_1 + \rho g y_1 + \tfrac{1}{2}\rho v_1^2 = p_2 + \rho g y_2 + \tfrac{1}{2}\rho v_2^2$$

$$p_1 = p_2 = p_{air}\,;\; v_1 \approx 0 \Rightarrow \tfrac{1}{2}\rho v_2^2 = \rho g(y_1 - y_2)$$

$$y_1 - y_2 = h_1 \Rightarrow v_2 = \sqrt{2g h_1}$$

$$v_D = v_2$$

$$v_C = 2 v_D = 2\sqrt{2gh_1}$$

h_2, that we want to calculate, is related to p_C as follows: Apply Bernoulli's eq. to the liquid in pipe E.

2, 1, h_2, C

$p_2 = p_C$

This liquid is stationary $\Rightarrow v_2 = v_1 = 0$.

$p_1 = p_{air}$

$$p_1 + \cancel{\tfrac{1}{2}\rho v_1^2} + \rho g y_1 = p_2 + \cancel{\tfrac{1}{2}\rho v_2^2} + \rho g y_2$$

$$\rho g(y_2 - y_1) = p_1 - p_2$$

$y_2 - y_1 = h_2$, $p_1 = p_{air}$, and $p_2 = p_C \Rightarrow \boxed{h_2 = \dfrac{p_{air} - p_C}{\rho g}}$

13-57 (cont)

Finally, apply Bernoulli's eq. to points C and D, to calculate p_c in terms of v_c:

$$p_c + \rho g y_c + \tfrac{1}{2}\rho v_c^2 = p_D + \rho g y_D + \tfrac{1}{2}\rho v_D^2$$

$y_c = y_D$; $p_D = p_{air} \Rightarrow p_{air} - p_c = \tfrac{1}{2}\rho(v_c^2 - v_D^2)$

$v_D = \tfrac{1}{2}v_c \Rightarrow p_{air} - p_c = \tfrac{1}{2}\rho(v_c^2 - \tfrac{1}{4}v_c^2) = \tfrac{3}{8}\rho v_c^2$

But $v_c = 2\sqrt{2gh_1} \Rightarrow p_{air} - p_c = \tfrac{3}{8}\rho(8gh_1) = 3\rho g h_1$

$$h_2 = \frac{p_{air} - p_c}{\rho g} = \frac{3\rho g h_1}{\rho g} = 3h_1 \quad ; \quad \boxed{h_2 = 3h_1}$$

13-59

a)

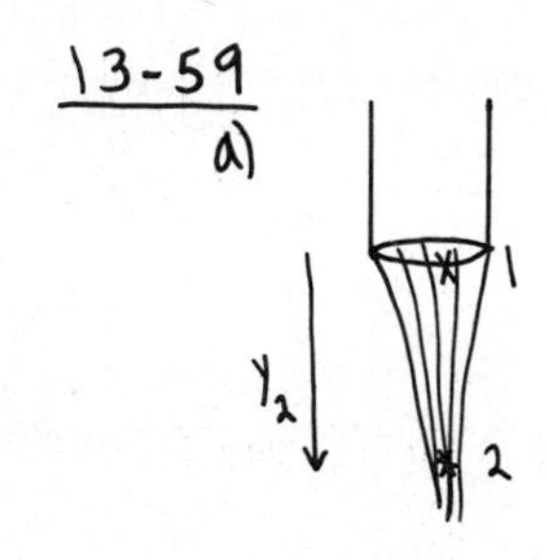

Let point 1 be at the end of the pipe, and point 2 be in the stream of liquid, a distance y_2 below the end of the tube.

free-fall $\Rightarrow a = g$

$$v_2^2 = v_1^2 + 2ay_2 = v_1^2 + 2gy_2$$

$$v_2 = \sqrt{v_1^2 + 2gy_2}$$

Equation of continuity $\Rightarrow v_1 A_1 = v_2 A_2$

$$A = \pi r^2 \Rightarrow v_2 = v_1\left(\frac{r_1^2}{r_2^2}\right)$$

Use in the above $\Rightarrow v_1\left(\frac{r_1^2}{r_2^2}\right) = \sqrt{v_1^2 + 2gy} \Rightarrow r_2 = \frac{\sqrt{v_1}\, r_1}{(v_1^2 + 2gy_2)^{1/4}}$

Note that this equation says that r_2 decreases the further the liquid is below the discharge end of the pipe.

b) $v_1 = 1.0\ \text{m}\cdot\text{s}^{-1}$

Want y_2 that makes $r_2 = \tfrac{1}{2}r_1 \Rightarrow \tfrac{1}{2}r_1 = \frac{\sqrt{v_1}\, r_1}{(v_1^2 + 2gy_2)^{1/4}}$

$$(v_1^2 + 2gy_2)^{1/4} = 2\sqrt{v_1}$$

$$v_1^2 + 2gy_2 = 16v_1^2$$

$$y_2 = \frac{15v_1^2}{2g} = \frac{15\,(1.0\ \text{m}\cdot\text{s}^{-1})^2}{2\,(9.8\ \text{m}\cdot\text{s}^{-2})} = \underline{0.765\ \text{m}}$$

13-61

a)

Note: The viscous drag force F is downward, since the bubble is traveling upward.

$B = mg + F$ ($a = 0$ at the terminal velocity)

$B = \rho' V g = \frac{4}{3}\pi r^3 \rho' g$, where ρ' is the density of the liquid

$mg = \rho V g = \frac{4}{3}\pi r^3 \rho g$, where ρ is the density of the air in the bubble

$F = 6\pi \eta r v_T$

Thus $B = mg + F \Rightarrow \frac{4}{3}\pi r^3 \rho' g = \frac{4}{3}\pi r^3 \rho g + 6\pi\eta r v_T$

$$6\eta v_T = \frac{4}{3} r^2 g(\rho' - \rho)$$

$$v_T = \frac{2r^2 g}{9\eta}(\rho' - \rho)$$

$r = 0.5\times10^{-3}\ \text{m}$, $\eta = 0.15\ \text{N·s·m}^{-2}$, $\rho' = 900\ \text{kg·m}^{-3}$, $\rho = 1.3\ \text{kg·m}^{-3}$

$$v_T = \frac{2(0.5\times10^{-3}\ \text{m})^2(9.8\ \text{m·s}^{-2})}{9(0.15\ \text{N·s·m}^{-2})}(900\ \text{kg·m}^{-3} - 1.3\ \text{kg·m}^{-3}) = 3.26\times10^{-3}\ \text{m·s}^{-1}$$

$$v_T = \underline{3.26\ \text{mm·s}^{-1}}$$

(Note: The precise value of $\rho = \rho_{air}$ that is used is unimportant, since $\rho' \gg \rho$.)

b) water at 20°C $\Rightarrow \rho' = 1000\ \text{kg·m}^{-3}$, $\eta = 1.005\times10^{-3}\ \text{N·s·m}^{-2}$

$$v_T = \frac{2r^2 g}{9\eta}(\rho' - \rho) = \frac{2(0.5\times10^{-3}\ \text{m})^2(9.8\ \text{m·s}^{-2})}{9(1.005\times10^{-3}\ \text{N·s·m}^{-2})}(1000\ \text{kg·m}^{-3} - 1.3\ \text{kg·m}^{-3})$$

$$v_T = \underline{0.541\ \text{m·s}^{-1}}$$

The bubble rises much faster in water, since water is much less viscous than the liquid in (a).

CHAPTER 14

Exercises 1, 3, 9, 11, 15, 17, 19

Problems 23, 25, 27, 31

Exercises

14-1

a) $T_C = 40°C$; $T_F = ?$

$T_F = \frac{9}{5}T_c + 32° = \frac{9}{5}(40°) + 32° = \underline{104°F}$; you are ill.

b) $T_F = 98.6°F$; $T_c = ?$

$T_c = \frac{5}{9}(T_F - 32°) = \frac{5}{9}(98.6° - 32°) = \underline{37.0°C}$

c) $T_F = \frac{9}{5}T_c + 32°$

$T_F = T_c = T \Rightarrow T = \frac{9}{5}T + 32°$

$\frac{4}{5}T = -32° \Rightarrow T = -40°$

$-40°C = -40°F$

14-3

$T_K = T_c + 273.15$

$T_c = -182.97°C \Rightarrow T_K = -182.97 + 273.15 = \underline{90.18\,K}$

$T_R = T_F + 459.7°$, so convert the given T_c to T_F

$T_F = \frac{9}{5}T_c + 32° = \frac{9}{5}(-182.97°) + 32° = -297.35°F$

Then $T_R = -297.35° + 459.7° = \underline{+162.3°R}$

14-9

Let $L_0 = 80.000\text{ cm}$; $T_0 = 20°C$

$T = 40°C \Rightarrow \Delta L = 0.024\text{ cm}$

Thus $\Delta L = \alpha L_0 \Delta T \Rightarrow \alpha = \frac{\Delta L}{L_0 \Delta T} = \frac{0.024\text{cm}}{(80.000\text{cm})(40°C - 20°C)} = \underline{1.5 \times 10^{-5}\,(C°)^{-1}}$

(Note: It was ok to leave the lengths in cm since in the calculation of α the length units cancel.)

14-11

The diameter of the hole undergoes linear expansion just as does a length of brass. α_{brass} is given in Table 14-1.

$$\Delta L = \alpha L_0 \Delta T = (2.0\times10^{-5}(C°)^{-1})(2.500 cm)(200°C - 20°C) = +0.009 cm$$

$$L = L_0 + \Delta L = 2.500 cm + 0.009 cm = \underline{2.509 cm}$$

14-15

Calculate ΔV for the ethanol. From Table 14-2, β for ethanol is $75\times10^{-5}(C°)^{-1}$.

$$\Delta V = \beta V_0 \Delta T = (75\times10^{-5}(C°)^{-1})(500 L)(10°C - 25°C) = -5.62 L$$

Thus the volume of the air space will be $\underline{5.62 L} = \underline{5.62\times10^{-3} m^3}$.

14-17

a) $\Delta L = \alpha L_0 \Delta T \Rightarrow \alpha = \dfrac{\Delta L}{L_0 \Delta T} = \dfrac{1.5\times10^{-2} m}{3 m (520°C - 20°C)} = \underline{1.0\times10^{-5}(C°)^{-1}}$

b) stress $\dfrac{F}{A} = -Y\alpha\Delta T$ (eq. 14-13)

$$\frac{F}{A} = -(2\times10^{11} Pa)(1.0\times10^{-5}(C°)^{-1})(20°C - 520°C) = \underline{+1.0\times10^{9} Pa}$$

Note: ΔT means final temperature minus initial temperature. Here ΔT is negative (the temperature decreases). This gives F/A positive, corresponding to a tensile (stretching) stress.

14-19

$$\Delta p = B\beta\Delta T \quad \text{(eq. 14-14)}$$

copper: $B = 1.4\times10^{11}$ Pa (Table 12-1)
$\beta = 5.1\times10^{-5}(C°)^{-1}$ (Table 14-2)

$$\Delta p = (1.4\times10^{11} Pa)(5.1\times10^{-5}(C°)^{-1})(30°C - 20°C) = \underline{+7.14\times10^{7} Pa}$$

Problems

14-23

a) Heat the ring to make its diameter equal to 3.002 in. The diameter of the ring undergoes linear expansion.

$$\Delta L = L_0 \alpha \Delta T \Rightarrow \Delta T = \frac{\Delta L}{L_0 \alpha} = \frac{+0.002 in}{(3.000 in)(1.2\times10^{-5}(C°)^{-1})} = +55.6 C°$$

14-23 (cont)

$$T = T_0 + \Delta T = 20°C + 55.6\,C° = \underline{75.6°C}$$

b) $\Delta L = L_0 \alpha \Delta T$

ΔT will be the same for both. Cool $\Rightarrow$ ΔL is negative. The brass shaft needs to contract 0.002 in more than the steel ring.

$\Rightarrow (-\Delta L_b) - (-\Delta L_s) = 0.002\text{ in}$ $\quad$ ($b \Rightarrow$ brass, $s \Rightarrow$ steel)

$\Delta L_s - \Delta L_b = +0.002\text{ in}$

$\Delta L_s = L_{os}\alpha_s \Delta T$

$\Delta L_b = L_{ob}\alpha_b \Delta T$ $\quad \Rightarrow (L_{os}\alpha_s - L_{ob}\alpha_b)\Delta T = 0.002\text{ in}$

$$\Delta T = \frac{0.002\text{ in}}{L_{os}\alpha_s - L_{ob}\alpha_b} = \frac{0.002\text{ in}}{(3.000\text{in})(1.2\times10^{-5}(C°)^{-1}) - (3.002\text{in})(2.0\times10^{-5}(C°)^{-1})} = -83.2\,C°$$

$$T = T_0 + \Delta T = 20°C - 83.2\,C° = \underline{-63.2°C}$$

14-25

Call the metals A and B. Use the data given to calculate α for each metal.

$$\Delta L = L_0 \alpha \Delta T \Rightarrow \alpha = \frac{\Delta L}{L_0 \Delta T}$$

metal A: $\alpha_A = \dfrac{0.075\text{cm}}{(30.0\text{cm})(100°C - 0°C)} = 2.50\times10^{-5}(C°)^{-1}$

metal B: $\alpha_B = \dfrac{0.045\text{cm}}{(30.0\text{cm})(100°C - 0°C)} = 1.50\times10^{-5}(C°)^{-1}$

Now consider the third (composite) rod. Let L_A be the length of metal A in this rod.

←L_A→←30cm−L_A→

A	B

For $\Delta T = +100\,C°$, $\Delta L = 0.065\text{ cm}$

$\Delta L = \Delta L_A + \Delta L_B$

$\Delta L_A = L_A \alpha_A \Delta T$

$\Delta L_B = L_B \alpha_B \Delta T = (30\text{ cm} - L_A)\alpha_B \Delta T$

$\Rightarrow \Delta L = [L_A\alpha_A + (30\text{cm} - L_A)\alpha_B]\Delta T$

$L_A(\alpha_A - \alpha_B) = \dfrac{\Delta L}{\Delta T} - (30\text{cm})\alpha_B$

$$L_A = \frac{\Delta L/\Delta T - (30\text{cm})\alpha_B}{\alpha_A - \alpha_B} = \frac{(0.065\text{cm}/100C°) - (30\text{cm})(1.50\times10^{-5}(C°)^{-1})}{2.50\times10^{-5}(C°)^{-1} - 1.50\times10^{-5}(C°)^{-1}} = \underline{20\text{cm}}$$

Then $L_B = \underline{10\text{ cm}}$.

14-27

$V_0 = 1000\ cm^3$ for the mercury and for the glass.
It must be that $\Delta V_{Hg} - \Delta V_{flask} = 15.2\ cm^3$.

mercury: $\Delta V_{Hg} = V_0 \beta_{Hg} \Delta T = (1\times10^3 cm^3)(18\times10^{-5}(C°)^{-1})(100\ C°) = 18.0\ cm^3$

glass: $\Delta V_{flask} = V_0 \beta_{glass} \Delta T = (1\times10^3 cm^3)\ \beta_{glass}\ (100C°)$

$\Delta V_{Hg} - \Delta V_{flask} = 15.2\ cm^3 \Rightarrow \Delta V_{flask} = \Delta V_{Hg} - 15.2\ cm^3 = 18.0\ cm^3 - 15.2\ cm^3 = 2.8\ cm^3$

$\beta_{glass} = \frac{\Delta V_{flask}}{(1\times10^3 cm^3)(100C°)} = \frac{2.8\ cm^3}{(1\times10^3 cm^3)(100C°)} = \underline{2.8\times10^{-5}\ (C°)^{-1}}$

14-31

Let V_0 be the initial volume of the water, and steel bomb.

The water temperature changes from 10°C to 75°C. With the density of water at these temperatures from Table 14-3, the change in volume of m grams of water is
$\Delta V = m\,(1.0258 - 1.0003)\ cm^3 \cdot g^{-1} = m\,(0.0255)\ cm^3 \cdot g^{-1}$

The initial volume of water, m grams at 10°C, is
$V_0 = m\,(1.0003)\ cm^3 \cdot g^{-1}$

Thus for the water $\frac{\Delta V}{V_0} = \frac{m(0.0255)\ cm^3 \cdot g^{-1}}{m\,(1.0003)\ cm^3 \cdot g^{-1}} = 0.0255$

But the thermal expansion of the steel causes its volume to increase:
$\frac{\Delta V}{V_0} = \beta \Delta T = (3.6\times10^{-5}(C°)^{-1})(65C°) = 0.00234$

The net $\frac{\Delta V}{V_0}$ is thus $0.0255 - 0.00234 = 0.0232$.

This $\frac{\Delta V}{V_0}$ increase corresponds to a thermal stress; the pressure must increase enough to make $\frac{\Delta V}{V_0} = -\frac{\Delta p}{B} = -0.0232$.

(Note: rigid bomb ⇒ B for the steel is much larger than B for the water: $B_{steel} = 1.6\times10^{11}\ Pa$ (Table 12-1) and $B_{water} = \frac{1}{k} = \frac{1}{45.8\times10^{-11}\ Pa^{-1}} = 2.18\times10^9\ Pa$ (Table 12-2), so this is in fact the case.)

$\frac{\Delta p}{B} = 0.0232 \Rightarrow \Delta p = (0.0232)(2.18\times10^9\ Pa) = 5.06\times10^7\ Pa$

$p = p_0 + \Delta p = 1.01\times10^5\ Pa + 5.06\times10^7\ Pa = \underline{5.07\times10^7\ Pa}$

CHAPTER 15

Exercises 1, 3, 9, 13, 15, 17, 21

Problems 25, 27, 31, 35

15-1

energy conservation ⇒ heat generated = kinetic energy decrease

$$Q = \tfrac{1}{2} m v^2 = \tfrac{1}{2}(1500\,\text{kg})(5\,\text{m}\cdot\text{s}^{-1})^2 = \underline{1.88\times10^4\,\text{J}}$$

15-3

a) $Q = mc(T_2 - T_1) = (0.20\,\text{kg})(4190\,\text{J}\cdot\text{kg}^{-1}\cdot(C^\circ)^{-1})(30^\circ C - 20^\circ C) = \underline{8380\,\text{J}}$

b) $Q = mc(T_2 - T_1) \Rightarrow T_2 - T_1 = \dfrac{Q}{mc} = \dfrac{8380\,\text{J}}{(0.20\,\text{kg})(138\,\text{J}\cdot\text{kg}^{-1}\cdot(C^\circ)^{-1})} = 304\,C^\circ$

$$T_2 = T_1 + 304\,C^\circ = 20^\circ C + 304\,C^\circ = \underline{324^\circ C}$$

($c_{Hg} = 138\,\text{J}\cdot\text{kg}^{-1}\cdot(C^\circ)^{-1}$ was obtained from Table 15-1.)

c) $Q = mc(T_2 - T_1)$

We must find the mass of mercury, that has the same volume as 0.20 kg of water.

$$V = \frac{m}{\rho} \Rightarrow \frac{m_{H_2O}}{\rho_{H_2O}} = \frac{m_{Hg}}{\rho_{Hg}} \Rightarrow m_{Hg} = \left(\frac{\rho_{Hg}}{\rho_{H_2O}}\right) m_{H_2O} = \left(\frac{13.6\times10^3\,\text{kg}\cdot\text{m}^{-3}}{1.0\times10^3\,\text{kg}\cdot\text{m}^{-3}}\right)(0.20\,\text{kg})$$

$$m_{Hg} = 2.72\,\text{kg}$$

$$T_2 - T_1 = \frac{Q}{mc} = \frac{8380\,\text{J}}{(2.72\,\text{kg})(138\,\text{J}\cdot\text{kg}^{-1}\cdot(C^\circ)^{-1})} = 22.3\,C^\circ$$

$$T_2 = T_1 + 22.3\,C^\circ = 20^\circ C + 22.3\,C^\circ = \underline{42.3^\circ C}$$

15-9

Heat must be added to do the following:

ice at $-10^\circ C$ to ice at $0^\circ C$: $Q_{ice} = mc_{ice}\,\Delta T = (1.0\times10^{-3}\,\text{kg})(2.0\times10^3\,\text{J}\cdot\text{kg}^{-1}\cdot(C^\circ)^{-1})(0^\circ C - (-10^\circ C))$

$$Q_{ice} = 20.0\,\text{J}$$

phase transition ice → liquid water: $Q_{ice\to liq.} = +mL_F = (1.0\times10^{-3}\,\text{kg})(334\times10^3\,\text{J}\cdot\text{kg}^{-1})$

$$Q_{ice\to liq.} = 334\,\text{J}$$

water at $0^\circ C$ (from melted ice) → water at $100^\circ C$: $Q_{water} = mc_{water}\,\Delta T$

$$Q_{water} = (1.0\times10^{-3}\,\text{kg})(4190\,\text{J}\cdot\text{kg}^{-1}\cdot(C^\circ)^{-1})(100^\circ C - 0^\circ C) = 419\,\text{J}$$

15-9 (cont)

phase transition liquid → gas (boil the water): $Q_{liq \to gas} = +mL_V$

$Q_{liq \to gas} = (1.0 \times 10^{-3}\ kg)(2256 \times 10^3\ J \cdot kg^{-1}) = 2256\ J$

The total Q is $20\ J + 334\ J + 419\ J + 2256\ J = \underline{3029\ J}$

$3029\ J \left(\frac{1\ cal}{4.186\ J}\right) = \underline{724\ cal}$

$3029\ J \left(\frac{1\ Btu}{1055\ J}\right) = \underline{2.87\ Btu}$

15-13

$Q = \frac{1}{2} m v^2$

Calculate the heat that must be added to a lead bullet of mass m and initial temperature 25°C to melt it:

$Q = \underbrace{mc\Delta T}_{\text{raise temperature to melting point}} + \underbrace{mL_F}_{\text{make phase transition}}$

Using data from Tables 15-1 and 15-2,

$Q = m\left[(130\ J \cdot kg^{-1} \cdot (C^\circ)^{-1})(327.3^\circ C - 25^\circ C) + 24.5 \times 10^3\ J \cdot kg^{-1}\right] = m\left[6.38 \times 10^4\ J \cdot kg^{-1}\right]$

Then $Q = \frac{1}{2} m v^2 \Rightarrow v = \sqrt{\frac{2Q}{m}} = \sqrt{\frac{2m[6.38 \times 10^4\ J \cdot kg^{-1}]}{m}} = \underline{357\ m \cdot s^{-1}}$

(Note: $1\ J \cdot kg^{-1} = 1\ N \cdot m \cdot kg^{-1} = 1\ kg \cdot m \cdot s^{-2} \cdot m \cdot kg^{-1} = 1\ m^2 \cdot s^{-2}$, so the expression used does give the correct mks units for v.)

15-15

$Q = mL_F$

$1\ ton = 2000\ lb \Rightarrow m = \frac{w}{g} = \frac{2000\ lb}{32\ ft \cdot s^{-2}} = 62.5\ slugs$

$m = 62.5\ slugs \left(\frac{1\ kg}{0.0685\ slugs}\right) = 912.4\ kg$ (The slug → kg conversion factor is from Appendix E.)

Then $Q = mL_F = (912.4\ kg)(334 \times 10^3\ J \cdot kg^{-1}) = 3.05 \times 10^8\ J.$

A one-ton air conditioner freezes one ton of ice in 24 hr

$\Rightarrow P = \frac{Q}{t} = \frac{3.05 \times 10^8\ J}{(24\ hr)(3600\ s/1\ hr)} = \underline{3.53 \times 10^3\ W}$

$3.53 \times 10^3\ W \left(\frac{1\ Btu \cdot hr^{-1}}{0.293\ W}\right) = \underline{1.20 \times 10^4\ Btu \cdot hr^{-1}}$ (using Appendix E)

15-17

$Q_{system} = 0$

Calculate Q for each component of the system:

aluminum can

$$Q_{can} = mc_{Al}\Delta T = (0.500\,kg)(910\,J\cdot kg^{-1}\cdot(C^\circ)^{-1})(T_2 - 20^\circ C) = (455\,J\cdot(C^\circ)^{-1})(T_2 - 20^\circ C)$$

$$Q_{can} = (455\,J\cdot(C^\circ)^{-1})T_2 - 9100\,J$$

water

$$Q_{H_2O} = mc_{H_2O}\Delta T = (0.118\,kg)(4190\,J\cdot kg^{-1}\cdot(C^\circ)^{-1})(T_2 - 20^\circ C) = (494\,J\cdot(C^\circ)^{-1})(T_2 - 20^\circ C)$$

$$Q_{H_2O} = (494\,J\cdot(C^\circ)^{-1})T_2 - 9880\,J$$

iron

$$Q_{Fe} = mc_{Fe}\Delta T = (0.200\,kg)(470\,J\cdot kg^{-1}\cdot(C^\circ)^{-1})(T_2 - 75^\circ C) = (94.0\,J\cdot(C^\circ)^{-1})(T_2 - 75^\circ C)$$

$$Q_{Fe} = (94\,J\cdot(C^\circ)^{-1})T_2 - 7050\,J$$

$$Q_{system} = 0 \Rightarrow Q_{can} + Q_{H_2O} + Q_{Fe} = 0$$

$$(455\,J\cdot(C^\circ)^{-1})T_2 - 9100\,J + (494\,J\cdot(C^\circ)^{-1})T_2 - 9880\,J + (94\,J\cdot(C^\circ)^{-1})T_2 - 7050\,J = 0$$

$$(1043\,J\cdot(C^\circ)^{-1})T_2 - 2.60\times10^4\,J = 0$$

$$T_2 = \underline{24.9^\circ C}$$

(Note: Q_{can} and Q_{H_2O} are positive; Q_{Fe} is negative.)

15-21

$Q_{system} = 0$

Calculate Q for each component of the system. (Beaker has small mass $\Rightarrow Q = mc\Delta T$ for the beaker can be neglected.)

0.500 kg of water

$$Q_{water} = mc\Delta T = (0.500\,kg)(4190\,J\cdot kg^{-1}\cdot(C^\circ)^{-1})(50^\circ C - 80^\circ C) = -6.28\times10^4\,J$$

m grams of ice, that becomes m grams of water at 50°C

$$Q_{ice} = mc_{ice}(0^\circ C - (-20^\circ C)) + mL_F + mc_{water}(50^\circ C - 0^\circ C)$$

$$Q_{ice} = m[(2.0\times10^3\,J\cdot kg^{-1}\cdot(C^\circ)^{-1})(20C^\circ) + 334\times10^3\,J\cdot kg^{-1} + (4190\,J\cdot kg^{-1}\cdot(C^\circ)^{-1})(50C^\circ)]$$

$$Q_{ice} = m[5.835\times10^5\,J\cdot kg^{-1}]$$

$$Q_{system} = 0 \Rightarrow Q_{water} + Q_{ice} = 0$$

$$-6.28\times10^4\,J + m[5.835\times10^5\,J\cdot kg^{-1}] = 0$$

$$m = 0.108\,kg = \underline{108\,grams}$$

Problems

15-25

Let m_{tot} be the total initial mass of the piece of ice.

The mass m_{melt} that melts is $m_{melt} = (0.5\times10^{-2})\, m_{tot}$.
The heat required to do this is

$$Q = m_{melt} L_F = (0.5\times10^{-2})\, m_{tot}\,(334\times10^{3}\ \mathrm{J\cdot kg^{-1}})$$

By conservation of energy $Q = U_{grav} = m_{tot}\, g h$, where h is the height from which the ice falls.
(mgh could be larger than Q, if some of this initial gravitational potential energy were converted to some form of energy other than heat added to the ice, but mgh can't be less than Q.)

$$Q = m_{tot}\, g h \Rightarrow (0.5\times10^{-2})\, \cancel{m_{tot}}\,(334\times10^{3}\ \mathrm{J\cdot kg^{-1}}) = \cancel{m_{tot}}\, g h$$

$$h = \frac{(0.5\times10^{-2})(334\times10^{3}\ \mathrm{J\cdot kg^{-1}})}{9.8\ \mathrm{m\cdot s^{-2}}} = \underline{170\ \mathrm{m}}$$

15-27

a) Eq. (15-2) $\Rightarrow c = \frac{1}{m}\frac{dQ}{dT}$

But the problem gives the molar heat capacity $C = Mc$.

$$C = \frac{M}{m}\frac{dQ}{dT} = \frac{1}{n}\frac{dQ}{dT} \quad \text{(n is the number of moles)}$$

$$\Rightarrow dQ = n\, C\, dT$$

$$Q = n\int_{T_1}^{T_2} C\, dT = n\int_{T_1}^{T_2} k\frac{T^3}{\theta^3}\, dT = \frac{nk}{\theta^3}\left(\frac{1}{4}T^4\Big/_{T_1}^{T_2}\right) = \frac{nk}{4\theta^3}(T_2^4 - T_1^4)$$

$$Q = \frac{(2\ \mathrm{mol})(1940\ \mathrm{J\cdot mol^{-1}\cdot K^{-1}})}{4\,(281\ \mathrm{K})^3}\left((50\mathrm{K})^4 - (10\mathrm{K})^4\right) = \underline{273\ \mathrm{J}}$$

b) The average molar heat capacity C_{av} is defined by

$$Q = n\, C_{av}\, \Delta T \Rightarrow C_{av} = \frac{Q}{n\Delta T} = \frac{273\ \mathrm{J}}{2\ \mathrm{mol}\,(50\mathrm{K} - 10\mathrm{K})} = \underline{3.41\ \mathrm{J\cdot mol^{-1}\cdot K^{-1}}}$$

c) $C = k\frac{T^3}{\theta^3}$

$$T = 50\mathrm{K} \Rightarrow C = (1940\ \mathrm{J\cdot mol^{-1}\cdot K^{-1}})\left(\frac{50\mathrm{K}}{281\mathrm{K}}\right)^3 = \underline{10.9\ \mathrm{J\cdot mol^{-1}\cdot K^{-1}}}$$

15-31

a) Calculate the mass of water required, and then the volume of water from this:

$$Q = mc\Delta T \Rightarrow m = \frac{Q}{c\Delta T} = \frac{4.2\times10^{9}\ \mathrm{J}}{(4190\ \mathrm{J\cdot kg^{-1}\cdot (C^\circ)^{-1}})(49^\circ\mathrm{C} - 27^\circ\mathrm{C})} = 4.56\times10^{4}\ \mathrm{kg}$$

15-31 (cont)

$$\rho = \frac{m}{V} \Rightarrow V = \frac{m}{\rho} = \frac{4.56 \times 10^4\,kg}{1.0 \times 10^3\,kg \cdot m^{-3}} = \underline{45.6\,m^3}$$

b) Repeat the above calculation, but now with Glauber salt in place of water. The essential difference is that Glauber salt undergoes a phase transition in this temperature range.

$$Q = m c_{solid}(32°C - 27°C) + m L_F + m c_{liq}(49°C - 32°C)$$

$$Q = m\left[(1930\,J \cdot kg^{-1} \cdot (C°)^{-1})(5\,C°) + 2.42 \times 10^5\,J \cdot kg^{-1} + (2850\,J \cdot kg^{-1} \cdot (C°)^{-1})(17\,C°)\right]$$

$$Q = m\left[3.00 \times 10^5\,J \cdot kg^{-1}\right]$$

$$m = \frac{Q}{3.00 \times 10^5\,J \cdot kg^{-1}} = \frac{4.2 \times 10^9\,J}{3.00 \times 10^5\,J \cdot kg^{-1}} = 1.40 \times 10^4\,kg$$

$$V = \frac{m}{\rho} = \frac{1.40 \times 10^4\,kg}{1.6 \times 10^3\,kg \cdot m^{-3}} = \underline{8.75\,m^3}$$

15-35

First try to determine what phases will be present after equilibrium has been reached:

heat required to melt all the ice: $Q = mL_F = (0.05\,kg)(334 \times 10^3\,J \cdot kg^{-1}) = 1.67 \times 10^4\,J$

heat required to heat water at 0°C from melted ice to 100°C:

$$Q = mc\,\Delta T = (0.05\,kg)(4190\,J \cdot kg^{-1} \cdot (C°)^{-1})(100\,C°) = 2.09 \times 10^4\,J$$

heat liberated if all the steam condenses:

$$Q = mL_V = (0.012\,kg)(2256 \times 10^3\,J \cdot kg^{-1}) = 2.71 \times 10^4\,J$$

$1.67 \times 10^4\,J < 2.71 \times 10^4\,J \Rightarrow$ all the ice will melt

$(1.67 \times 10^4\,J + 2.09 \times 10^4\,J) > 2.71 \times 10^4\,J \Rightarrow$ all the steam will condense

thus at equilibrium the only phase will be liquid water, and the final temperature of the system will be between 0°C and 100°C.

$$Q_{system} = 0$$

for the can

$$Q_{can} = m c_{Cu}\,\Delta T = (0.322\,kg)(390\,J \cdot kg^{-1} \cdot (C°)^{-1})(T_2 - 0°C) = [125.6\,J \cdot (C°)^{-1}]\,T_2$$

for the ice

$$Q_{ice} = mL_F + m c_{water}\,\Delta T = (0.05\,kg)(334 \times 10^3\,J \cdot kg^{-1}) + (0.05\,kg)(4190\,J \cdot kg^{-1} \cdot (C°)^{-1}) \cdot (T_2 - 0°C)$$

15-35 (cont)

$$Q_{ice} = 1.67\times10^4\,J + (209.5\,J\cdot(C^\circ)^{-1})T_2$$

for the steam

$$Q_{steam} = -mL_v + mc_{water}\Delta T$$

$$Q_{steam} = -(0.012\,kg)(2256\times10^3\,J\cdot kg^{-1}) + (0.012\,kg)(4190\,J\cdot kg^{-1}\cdot(C^\circ)^{-1})(T_2 - 100^\circ C)$$

$$Q_{steam} = -2.71\times10^4\,J + (50.3\,J\cdot(C^\circ)^{-1})T_2 - 5.03\times10^3\,J$$

$$Q_{steam} = (50.3\,J\cdot(C^\circ)^{-1})T_2 - 3.21\times10^4\,J$$

$$Q_{system} = 0 \Rightarrow Q_{can} + Q_{ice} + Q_{steam} = 0$$

$$[125.6\,J\cdot(C^\circ)^{-1}]T_2 + 1.67\times10^4\,J + [209.5\,J\cdot(C^\circ)^{-1}]T_2 + [50.3\,J\cdot(C^\circ)^{-1}]T_2 - 3.21\times10^4\,J = 0$$

$$[385.4\,J\cdot(C^\circ)^{-1}]T_2 = 1.54\times10^4\,J \Rightarrow T_2 = \underline{40.0^\circ C}$$

CHAPTER 16

Exercises 3, 7, 11, 13, 15, 17

Problems 23, 25, 29, 31

Exercises

16-3

a)

T₁ = −10°C | wood | styrofoam | T₂ = 20°C ; T at the interface; 3cm, 3cm

heat current through the wood:

$$H_w = k_w \frac{A(T-T_1)}{L_w}$$

heat current through the styrofoam:

$$H_s = k_s \frac{A(T_2-T)}{L_s}$$

$$H_w = H_s \Rightarrow k_w \frac{A(T-T_1)}{L_w} = k_s \frac{A(T_2-T)}{L_s} \qquad (L_s = L_w = 3\,cm)$$

$$k_w(T-T_1) = k_s(T_2-T)$$

$$(k_s+k_w)T = k_s T_2 + k_w T_1$$

$$T = \frac{k_s T_2 + k_w T_1}{k_s+k_w} = \left(\frac{k_s}{k_s+k_w}\right)T_2 + \left(\frac{k_w}{k_s+k_w}\right)T_1$$

(Note: The units for k_w and k_s will cancel, if the same units are used for both.)

From Table 16-1, $k_s = 2\times10^{-5}\,cal\cdot s^{-1}\cdot cm^{-1}\cdot(C°)^{-1}$

$$\Rightarrow T = \left(\frac{2\times10^{-5}}{2\times10^{-5}+9.5\times10^{-5}}\right)(20°C) + \left(\frac{9.5\times10^{-5}}{2\times10^{-5}+9.5\times10^{-5}}\right)(-10°C)$$

$$T = 3.48°C - 8.26°C = \underline{-4.78°C}$$

b) $$H_w = k_w \frac{A(T-T_1)}{L_w} = \left[9.5\times10^{-5}\,cal\cdot s^{-1}\cdot cm^{-1}\cdot(C°)^{-1}\right]\frac{(1\times10^4\,cm^2)(-4.78°C-(-10°C))}{3\,cm}$$

$$H_w = 1.65\,cal\cdot s^{-1} = 1.65\,cal\cdot s^{-1}\left(\frac{4.186\,J}{1\,cal}\right) = \underline{6.91\,J\cdot s^{-1}}, \text{ for } A = 1\,m^2$$

or

$$H_s = k_s \frac{A(T_2-T)}{L_s} = \left[2\times10^{-5}\,cal\cdot s^{-1}\cdot cm^{-1}\cdot(C°)^{-1}\right]\frac{(1\times10^4\,cm^2)(20°C-(-4.78°C))}{3\,cm}$$

$H_s = 1.65\,cal\cdot s^{-1}$, which equals H_w, as it should.

16-7

a) $$\text{temperature gradient} = \frac{T_2-T_1}{L} = \frac{100°C-0°C}{0.10\,m} = \underline{1000\,C°\cdot m^{-1}}$$

b) $$H = k\frac{A(T_2-T_1)}{L} = \left[385\,J\cdot s^{-1}\cdot m^{-1}\cdot(C°)^{-1}\right]\frac{(1\times10^{-4}\,m^2)(100°C-0°C)}{0.10\,m} = \underline{38.5\,J\cdot s^{-1}}$$

16-7 (cont)

c) $H = 38.5\ J\cdot s^{-1}$ for all sections of the rod

T_2 [2cm, T] T_1

$$H = k\frac{A(T_2 - T)}{L} \Rightarrow T_2 - T = \frac{LH}{kA}$$

$$T = T_2 - \frac{LH}{kA}$$

$$T = 100°C - \frac{(0.02m)(38.5\ J\cdot s^{-1})}{(385\ J\cdot s^{-1}\cdot m^{-1}\cdot (C°)^{-1})(1\times 10^{-4} m^2)} = 100°C - 20C° = \underline{80°C}$$

16-11

From Table 16-2, for a vertical pipe

$h = 1.32\left(\frac{\Delta T}{D}\right)^{1/4}\ J\cdot s^{-1} m^{-2}\cdot (C°)^{-1}$, D in meters

$D = 7.5\times 10^{-2} m$, $\Delta T = 95°C - 20°C = 75°C$

$$\Rightarrow h = 1.32\left(\frac{75}{7.5\times 10^{-2}}\right)^{1/4} J\cdot s^{-1}\cdot m^{-2}\cdot (C°)^{-1} = 7.42\ J\cdot s^{-1}\cdot m^{-2}\cdot (C°)^{-1}$$

$H = hA\Delta T$

A = surface area of pipe $= \pi DL = \pi(7.5\times 10^{-2} m)(4m) = 0.942 m^2$

$H = [7.42\ J\cdot s^{-1}\cdot m^{-2}\cdot (C°)^{-1}][0.942 m^2][75C°] = 524\ J\cdot s^{-1}$

$H = \frac{Q}{t} \Rightarrow Q = Ht = (524\ J\cdot s^{-1})(3600 s) = \underline{1.89\times 10^6 J}$

[D, L]

16-13

$$H_{net} = Ae\sigma(T_1^4 - T_2^4)$$

$$H_{net} = (1.2m^2)(1)(5.67\times 10^{-8} W\cdot m^{-2}\cdot K^{-4})[(303K)^4 - (273K)^4] = \underline{196W}$$

This is larger than the result in Example 16-7. The lower surrounding temperature increases the rate of heat loss by radiation.

16-15

$$H = Ae\sigma T^4 \Rightarrow A = \frac{H}{e\sigma T^4}$$

25-W lamp $\Rightarrow H = 25W$

$$A = \frac{25W}{(0.30)(5.67\times 10^{-8} W\cdot m^{-2}\cdot K^{-4})(2450K)^4} = 4.08\times 10^{-5} m^2 = \underline{0.408\ cm^2}$$

16-17

$H = 20kW$

The heat energy for a week is $Q = Ht = (20\times 10^3 W)(7da)\left(\frac{24hr}{1da}\right)\left(\frac{3600s}{1hr}\right)$

$Q = 1.21\times 10^{10} J$

16-17 (cont)

mass of water required:

$$Q = mc\,\Delta T \Rightarrow m = \frac{Q}{c\Delta T} = \frac{1.21 \times 10^{10}\,\text{J}}{(4190\,\text{J}\cdot\text{kg}^{-1}\cdot(\text{C}°)^{-1})(80°\text{C} - 40°\text{C})} = 7.22 \times 10^{4}\,\text{kg}$$

volume of water required:

$$\rho = \frac{m}{V} \Rightarrow V = \frac{m}{\rho} = \frac{7.22 \times 10^{4}\,\text{kg}}{1.0 \times 10^{3}\,\text{kg}\cdot\text{m}^{-3}} = \underline{72.2\,\text{m}^3}, \text{ volume of water.}$$

Problems

16-23

heat current through the walls of the icebox:

$$H = k\frac{A\Delta T}{L} = (0.05\,\text{J}\cdot\text{s}^{-1}\cdot\text{m}^{-1}\cdot(\text{C}°)^{-1})\frac{(2\,\text{m}^2)(20°\text{C} - 5°\text{C})}{0.05\,\text{m}} = 30\,\text{J}\cdot\text{s}^{-1}$$

The heat carried through the walls in 1 hr is thus

$$Q = Ht = (30\,\text{J}\cdot\text{s}^{-1})(3600\,\text{s}) = 1.08 \times 10^{5}\,\text{J}$$

This heat melts ice; the ice at 0°C is converted to water at 5°C:

$$Q = mL_F + mc_{water}\Delta T$$

$$\Rightarrow mL_F + mc_{water}\Delta T = 1.08 \times 10^{5}\,\text{J}$$

$$m = \frac{1.08 \times 10^{5}\,\text{J}}{334 \times 10^{3}\,\text{J}\cdot\text{kg}^{-1} + (4190\,\text{J}\cdot\text{kg}^{-1}\cdot(\text{C}°)^{-1})(5\,\text{C}°)} = 0.304\,\text{kg}$$

cost of this much ice is (0.304)(25¢) = 7.6¢, which is cost per hour

16-25

$T_2 =$ 100°C	aluminum (Al)	steel (s)	copper (cu)	$T_1 =$ 0°C
	L	L	L	

] A

$T_{Al,s}$ (Al–steel junction), $T_{s,cu}$ (steel–copper junction)

Treat the rod as a whole, to calculate the heat current H:

$$H = \frac{A(T_2 - T_1)}{R}$$

From problem 16-22, $R = R_{Al} + R_s + R_{cu}$

$$R = \frac{L}{k} \Rightarrow R = L\left(\frac{1}{k_{Al}} + \frac{1}{k_s} + \frac{1}{k_{cu}}\right) = L\left(\frac{1}{205} + \frac{1}{50.2} + \frac{1}{385}\right)\text{J}^{-1}\cdot\text{s}\cdot\text{m}\cdot\text{C}°$$

$$R = (0.0274\,\text{J}^{-1}\cdot\text{s}\cdot\text{m}\cdot\text{C}°)L$$

$$H = \frac{A(100°\text{C} - 0°\text{C})}{L(0.0274\,\text{J}^{-1}\cdot\text{s}\cdot\text{m}\cdot\text{C}°)} = \frac{A}{L}(3650\,\text{J}\cdot\text{s}^{-1}\cdot\text{m}^{-1})$$

16-25 (cont)

Any segment of the rod has this same heat current.

copper segment:

$$H_{Cu} = \frac{k_{Cu} A(T_{s,Cu} - 0°C)}{L} = \frac{A}{L}(3650 \, J \cdot s^{-1} \cdot m^{-1})$$

$$T_{s,Cu} = 0°C + \frac{3650 \, J \cdot s^{-1} \cdot m^{-1}}{385 \, J \cdot s^{-1} \cdot m^{-1} \cdot (C°)^{-1}} = \underline{9.48°C}$$ temperature at the steel-copper junction

aluminum segment:

$$H_{Al} = \frac{k_{Al} A(100°C - T_{Al,s})}{L} = \frac{A}{L}(3650 \, J \cdot s^{-1} \cdot m^{-1})$$

$$T_{Al,s} = 100°C - \frac{3650 \, J \cdot s^{-1} \cdot m^{-1}}{205 \, J \cdot s^{-1} \cdot m^{-1} \cdot (C°)^{-1}} = \underline{82.2°C}$$ temperature of the aluminum-steel junction

As a check use the above two junction temperatures to calculate the heat current through the steel segment:

$$H_s = \frac{k_s A(T_{Al,s} - T_{s,Cu})}{L} = \frac{A}{L}(50.2 \, J \cdot s^{-1} \cdot m^{-1} \cdot (C°)^{-1})(82.2°C - 9.5°C)$$

$$H_s = \frac{A}{L}(3650 \, J \cdot s^{-1} \cdot m^{-1}),$$ as it should.

16-29

$A = \pi r^2 = \pi(0.30 \, m)^2 = 0.283 \, m^2$

$A = 2\pi r h = 2\pi(0.30 \, m)(1.0 \, m) = 1.88 \, m^2$

$A = 0.283 \, m^2$

$H = hA\Delta T$ for each part of the tank surface. Obtain h from Table 16-2.

top: horizontal plate facing upward

$h = 2.49(\Delta T)^{1/4} \, J \cdot s^{-1} \cdot m^{-2} \cdot (C°)^{-1}$, with ΔT in °C

$$H_t = [2.49(\Delta T)^{1/4} \, J \cdot s^{-1} \cdot m^{-2} \cdot (C°)^{-1}](0.283 \, m^2)\Delta T = (0.705(\Delta T)^{5/4}) \, J \cdot s^{-1}$$

bottom: horizontal plate facing downward

$h = 1.31(\Delta T)^{1/4} \, J \cdot s^{-1} \cdot m^{-2} \cdot (C°)^{-1}$, with ΔT in °C

$$H_b = [1.31(\Delta T)^{1/4} \, J \cdot s^{-1} \cdot m^{-2} \cdot (C°)^{-1}](0.283 \, m^2)\Delta T = (0.371(\Delta T)^{5/4}) \, J \cdot s^{-1}$$

16-29 (cont)

sides: vertical pipe (diameter $D = 0.60\,m$)

$$h = 1.32\left(\frac{\Delta T}{D}\right)^{1/4} J\cdot s^{-1}\cdot m^{-2}\cdot(C°)^{-1} = \frac{1.32}{(0.60)^{1/4}}(\Delta T)^{1/4} J\cdot s^{-1}\cdot m^{-2}\cdot(C°)^{-1}$$

$$h = 1.50(\Delta T)^{1/4} J\cdot s^{-1}\cdot m^{-2}\cdot(C°)^{-1}$$

$$H_s = [1.50(\Delta T)^{1/4} J\cdot s^{-1}\cdot m^{-2}\cdot(C°)^{-1}][1.88\,m^2]\,\Delta T = (2.82(\Delta T)^{5/4})\,J\cdot s^{-1}$$

The total H is $H_{tot} = H_t + H_b + H_s$.

$H_{tot} = 1\,kW = 1\times10^3\,J\cdot s^{-1}$

$$\Rightarrow [0.705 + 0.371 + 2.82]\,\Delta T^{5/4}\,J\cdot s^{-1} = 1\times10^3\,J\cdot s^{-1}$$

$$\Delta T^{5/4} = \frac{1\times10^3}{3.90} = 256\ ;\ \Delta T = (256)^{4/5} = 84.4\,K = \underline{84.4\,C°}$$

The tank surface will be this much hotter than the surrounding air.

16-31

$$H = Ae\sigma(T_2^4 - T_1^4)$$

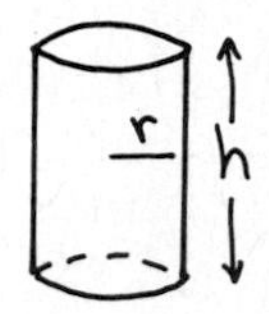

The surface area of the cylindrical can is

$A = 2\pi rh + 2\pi r^2 = 2\pi r(h+r)$

$A = 2\pi(0.025\,m)(0.10\,m + 0.025\,m) = 0.0196\,m^2$

$$H = (0.0196\,m^2)(0.2)(5.67\times10^{-8}\,W\cdot m^{-2}\cdot K^{-4})([80K]^4 - [4K]^4) = 9.10\times10^{-3}\,J\cdot s^{-1}$$

This is the net heat current due to radiation. Calculate the net amount of heat absorbed in 1 hr.

$$Q = Ht = (9.10\times10^{-3}\,J\cdot s^{-1})(3600\,s) = 32.8\,J$$

This heat causes the helium at 4K to undergo the liquid → gas phase transition.

$$Q = mL_v \Rightarrow m = \frac{Q}{L_v} = \frac{32.8\,J}{2.0\times10^4\,J\cdot kg^{-1}} = 1.64\times10^{-3}\,kg = \underline{1.64\,grams}$$

CHAPTER 17

Exercises 3, 7, 9, 13

Problems 17, 19, 21

Exercises

17-3

a) $n = \frac{m}{M} = \frac{0.2\,kg}{4.003 \times 10^{-3}\,kg \cdot mol^{-1}} = \underline{50.0\ moles}$

(M is from Appendix D; note that the atomic masses given there are in $g \cdot mol^{-1}$.)

b) $pV = nRT \Rightarrow p = \frac{nRT}{V} = \frac{(50\,mol)(8.314\,J \cdot mol^{-1} \cdot K^{-1})([27+273]\,K)}{20 \times 10^{-3}\,m^3} = \underline{6.24 \times 10^6\ Pa}$

(Note that T **must** be in kelvins, and that V = 20 L was converted to mks units (m^3).)

$p = (6.24 \times 10^6\,\cancel{Pa})\left(\frac{1\,atm}{1.013 \times 10^5\,\cancel{Pa}}\right) = \underline{61.6\ atm}$

17-7

$pV = nRT$

nR constant $\Rightarrow \frac{pV}{T} = nR = \text{constant}$

$\Rightarrow \frac{p_1 V_1}{T_1} = \frac{p_2 V_2}{T_2} \Rightarrow T_2 = T_1 \left(\frac{p_2}{p_1}\right)\left(\frac{V_2}{V_1}\right)$

$T_1 = 27 + 273 = 300\,K$

$p_1 = 1.01 \times 10^5\,Pa$ (air pressure)

$p_2 = 40 \times 10^5\,Pa + 1 \times 10^5\,Pa = 41 \times 10^5\,Pa$, since the final gauge pressure is $40 \times 10^5\,Pa$

$T_2 = 300\,K \left(\frac{41 \times 10^5\,Pa}{1.01 \times 10^5\,Pa}\right)\left(\frac{50\,cm^3}{800\,cm^3}\right) = 761\,K$

$T_2 = (761 - 273)\,^\circ C = \underline{488\,^\circ C}$

(Note that the units cancel in the $\frac{V_2}{V_1}$ volume ratio, so it was not necessary to convert the volumes given in cm^3 to m^3.)

17-9

a) $pV = nRT$

$n = \frac{m}{M} \Rightarrow pV = \frac{m}{M} RT$

$\rho = \frac{m}{V} = \frac{pM}{RT}$

17-9 (cont)

b)

$$p = 1\,\text{atm} = 1.01\times10^5\,\text{Pa}$$
$$M = 28.8\times10^{-3}\,\text{kg}\cdot\text{mol}^{-1}$$
$$T = 20^\circ\text{C} = 293\,\text{K}$$

$$\rho = \frac{(1.01\times10^5\,\text{Pa})(28.8\times10^{-3}\,\text{kg}\cdot\text{mol}^{-1})}{(8.314\,\text{J}\cdot\text{mol}^{-1}\cdot\text{K}^{-1})(293\,\text{K})} = \underline{1.19\,\text{kg}\cdot\text{m}^{-3}}$$

17-13

The phase diagram is as in Fig. 17-4.

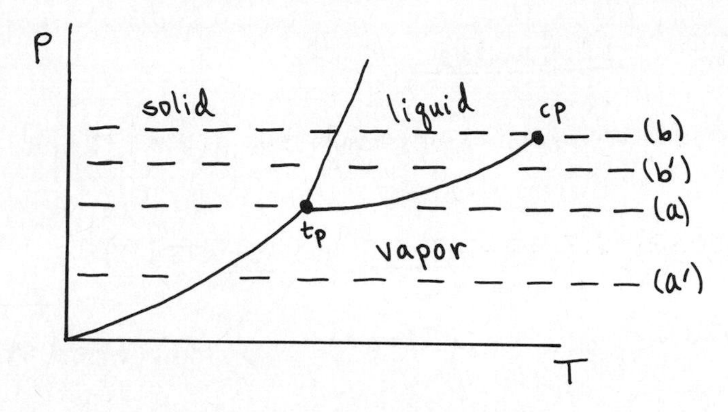

a) Increase T from a low value at constant pressure ⇒ move along a horizontal line in the phase diagram. To have a solid → liquid transition this line must be above the dashed line marked (a) on the phase diagram. This implies that the pressure must be above the triple point pressure p_{tp}. From Table 17-1, $p_{tp} = \underline{0.125\times10^5\,\text{Pa}}$.

For $p < p_{tp}$ the system moves along the dashed line marked (a′) as T is increased from a low value. The solid → vapor transition (sublimation) is observed.

b) To observe the liquid → vapor (boiling) transition the system must move along a line below the dashed line (b) as T is increased from a low value. This implies that the pressure must be below the critical point pressure p_{cp}. From Table 17-2, $p_{cp} = \underline{33.9\times10^5\,\text{Pa}}$.

For $p_1 < p < p_2 \Rightarrow p_{tp} < p < p_{cp} \Rightarrow 0.125\times10^5\,\text{Pa} < p < 33.9\times10^5\,\text{Pa}$, the system moves along the line (b′), and the solid → liquid (melting) and liquid → vapor (boiling) phase transitions are observed.

Problems

17-17

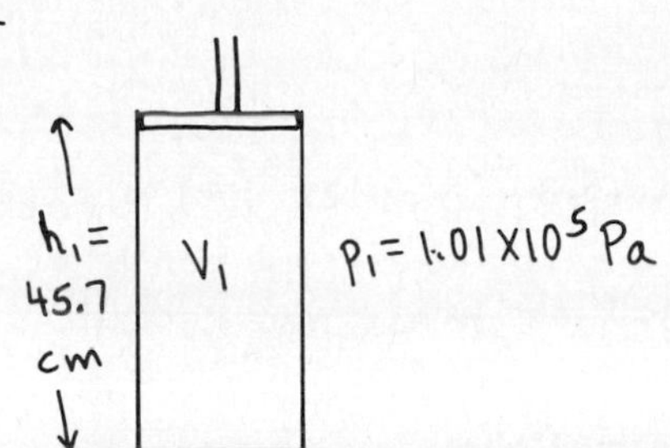

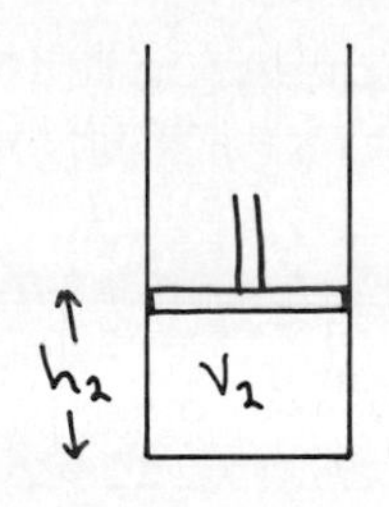

$p_2 = 2.76\times10^5\ \text{Pa} + 1.01\times10^5\ \text{Pa}$

$p_2 = 3.77\times10^5\ \text{Pa}$

$pV = nRT$

n, R, T constant $\Rightarrow p_1V_1 = p_2V_2$

$V = hA$, where A is the cross sectional area of the cylinder

$p_1h_1A = p_2h_2A$

$$h_2 = h_1\left(\frac{p_1}{p_2}\right) = 45.7\text{cm}\left(\frac{1.01\times10^5\ \text{Pa}}{3.77\times10^5\ \text{Pa}}\right) = \underline{12.2\ \text{cm}}$$

17-19

a) $pV = nRT$

Consider the cooling, after the stopcock is closed:

n, R, V constant $\Rightarrow \frac{p}{T} = \frac{nR}{V} =$ constant $\Rightarrow \frac{p_1}{T_1} = \frac{p_2}{T_2}$

At $T_1 = 400\text{K}$ the flask is open to the air $\Rightarrow p_1 = 1.01\times10^5$ Pa

$$p_2 = p_1\left(\frac{T_2}{T_1}\right) = 1.01\times10^5\text{Pa}\left(\frac{300\text{K}}{400\text{K}}\right) = \underline{7.58\times10^4\ \text{Pa}}$$

b) $pV = nRT$; apply to the flask at the end of the procedure:

$n = \frac{m}{M} \Rightarrow pV = \frac{m}{M}RT$

$$m = \frac{pVM}{RT} = \frac{(7.58\times10^4\text{Pa})(2\times10^{-3}\text{m}^3)(32.0\times10^{-3}\text{kg}\cdot\text{mol}^{-1})}{(8.314\ \text{J}\cdot\text{mol}^{-1}\cdot\text{K}^{-1})(300\text{K})} = 1.94\times10^{-3}\text{kg} = \underline{1.94\text{grams}}$$

Note: The atomic mass of oxygen is given in Appendix D as $15.999\text{g}\cdot\text{mol}^{-1}$. But oxygen gas is diatomic, and we need M in mks units

$\Rightarrow M = 2(15.999\times10^{-3}\text{kg}\cdot\text{mol}^{-1}) = 32.0\times10^{-3}\ \text{kg}\cdot\text{mol}^{-1}$.

17-21

a) Consider the volume occupied by the gas in one cylinder, after its pressure is changed from 15×10^5 Pa to 1.01×10^5 Pa.

$pV = nRT$

n, R, T constant $\Rightarrow pV$ constant $\Rightarrow p_1V_1 = p_2V_2$

$$V_2 = V_1\left(\frac{p_1}{p_2}\right) = (2.5\text{m}^3)\left(\frac{15\times10^5\ \text{Pa}}{1.01\times10^5\text{Pa}}\right) = 37.1\ \text{m}^3, \text{ from 1 cylinder}$$

17-21 (cont)

The number of cylinders required is thus

$$\frac{500\ m^3}{37.1 m^3\cdot cylinder^{-1}} = \underline{13.5\ cylinders}$$

b) This is a buoyancy (Archimedes' Principle, Chapter 13) problem.

Force diagram for the balloon plus a load of mass m_L:

$B = \rho_{air} V_{balloon}\, g$

$a = 0$

$m_{hyd} + m_L$

$m_L g$ $\quad m_{hyd}\, g$

(m_{hyd} is the mass of hydrogen gas in the balloon)

$$\Sigma F_y = ma_y$$

$$B - m_L g - m_{hyd}\, g = 0$$

$$m_{hyd} = \rho_{hyd} V_{balloon}\, g$$

$$\rho_{air} V_{balloon}\, g - m_L g - \rho_{hyd} V_{balloon}\, g = 0$$

$$m_L = (\rho_{air} - \rho_{hyd}) V_{balloon}$$

From Example 17-1 the volume occupied by 1 mole of gas at STP is $0.0224 m^3$. We can use this to calculate ρ_{air} and ρ_{hyd} at STP:

From Example 17-2, $M_{air} = 28.8\times10^{-3} kg\cdot mol^{-1}$

$$\Rightarrow \rho_{air} = \frac{28.8\times10^{-3}\ kg\cdot mol^{-1}}{0.0224\ m^3\cdot mol^{-1}} = 1.29\ kg\cdot m^{-3}$$

Hydrogen is diatomic and from Appendix D its atomic mass is $1.008\times10^{-3}\ kg\cdot mol^{-1}$

$$\Rightarrow M_{hyd} = 2(1.008\times10^{-3} kg\cdot mol^{-1}) = 2.02\times10^{-3} kg\cdot mol^{-1}$$

Thus $$\rho_{hyd} = \frac{2.02\times10^{-3}\ kg\cdot mol^{-1}}{0.0224\ m^3\cdot mol^{-1}} = 0.0902\ kg\cdot m^{-3}.$$

$$m_L = (1.29\ kg\cdot m^{-3} - 0.090\ kg\cdot m^{-3})(500 m^3) = 600 kg$$

$$w_L = m_L g = (600 kg)(9.8 m\cdot s^{-2}) = \underline{5880 N}$$

c) Helium is monatomic.

From Appendix D, $M_{He} = 4.003\times10^{-3} kg\cdot mol^{-1}$

$$\Rightarrow \rho_{He} = \frac{4.003\times10^{-3}\ kg\cdot mol^{-1}}{0.0224\ m^3\cdot mol^{-1}} = 0.179\ kg\cdot m^{-3}$$

$$m_L = (\rho_{air} - \rho_{He}) V_{balloon} = (1.29\ kg\cdot m^{-3} - 0.179\ kg\cdot m^{-3})(500\ m^3) = 556\ kg$$

$$w_L = m_L g = (556\ kg)(9.8\ m\cdot s^{-2}) = \underline{5450 N}$$

CHAPTER 18

Exercises 3, 5, 9, 11, 15, 17

Problems 19, 23, 25

Exercises

18-3

$W = \int_1^2 p\,dV$

$pV = nRT$; pressure constant $\Rightarrow p\,dV = nR\,dT$

$W = \int_{T_1}^{T_2} nR\,dT = nR(T_2 - T_1) = (3\,\text{mol})(8.314\,\text{J}\cdot\text{mol}^{-1}\cdot\text{K}^{-1})(400\text{K} - 300\text{K}) = \underline{2494\,\text{J}}$

18-5

a) $\Delta U = Q - W$

$Q = mc\,\Delta T$

$\Delta T > 0 \Rightarrow Q > 0 \Rightarrow$ heat energy has been added to the water

This heat comes from the burning fuel-oxygen mixture

$\Rightarrow Q$ for the system is negative.

b) $W = \int p\,dV$

constant volume $\Rightarrow W = 0$.

c) $\Delta U = Q - W$

$Q < 0,\ W = 0 \Rightarrow \Delta U < 0$

The internal energy of the fuel-oxygen mixture decreased.

18-9

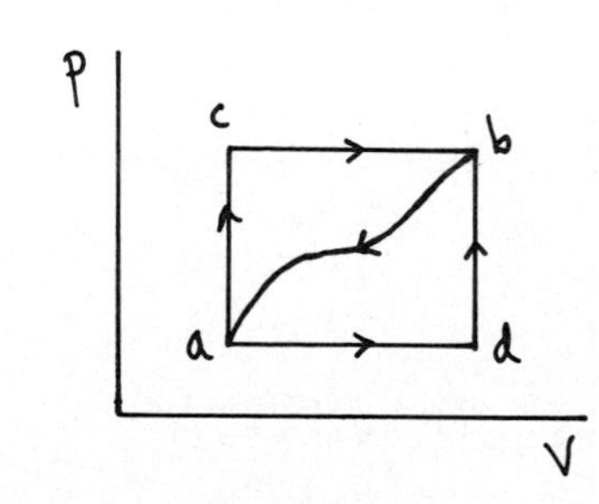

$Q_{acb} = +80\,\text{J}$

$W_{acb} = +30\,\text{J}\ \ (\Delta V > 0 \Rightarrow W > 0)$

a) $\Delta U = Q - W$

ΔU is path independent ; Q and W depend on the path

$\Delta U = U_b - U_a$

This can be calculated for any path from a to b, in particular for path acb:

$\Delta U_{a \to b} = Q_{acb} - W_{acb} = 80\,\text{J} - 30\,\text{J} = 50\,\text{J}$

18-9 (cont)

Now consider path adb: (ΔU is the same as for acb)

$\Delta U_{a\to b} = Q_{adb} - W_{adb} \Rightarrow Q_{adb} = \Delta U_{a\to b} + W_{adb}$

$W_{adb} = +10\,J$ ($\Delta V > 0$ for this path $\Rightarrow W > 0$)

Thus

$$Q_{adb} = 50\,J + 10\,J = \underline{60\,J}$$

b) for path ba:

$\Delta U_{b\to a} = Q_{ba} - W_{ba}$

$|W_{ba}| = 20\,J$. But for $b \to a$, $\Delta V < 0 \Rightarrow W < 0 \Rightarrow W_{ba} = -20\,J$

$\Delta U_{b\to a} = U_a - U_b = -(U_b - U_a) = -\Delta U_{a\to b} = -50\,J$

$Q_{ba} = \Delta U_{b\to a} + W_{ba} = -50\,J - 20\,J = \underline{-70\,J}$

$Q_{ba} < 0 \Rightarrow$ heat is liberated from (goes out of) the system

c) $U_a = 0$, $U_d = 40\,J$

process $a \to d$

$\Delta U_{a\to d} = U_d - U_a = 40\,J$

$W_{adb} = +10\,J$

$W_{adb} = W_{ad} + W_{db}$, but $W_{db} = 0$ since $\Delta V = 0$ for this process.

Thus $W_{ad} = +10\,J$.

$\Delta U_{a\to d} = Q_{ad} - W_{ad}$

$\Rightarrow Q_{ad} = \Delta U_{a\to b} + W_{ad} = 40\,J + 10\,J = \underline{+50\,J}$ ($+ \Rightarrow$ absorbed by the system)

process $d \to b$

$\Delta U_{d\to b} = Q_{db} - W_{db} \Rightarrow Q_{db} = \Delta U_{d\to b} + W_{db}$

$\Delta U_{d\to b} = U_b - U_d$

$\Delta U_{a\to b} = U_b - U_a = 50\,J$; $U_a = 0 \Rightarrow U_b = 50\,J$

Thus

$\Delta U_{d\to b} = 50\,J - 40\,J = +10\,J$

$W_{db} = 0$, since $\Delta V = 0$.

$Q_{db} = \Delta U_{d\to b} + W_{db} = 10\,J + 0 = +10\,J$ ($+ \Rightarrow$ absorbed by system)

18-11

a) ideal gas $\Rightarrow pV = nRT$

$W = \int_1^2 p\,dV$

T constant (isothermal) $\Rightarrow p = \frac{nRT}{V}$

$$W = \int_{V_1}^{V_2} \left(\frac{nRT}{V}\right) dV = nRT \int_{V_1}^{V_2} \frac{dV}{V} = nRT \ln\left(\frac{V_2}{V_1}\right)$$

$$W = (0.10\,\text{mol})(8.314\ \text{J}\cdot\text{mol}^{-1}\cdot\text{K}^{-1})(273\,\text{K}) \ln\left(\frac{\frac{1}{5}V_1}{V_1}\right) = \underline{-365\ \text{J}}$$

(since $\Delta V < 0$, $W < 0$ as we calculated)

b) $\Delta U = nC_V \Delta T$, for any ideal gas process

$\Delta T = 0 \Rightarrow \underline{\Delta U = 0}$.

c) $\Delta U = Q - W$

$\Delta U = 0 \Rightarrow Q = W = \underline{-365\ \text{J}}$

(negative $\Rightarrow$ the gas liberates 365 J of heat to the surroundings)

18-15

For an adiabatic process we have that

$T_1 V_1^{\gamma-1} = T_2 V_2^{\gamma-1}$, $p_1 V_1^{\gamma} = p_2 V_2^{\gamma}$, and $pV = nRT$

Air is mostly diatomic $(O_2, N_2, H_2) \Rightarrow \gamma = 1.4$

$$T_1 V_1^{\gamma-1} = T_2 V_2^{\gamma-1} \Rightarrow T_2 = T_1 \left(\frac{V_1}{V_2}\right)^{\gamma-1} = (293\,\text{K})\left(\frac{V_1}{\frac{1}{10}V_1}\right)^{1.4-1}$$

$$T_2 = (293\,\text{K})(10)^{0.4} = \underline{736\,\text{K}} = \underline{463^\circ\text{C}}$$

(Note: In the relation $T_1 V_1^{\gamma-1} = T_2 V_2^{\gamma-1}$ the temperature must be in kelvins.)

$$p_1 V_1^{\gamma} = p_2 V_2^{\gamma} \Rightarrow p_2 = p_1 \left(\frac{V_1}{V_2}\right)^{\gamma} = (1\,\text{atm})\left(\frac{V_1}{\frac{1}{10}V_1}\right)^{1.4} = (1\,\text{atm})(10)^{1.4} = \underline{25.1\,\text{atm}}$$

18-17

a) adiabatic $\Rightarrow Q = 0$

$\Delta U = Q - W \Rightarrow W = -\Delta U$

For any ideal gas process $\Delta U = nC_V \Delta T$.

From Table 18-1, for oxygen $C_V = 21.10\ \text{J}\cdot\text{mol}^{-1}\cdot\text{K}^{-1}$

$$\Rightarrow \Delta U = (0.10\,\text{mol})(21.10\ \text{J}\cdot\text{mol}^{-1}\cdot\text{K}^{-1})(283\,\text{K} - 303\,\text{K}) = -42.2\ \text{J}$$

Thus $W = -\Delta U = \underline{+42.2\ \text{J}}$.

(expansion $\Rightarrow \Delta V > 0 \Rightarrow W > 0$)

18-17 (cont)

b) adiabatic $\Rightarrow$ $\underline{Q = 0}$

Problems

18-19

$\Delta U = Q - W$

$Q = +2.11 \times 10^5 \text{ J}$

$\Delta U = 0 \Rightarrow Q = W$

constant pressure $\Rightarrow W = p\Delta V$

Thus $p\Delta V = Q$

$$\Delta V = \frac{Q}{p} = \frac{2.11 \times 10^5 \text{ J}}{6.89 \times 10^5 \text{ Pa}} = \underline{0.306 \text{ m}^3}$$

18-23

a) isothermal $\Rightarrow \Delta T = 0$

$\Delta U = nC_V \Delta T$ for any ideal gas process $\Rightarrow$ $\underline{\Delta U = 0}$

Thus $\Delta U = Q - W \Rightarrow Q = W = \underline{+500 \text{ J}}$

b) adiabatic $\Rightarrow$ $\underline{Q = 0}$

$\Delta U = Q - W = -W = -\underline{500 \text{ J}}$ ($\Delta U < 0 \Rightarrow$ the gas cools)

18-25

a) $pV = nRT$

constant pressure for $1 \to 2 \Rightarrow \frac{V}{T} = \frac{nR}{p}$ = constant

Thus $\frac{V_1}{T_1} = \frac{V_2}{T_2} \Rightarrow T_2 = T_1 \left(\frac{V_2}{V_1}\right)$

$T_1 = 300 \text{ K}$, $V_1 = 0.020 \text{ m}^3$, $V_2 = 0.040 \text{ m}^3$ $\Rightarrow T_2 = 300 \text{ K} \left(\frac{0.040 \text{ m}^3}{0.020 \text{ m}^3}\right) = 600 \text{ K}$

adiabatic process $2 \to 3$:

$T_2 = 600 \text{ K}$, $T_3 = 300 \text{ K}$

expansion $\Rightarrow \Delta V > 0$

18-25 (cont)

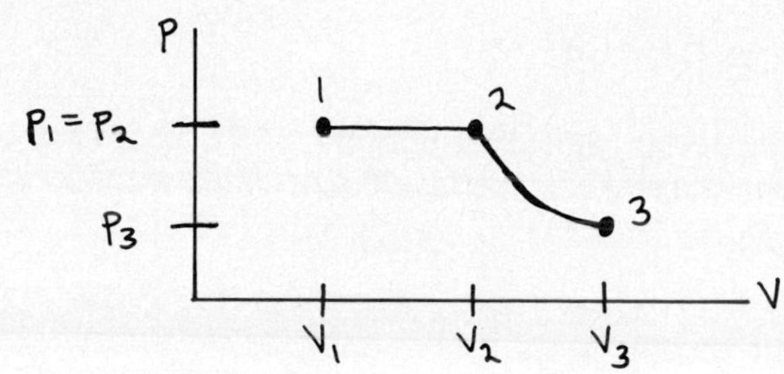

b) $Q_{12} = nC_p \Delta T$, since $\Delta p = 0$.

From Table 18-1, $C_p = 20.78 \text{ J·mol}^{-1}\text{·K}^{-1}$

$$Q_{12} = (2\text{ mol})(20.78 \text{ J·mol}^{-1}\text{·K}^{-1})(600\text{K} - 300\text{K}) = 12{,}468\text{ J}$$

$Q_{23} = 0$, since this process is adiabatic.

$$Q_{tot} = Q_{12} + Q_{23} = \underline{12{,}468\text{ J}}$$

c) $\Delta U_{12} = nC_v \Delta T = nC_v(T_2 - T_1)$

$\Delta U_{23} = nC_v \Delta T = nC_v(T_3 - T_2)$

$$\Rightarrow \Delta U_{tot} = \Delta U_{12} + \Delta U_{23} = nC_v(T_2 - T_1 + T_3 - T_2) = nC_v(T_3 - T_1) = 0,$$

since $T_1 = T_3 = 300\text{K}$.

d) $\Delta U = Q - W \Rightarrow W = Q - \Delta U$

$\Delta U = 0 \Rightarrow W = \underline{Q = 12{,}468\text{ J}}$ for the complete process

e) $2 \to 3$ is an adiabatic process $\Rightarrow T_2 V_2^{\gamma-1} = T_3 V_3^{\gamma-1}$

$$V_3^{\gamma-1} = V_2^{\gamma-1}\left(\frac{T_2}{T_3}\right) \Rightarrow V_3 = V_2\left(\frac{T_2}{T_3}\right)^{1/(\gamma-1)}$$

$V_2 = 0.040\text{ m}^3$, $T_2 = 600\text{K}$, $T_3 = 300\text{K}$ (all from the above)

For helium, $\gamma = 1.67$ (from Table 18-1)

$$V_3 = 0.040\text{m}^3\left(\frac{600\text{K}}{300\text{K}}\right)^{1/(1.67-1)} = (0.040\text{m}^3)(2)^{1.49} = \underline{0.112\text{m}^3}$$

CHAPTER 19

Exercises 1, 5, 7, 11, 13, 17

Problems 21, 23

Exercises

19-1

a) $\text{eff} = \dfrac{\text{work output}}{\text{heat energy input}} = \dfrac{W}{Q_H} = \dfrac{3000\text{ J}}{8000\text{ J}} = 0.375 \Rightarrow \underline{37.5\%}$

b) $\Delta U = Q - W$

cycle $\Rightarrow \Delta U = 0 \Rightarrow Q = W$

$Q = Q_H + Q_C \Rightarrow Q_C = W - Q_H$

$Q_C = 3000\text{ J} - 8000\text{ J} = \underline{-5000\text{ J}}$

c) $Q = mL_C$, where Q equals the 8000 J of heat input and L_C is the heat of combustion.

$m = \dfrac{Q}{L_C} = \dfrac{8000\text{ J}}{5.0\times10^4\text{ J}\cdot\text{g}^{-1}} = \underline{0.16\text{ g}}$

d) 50 cycles $\Rightarrow W = (50)(3000\text{ J}) = 1.5\times10^5\text{ J}$

$P = \dfrac{W}{t} = \dfrac{1.5\times10^5\text{ J}}{1\text{ s}} = \underline{1.5\times10^5\text{ W}}$

$P = (1.5\times10^5\text{ W})\left(\dfrac{1\text{ hp}}{746\text{ W}}\right) = \underline{201\text{ hp}}$

19-5

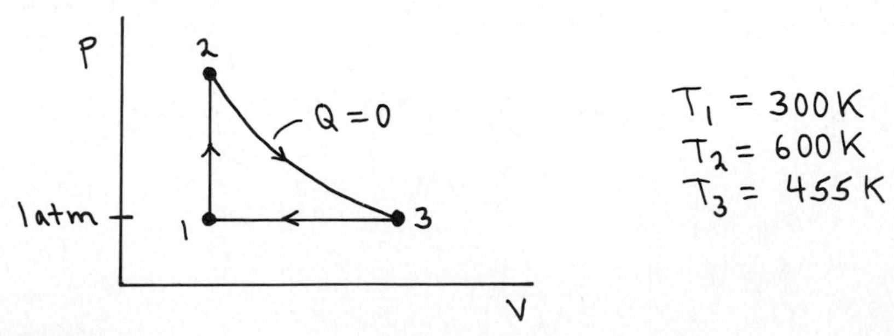

$T_1 = 300\text{ K}$
$T_2 = 600\text{ K}$
$T_3 = 455\text{ K}$

a) point 1

$p_1 = 1\text{ atm} = 1.01\times10^5\text{ Pa}$ (from the figure)

$pV = nRT \Rightarrow V_1 = \dfrac{nRT_1}{p_1} = \dfrac{(0.1\text{ mol})(8.314\text{ J}\cdot\text{mol}^{-1}\cdot\text{K}^{-1})(300\text{ K})}{1.01\times10^5\text{ Pa}} = \underline{2.47\times10^{-3}\text{ m}^3}$

19-5 (cont)

point 2

from the figure, $V_2 = V_1 = \underline{2.47\times10^{-3}\,m^3}$

$pV = nRT$; n, R, V constant $\Rightarrow \frac{p}{T} = \frac{nR}{V}$ = constant

$$\frac{p_1}{T_1} = \frac{p_2}{T_2} \Rightarrow p_2 = p_1\left(\frac{T_2}{T_1}\right) = (1\,atm)\left(\frac{600K}{300K}\right) = \underline{2atm} = 2.03\times10^5\,Pa$$

point 3

from the figure, $p_3 = p_1 = \underline{1atm} = 1.01\times10^5\,Pa$

$pV = nRT$; apply to states 1 and 3 $\Rightarrow p, n, R$ constant

$$\frac{V}{T} = \frac{nR}{p} = \text{constant} \Rightarrow \frac{V_1}{T_1} = \frac{V_3}{T_3} \Rightarrow V_3 = V_1\left(\frac{T_3}{T_1}\right) = (1atm)\left(\frac{455K}{300K}\right)$$

$$V_3 = \underline{3.73\times10^{-3}\,m^3}$$

b) process $1\rightarrow2$

$$\Delta V = 0 \Rightarrow W = 0$$

process $2\rightarrow3$

adiabatic $\Rightarrow Q = 0$

From Eq. (18-27), $W = \left(\frac{1}{\gamma-1}\right)(p_2V_2 - p_3V_3)$

$$W = \left(\frac{1}{\frac{5}{3}-1}\right)\left[(2.03\times10^5\,Pa)(2.47\times10^{-3}m^3) - (1.01\times10^5\,Pa)(3.73\times10^{-3}m^3)\right]$$

$$W = (1.5)(499J - 377J) = 183J \quad (\Delta V > 0 \Rightarrow W > 0)$$

process $3\rightarrow1$

$$W = \int_3^1 p\,dV = p\int_{V_3}^{V_1} dV = p(V_1 - V_3) \quad \text{(since } p \text{ is constant)}$$

$$W = (1.01\times10^5\,Pa)(2.47\times10^{-3}m^3 - 3.73\times10^{-3}m^3) = -127J \quad (\Delta V < 0 \Rightarrow W < 0)$$

$$W_{tot} = W_{1\rightarrow2} + W_{2\rightarrow3} + W_{3\rightarrow1} = 0 + 183J - 127J = \underline{+56J}$$

19-7

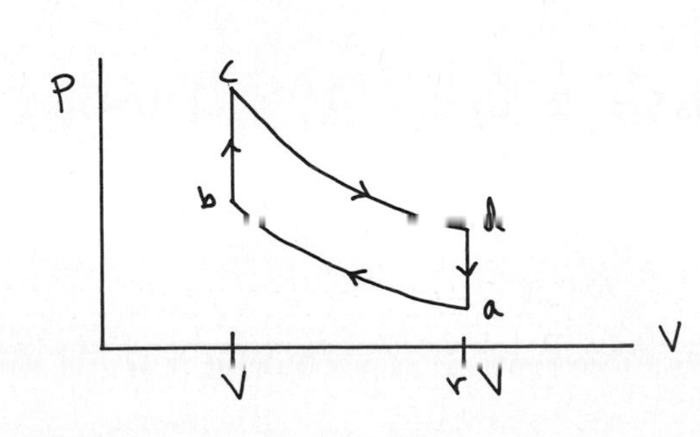

ab is an adiabatic process

$\Rightarrow T_a V_a^{\gamma-1} = T_b V_b^{\gamma-1}$

$V_a = rV,\ V_b = V \Rightarrow V_a = rV_b$

$T_a = 22°C = 295K$

19-7 (cont)

$$T_b = T_a\left(\frac{V_a}{V_b}\right)^{\gamma-1} = T_a\left(\frac{rV_b}{V_b}\right)^{\gamma-1} = T_a r^{\gamma-1}$$

$$T_b = 295\,\text{K}\,(8)^{1.4-1} = \underline{678\,\text{K}}$$

19-11

a) Carnot cycle $\Rightarrow \frac{Q_C}{Q_H} = -\frac{T_C}{T_H}$

$$T_C = -T_H\left(\frac{Q_C}{Q_H}\right) = -(400\text{K})\left(\frac{-335\,\text{J}}{420\,\text{J}}\right) = \underline{319\,\text{K}}$$

b) $e = 1 - \frac{T_C}{T_H} = 1 - \frac{319\text{K}}{400\text{K}} = 0.202;\ \underline{20.2\%}$

or

$$e = \frac{W}{Q_H} = \frac{Q_C + Q_H}{Q_H} = 1 + \frac{Q_C}{Q_H} = 1 + \frac{-335\,\text{J}}{420\,\text{J}} = 0.202;\ \underline{20.2\%}$$

19-13

a) Carnot cycle $\Rightarrow -\frac{T_C}{T_H} = \frac{Q_C}{Q_H} \Rightarrow Q_H = -Q_C\left(\frac{T_H}{T_C}\right)$

$T_H = 27°\text{C} = 300\,\text{K}$; $T_C = 0°\text{C} = 273\,\text{K}$

Q_C = heat absorbed by the refrigerator at 0°C = heat removed from the water

$Q_C = mL_F = (50\,\text{kg})(334\times10^3\,\text{J}\cdot\text{kg}^{-1}) = 1.67\times10^7\,\text{J}$

(positive; this heat is absorbed by the refrigerator)

$$Q_H = -Q_C\left(\frac{T_H}{T_C}\right) = -(1.67\times10^7\,\text{J})\left(\frac{300\,\text{K}}{273\,\text{K}}\right) = \underline{-1.84\times10^7\,\text{J}}$$

(negative because this heat is rejected by the refrigerator)

b) cycle $\Rightarrow \Delta U = 0 \Rightarrow W = Q$

$$W = Q_C + Q_H = 1.67\times10^7\,\text{J} + (-1.84\times10^7\,\text{J}) = \underline{-1.7\times10^6\,\text{J}}$$

(W is negative because this much mechanical (or electrical) energy must be supplied to the refrigerator, rather than being obtained from it.)

19-17

reversible $\Rightarrow \Delta S = \int \frac{dQ}{T}$

isothermal $\Rightarrow$ T is constant $\Rightarrow \Delta S = \frac{1}{T}\int dQ = \frac{Q}{T}$

$\Delta T = 0 \Rightarrow \Delta U = nC_V\Delta T = 0 \Rightarrow Q = W$

$W = \int_1^2 p\,dV$; $p = \frac{nRT}{V}$

19-17 (cont)

$$W = nRT\int_{V_1}^{V_2} \frac{dV}{V} = nRT\ln\left(\frac{V_2}{V_1}\right) = (2\,\text{mol})(8.314\,\text{J}\cdot\text{mol}^{-1}\cdot\text{K}^{-1})(300\,\text{K})\ln\left(\frac{0.04\,\text{m}^3}{0.02\,\text{m}^3}\right)$$

$$W = 3458\,\text{J} \Rightarrow Q = 3458\,\text{J}$$

$$\Delta S = \frac{Q}{T} = \frac{3458\,\text{J}}{300\,\text{K}} = \underline{11.5\,\text{J}\cdot\text{K}^{-1}}$$

Problems

19-21

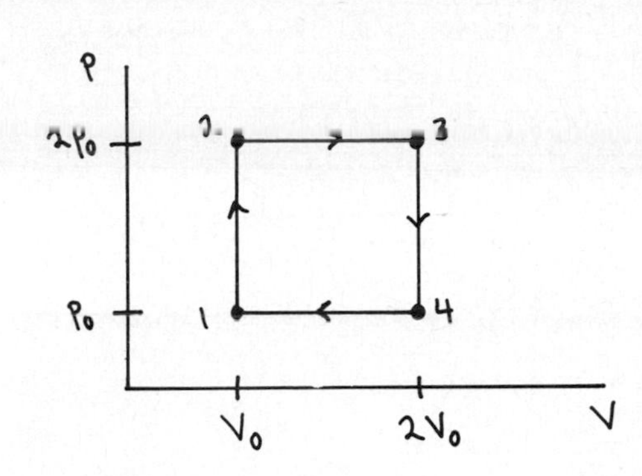

$$C_V = 12\,\text{J}\cdot\text{mol}^{-1}\cdot\text{K}^{-1}$$

$$C_p = C_V + R = (12 + 8.314)\,\text{J}\cdot\text{mol}^{-1}\cdot\text{K}^{-1}$$

$$C_p = 20.3\,\text{J}\cdot\text{mol}^{-1}\cdot\text{K}^{-1}$$

Calculate Q and W for each process:

process 1→2

$$\Delta V = 0 \Rightarrow W = 0$$

$\Delta V = 0 \Rightarrow Q = nC_V\Delta T$; $pV = nRT \Rightarrow V\Delta p = nR\Delta T$ (true when V is constant)

$$Q = nC_V\frac{V\Delta p}{nR} = \frac{C_V}{R}V\Delta p = \frac{C_V}{R}V_0(2p_0 - p_0) = \frac{C_V}{R}p_0V_0$$

$Q > 0 \Rightarrow$ heat is absorbed by the gas

process 2→3

$$\Delta p = 0 \Rightarrow W = p\Delta V = 2p_0(2V_0 - V_0) = 2p_0V_0$$

$\Delta p = 0 \Rightarrow Q = nC_p\Delta T$; $pV = nRT \Rightarrow p\Delta V = nR\Delta T$ (true when p is constant)

$$Q = nC_p\frac{p\Delta V}{nR} = \frac{C_p}{R}p\Delta V = \frac{C_p}{R}2p_0(2V_0 - V_0) = 2\frac{C_p}{R}p_0V_0$$

$Q > 0 \Rightarrow$ heat is absorbed by the gas

process 3→4

$$\Delta V = 0 \Rightarrow W = p\Delta V = 0$$

$$\Delta V = 0 \Rightarrow Q = nC_V\Delta T = \frac{C_V}{R}V\Delta p = \frac{C_V}{R}(2V_0)(p_0 - 2p_0) = -2\frac{C_V}{R}p_0V_0$$

$Q < 0 \Rightarrow$ heat is rejected by the gas

19-21 (cont)

process 4→1

$\Delta p = 0 \Rightarrow W = p\Delta V = p_0(V_0 - 2V_0) = -p_0 V_0$

$\Delta p = 0 \Rightarrow Q = nC_p\Delta T = \frac{C_p}{R} p\Delta V = \frac{C_p}{R} p_0(V_0 - 2V_0) = -\frac{C_p}{R} p_0 V_0$

$Q < 0 \Rightarrow$ heat is rejected by the gas

total work

$$W = W_{1\to2} + W_{2\to3} + W_{3\to4} + W_{4\to1} = 0 + 2p_0V_0 + 0 - p_0V_0 = p_0V_0$$

total heat absorbed

$$Q_H = Q_{1\to2} + Q_{2\to3} = \frac{C_v}{R} p_0 V_0 + 2\frac{C_p}{R} p_0 V_0 = \left(\frac{C_v + 2C_p}{R}\right) p_0 V_0$$

$$C_p = C_v + R \Rightarrow \frac{C_v + 2C_p}{R} = \frac{3C_v + 2R}{R} = 3\frac{C_v}{R} + 2$$

$$Q_H = \left(3\frac{C_v}{R} + 2\right) p_0 V_0$$

total heat rejected

$$Q_C = Q_{3\to4} + Q_{4\to1} = -2\frac{C_v}{R} p_0 V_0 - \frac{C_p}{R} p_0 V_0 = -\left(\frac{2C_v + C_p}{R}\right) p_0 V_0$$

$$C_p = C_v + R \Rightarrow \frac{2C_v + C_p}{R} = \frac{3C_v + R}{R} = 3\frac{C_v}{R} + 1$$

$$Q_C = -\left(3\frac{C_v}{R} + 1\right) p_0 V_0$$

(Note: Net heat flow $Q = Q_C + Q_H = \left(3\frac{C_v}{R} + 2\right)p_0V_0 - \left(3\frac{C_v}{R} + 1\right)p_0V_0 = p_0V_0 = W$, so $Q = W$ as it should for a cycle.)

efficiency

$$e = \frac{W}{Q_H} = \frac{p_0V_0}{\left(3\frac{C_v}{R} + 2\right)p_0V_0} = \frac{1}{\left(3\frac{C_v}{R} + 2\right)} = \frac{1}{\left(3\left[\frac{12\,\text{J}\cdot\text{mol}^{-1}\cdot\text{K}^{-1}}{8.314\,\text{J}\cdot\text{mol}^{-1}\cdot\text{K}^{-1}}\right] + 2\right)} = 0.158$$

$e = \underline{15.8\%}$

19-23

First use the methods of Chapter 15 to find the final temperature of the system:

$Q_{system} = 0$

Q for the 0.50 kg of water:

$Q_{water} = mc\Delta T = (0.50\,\text{kg})(4190\,\text{J}\cdot\text{kg}^{-1}\cdot(\text{C}°)^{-1})(T_2 - 60°\text{C}) = (2095\,\text{J}\cdot(\text{C}°)^{-1})T_2 - 1.257\times10^5\,\text{J}$

19-23 (cont)

Q for the 0.08 kg of ice (warms as ice to 0°C, melts, and warms as water to T_2):

$$Q_{ice} = m c_{ice} \Delta T + m L_F + m c_{water} \Delta T$$

$$Q_{ice} = (0.08\,kg)(2.0\times10^3\,J\cdot kg^{-1}\cdot(C^\circ)^{-1})(0^\circ C - (-15^\circ C)) + (0.08\,kg)(334\times10^3\,J\cdot kg^{-1}) + (0.08\,kg)(4190\,J\cdot kg^{-1}\cdot(C^\circ)^{-1})(T_2 - 0^\circ C)$$

$$Q_{ice} = 2400\,J + 26{,}720\,J + (335\,J\cdot(C^\circ)^{-1})T_2 = 2.912\times10^4\,J + (335\,J\cdot(C^\circ)^{-1})\,T_2$$

$$Q_{system} = 0 \Rightarrow Q_{water} + Q_{ice} = 0$$

$$(2095\,J\cdot(C^\circ)^{-1})T_2 - 1.257\times10^5\,J + 2.912\times10^4\,J + (335\,J\cdot(C^\circ)^{-1})\,T_2 = 0$$

$$(2430\,J\cdot(C^\circ)^{-1})T_2 = 9.658\times10^4\,J \Rightarrow T_2 = 39.4^\circ C$$

Now can calculate the entropy changes:

ice

The process takes ice at −15°C and produces water at 39.4°C. Calculate ΔS as if this were done reversibly. ΔS is path independent, so ΔS for a reversible process will be the same as ΔS for the actual (irreversible) process, as long as the initial and final states are the same.

$$\Delta S = \int \frac{dQ}{T}$$

For a temperature change $dQ = mc\,dT \Rightarrow \Delta S = \int_{T_1}^{T_2} \frac{mc\,dT}{T} = mc \int_{T_1}^{T_2} \frac{dT}{T}$

$$\Delta S = mc \ln\left(\frac{T_2}{T_1}\right)$$ (T_1 and T_2 must be in kelvins)

For a phase change, since it occurs at constant T,

$$\Delta S = \int \frac{dQ}{T} = \frac{1}{T}\int dQ = \frac{mL}{T}$$

Therefore

$$\Delta S_{ice} = m c_{ice} \ln\left(\frac{273\,K}{258\,K}\right) + \frac{mL_F}{273\,K} + m c_{water} \ln\left(\frac{312.4\,K}{273\,K}\right)$$

(Note: The units of c are $J\cdot kg^{-1}\cdot(C^\circ)^{-1}$, but one C° = one K, so the units of c may just as well be written as $J\cdot kg^{-1}\cdot K^{-1}$.)

$$\Delta S_{ice} = (0.08\,kg)(2000\,J\cdot kg^{-1}\cdot K^{-1})\ln\left(\frac{273}{258}\right) + \frac{(0.08\,kg)(334\times10^3\,J\cdot kg^{-1})}{273\,K} + (0.08\,kg)(4190\,J\cdot kg^{-1}\cdot K^{-1})\ln\left(\frac{312.4\,K}{273\,K}\right)$$

$$\Delta S_{ice} = 9.0\,J\cdot K^{-1} + 97.9\,J\cdot K^{-1} + 45.2\,J\cdot K^{-1} = 152\,J\cdot K^{-1}$$

water

The process takes water at 60°C and produces water at 39.4°C.

$T_2 = 39.4 + 273 = 312.4\,K$

$T_1 = 60 + 273 = 333\,K$

19-23 (cont)

$$\Delta S_{water} = m c_{water} \ln\left(\frac{T_2}{T_1}\right) = (0.50\,\text{kg})(4190\,\text{J}\cdot\text{kg}^{-1}\cdot\text{K}^{-1}) \ln\left(\frac{312.4\,\text{K}}{333\,\text{K}}\right)$$

$$\Delta S_{water} = -134\,\text{J}\cdot\text{K}^{-1}$$

$$\Delta S_{system} = \Delta S_{ice} + \Delta S_{water} = 152\,\text{J}\cdot\text{K}^{-1} - 134\,\text{J}\cdot\text{K}^{-1} = \underline{+18\,\text{J}\cdot\text{K}^{-1}}$$

Note: $\Delta S_{system} > 0$, as it must be for an irreversible process.

CHAPTER 20

Exercises 3, 5, 9, 11, 13

Problems 17, 21

20-3

a) $\rho = \frac{m}{V} \Rightarrow V = \frac{m}{\rho}$

for water $M_{H_2O} = 2M_H + M_O = 2(1.008\,g\cdot mol^{-1}) + 15.999\,g\cdot mol^{-1} = 18.015\,g\cdot mol^{-1}$
(Using Appendix D.)

$1\,mol \Rightarrow m = 18.015\,g$

$\rho = 1.0\,g\cdot cm^{-3} \Rightarrow V = \frac{18.015\,g}{1.0\,g\cdot cm^{-3}} = \underline{18.0\,cm^3}$

b) Calculate the volume per molecule:

$1\,mole \Rightarrow V = 18.0\,cm^3 = 18\times10^{-6}\,m^3$

$1\,mole \Rightarrow N_A = 6.02\times10^{23}$ molecules

$$\text{volume per molecule} = \frac{18\times10^{-6}\,m^3}{6.02\times10^{23}\,\text{molecules}} = 2.99\times10^{-29}\,m^3\cdot \text{molecule}^{-1}$$

$V = a^3$, where a is the length of each side of the cube occupied by a molecule

$a^3 = 2.99\times10^{-29}\,m^3 \Rightarrow a = \underline{3.10\times10^{-10}\,m}$

c) atoms and molecules are on the order of $10^{-10}\,m$ in diameter, in agreement with the above estimate

20-5

$$v_{rms} = \sqrt{\frac{3RT}{M}} \Rightarrow \frac{T}{M} = \frac{v_{rms}^2}{3R}$$

$$v_{rms}\ \text{same} \Rightarrow \frac{T_{O_2}}{M_{O_2}} = \frac{T_{H_2}}{M_{H_2}} \Rightarrow T_{O_2} = T_{H_2}\left(\frac{M_{O_2}}{M_{H_2}}\right)$$

$M_{O_2} = (32)(1.66\times10^{-27}\,kg) = 5.31\times10^{-26}\,kg$

$M_{H_2} = (2)(1.67\times10^{-27}\,kg) = 3.34\times10^{-27}\,kg$

$$T_{O_2} = (273\,K)\left(\frac{5.31\times10^{-26}\,kg}{3.34\times10^{-27}\,kg}\right) = \underline{4340\,K} = 4097^\circ C$$

20-9

a) $\frac{1}{2}m(v^2)_{av} = \frac{3}{2}kT = \frac{3}{2}(1.38\times10^{-23}\,J\cdot K^{-1})(300\,K) = \underline{6.21\times10^{-21}\,J}$

20-9 (cont)

b) $(v^2)_{av} = \frac{2(6.21\times10^{-21}\,J)}{m} = \frac{2(6.21\times10^{-21}\,J)}{32(1.66\times10^{-27}\,kg)} = \underline{2.34\times10^{5}\,m^2\cdot s^{-2}}$

c) $v_{rms} = \sqrt{(v^2)_{av}} = \sqrt{2.34\times10^{5}\,m^2\cdot s^{-2}} = \underline{484\,m\cdot s^{-1}}$

d) $p = mv = 32(1.66\times10^{-27}\,kg)(484\,m\cdot s^{-1}) = \underline{2.57\times10^{-23}\,kg\cdot m\cdot s^{-1}}$

e) time between collisions with a given wall is

$t = \frac{0.2\,m}{v_{rms}} = \frac{0.2\,m}{484\,m\cdot s^{-1}} = 4.13\times10^{-4}\,s$

$F = \frac{dp}{dt}$; $(F)_{av} = \frac{\Delta p}{\Delta t}$

In a collision $\vec{v}$ changes direction $\Rightarrow \Delta p = 2m v_{rms} = 2p$

$(F)_{av} = \frac{2(2.57\times10^{-23}\,kg\cdot m\cdot s^{-1})}{4.13\times10^{-4}\,s} = \underline{1.24\times10^{-19}\,N}$

f) $\text{pressure} = \frac{F}{A} = \frac{1.24\times10^{-19}\,N}{(0.10\,m)^2} = \underline{1.24\times10^{-17}\,Pa}$

g) $\text{pressure} = 1\,atm = 1.01\times10^{5}\,Pa$

$\Rightarrow \frac{1.01\times10^{5}\,Pa}{1.24\times10^{-17}\,Pa} = \underline{8.15\times10^{21}\text{ molecules}}$

h) $pV = nRT = NkT$

$N = \frac{pV}{kT} = \frac{(1.01\times10^{5}\,Pa)(0.1\,m)^3}{(1.38\times10^{-23}\,J\cdot K^{-1})(300\,K)} = \underline{2.44\times10^{22}}\text{ molecules}$

i) $\frac{\text{actual}}{\text{needed}} = \frac{2.44\times10^{22}}{8.15\times10^{21}} = \underline{3.0}$

This factor arises from the relation $(v_x^2)_{av} = \frac{1}{3}(v^2)_{av}$.

20-11

$\frac{1}{2}R$ contribution to C_v for each degree of freedom

$\Rightarrow C_v = 6(\frac{1}{2}R) = 3R = 3(8.314\,J\cdot mol^{-1}\cdot K^{-1}) = \underline{24.9\,J\cdot mol^{-1}\cdot K^{-1}}$

actual C_v

$c_v = 2000\,J\cdot kg^{-1}\cdot K^{-1}$

$C_v = Mc_v$; for water $M = 2M_H + M_O = 2(1.008\times10^{-3}\,kg\cdot mol^{-1}) + 15.999\times10^{-3}\,kg\cdot mol^{-1}$

$M = 0.0180\,kg\cdot mol^{-1}$

$C_v = (0.0180\,kg\cdot mol^{-1})(2000\,J\cdot kg^{-1}\cdot K^{-1}) = 36.0\,J\cdot mol^{-1}\cdot K^{-1}$

actual C_v − calculated $C_v = 36.0\,J\cdot mol^{-1}\cdot K^{-1} - 24.9\,J\cdot mol^{-1}\cdot K^{-1}$

$= 11.1\,J\cdot mol^{-1}\cdot K^{-1} \simeq 1.4R = 2.8(\frac{R}{2})$

20-11 (cont)

Thus the equivalent of nearly 3 vibrational degrees of freedom contribute to C_V.

20-13

Eq. (20-18) $\Rightarrow f(v) = f(\epsilon) = \frac{8\pi}{m}\left(\frac{m}{2\pi kT}\right)^{3/2} \epsilon e^{-\epsilon/kT}$

At the maximum of $f(\epsilon)$, $\frac{df}{d\epsilon} = 0$.

$$\frac{df}{d\epsilon} = \frac{8\pi}{m}\left(\frac{m}{2\pi kT}\right)^{3/2} \frac{d}{d\epsilon}\left(\epsilon e^{-\epsilon/kT}\right) = 0$$

$$\frac{d}{d\epsilon}\left(\epsilon e^{-\epsilon/kT}\right) = 0 \Rightarrow e^{-\epsilon/kT} - \frac{\epsilon}{kT} e^{-\epsilon/kT} = 0$$

Thus $\frac{\epsilon}{kT} = 1 \Rightarrow \epsilon = kT$, as was to be shown.

Problems

20-17

a) $v_{rms} = \sqrt{\frac{3RT}{M}}$

$M_{N_2} = 2(14.007 \times 10^{-3}\ kg\cdot mol^{-1}) = 28.01 \times 10^{-3}\ kg\cdot mol^{-1}$ (from Appendix D).

$$v_{rms} = \sqrt{\frac{3(8.314\ J\cdot mol^{-1}\cdot K^{-1})(300K)}{28.01 \times 10^{-3}\ kg\cdot mol^{-1}}} = \underline{517\ m\cdot s^{-1}}$$; larger than the sound velocity

b) $(v_x^2)_{av} = \frac{1}{3}(v^2)_{av} \Rightarrow (v_x)_{rms} = \sqrt{(v_x^2)_{av}} = \frac{1}{\sqrt{3}}\sqrt{(v^2)_{av}} = \frac{1}{\sqrt{3}} v_{rms}$

$$(v_x)_{rms} = \frac{1}{\sqrt{3}} v_{rms} = \frac{517\ m\cdot s^{-1}}{\sqrt{3}} = \underline{298\ m\cdot s^{-1}}$$

The sound velocity is close to this.

20-21

a) (m) ↑ v_1

The only work is that done by gravity $\Rightarrow K_1 + U_1 = K_2 + U_2$, with $U = -G\frac{mm_E}{r}$

"Escape" $\Rightarrow v_2 = 0$ when $r_2 \to \infty \Rightarrow K_2 = 0,\ U_2 = 0$

$r_1 = R$ = radius of the earth

Thus $\frac{1}{2}mv_1^2 - G\frac{mm_E}{R} = 0 \Rightarrow \frac{1}{2}mv_1^2 = \left(\frac{Gm_E}{R}\right)m$

At the earth's surface $w = F_{grav} = G\frac{mm_E}{R^2} = mg \Rightarrow g = G\frac{m_E}{R^2}$

Thus $\frac{1}{2}mv_1^2 = mgR$, as was to be shown.

20-21 (cont)

b) $\frac{1}{2}m(v^2)_{av} = \frac{3}{2}kT = mgR \Rightarrow T = \frac{2mgR}{3k}$

oxygen

$$T = \frac{2(32 \times 1.66 \times 10^{-27}\ kg)(9.8\ m \cdot s^{-2})(6.38 \times 10^{6}\ m)}{3(1.38 \times 10^{-23}\ J \cdot K^{-1})} = 1.60 \times 10^{5}\ K$$

hydrogen

$$T = \frac{2(2 \times 1.67 \times 10^{-27}\ kg)(9.8\ m \cdot s^{-2})(6.38 \times 10^{6}\ m)}{3(1.38 \times 10^{-23}\ J \cdot K^{-1})} = 1.01 \times 10^{4}\ K$$

(Result for H_2 is smaller than that for O_2 by a factor of $\frac{m_{H_2}}{m_{O_2}} \approx \frac{2}{32} = \frac{1}{16}$)

CHAPTER 21

Exercises 3, 5, 9, 15, 17

Problems 19, 23, 25

Exercises

21-3

a) $c = 345\ m\cdot s^{-1}$

$c = f\lambda \Rightarrow \lambda = \frac{c}{f}$

$f = 20\ Hz \Rightarrow \lambda = \frac{345\ m\cdot s^{-1}}{20\ s^{-1}} = \underline{17.2\ m}$

$f = 20{,}000\ Hz \Rightarrow \lambda = \frac{345\ m\cdot s^{-1}}{20{,}000\ s^{-1}} = \underline{0.0172\ m}$

b) $c = 1480\ m\cdot s^{-1}$

$f = 20\ Hz \Rightarrow \lambda = \frac{1480\ m\cdot s^{-1}}{20 s^{-1}} = \underline{74.0\ m}$

$f = 20{,}000\ Hz \Rightarrow \lambda = \frac{1480\ m\cdot s^{-1}}{20{,}000\ s^{-1}} = \underline{0.0740 m}$

21-5

Eq. (21-4) is $y(x,t) = A \sin 2\pi\left(\frac{t}{\tau} - \frac{x}{\lambda}\right)$

$\sin\theta = -\sin(-\theta) \Rightarrow y = -A\sin 2\pi\left(\frac{x}{\lambda} - \frac{t}{\tau}\right) = -A\sin \frac{2\pi}{\lambda}\left(x - \frac{\lambda t}{\tau}\right)$

$\tau = \frac{1}{f} \Rightarrow \frac{\lambda}{\tau} = \lambda f = c \Rightarrow y = -A\sin\left[\frac{2\pi}{\lambda}(x - ct)\right]$, as was to be shown.

21-9

a) $c = \sqrt{\frac{S}{\mu}}$

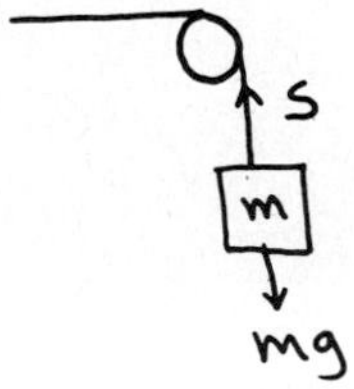

$S = mg = (5\ kg)(9.8\ m\cdot s^{-2}) = 49\ N$ (tension in the string)

$c = \sqrt{\frac{49\ N}{0.02\ kg\cdot m^{-1}}} = \underline{49.5\ m\cdot s^{-1}}$

b) $c = f\lambda \Rightarrow \lambda = \frac{c}{f} = \frac{49.5\ m\cdot s^{-1}}{240\ s^{-1}} = \underline{0.206\ m}$

21-15

a) Eq. (21-22) $\Rightarrow c = \sqrt{\frac{B}{\rho}}$

Eq. (21-27) $\Rightarrow B_{iso} = p \Rightarrow c = \sqrt{\frac{p}{\rho}}$

But for an ideal gas, $pV = nRT \Rightarrow p = \frac{nRT}{V} = \frac{mRT}{MV}$

$\rho = \frac{m}{V} \Rightarrow \frac{p}{\rho} = \left(\frac{mRT}{MV}\right)\left(\frac{V}{m}\right) = \frac{RT}{M}$

Thus $c = \sqrt{\frac{RT}{M}}$, as was to be shown. (Compare this to Eq. (21-29).)

b) For air, $M = 28.8 \times 10^{-3}\ \text{kg}\cdot\text{mol}^{-1}$

$$c = \sqrt{\frac{(8.314\ \text{J}\cdot\text{mol}^{-1}\cdot\text{K}^{-1})(300\text{K})}{28.8 \times 10^{-3}\ \text{kg}\cdot\text{mol}^{-1}}} = \underline{294\ \text{m}\cdot\text{s}^{-1}}$$

(Note: Example 21-5 shows that if the wave is adiabatic, then $c = \sqrt{\frac{\gamma RT}{M}} = 348\ \text{m}\cdot\text{s}^{-1}$.)

c) To be isothermal the heat transport must be fast compared to the frequency of the wave motion $\Rightarrow$ gas must have large thermal conductivity and/or the wave must have large wavelength, low frequency.

21-17

Eq. (21-35) $P_{av} = \frac{1}{2}\sqrt{\mu S}\,\omega^2 A^2$

$c = \sqrt{\frac{S}{\mu}} \Rightarrow \sqrt{\mu} = \frac{\sqrt{S}}{c} \Rightarrow P_{av} = \frac{1}{2}\frac{S}{c}\omega^2 A^2$

$\omega = 2\pi f \Rightarrow \frac{\omega}{c} = \frac{2\pi f}{c} = \frac{2\pi}{\lambda} = k$

Thus $P_{av} = \frac{1}{2} S\omega k A^2$, as was to be shown.

Problems

21-19

$A = 0.10\text{m}$, $\lambda = 2\text{m}$, $c = 1\ \text{m}\cdot\text{s}^{-1}$; propagates to the right.

a) $c = f\lambda \Rightarrow f = \frac{c}{\lambda} = \frac{1\ \text{m}\cdot\text{s}^{-1}}{2\text{m}} = \underline{0.50\ \text{Hz}}$

b) $\omega = 2\pi f = 2\pi(0.5\ \text{s}^{-1}) = \underline{3.14\ \text{rad}\cdot\text{s}^{-1}}$

c) $k = \frac{2\pi}{\lambda} = \frac{2\pi}{2\text{m}} = \underline{3.14\ \text{m}^{-1}}$

d) traveling to the right $\Rightarrow y(x,t) = \pm A\sin(\omega t - kx)$

21-19 (cont)

$$v = \frac{\partial y}{\partial t} = \pm A\omega \cos(\omega t - kx)$$

$t = 0, x = 0 \Rightarrow v = \pm A\omega$ for left-hand end of string $(x=0)$ at $t = 0$.
It is specified that this velocity is downward

$$\Rightarrow v = -A\omega \Rightarrow y(x,t) = -A\sin(\omega t - kx)$$

Thus $y(x,t) = -(0.10\text{m})\sin(\pi(t-x))$, for x in m, t in seconds.

e) left end $\Rightarrow x = 0$

$$y(0,t) = -(0.10\text{m})\sin(\pi t), \quad t \text{ in seconds.}$$

f) 1.5 m to right of origin $\Rightarrow x = 1.5$ m

$$y(t) = -(0.10\text{m})\sin(\pi t - \tfrac{3}{2}\pi) = -(0.10\text{m})\sin(\pi t - 2\pi + \tfrac{1}{2}\pi)$$

But $\sin\theta = \sin(\theta - 2\pi)$

$$\Rightarrow y(t) = -(0.10\text{m})\sin(\pi t + \tfrac{1}{2}\pi)$$

But $\sin(\theta + \frac{\pi}{2}) = \cos\theta \Rightarrow \sin(\pi t + \frac{1}{2}\pi) = \cos(\pi t)$ (Appendix B)

$$\Rightarrow y(t) = -(0.10\text{m})\cos(\pi t)$$

g) $y(x,t) = -(0.10\text{m})\sin(\pi[t-x])$

$$v_y = \frac{\partial y}{\partial t} = -(0.10\text{m})(\pi\,\text{s}^{-1})\cos(\pi(t-x))$$

max v_y is when $\cos(\pi[t-x]) = \pm 1 \Rightarrow (v_y)_{max} = 0.10\pi\ \text{m}\cdot\text{s}^{-1} = \underline{0.314\ \text{m}\cdot\text{s}^{-1}}$

h) $y(t)$ for this x was derived in part (f)

$t = 3.25\,\text{s} \Rightarrow y = -(0.10\text{m})\cos(3.25\pi) = \underline{+0.0707\ \text{m}}$

$$v = \frac{dy}{dt} = -(0.10\text{m})\frac{d}{dt}\cos(\pi t) = -(0.10\text{m})(-\pi\,\text{s}^{-1})(\sin\pi t) = +(0.314\,\text{m}\cdot\text{s}^{-1})\sin\pi t$$

$t = 3.25\,\text{s} \Rightarrow v = \underline{-0.222\ \text{m}\cdot\text{s}^{-1}}$

(At $t = 3.25$ s this point in the string is displaced up and is moving downward.)

i) From (d), $y(x,t) = -(0.10\text{m})\sin(\pi(t-x))$

$t = 3.25\,\text{s} = (2 + \frac{5}{4})\,\text{s} \Rightarrow \sin(\pi t - \pi x) = \sin(2\pi + \frac{5}{4}\pi - \pi x) = \sin[\pi(\frac{5}{4} - x)]$

$$y(x) = -(0.10\text{m})\sin[\pi(\tfrac{5}{4} - x)]$$

21-19 (cont)

For the sketch calculate y(x) at several x:

x	y(x)	x	y(x)	x	y(x)	x	y(x)
0	+0.071 m	$1\frac{1}{4}$ m	0	$2\frac{1}{4}$ m	0	$3\frac{1}{4}$ m	0
$\frac{1}{4}$ m	0	$1\frac{1}{2}$ m	0.071 m	$2\frac{1}{2}$ m	−0.071 m	$3\frac{1}{2}$ m	0.071 m
$\frac{1}{2}$ m	−0.071 m	$1\frac{3}{4}$ m	0.10 m	$2\frac{3}{4}$ m	−0.10 m	$3\frac{3}{4}$ m	0.10 m
$\frac{3}{4}$ m	−0.10 m	2 m	0.071 m	3 m	−0.071 m	4 m	0.071 m
1 m	−0.071 m						

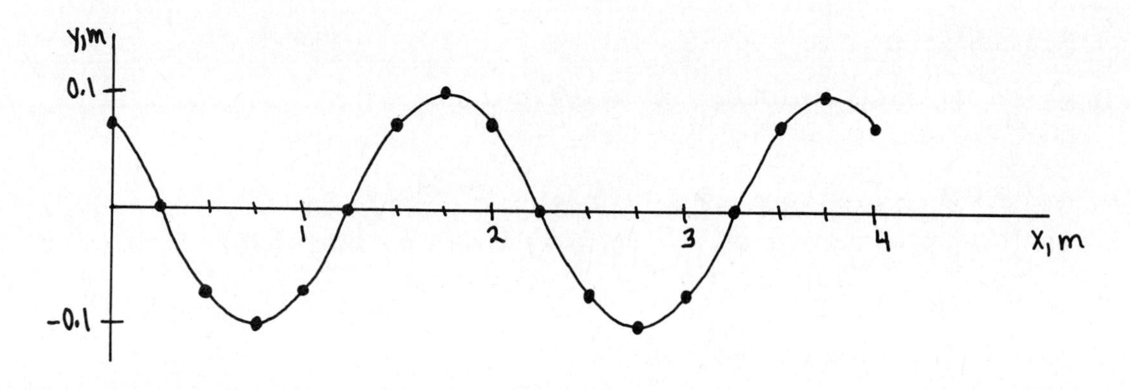

21-23

transverse wave speed $c_t = \sqrt{\frac{S}{\mu}} = \sqrt{\frac{S}{m/L}} = \sqrt{\frac{SL}{m}}$

longitudinal wave speed $c_l = \sqrt{\frac{Y}{\rho}} = \sqrt{\frac{Y}{m/V}} = \sqrt{\frac{Y}{m/LA}} = \sqrt{\frac{YLA}{m}}$

$$c_l = 10c_t \Rightarrow \sqrt{\frac{YLA}{m}} = 10\sqrt{\frac{SL}{m}}$$

$$\sqrt{YA} = 10\sqrt{S} \Rightarrow YA = 100S$$

stress $\frac{S}{A} = \frac{Y}{100}$

(S is the tension in the wire, so plays the role of F in the notation of Chapter 12.)

21-25

$$c = \sqrt{\frac{\gamma RT}{M}} \quad (\text{diatomic} \Rightarrow \gamma = 1.4)$$

$$v_{rms} = \sqrt{\frac{3RT}{M}}$$

$$\frac{c}{v_{rms}} = \sqrt{\frac{\gamma RT}{M}}\sqrt{\frac{M}{3RT}} = \sqrt{\frac{\gamma}{3}} = \sqrt{\frac{1.4}{3}} = 0.683$$

$$\underline{c = 0.683\, v_{rms}}$$

CHAPTER 22

Exercises 3, 5, 7, 11

Problems 13, 15

Exercises

22-3

The wave equation is $\frac{\partial^2 y}{\partial t^2} = c^2 \frac{\partial^2 y}{\partial x^2}$

$y(x,t) = -[2A\cos\omega t]\sin kx$

$\frac{\partial y}{\partial t} = -2A\sin kx(-\omega\sin\omega t) = 2\omega A\sin kx\sin\omega t$

$\frac{\partial^2 y}{\partial t^2} = 2\omega A\sin kx(\omega\cos\omega t) = 2\omega^2 A\cos\omega t\sin kx$

$\frac{\partial y}{\partial x} = -(2A\cos\omega t)k\cos kx$

$\frac{\partial^2 y}{\partial x^2} = -(2A\cos\omega t)(-k^2\sin kx) = +2k^2 A\cos\omega t\sin kx$

Put these two second partial derivatives into the wave equation

$\Rightarrow 2\omega^2 A\cos\omega t\sin kx = c^2 2k^2 A\cos\omega t\sin kx$

$\omega^2 = k c^2$

$\omega = 2\pi f$

$k = \frac{2\pi}{\lambda} \Rightarrow (2\pi f)^2 = \left(\frac{2\pi}{\lambda}\right)^2 c^2$

$(f\lambda)^2 = c^2$

$c^2 = c^2$ ✓

22-5

a)

fundamental

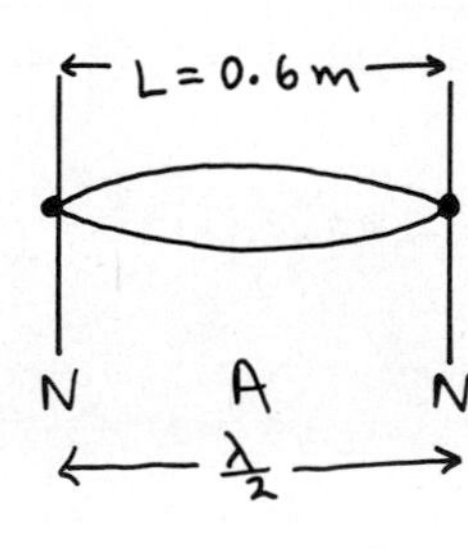

$f = 30\,\text{Hz}$

From the sketch, $\frac{\lambda}{2} = L \Rightarrow \lambda = 2L = 1.2\,\text{m}$

$c = f\lambda = (30\,\text{s}^{-1})(1.2\,\text{m}) = \underline{36.0\,\text{m}\cdot\text{s}^{-1}}$

b) $c = \sqrt{\frac{S}{\mu}} \Rightarrow S = \mu c^2$

$\mu = \frac{m}{L} = \frac{0.030\,\text{kg}}{0.60\,\text{m}} = 0.05\,\text{kg}\cdot\text{m}^{-1}$

$S = (0.05\,\text{kg}\cdot\text{m}^{-1})(36.0\,\text{m}\cdot\text{s}^{-1})^2 = \underline{64.8\,\text{N}}$

22-7

a) open at both ends $\Rightarrow$ displacement antinode at each end

fundamental

A N A ($\leftarrow L \rightarrow$, $\leftarrow \frac{\lambda}{2} \rightarrow$)

$\frac{\lambda}{2} = L \Rightarrow \lambda = 2L = 2(4.88\,m) = 9.76\,m$

$f = \frac{c}{\lambda} = \frac{345\,m\cdot s^{-1}}{9.76\,m} = \underline{35.3\,Hz}$

b) closed at one end $\Rightarrow$ displacement node at closed end

fundamental

A N ($\leftarrow L \rightarrow$, $\leftarrow \frac{\lambda}{4} \rightarrow$)

$\frac{\lambda}{4} = L \Rightarrow \lambda = 4L = 4(4.88\,m) = 19.5\,m$

$f = \frac{c}{\lambda} = \frac{345\,m\cdot s^{-1}}{19.5\,m} = \underline{17.7\,Hz}$

22-11

a) rod clamped at its center $\Rightarrow$ node at center, antinodes at ends

A N A ($\leftarrow L \rightarrow$, $\leftarrow \frac{\lambda}{2} \rightarrow$)

$L = \frac{\lambda}{2} \Rightarrow \lambda = 2L = 2\,m$

$c = f\lambda = (2480\,s^{-1})(2\,m) = \underline{4960\,m\cdot s^{-1}}$

b) The powder heaps are at the displacement nodes, so the distance between heaps is $\frac{\lambda}{2}$.

Thus $\frac{\lambda}{2} = 0.069\,m \Rightarrow \lambda = 2(0.069\,m) = 0.138\,m$

Then $c = f\lambda = (2480\,s^{-1})(0.138\,m) = \underline{342\,m\cdot s^{-1}}$

Problems

22-13

a) fundamental

N A N ($\leftarrow L \rightarrow$, $\leftarrow \frac{\lambda}{2} \rightarrow$)

$\frac{\lambda}{2} = L \Rightarrow \lambda = 2L = 2(0.68\,m) = 1.36\,m$

$c = f\lambda = (220\,Hz)(1.36\,m) = 299\,m\cdot s^{-1}$

$c = \sqrt{\frac{S}{\mu}} \Rightarrow S = \mu c^2 = \left(\frac{m}{L}\right)c^2 = \left(\frac{1.29\times10^{-3}\,kg}{0.68\,m}\right)(299\,m\cdot s^{-1})^2 = \underline{170\,N}$

b) $S = \mu c^2 = \mu f^2 \lambda^2$; $\lambda = 2L$ stays the same

$S_1 = \mu f_1^2 \lambda^2,\ S_2 = \mu f_2^2 \lambda^2 \Rightarrow \frac{S_2}{S_1} = \left(\frac{f_2}{f_1}\right)^2$

$\Rightarrow S_2 = S_1\left(\frac{f_2}{f_1}\right)^2 = 170\,N\left(\frac{233\,Hz}{220\,Hz}\right)^2 = 190.7\,N$

The percentage increase is $\frac{S_2 - S_1}{S_1} \times 100\% = \frac{190.7\,N - 170\,N}{170\,N} \times 100\% = \underline{12.2\%}$

22-15

a)

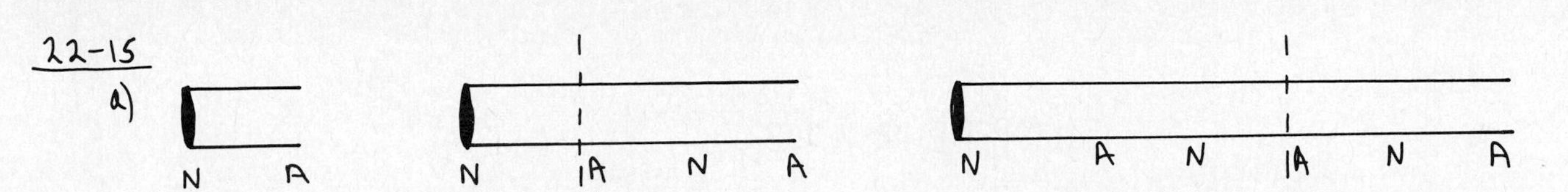

The successive lengths differ by an additional $\frac{\lambda}{2}$ length (the A to A distance)

$$\Rightarrow \frac{\lambda}{2} = 55.5\text{ cm} - 18.0\text{ cm} = 93.0\text{ cm} - 55.5\text{ cm} = 37.5\text{ cm}$$

$$\lambda = 2(37.5\text{ cm}) = 75\text{ cm} = 0.75\text{ m}$$

$$c = f\lambda = (500\text{Hz})(0.75\text{m}) = \underline{375\text{ m}\cdot\text{s}^{-1}}$$

b) $c = \sqrt{\frac{\gamma RT}{M}} \Rightarrow \gamma = \frac{Mc^2}{RT}$

For air, $M = 28.8 \times 10^{-3}\text{ kg}\cdot\text{mol}^{-1}$

$$\Rightarrow \gamma = \frac{(28.8 \times 10^{-3}\text{ kg}\cdot\text{mol}^{-1})(375\text{ m}\cdot\text{s}^{-1})^2}{(8.314\text{ J}\cdot\text{mol}^{-1}\cdot\text{K}^{-1})(350\text{ K})} = \underline{1.39}$$

CHAPTER 23

Exercises 3, 7, 9, 11

Problems 13, 17, 19

Exercises

23-3

a) intensity in water is [eq. 23-10]

$$I_w = \frac{p^2_{max,w}}{2\sqrt{\rho_w B_w}}$$

and in air is $I_a = \dfrac{p^2_{max,a}}{2\sqrt{\rho_a B_a}}$

$$I_w = I_a \Rightarrow \frac{p^2_{max,w}}{2\sqrt{\rho_w B_w}} = \frac{p^2_{max,a}}{2\sqrt{\rho_a B_a}}$$

$$\frac{p_{max,w}}{p_{max,a}} = \left(\frac{\rho_w B_w}{\rho_a B_a}\right)^{1/4}$$

$\rho_w = 1.0\times 10^3\,kg\cdot m^{-3}$; $B_w = \frac{1}{k_w} = \frac{1}{45.8\times 10^{-11}}\,Pa = 2.18\times 10^9\,Pa$ (Table 12-2)

$\rho_a = 1.3\,kg\cdot m^{-3}$; $B_a = 1.42\times 10^5\,Pa$ (Example 23-1)

$$\frac{p_{max,w}}{p_{max,a}} = \left(\frac{(1.0\times 10^3\,kg\cdot m^{-3})(2.18\times 10^9\,Pa)}{(1.3\,kg\cdot m^{-3})(1.42\times 10^5\,Pa)}\right)^{1/4} = \underline{58.6}$$

b) $p^2_{max} = 2I\sqrt{\rho B}$

$$p^2_{max,w} = p^2_{max,a} \Rightarrow 2I_w\sqrt{\rho_w B_w} = 2I_a\sqrt{\rho_a B_a}$$

$$\frac{I_w}{I_a} = \sqrt{\frac{\rho_a B_a}{\rho_w B_w}} = \left(\frac{1}{58.6}\right)^2 = \underline{2.91\times 10^{-4}}$$

23-7

a) $\beta_2 = 10\log\left(\frac{I_2}{I_0}\right)$; $\beta_1 = 10\log\left(\frac{I_1}{I_0}\right)$

$$\beta_2 - \beta_1 = 10\left[\log\left(\frac{I_2}{I_0}\right) - \log\left(\frac{I_1}{I_0}\right)\right] = 10\log\left(\frac{I_2}{I_0}\frac{I_0}{I_1}\right) = 10\log\left(\frac{I_2}{I_1}\right)$$

(We have used that $\log\left(\frac{a}{b}\right) = \log a - \log b$.)

b) $I_1 = \dfrac{(p_{max})_1^2}{2\rho c}$; $I_2 = \dfrac{(p_{max})_2^2}{2\rho c}$ (eq. 23-10)

23-7 (cont)

From the result derived in (a)

$$\beta_2-\beta_1 = 10\log\left(\frac{I_2}{I_1}\right) = 10\log\left[\frac{(p_{max})_2^2}{2\rho c}\cdot\frac{2\rho c}{(p_{max})_1^2}\right] = 10\log\left[\frac{(p_{max})_2}{(p_{max})_1}\right]^2$$

$$\beta_2-\beta_1 = 20\log\left(\frac{(p_{max})_2}{(p_{max})_1}\right)$$

(We have used that $\log x^a = a\log x$.)

c) $\beta = 10\log\left(\frac{I}{I_0}\right) = 10[\log I - \log I_0]$

$I_0 = 10^{-12} \Rightarrow \log I_0 = -12$

$\beta = 10\log I - 10(-12) = 120 + 10\log I$

23-9

$$f_{beat} = f_1 - f_2$$

$3.6\ \text{beats}\cdot s^{-1} \Rightarrow f_1 - f_2 = 3.6\ \text{Hz}$

one frequency is 440 Hz $\Rightarrow$ the other is 440 Hz $\pm$ 3.6 Hz

$\Rightarrow$ 443.6 Hz or 436.4 Hz

23-11

a)

$\rightarrow v_s = 30\ m\cdot s^{-1}$

$f_s = 500\ \text{Hz}$ (S) (S) (L)

$v_s t$

position of source after time t

ct (distance wave emitted at $t=0$ has traveled in time t)

$$\lambda = \frac{\text{distance over which waves are spread}}{\text{number of waves emitted}} = \frac{ct - v_s t}{f_s t} = \frac{c - v_s}{f_s} \quad \text{(eq. 23-15)}$$

$$\lambda = \frac{345\ m\cdot s^{-1} - 30\ m\cdot s^{-1}}{500\ \text{Hz}} = \underline{0.63\ m}$$

b)

$v_s = 30\ m\cdot s^{-1}$

(L) $f_s = 500\ \text{Hz}$ (S) (S)

ct $v_s t$

$$\lambda = \frac{ct + v_s t}{f_s t} = \frac{c + v_s}{f_s} \quad \text{(eq. 23-16)}$$

23-11 (cont)

$$\lambda = \frac{345\,m\cdot s^{-1} + 30\,m\cdot s^{-1}}{500\ Hz} = \underline{0.75\,m}$$

c) $\dfrac{f_L}{c+v_L} = \dfrac{f_S}{c+v_S}$

(S) $\rightarrow v$ (+ ←) $\ddot{\chi}$ L

$$f_L = \left(\frac{c}{c-v_S}\right) f_S = \left(\frac{345}{345-30}\right)(500\,Hz) = \underline{548\,Hz}$$

or

$$f = \frac{c}{\lambda} = \frac{345\,m\cdot s^{-1}}{0.63\ m} = 548\,Hz$$

d) $\ddot{\chi}$ (+ →) (S) $\rightarrow v$

$$f_L = \left(\frac{c}{c+v_S}\right) f_S = \left(\frac{345}{345+30}\right)(500\,Hz) = \underline{460\,Hz}$$

or

$$f = \frac{c}{\lambda} = \frac{345\,m\cdot s^{-1}}{0.75\,m} = 460\,Hz$$

e) $30\,m\cdot s^{-1}$ → (S) (+ ←) (L) ← $15\,m\cdot s^{-1}$

$v_S = -30\,m\cdot s^{-1}$

$v_L = +15\,m\cdot s^{-1}$

$$f_L = \left(\frac{c+v_L}{c+v_S}\right) f_S = \left(\frac{345+15}{345-30}\right)(500\,Hz) = \underline{571\,Hz}$$

f) $15\,m\cdot s^{-1}$ ← (L) (+ →) (S) → $30\,m\cdot s^{-1}$

$v_S = +30\,m\cdot s^{-1}$

$v_L = -15\,m\cdot s^{-1}$

$$f_L = \left(\frac{c+v_L}{c+v_S}\right) f_S = \left(\frac{345-15}{345+30}\right)(500\,Hz) = \underline{440\,Hz}$$

g) wind $\rightarrow v_w = 10\,m\cdot s^{-1}$

v_L and v_S in the Doppler formulas are relative to the air.

(a) $v \rightarrow 30\,m\cdot s^{-1}$, $v_w \rightarrow 10\,m\cdot s^{-1}$ $\Rightarrow v_S = 20\,m\cdot s^{-1}$

$$\lambda = \frac{345\,m\cdot s^{-1} - 20\,m\cdot s^{-1}}{500\,Hz} = \underline{0.65\,m}$$

(b) $$\lambda = \frac{345\,m\cdot s^{-1} + 20\,m\cdot s^{-1}}{500\ Hz} = \underline{0.73\,m}$$

(c) $v_S = -20\,m\cdot s^{-1}$ → (S) (+ ←) (L) ← $v_L = +10\,m\cdot s^{-1}$

$$f_L = \left(\frac{c+v_L}{c+v_S}\right) f_S = \left(\frac{345\,m\cdot s^{-1} + 10\,m\cdot s^{-1}}{345\,m\cdot s^{-1} - 20\,m\cdot s^{-1}}\right)(500\,Hz)$$

$$f_L = \underline{546\,Hz}$$

23-11 (cont)

(d) $v_L = -10\ m\cdot s^{-1}$ $v_S = +20\ m\cdot s^{-1}$

L (+ →) S

$$f_L = \left(\frac{c+v_L}{c+v_S}\right) f_S = \left(\frac{345-10}{345+20}\right)(500\,Hz)$$

$$f_L = \underline{459\,Hz}$$

(e) $v_S = -20\ m\cdot s^{-1}$ $v_L = +25\ m\cdot s^{-1}$

S (← +) L

$$f_L = \left(\frac{c+v_L}{c+v_S}\right) f_S = \left(\frac{345+25}{345-20}\right)(500\,Hz) = \underline{569\,Hz}$$

(f) $v_L = -25\ m\cdot s^{-1}$ $v_S = +20\ m\cdot s^{-1}$

L (+ →) S

$$f_L = \left(\frac{c+v_L}{c+v_S}\right) f_S = \left(\frac{345-25}{345+20}\right)(500\,Hz) = \underline{438\,Hz}$$

Problems

23-13

a) Calculate I at this distance:

$$\beta = 10 \log\left(\frac{I}{I_0}\right)$$

$$80 = 10 \log\left(\frac{I}{10^{-12}\,W\cdot m^{-2}}\right) \Rightarrow \log\left(\frac{I}{10^{-12}\,W\cdot m^{-2}}\right) = 8$$

$$\frac{I}{10^{-12}\,W\cdot m^{-2}} = 10^8 \Rightarrow I = 1\times 10^{-4}\ W\cdot m^{-2}$$

$$\text{eq. (23-9)} \Rightarrow I = \frac{c\,p_{max}^2}{2B} \Rightarrow p_{max} = \sqrt{\frac{2BI}{c}}$$

$B_{air} = 1.42\times 10^5\ Pa$ (Example 23-1)

$$p_{max} = \sqrt{\frac{2(1.42\times10^5\,Pa)(1\times10^{-4}\,W\cdot m^{-2})}{345\,m\cdot s^{-1}}} = \underline{0.287\,Pa}$$

b) eq (23-5) $p_{max} = BkA \Rightarrow A = \dfrac{p_{max}}{Bk}$

$$k = \frac{2\pi}{\lambda} = \frac{2\pi f}{c} = \frac{2\pi(440\,Hz)}{345\,m\cdot s^{-1}} = 8.01\ m^{-1}$$

$$A = \frac{0.287\,Pa}{(1.42\times10^5\,Pa)(8.01\,m^{-1})} = \underline{2.52\times10^{-7}\,m}$$

c) Consider concentric spheres, of radii r_1 and r_2, with the source at the center.

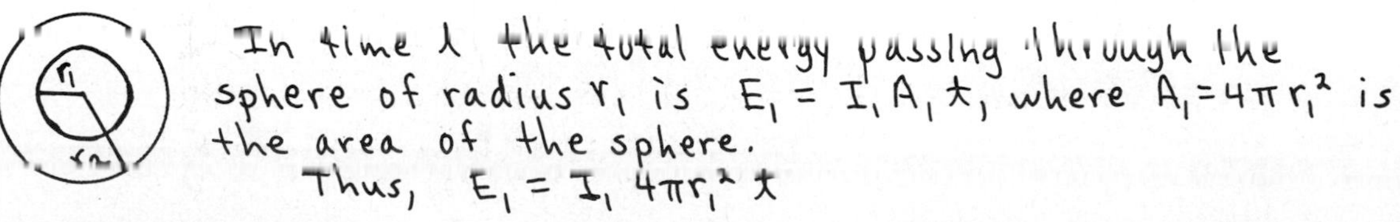

In time t the total energy passing through the sphere of radius r_1 is $E_1 = I_1 A_1 t$, where $A_1 = 4\pi r_1^2$ is the area of the sphere.

Thus, $E_1 = I_1\, 4\pi r_1^2\, t$

23-13 (cont)

But by energy conservation this must equal the energy passing through the sphere of radius r_2, $E_2 = I_2 4\pi r_2^2 t$.

$$E_1 = E_2 \Rightarrow I_1 4\pi r_1^2 t = I_2 4\pi r_2^2 t$$

$$r_2 = r_1 \left(\frac{I_1}{I_2}\right)^{1/2}$$

$\beta_1 = 80\,dB \Rightarrow I_1 = 1\times10^{-4}\ W\cdot m^{-2}$ (part (a))

$\beta_2 = 60\,dB \Rightarrow I_2 = 1\times10^{-6}\ W\cdot m^{-2}$

$$\frac{I_1}{I_2} = \frac{1\times10^{-4}\ W\cdot m^{-2}}{1\times10^{-6}\ W\cdot m^{-2}} = 1\times10^{2}$$

$$r_2 = 5m\,(1\times10^{2})^{1/2} = \underline{50m}$$

23-17

car approaching:

v_{car} → (S) (← +) $v_L = 0$

f_s f_{L1}

Positive direction is from the listener to the source $\Rightarrow v_s = -v_{car}$.

$$\frac{f_L}{c+v_L} = \frac{f_s}{c+v_s} \Rightarrow f_L = f_s\left(\frac{c+v_L}{c+v_s}\right)$$

$$f_{L1} = f_s\left(\frac{c}{c-v_{car}}\right)$$

car moving away:

$v_L = 0$ (+ →) → v_{car} (S)

f_{L2} f_s

Positive direction is from listener to the source $\Rightarrow v_s = +v_{car}$

$$f_{L2} = f_s\left(\frac{c}{c+v_{car}}\right)$$

$$\frac{f_{L1}}{f_{L2}} = \frac{f_s\left(\frac{c}{c-v_{car}}\right)}{f_s\left(\frac{c}{c+v_{car}}\right)} = \frac{c+v_{car}}{c-v_{car}}$$

half-tone drop $\Rightarrow \frac{f_{L1}}{f_{L2}} = 1.059 = \frac{c+v_{car}}{c-v_{car}}$

$$1.059c - 1.059\,v_{car} = c + v_{car}$$

$$2.059\,v_{car} = 0.059c$$

$$v_{car} = \frac{0.059}{2.059}(345\,m\cdot s^{-1}) = \underline{9.89\,m\cdot s^{-1}}$$

23-19

a) The wavelength of the waves is $\lambda_0 = \frac{c}{f_0}$.

The waves move at velocity $c+v$ relative to the surface.

$$\Rightarrow f = \frac{\text{velocity}}{\text{wavelength}} = \frac{c+v}{c/f_0} = f_0\left(\frac{c+v}{c}\right)$$

The number of waves that strike the surface in time t is

$$ft = f_0 t\left(\frac{c+v}{c}\right) = t(c+v)/\lambda_0 .$$

b)

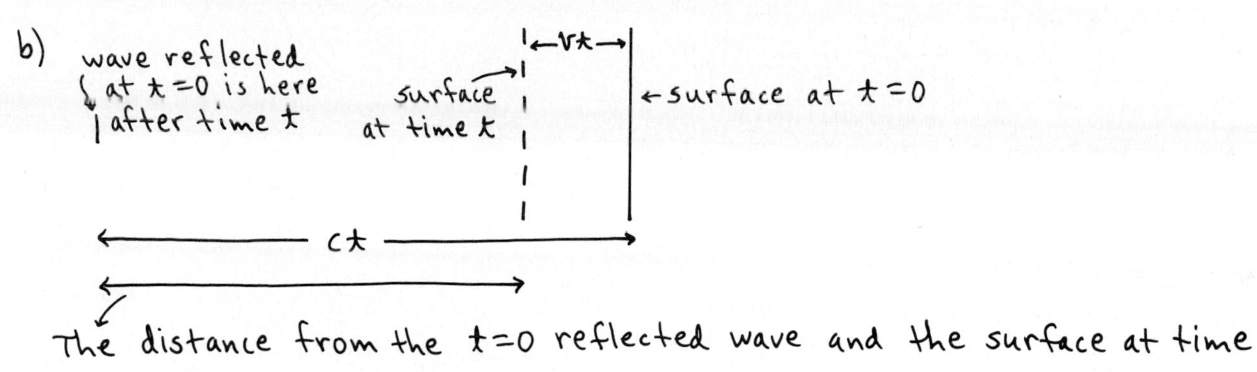

The distance from the $t=0$ reflected wave and the surface at time t is $ct - vt = \underline{(c-v)t}$.

c) $$\lambda = \frac{\text{distance over which waves are spread}}{\text{number of waves}} = \frac{(c-v)t}{ft}$$

$\lambda = (c-v)t\left[\frac{\lambda_0}{(c+v)t}\right]$, using the expression for ft from (a).

$$\Rightarrow \lambda = \lambda_0\left(\frac{c-v}{c+v}\right)$$

d) $$f = \frac{c}{\lambda} = \frac{c}{\lambda_0}\left(\frac{c+v}{c-v}\right) = f_0\left(\frac{c+v}{c-v}\right)$$

Note: this problem can also be analyzed by treating the wall as a moving listener, and then again as a moving source.

$v \leftarrow$ (S) (+ ←) (L) original source, $f_s = f_0$; wall

$v_s = 0$

$v_L = +v$

$$\frac{f_L}{c+v_L} = \frac{f_s}{c+v_s} \Rightarrow f_L = f_0\left(\frac{c+v}{c}\right)$$

The above f_L becomes f_s as the source re-emits (reflects) the waves:

$\leftarrow v$ (L) (+ →) (S) wall

$$f_s = f_0\left(\frac{c+v}{c}\right)$$

$v_L = 0$

$v_s = -v$

23-19 (cont)

$$\frac{f_L}{c+v_L} = \frac{f_S}{c+v_S} \Rightarrow f_L = f_S\left(\frac{c}{c-v}\right) = f_0\left(\frac{c+v}{c}\right)\left(\frac{c}{c-v}\right) = f_0\left(\frac{c+v}{c-v}\right)$$

This result agrees with the preceeding careful analysis.

e) She hears the frequencies f_0 and $f_L = f_0\left(\frac{c+v}{c-v}\right)$

$$f_{beat} = f_L - f_0 = f_0\left(\frac{c+v}{c-v}\right) - f_0 = f_0\left(\frac{c+v-c+v}{c-v}\right) = \frac{2vf_0}{c-v}$$

CHAPTER 24

Exercises 1, 5, 7, 13, 15, 17

Problems 19, 23, 25

Exercises

24-1

1 mole $\Rightarrow N_A = 6.02 \times 10^{23}$ atoms

There is one proton in a hydrogen atom (atomic number = 1 for hydrogen) $\Rightarrow 6.02 \times 10^{23}$ protons in 1 mole.

A proton has charge $e = 1.60 \times 10^{-19}$ C, so the charge of 1 mole of protons is $Q = (6.02 \times 10^{23})(1.60 \times 10^{-19}\,C) = \underline{9.65 \times 10^{4}\,C}$. (A huge amount of charge.)

24-5

a) $q_1 = q_2 = q$

$$F = \frac{1}{4\pi\epsilon_0}\frac{|q_1 q_2|}{r^2} = \frac{1}{4\pi\epsilon_0}\frac{q^2}{r^2} \Rightarrow q = r\sqrt{\frac{F}{1/4\pi\epsilon_0}} = 0.05\,m\sqrt{\frac{0.10\,N}{9 \times 10^9\,N\cdot m^2\cdot C^{-2}}}$$

$$q = \underline{1.67 \times 10^{-7}\,C}, \text{ on each}$$

b) $q_2 = 2q_1$

$$F = \frac{1}{4\pi\epsilon_0}\frac{|q_1 q_2|}{r^2} = \frac{1}{4\pi\epsilon_0}\frac{2q_1^2}{r^2} \Rightarrow q_1 = r\sqrt{\frac{F}{2(1/4\pi\epsilon_0)}} = \frac{1}{\sqrt{2}}(1.67 \times 10^{-7}\,C)$$

$$q_1 = \underline{1.18 \times 10^{-7}\,C}$$

$$q_2 = 2q_1 = 2(1.18 \times 10^{-7}\,C) = \underline{2.36 \times 10^{-7}\,C}$$

24-7

$$m_H \text{ (mass of 1 hydrogen atom)} = \frac{1 \times 10^{-3}\,kg}{6.02 \times 10^{23}} = 1.66 \times 10^{-27}\,kg$$

$$w_H = m_H g$$

The force of attraction between a proton and an electron separated by a distance r is

$$F = \frac{1}{4\pi\epsilon_0}\frac{|q_1 q_2|}{r^2} = \frac{1}{4\pi\epsilon_0}\frac{e^2}{r^2}$$

$$F = w_H \rightarrow \frac{1}{4\pi\epsilon_0}\frac{e^2}{r^2} = m_H g$$

24-7 (cont)

$$r = e\sqrt{\frac{1/4\pi\epsilon_0}{m_H g}} = 1.60\times10^{-19}\,C\sqrt{\frac{9\times10^9\,N\cdot m^2\cdot C^{-2}}{(1.66\times10^{-27}kg)(9.8\,m\cdot s^{-2})}} = 0.119\,m$$

(This is a very large distance compared to atomic dimensions.)

24-13

a)

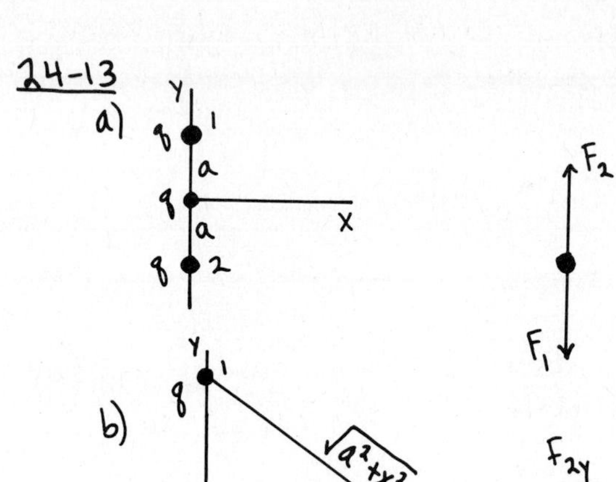

Forces on q at the origin due to charges 1 and 2 are:

$$F_1 = F_2 = \frac{1}{4\pi\epsilon_0}\frac{q^2}{a^2}$$

$\vec{F}_1$ and $\vec{F}_2$ are equal and opposite, so the total force $\vec{F} = \vec{F}_1 + \vec{F}_2 = 0$.

b)

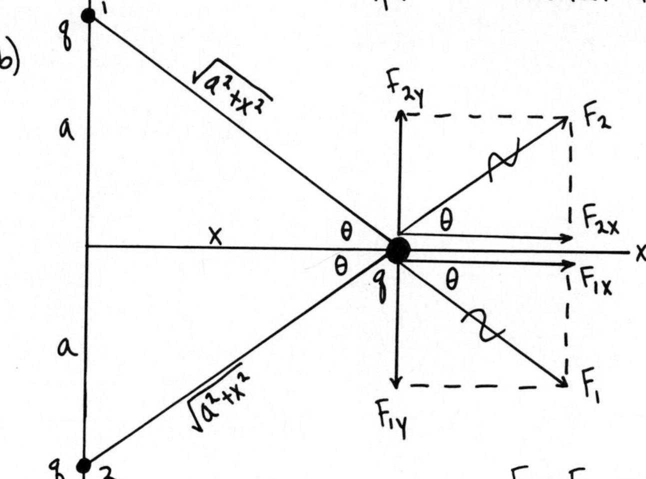

$$F_1 = F_2 = \frac{1}{4\pi\epsilon_0}\frac{q^2}{a^2+x^2}$$

$$\cos\theta = \frac{x}{\sqrt{a^2+x^2}}$$

$$\sin\theta = \frac{a}{\sqrt{a^2+x^2}}$$

$$F_y = F_{2y} - F_{1y} = F_1\sin\theta - F_2\sin\theta = 0$$

$$F_x = F_{1x} + F_{2x} = F_1\cos\theta + F_2\cos\theta$$

$$F = F_x = \frac{1}{2\pi\epsilon_0}\frac{xq^2}{(a^2+x^2)^{3/2}} \quad \text{(in the +x-direction)}$$

c) Note: $F(x) = -F(-x)$

$F(0) = 0$

$F(x) \to 0$ as $x \to \infty$.

Thus $F(x)$ has a maximum, that we can locate by $\frac{dF}{dx} = 0$

$$\Rightarrow \frac{d}{dx}\left(\frac{x}{(a^2+x^2)^{3/2}}\right) = 0$$

$$\frac{1}{(a^2+x^2)^{3/2}} - \frac{3}{2}\frac{x}{(a^2+x^2)^{5/2}}(2x) = 0$$

$$a^2 + x^2 - 3x^2 = 0 \Rightarrow x = \pm\frac{a}{\sqrt{2}}$$

The sketch of $F(x)$ must therefore be qualitatively

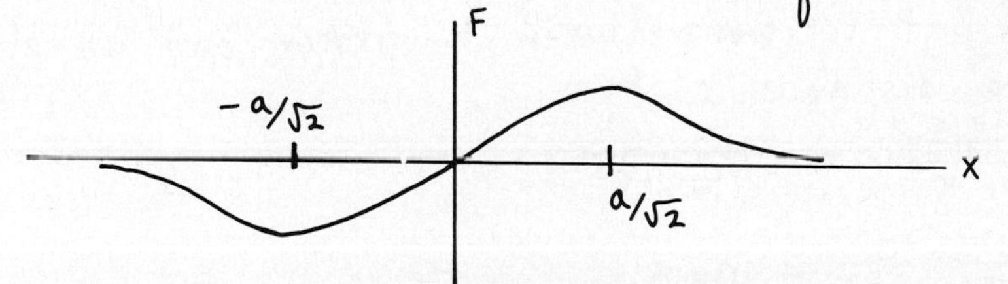

24-13 (cont)

d) x large $\Rightarrow a^2+x^2 \simeq x^2$

$$F(x) \longrightarrow \frac{1}{2\pi\epsilon_0}\frac{q^2 x}{x^3} = \frac{q^2}{2\pi\epsilon_0 x^2} = \frac{1}{4\pi\epsilon_0}\frac{q(2q)}{x^2}$$

Thus the two charges at $y = \pm a$ act like a point charge of charge $2q$ when the third charge is very far away.

24-15

a)

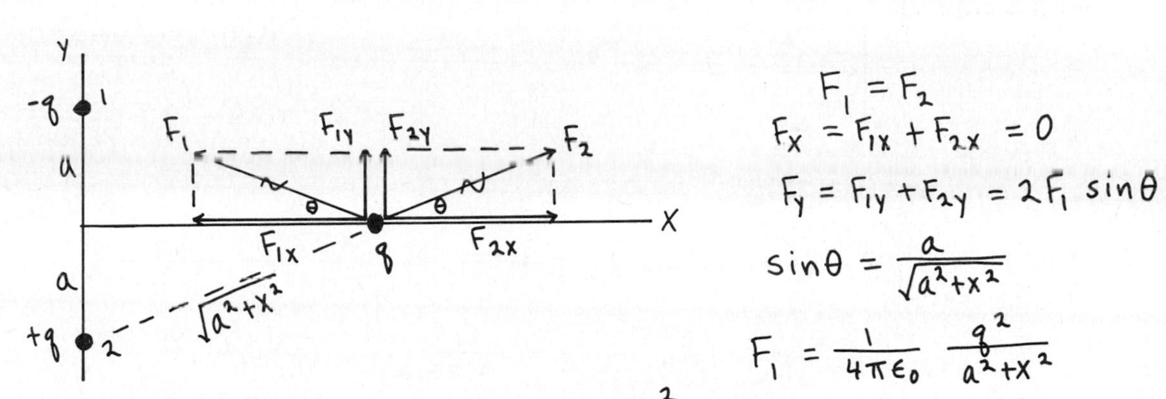

$$F_1 = F_2$$

$$F_x = F_{1x} + F_{2x} = 0$$

$$F_y = F_{1y} + F_{2y} = 2F_1 \sin\theta$$

$$\sin\theta = \frac{a}{\sqrt{a^2+x^2}}$$

$$F_1 = \frac{1}{4\pi\epsilon_0}\frac{q^2}{a^2+x^2}$$

Resultant force $F = F_y = \dfrac{1}{2\pi\epsilon_0}\dfrac{aq^2}{(a^2+x^2)^{3/2}}$

$$x \gg a \Rightarrow (a^2+x^2)^{3/2} \rightarrow x^3 \Rightarrow F \rightarrow \frac{aq^2}{2\pi\epsilon_0 x^3} \sim x^{-3}$$

b)

F_2, q, F_1, $-q$ 1, a, a, $+q$ 2; distances y, $y-a$, $y+a$

$$F = F_1 - F_2$$

$$F_1 = \frac{1}{4\pi\epsilon_0}\frac{q^2}{(y-a)^2}$$

$$F_2 = \frac{1}{4\pi\epsilon_0}\frac{q^2}{(y+a)^2}$$

$$\Rightarrow F = \frac{q^2}{4\pi\epsilon_0}\left(\frac{1}{(y-a)^2} - \frac{1}{(y+a)^2}\right)$$

$$F = \frac{q^2}{4\pi\epsilon_0 y^2}\left[\frac{1}{(1-a/y)^2} - \frac{1}{(1+a/y)^2}\right]$$

The binomial expansion (Appendix B)

$$\Rightarrow \frac{1}{(1-a/y)^2} = (1-a/y)^{-2} = 1 + \frac{2a}{y} + \ldots$$

and

$$\frac{1}{(1+a/y)^2} = (1+a/y)^{-2} = 1 - \frac{2a}{y} + \ldots$$

Thus

$$\frac{1}{(1-a/y)^2} - \frac{1}{(1+a/y)^2} \simeq 1 + \frac{2a}{y} - \left(1 - \frac{2a}{y}\right) = \frac{4a}{y}$$

24-15 (cont)

$$F \to \frac{q^2}{4\pi\epsilon_0 y^2}\,\frac{4a}{y} = \frac{aq^2}{\pi\epsilon_0 y^3} \sim y^{-3}$$

24-17

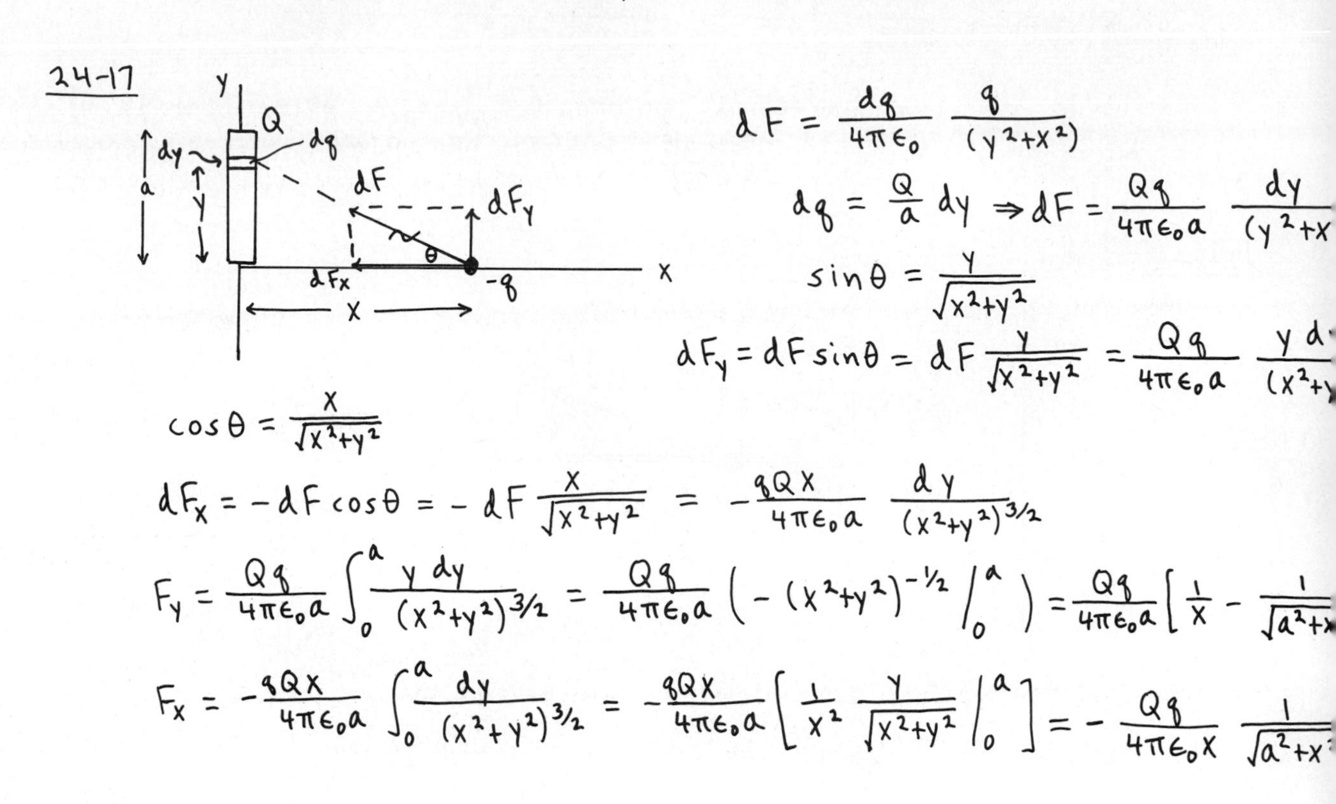

$$dF = \frac{dq}{4\pi\epsilon_0}\,\frac{q}{(y^2+x^2)}$$

$$dq = \frac{Q}{a}\,dy \Rightarrow dF = \frac{Qq}{4\pi\epsilon_0 a}\,\frac{dy}{(y^2+x$$

$$\sin\theta = \frac{y}{\sqrt{x^2+y^2}}$$

$$dF_y = dF\sin\theta = dF\frac{y}{\sqrt{x^2+y^2}} = \frac{Qq}{4\pi\epsilon_0 a}\,\frac{y\,d}{(x^2+y}$$

$$\cos\theta = \frac{x}{\sqrt{x^2+y^2}}$$

$$dF_x = -dF\cos\theta = -dF\frac{x}{\sqrt{x^2+y^2}} = -\frac{qQx}{4\pi\epsilon_0 a}\,\frac{dy}{(x^2+y^2)^{3/2}}$$

$$F_y = \frac{Qq}{4\pi\epsilon_0 a}\int_0^a \frac{y\,dy}{(x^2+y^2)^{3/2}} = \frac{Qq}{4\pi\epsilon_0 a}\left(-(x^2+y^2)^{-1/2}\Big|_0^a\right) = \frac{Qq}{4\pi\epsilon_0 a}\left[\frac{1}{x} - \frac{1}{\sqrt{a^2+x}}\right.$$

$$F_x = -\frac{qQx}{4\pi\epsilon_0 a}\int_0^a \frac{dy}{(x^2+y^2)^{3/2}} = -\frac{qQx}{4\pi\epsilon_0 a}\left[\frac{1}{x^2}\,\frac{y}{\sqrt{x^2+y^2}}\Big|_0^a\right] = -\frac{Qq}{4\pi\epsilon_0 x}\,\frac{1}{\sqrt{a^2+x}}$$

Problems

24-19

a)

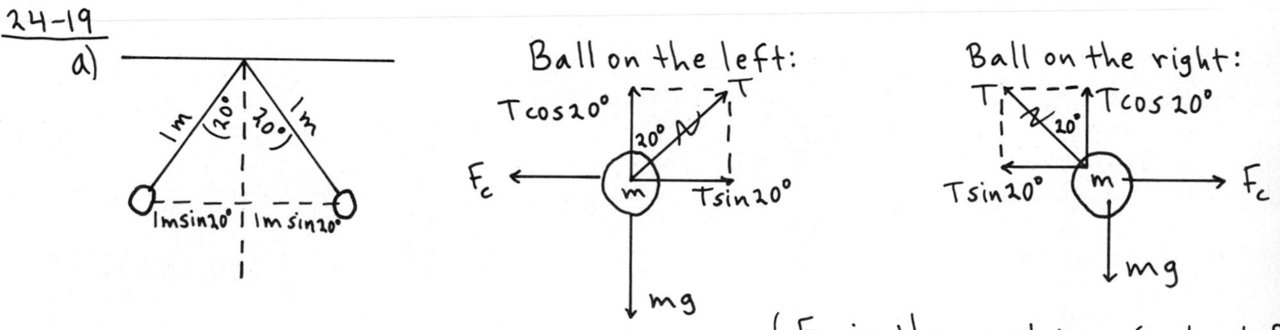

(F_c is the repulsive Coulomb fo exerted by one ball on the other.)

b) $\sum F_y = 0 \Rightarrow T\cos 20° - mg = 0$

$$T = \frac{mg}{\cos 20°}$$

$\sum F_x = 0 \Rightarrow T\sin 20° - F_c = 0$

$$F_c = \left(\frac{mg}{\cos 20°}\right)\sin 20° = mg\tan 20°$$

24-19 (cont)

$$F_c = \frac{1}{4\pi\epsilon_0}\frac{|q_1 q_2|}{r^2} = \frac{1}{4\pi\epsilon_0}\frac{q^2}{(2m\sin 20°)^2}$$

$$F_c = mg\tan 20° \Rightarrow \frac{1}{4\pi\epsilon_0}\frac{q^2}{(2m\sin 20°)^2} = mg\tan 20°$$

$$q = (2m\sin 20°)\sqrt{\frac{mg\tan 20°}{1/4\pi\epsilon_0}} = (2m\sin 20°)\sqrt{\frac{(10\times10^{-3}kg)(9.8m\cdot s^{-2})\tan 20°}{9\times10^{9}\,N\cdot m^2\cdot C^{-2}}}$$

$q = \underline{1.36\times10^{-6}C}$

c) The separation r between the balls now is $2l\sin\theta$

$$\Rightarrow F_c = \frac{1}{4\pi\epsilon_0}\frac{q^2}{(2l\sin\theta)^2}$$

$$F_c = mg\tan\theta \Rightarrow \frac{1}{4\pi\epsilon_0}\frac{q^2}{4l^2\sin^2\theta} = mg\tan\theta$$

$$\sin^2\theta\tan\theta = \left(\frac{1}{4\pi\epsilon_0}\right)\frac{q^2}{4l^2mg} = (9\times10^9 N\cdot m^2\cdot C^{-2})\frac{(1.36\times10^{-6}C)^2}{4(0.5m)^2(10\times10^{-3}kg)(9.8m\cdot s^{-2})}$$

$$\boxed{\sin^2\theta\tan\theta = 0.1699}$$

Solve this equation by trial and error. This will go quicker if we can make a good estimate of the value of θ that solves this equation:

θ small $\Rightarrow \tan\theta \approx \sin\theta$.

The equation then becomes $\sin^3\theta = 0.1699 \Rightarrow \sin\theta = 0.554 \Rightarrow \theta = 0.587\,rad$, or $\theta = 33.6°$

Now refine this guess:

θ	$\sin^2\theta\tan\theta$
33.6°	0.2035
33.0°	0.1926
30.0°	0.1443
31.0°	0.1594

θ	$\sin^2\theta\tan\theta$
32.0°	0.1755
31.5°	0.1673
31.6°	0.1689
31.7°	0.1705

θ	$\sin^2\theta\tan\theta$
31.65°	0.1697
31.66°	0.1699

This rounds to $\underline{\theta = 31.7°}$.

24-23

a)

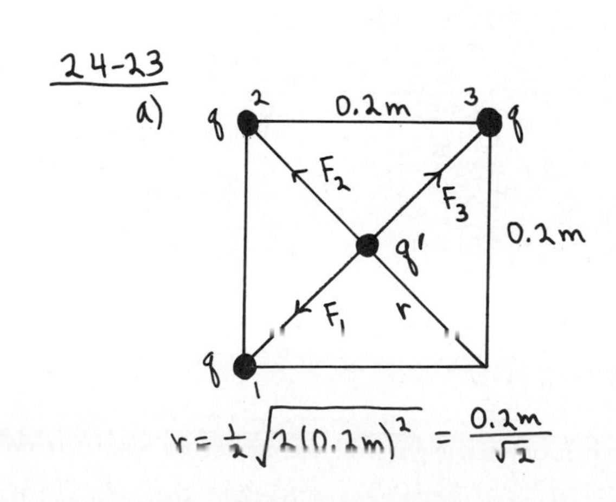

$r = \frac{1}{2}\sqrt{2(0.2m)^2} = \frac{0.2m}{\sqrt{2}}$

$q>0,\ q'<0 \Rightarrow$ The forces on q' are all attractive.

$F_1 = F_2 = F_3$

$\vec{F}_1 + \vec{F}_3 = 0$

Thus total $\vec{F} = \vec{F}_2$; direction is away from the vacant corner.

$$F_2 = \frac{1}{4\pi\epsilon_0}\frac{|qq'|}{r^2} = (9\times10^9 N\cdot m^2\cdot C^{-2})\frac{(2\times10^{-9}C)(1\times10^{-9}C)}{(0.2m/\sqrt{2})^2}$$

$F_2 = \underline{9.0\times10^{-7}C}$

24-23 (cont)

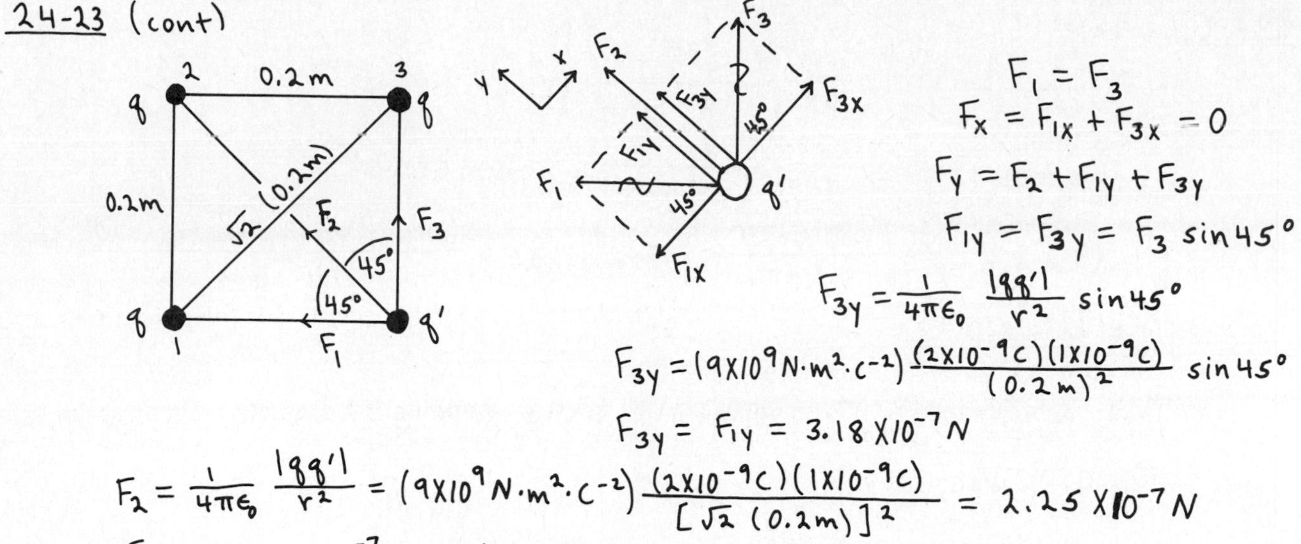

$$F_1 = F_3$$

$$F_x = F_{1x} + F_{3x} = 0$$

$$F_y = F_2 + F_{1y} + F_{3y}$$

$$F_{1y} = F_{3y} = F_3 \sin 45°$$

$$F_{3y} = \frac{1}{4\pi\epsilon_0}\frac{|qq'|}{r^2}\sin 45°$$

$$F_{3y} = (9\times10^9\ \mathrm{N\cdot m^2\cdot C^{-2}})\frac{(2\times10^{-9}\ \mathrm{C})(1\times10^{-9}\ \mathrm{C})}{(0.2\ \mathrm{m})^2}\sin 45°$$

$$F_{3y} = F_{1y} = 3.18\times10^{-7}\ \mathrm{N}$$

$$F_2 = \frac{1}{4\pi\epsilon_0}\frac{|qq'|}{r^2} = (9\times10^9\ \mathrm{N\cdot m^2\cdot C^{-2}})\frac{(2\times10^{-9}\ \mathrm{C})(1\times10^{-9}\ \mathrm{C})}{[\sqrt{2}\,(0.2\ \mathrm{m})]^2} = 2.25\times10^{-7}\ \mathrm{N}$$

$F_y = 2.25\times10^{-7}\ \mathrm{N} + 2(3.18\times10^{-7}\ \mathrm{N}) = \underline{8.61\times10^{-7}\ \mathrm{N}}$, directed toward the opposite corner

24-25

a)

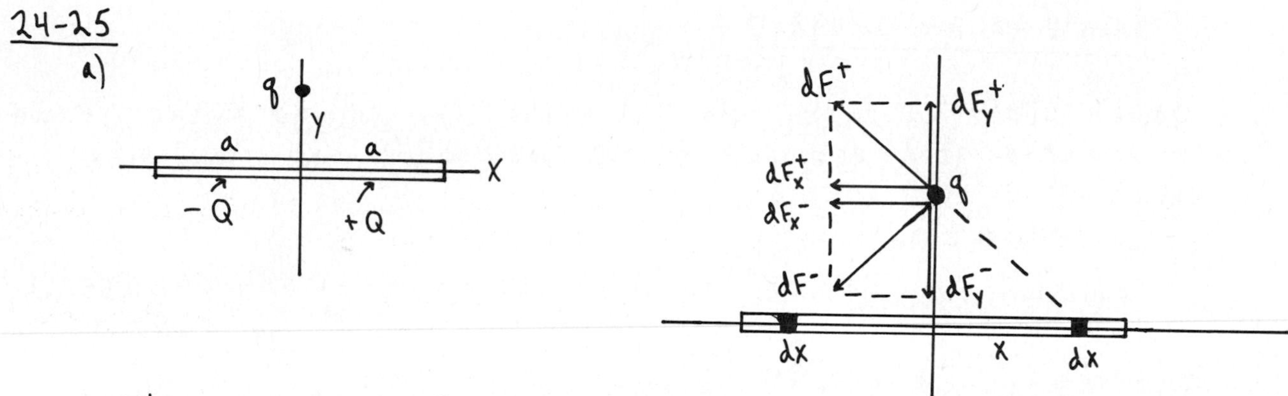

The y-components from the +Q and −Q segments cancel ⇒ $F_y = 0$

The x-components add, so calculate F_x from the +Q half and then multiply by 2:

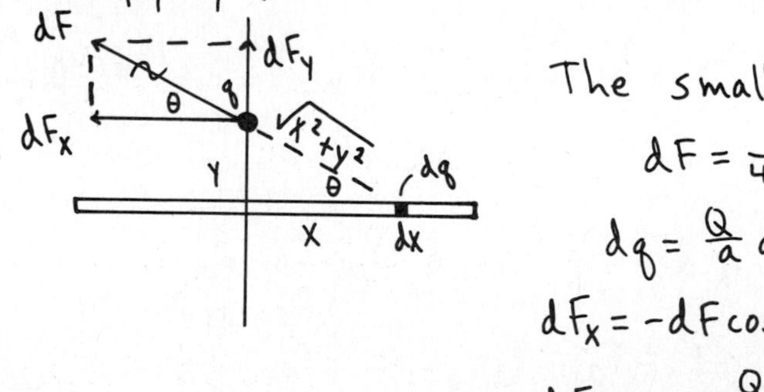

The small segment dx exerts a force

$$dF = \frac{1}{4\pi\epsilon_0}\frac{q\,dq}{x^2+y^2}$$

$$dq = \frac{Q}{a}dx \Rightarrow dF = \frac{Qq}{4\pi\epsilon_0 a}\frac{dx}{x^2+y^2}$$

$$dF_x = -dF\cos\theta\ ;\quad \cos\theta = \frac{x}{\sqrt{x^2+y^2}}$$

$$dF_x = -\frac{Qq}{4\pi\epsilon_0 a}\frac{x\,dx}{(x^2+y^2)^{3/2}}$$

$$F_x = 2\int_0^a dF_x = -\frac{Qq}{2\pi\epsilon_0 a}\int_0^a \frac{x\,dx}{(x^2+y^2)^{3/2}} = -\frac{Qq}{2\pi\epsilon_0 a}\left(-(x^2+y^2)^{-1/2}\Big|_0^a\right)$$

$$F_x = -\frac{Qq}{2\pi\epsilon_0 a}\left[\frac{1}{y} - \frac{1}{\sqrt{a^2+y^2}}\right]$$

24-25 (cont)

$$F = \frac{Qq}{2\pi\epsilon_0 a}\left[\frac{1}{y} - \frac{1}{\sqrt{a^2+y^2}}\right], \text{ in the } -x\text{-direction}$$

limit as y becomes large:

$$F = \frac{Qq}{2\pi\epsilon_0 a y}\left[1 - \frac{1}{\sqrt{1+a^2/y^2}}\right]$$

$$\frac{1}{\sqrt{1+a^2/y^2}} = (1+a^2/y^2)^{-1/2} = 1 - \frac{a^2}{2y^2} + \cdots \quad \text{(binomial theorem, Appendix B)}$$

Thus

$$F \to \frac{Qq}{2\pi\epsilon_0 a y}\left[1-\left(1-\frac{a^2}{2y^2}\right)\right] = \frac{qQa}{4\pi\epsilon_0 y^3} \sim y^{-3}, \text{ as was to be shown.}$$

b)

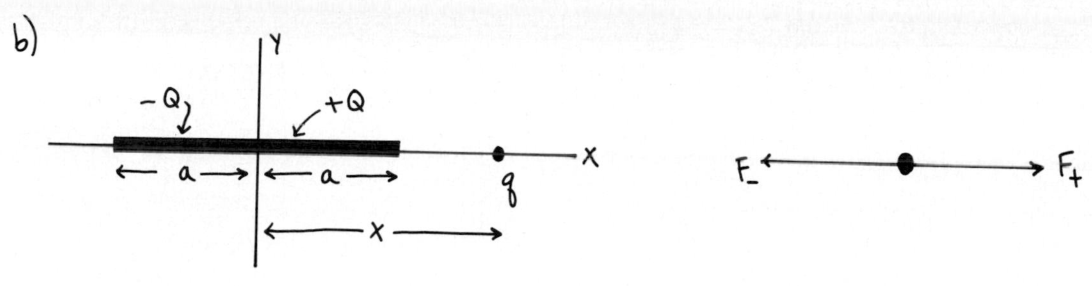

Calculate F_+:

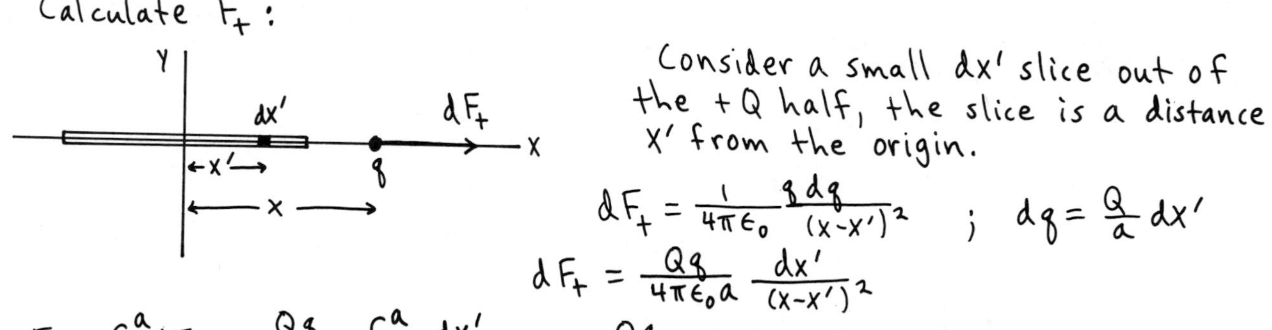

Consider a small dx' slice out of the $+Q$ half, the slice is a distance x' from the origin.

$$dF_+ = \frac{1}{4\pi\epsilon_0}\frac{q\,dq}{(x-x')^2} \quad ; \quad dq = \frac{Q}{a}dx'$$

$$dF_+ = \frac{Qq}{4\pi\epsilon_0 a}\frac{dx'}{(x-x')^2}$$

$$F_+ = \int_0^a dF_+ = \frac{Qq}{4\pi\epsilon_0 a}\int_0^a \frac{dx'}{(x-x')^2} = \frac{Qq}{4\pi\epsilon_0 a}\left(\frac{1}{x-x'}\Big|_0^a\right) = \frac{Qq}{4\pi\epsilon_0 a}\left[\frac{1}{x-a} - \frac{1}{x}\right]$$

Calculate F_-:

$$dF_- = \frac{1}{4\pi\epsilon_0}\frac{q\,dq}{(x+x')^2} \quad ; \quad dq = \frac{Q}{a}dx'$$

$$dF_- = \frac{Qq}{4\pi\epsilon_0 a}\frac{dx'}{(x+x')^2}$$

$$F_- = \int_0^a dF_- = \frac{Qq}{4\pi\epsilon_0 a}\int_0^a \frac{dx'}{(x+x')^2}$$

$$F_- = \frac{Qq}{4\pi\epsilon_0 a}\left(-\frac{1}{x+x'}\Big|_0^a\right) = \frac{Qq}{4\pi\epsilon_0 a}\left[\frac{1}{x} - \frac{1}{x+a}\right]$$

$$F = F_+ - F_- = \frac{Qq}{4\pi\epsilon_0 a}\left[\frac{1}{x-a} - \frac{1}{x} - \frac{1}{x} + \frac{1}{x+a}\right]$$

$$F = \frac{Qq}{4\pi\epsilon_0 a}\left[\frac{1}{x-a} + \frac{1}{x+a} - \frac{2}{x}\right], \text{ in the } +x\text{-direction.}$$

(Since $F_+ > F_-$.)

24-25 (cont)

Limit as $x \rightarrow$ large

$$\frac{1}{x-a} = \frac{1}{x}\left(1-\frac{a}{x}\right)^{-1} = \frac{1}{x}\left(1+\frac{a}{x}+\frac{a^2}{x^2}+\ldots\right) \rightarrow \frac{1}{x}+\frac{a}{x^2}+\frac{a^2}{x^3}$$

$$\frac{1}{x+a} = \frac{1}{x}\left(1+\frac{a}{x}\right)^{-1} = \frac{1}{x}\left(1-\frac{a}{x}+\frac{a^2}{x^2}-\ldots\right) \rightarrow \frac{1}{x}-\frac{a}{x^2}+\frac{a^2}{x^3}$$

Thus

$$\left[\frac{1}{x-a}+\frac{1}{x+a}-\frac{2}{x}\right] \rightarrow \frac{1}{x}+\frac{a}{x^2}+\frac{a^2}{x^3}+\frac{1}{x}-\frac{a}{x^2}+\frac{a^2}{x^3}-\frac{2}{x} = \frac{2a^2}{x^3}$$

And

$$F \rightarrow \frac{Qq}{4\pi\epsilon_0 a}\left(\frac{2a^2}{x^3}\right) = \frac{qQa}{2\pi\epsilon_0 x^3} \sim x^{-3}, \text{ as desired}$$

CHAPTER 25

Exercises 5, 7, 9, 13, 17, 19, 23

Problems 27, 29, 31, 35

Exercises

25-5

$\vec{E} = \frac{\vec{F}}{q}$. Thus $\vec{F}\downarrow$ but q negative $\Rightarrow \vec{E}\uparrow$ (upward)

$$E = \frac{F}{|q|} = \frac{20\times10^{-9}\,N}{5\times10^{-9}\,C} = \underline{4.0\,N\cdot C^{-1}}$$

25-7

$F = qE \Rightarrow F = eE$

$F = mg \Rightarrow eE = mg$

$$E = \frac{mg}{e} = \frac{(9.11\times10^{-31}\,kg)(9.8\,m\cdot s^{-2})}{1.60\times10^{-19}\,C} = \underline{5.58\times10^{-11}\,N\cdot C^{-1}}$$

25-9

The force is opposite to the electric field, since q is negative. The acceleration of the electron is thus upward and equal to $a = \frac{F}{m} = \frac{eE}{m}$

Use the kinematic information to solve for the acceleration:

x-component	y-component
$a_x = 0$	$a_y = a$
$x - x_0 = 0.02\,m$	$y - y_0 = 0.005\,m$
$v_{0x} = v_0 = 1.0\times10^{7}\,m\cdot s^{-1}$	$v_{0y} = 0$

This is very similar to a projectile motion problem, except that the acceleration is upward.

Use the x-component motion to solve for t:

$t = ?$

$x - x_0 = v_{0x}t + \frac{1}{2}a_x t^2$ (with $a_x = 0$)

$$t = \frac{x - x_0}{v_{0x}} = \frac{0.02\,m}{1.0\times10^{7}\,m\cdot s^{-1}} = 2.0\times10^{-9}\,s$$

Now use the y-component to solve for a:

$a = ?$

$t = 2.0\times10^{-9}\,s$

$y - y_0 = v_{0y}t + \frac{1}{2}a_y t^2$ (with $v_{0y} = 0$)

$$a_y = \frac{2(y - y_0)}{t^2} = \frac{2(0.005\,m)}{(2.0\times10^{-9}\,s)^2} = 2.5\times10^{15}\,m\cdot s^{-2}$$

25-9 (cont)

$$a = \frac{eE}{m} \Rightarrow E = \frac{ma}{e} = \frac{(9.11\times10^{-31}\,kg)(2.5\times10^{15}\,m\cdot s^{-2})}{1.60\times10^{-19}\,C} = \underline{1.42\times10^{4}\,N\cdot C^{-1}}$$

25-13

a)

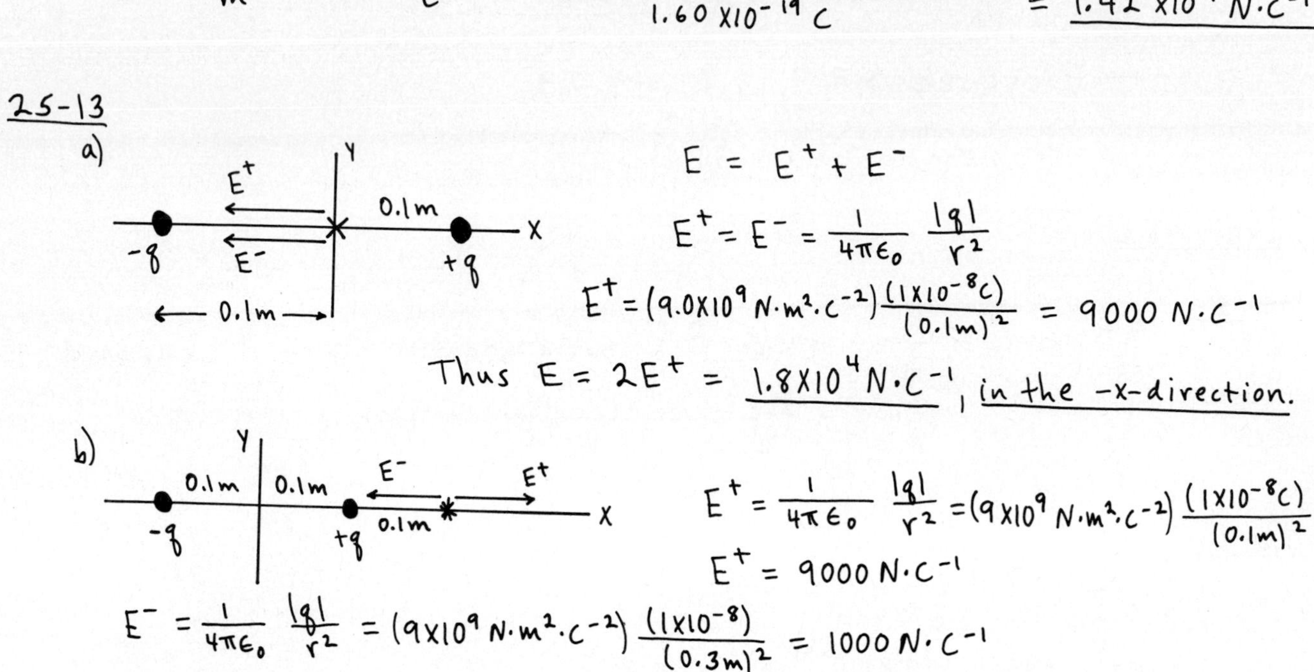

$$E = E^+ + E^-$$

$$E^+ = E^- = \frac{1}{4\pi\epsilon_0}\frac{|q|}{r^2}$$

$$E^+ = (9.0\times10^9\,N\cdot m^2\cdot C^{-2})\frac{(1\times10^{-8}C)}{(0.1m)^2} = 9000\,N\cdot C^{-1}$$

Thus $E = 2E^+ = \underline{1.8\times10^4\,N\cdot C^{-1}}$, in the $-x$-direction.

b)

$$E^+ = \frac{1}{4\pi\epsilon_0}\frac{|q|}{r^2} = (9\times10^9\,N\cdot m^2\cdot C^{-2})\frac{(1\times10^{-8}C)}{(0.1m)^2}$$

$$E^+ = 9000\,N\cdot C^{-1}$$

$$E^- = \frac{1}{4\pi\epsilon_0}\frac{|q|}{r^2} = (9\times10^9\,N\cdot m^2\cdot C^{-2})\frac{(1\times10^{-8})}{(0.3m)^2} = 1000\,N\cdot C^{-1}$$

$$E = E^+ - E^- = 9000\,N\cdot C^{-1} - 1000\,N\cdot C^{-1} = \underline{8000\,N\cdot C^{-1}}$$, in the $+x$-direction

c)

$$\sin\theta = \frac{0.2m}{0.25m} \Rightarrow \theta = 53.1^\circ$$

$$E^+ = \frac{1}{4\pi\epsilon_0}\frac{|q|}{r^2} = (9\times10^9\,N\cdot m^2\cdot C^{-2})\frac{(1\times10^{-8}C)}{(0.15m)^2} = 4000\,N\cdot C^{-1}$$

$$E^- = \frac{1}{4\pi\epsilon_0}\frac{|q|}{r^2} = (9\times10^9\,N\cdot m^2\cdot C^{-2})\frac{(1\times10^{-8}C)}{(0.25m)^2} = 1440\,N\cdot C^{-1}$$

$$E_x^- = -E^-\sin\theta = -(1440\,N\cdot C^{-1})\sin 53.1^\circ = -1152\,N\cdot C^{-1}$$

$$E_y^- = -E^-\cos\theta = -(1440\,N\cdot C^{-1})\cos 53.1^\circ = -864\,N\cdot C^{-1}$$

$$E_x = E_x^+ + E_x^- = 0 - 1152\,N\cdot C^{-1} = -1152\,N\cdot C^{-1}$$

$$E_y = E_y^+ + E_y^- = 4000\,N\cdot C^{-1} - 864\,N\cdot C^{-1} = 3136\,N\cdot C^{-1}$$

25-13 (cont)

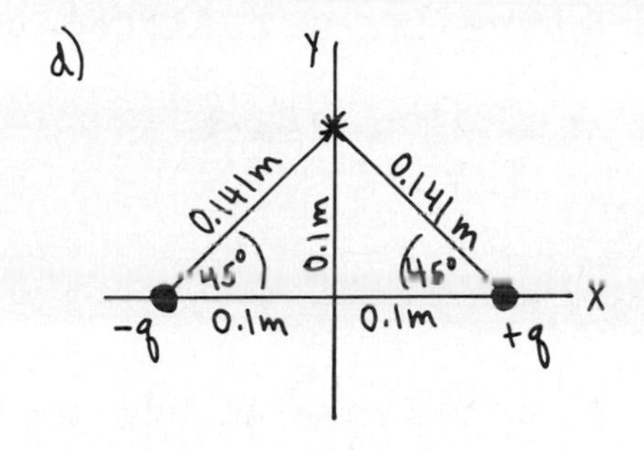

$$E = \sqrt{E_x^2 + E_y^2}$$

$$E = \sqrt{(-1152\ N\cdot C^{-1})^2 + (3136\ N\cdot C^{-1})^2} = \underline{3340\ N\cdot C^{-1}}$$

$$\tan\theta = \frac{3136\ N\cdot C^{-1}}{1152\ N\cdot C^{-1}} = 2.72$$

$\Rightarrow \theta = \underline{69.8°}$, $\underline{\text{above the } -x\text{-axis}}$

d)

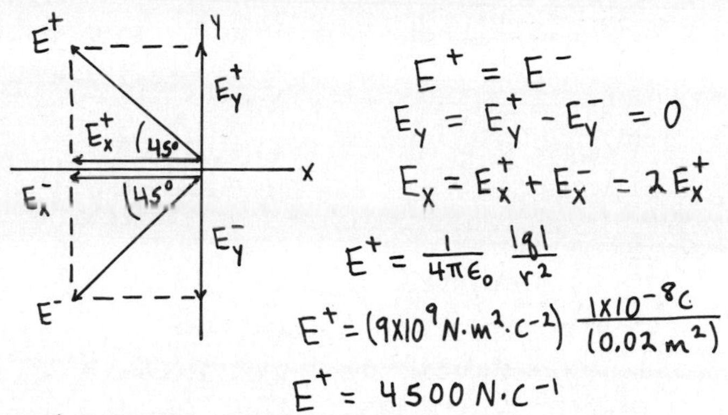

$$E^+ = E^-$$

$$E_y = E_y^+ - E_y^- = 0$$

$$E_x = E_x^+ + E_x^- = 2E_x^+$$

$$E^+ = \frac{1}{4\pi\epsilon_0}\frac{|q|}{r^2}$$

$$E^+ = (9\times10^9\ N\cdot m^2\cdot C^{-2})\frac{1\times10^{-8}C}{(0.02\ m^2)}$$

$$E^+ = 4500\ N\cdot C^{-1}$$

$$E_x^+ = -E^+\cos 45° = -3180\ N\cdot C^{-1}$$

$$\Rightarrow E_x = 2(-3180\ N\cdot C^{-1}) = -6360\ N\cdot C^{-1}$$

Thus $E = \underline{6360\ N\cdot C^{-1}}$, in the $\underline{-x\text{- direction.}}$

25-17

For a sphere of radius R and charge q, $E = \frac{1}{4\pi\epsilon_0}\frac{q}{r^2}$ for $r>R$ (Example 25-12).

Thus just outside the surface the electric field is

$$E = \frac{1}{4\pi\epsilon_0}\frac{q}{R^2}.$$

$$\Rightarrow q = \frac{ER^2}{(1/4\pi\epsilon_0)} = \frac{(1300\ N\cdot C^{-1})(0.05\ m)^2}{9\times10^9\ N\cdot m^2\cdot C^{-2}} = 3.61\times10^{-10}\ C$$

One electron has a charge of 1.60×10^{-19} C, so need

$$\frac{3.61\times10^{-10}\ C}{1.60\times10^{-19}\ C} = \underline{2.26\times10^9 \text{ electrons}}$$

25-19

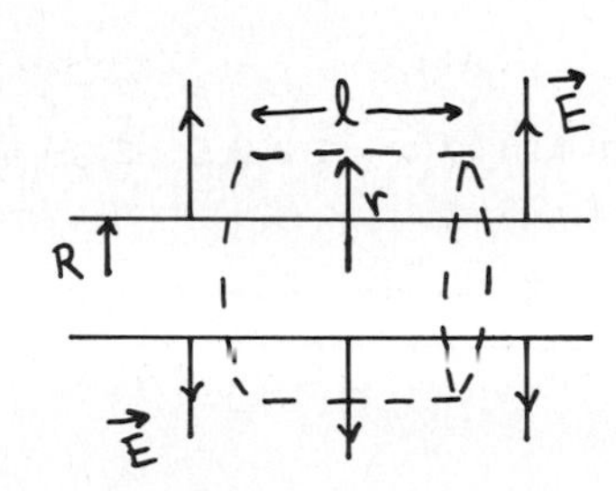

Apply Gauss's Law to a cylindrical surface coaxial with the conductor, of radius r and length l.

By symmetry the electric field is perpendicular to the cylinder axis and is directed outward.

25-19 (cont)

$$\oint \vec{E}\cdot d\vec{A} = \int_{\substack{\text{end}\\\text{faces}}} \vec{E}\cdot d\vec{A} + \int_{\substack{\text{curved}\\\text{surface}}} \vec{E}\cdot d\vec{A}$$

On the end faces $\vec{E}\cdot d\vec{A} = 0$ and on the curved surface $\vec{E}\cdot d\vec{A} = E\,dA$

Also, E by symmetry must depend only on the distance from the axis $\Rightarrow$ E is constant on the curved surface.

Thus $\oint \vec{E}\cdot d\vec{A} = 0 + E\int_{\substack{\text{curved}\\\text{surface}}} dA = E(2\pi r \ell)$

The enclosed charge Q is $\lambda\ell$, where λ is the charge per unit length on the conducting cylinder.
(λ is related to the surface charge density by $\lambda = \sigma(2\pi R)$, where R is the radius of the conducting cylinder.)

Apply Gauss's Law: $\epsilon_0 \oint \vec{E}\cdot d\vec{A} = Q$

$$\epsilon_0 E(2\pi r\ell) = \lambda\ell \Rightarrow E = \frac{\lambda}{2\pi\epsilon_0 r}$$

This agrees with eq. (25-22) for the electric field due to a line of charge.

25-23

a) Apply Gauss's Law to a sphere of radius r, where $a<r<b$.

The electric field is radially outward, so $\vec{E}\cdot d\vec{A} = E\,dA$, and E is constant on the surface.

Thus $\oint \vec{E}\cdot d\vec{A} = \oint E\,dA = E\oint dA = E(4\pi r^2)$

$Q = q$ is the enclosed charge.

$\epsilon_0 \oint \vec{E}\cdot d\vec{A} = Q \Rightarrow \epsilon_0 E\, 4\pi r^2 = q \Rightarrow E = \frac{q}{4\pi\epsilon_0 r^2}$, for $a<r<b$.

b) In a static situation, $E=0$ inside a conductor. Thus $E=0$ for $b<r<c$ and $E=0$ for $r<a$.

The only region left is $r>c$. Apply Gauss's Law to a sphere of radius $r>c$:

$\oint \vec{E}\cdot d\vec{A} = E(4\pi r^2)$

$Q=q$ (The outer hollow sphere carries no net charge.)

25-23 (cont)

$$\epsilon_0 \oint \vec{E}\cdot d\vec{A} = Q \Rightarrow \epsilon_0 E(4\pi r^2) q$$

$$E = \frac{q}{4\pi\epsilon_0 r^2}$$

c)

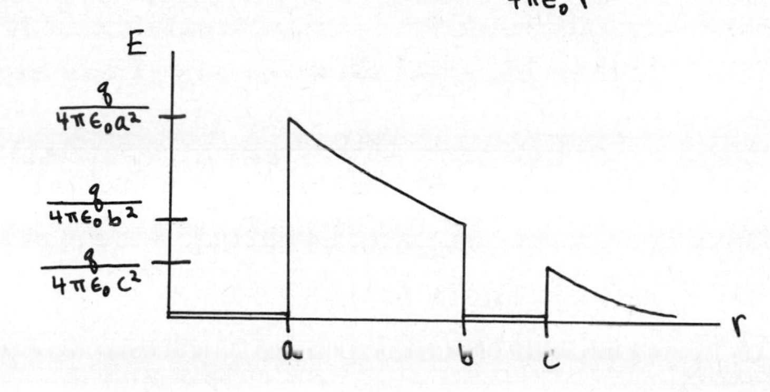

d) Apply Gauss's Law to a sphere of radius r, where $b < r < c$:

$E = 0$ inside conductor $\Rightarrow Q = 0$. Thus a charge $-q$ must be on the inside surface of the hollow sphere. The hollow sphere has no net charge, so it must have a charge $+q$ on its outer surface.

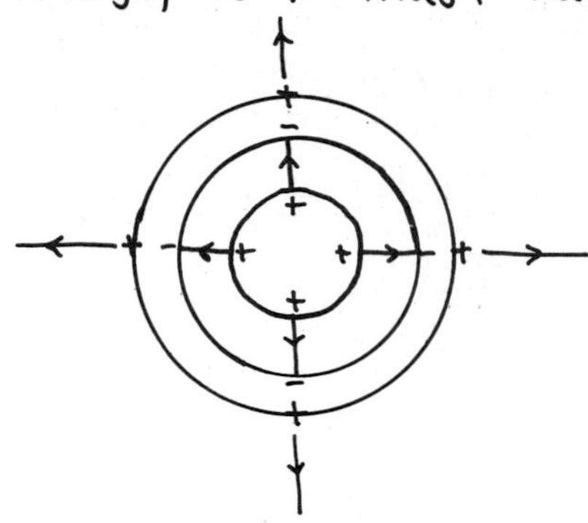

Problems

25-27

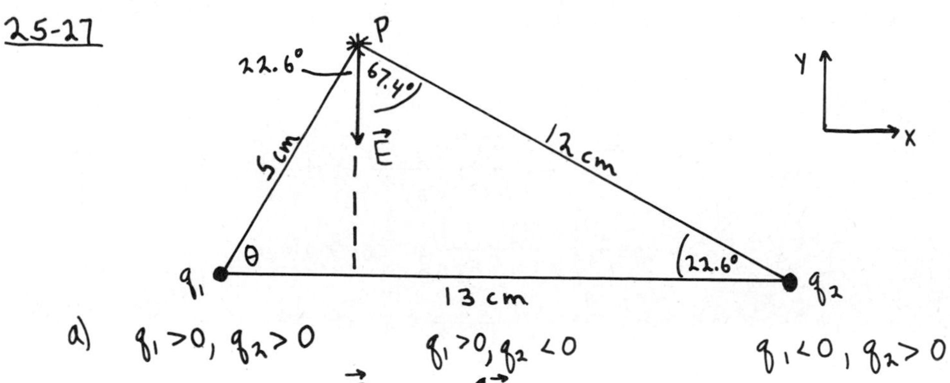

$$\sin\theta = \frac{12\text{cm}}{13\text{cm}} \Rightarrow \theta = 67.4°$$

$$|q_1| = 5\times10^{-6}\,C$$

a) $q_1 > 0,\ q_2 > 0$ $\quad$ $q_1 > 0,\ q_2 < 0$ $\quad$ $q_1 < 0,\ q_2 > 0$ $\quad$ $q_1 < 0,\ q_2 < 0$

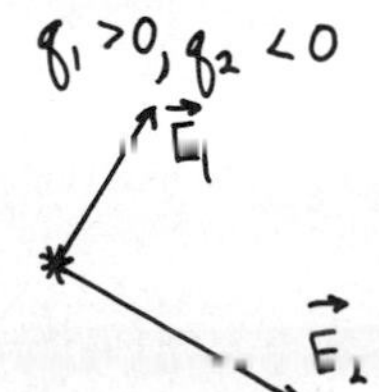

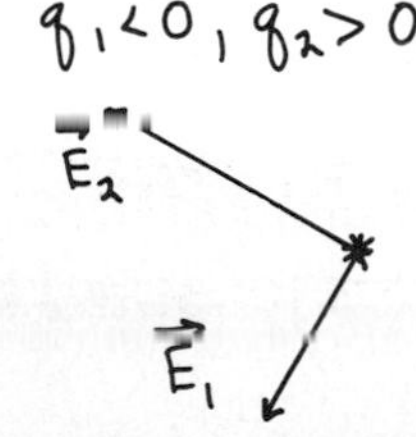

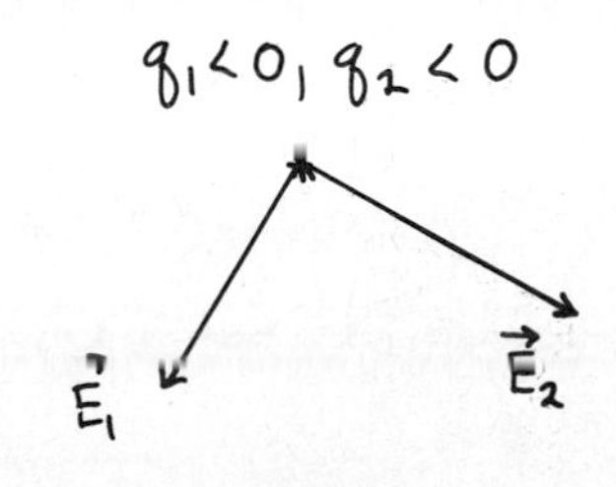

25-27 (cont)

b) Only in the $q_1 < 0$, $q_2 < 0$ sketch is it possible to have $E_x = E_{1x} + E_{2x} = 0$ and $E_y = E_{1y} + E_{2y}$ negative $\Rightarrow q_1 = -5\times10^{-6}\,C$ and $q_2 = -|q_2|$.

c)

$$E_1 = \frac{1}{4\pi\epsilon_0}\frac{|q_1|}{r_1^2} = (9\times10^9\,N\cdot m^2\cdot C^{-2})\frac{5\times10^{-6}C}{(0.05m)^2} = 1.80\times10^7\,N\cdot C^{-1}$$

$$E_2 = \frac{1}{4\pi\epsilon_0}\frac{|q_2|}{r_2^2} = (9\times10^9\,N\cdot m^2\cdot C^{-2})\frac{|q_2|}{(0.12m)^2}$$

$$E_2 = (6.25\times10^{11}\,N\cdot C^{-2})\,|q_2|$$

$$E_{2x} = E_2 \sin 67.4^\circ = |q_2|\,(5.77\times10^{11}\,N\cdot C^{-2})$$

$$E_{1x} = -E_1 \sin 22.6^\circ = -6.92\times10^6\,N\cdot C^{-2}$$

$$E_x = E_{1x} + E_{2x} = 0 \Rightarrow -6.92\times10^6\,N\cdot C^{-2} + |q_2|\,(5.77\times10^{11}\,N\cdot C^{-2}) = 0$$

$|q_2| = 1.20\times10^{-5}\,C$ and $q_2 = -1.20\times10^{-5}\,C$

With $|q_2|$ known, $E_2 = (6.25\times10^{11}\,N\cdot C^{-2})(1.20\times10^{-5}C) = 7.50\times10^6\,N\cdot C^{-1}$

$$E = E_y = E_{1y} + E_{2y} = -E_1\cos 22.6^\circ - E_2 \cos 67.4^\circ$$

$$E_y = -(1.80\times10^7\,N\cdot C^{-1})\cos 22.6^\circ - (7.50\times10^6\,N\cdot C^{-1})\cos 67.4^\circ = -1.95\times10^7\,N\cdot C^{-1}$$

Thus $E = 1.95\times10^7\,N\cdot C^{-1}$, in the $-y$-direction.

25-29

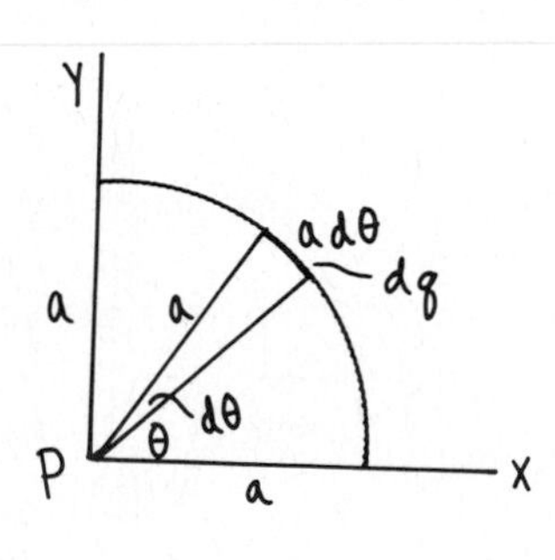

Divide the charge distribution into small segments, use the point charge formula for the electric field due to each small segment, and integrate over the charge distribution to find the total x and y components of the field.

$$dq = \frac{Q}{\frac{1}{4}(2\pi a)}(a\,d\theta) = \frac{2Q}{\pi}\,d\theta$$

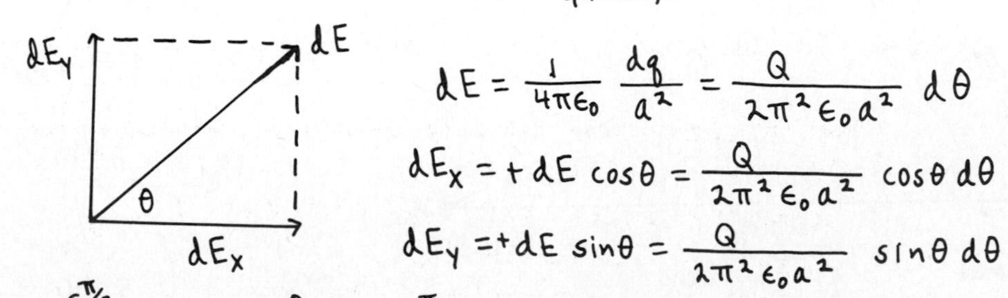

$$dE = \frac{1}{4\pi\epsilon_0}\frac{dq}{a^2} = \frac{Q}{2\pi^2\epsilon_0 a^2}\,d\theta$$

$$dE_x = +dE\cos\theta = \frac{Q}{2\pi^2\epsilon_0 a^2}\cos\theta\,d\theta$$

$$dE_y = +dE\sin\theta = \frac{Q}{2\pi^2\epsilon_0 a^2}\sin\theta\,d\theta$$

$$E_x = \int_0^{\pi/2} dE_x = \frac{Q}{2\pi^2\epsilon_0 a^2}\int_0^{\pi/2}\cos\theta\,d\theta = \frac{Q}{2\pi^2\epsilon_0 a^2}\left(\sin\theta\Big|_0^{\pi/2}\right) = \frac{Q}{2\pi^2\epsilon_0 a^2}$$

$$E_y = \int_0^{\pi/2} dE_y = \frac{Q}{2\pi^2\epsilon_0 a^2}\int_0^{\pi/2}\sin\theta\,d\theta = \frac{Q}{2\pi^2\epsilon_0 a^2}\left(-\cos\theta\Big|_0^{\pi/2}\right) = \frac{Q}{2\pi^2\epsilon_0 a^2}$$

25-31

Find the flux through the parallelopiped, and then apply Gauss's Law to find the net charge within:

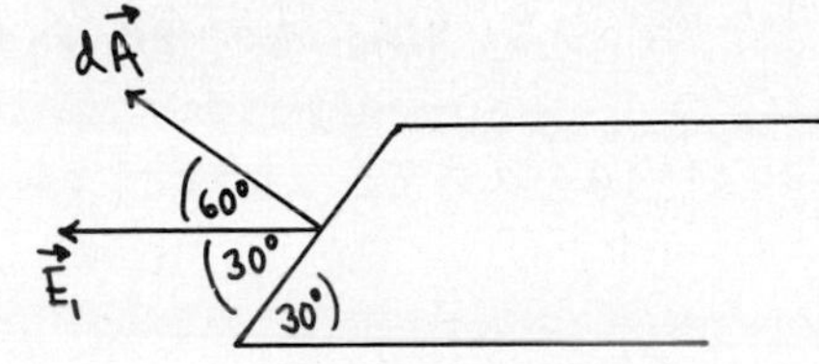

$\vec{E}_1 \cdot d\vec{A} = E_1 \cos 60° \, dA$

Flux through this face is $\Phi_1 = \int \vec{E}_1 \cdot d\vec{A}$

$\Phi_1 = \int E_1 \cos 60° \, dA$

E_1 is constant $\Rightarrow \Phi_1 = E_1 \cos 60° \int dA = E_1 \cos 60° A$, where $A = (0.06\text{m})(0.05\text{m}) = 0.003\text{m}^2$

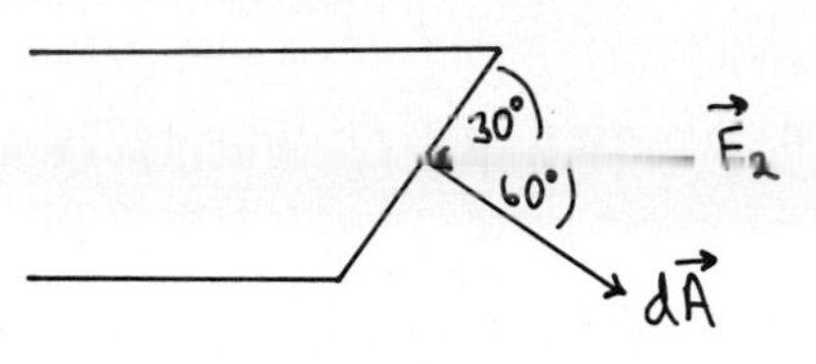

$\vec{E}_2 \cdot d\vec{A} = -E_2 \cos 60° \, dA$

Flux through this face is $\Phi_2 = \int \vec{E}_2 \cdot d\vec{A} = \int E_2 \cos 60° \, dA$

E_2 is constant $\Rightarrow \Phi_2 = -E_2 \cos 60° \int dA = -E_2 \cos 60° A$, $A = 0.003\text{m}^2$

The net flux is $\Phi = \Phi_1 + \Phi_2 = (E_1 - E_2) \cos 60° A$

$\Phi = (9 \times 10^4 \text{N}\cdot\text{C}^{-1} - 11 \times 10^4 \text{N}\cdot\text{C}^{-1}) \cos 60° (0.003\text{m}^2) = -30 \text{ N}\cdot\text{m}^2\cdot\text{C}^{-1}$

Gauss's Law: $\epsilon_0 \oint \vec{E} \cdot d\vec{A} = Q$

$\oint \vec{E} \cdot d\vec{A} = \Phi = -30 \text{N}\cdot\text{m}^2\cdot\text{C}^{-1}$

$\Rightarrow Q = (8.854 \times 10^{-12} \text{C}^2\cdot\text{N}^{-1}\cdot\text{m}^{-2})(-30 \text{ N}\cdot\text{m}^2\cdot\text{C}^{-1}) = \underline{-2.66 \times 10^{-10} \text{C}}$

(The enclosed charge is negative; the flux into the parallelopiped is larger than the flux out.)

25-35

$\rho(r) = \rho_0 (1 - r/R)$ for $r \le R$, $\rho_0 = \frac{3Q}{\pi R^3}$

$\rho = 0$ for $r \ge R$

a) The charge density varies with r inside the spherical volume.

Divide the volume up into thin concentric shells, of radius r and thickness dr.

The volume of such a shell is $dV = 4\pi r^2 dr$.

The charge in the shell is

$dq = \rho(r) dV = 4\pi r^2 \rho_0 (1 - r/R) dr$

The total charge Q in the charge distribution is obtained by integrating dq over all such shells into which the sphere can be subdivided:

25-35 (cont)

$$Q = \int dq = \int_0^R 4\pi r^2 \rho_0 (1 - r/R)\,dr = 4\pi\rho_0 \int_0^R \left(r^2 - \tfrac{1}{R} r^3\right) dr = 4\pi\rho_0 \left(\frac{r^3}{3} - \frac{r^4}{4R}\right)\Big|_0^R$$

$$Q = 4\pi\rho_0 \left(\frac{R^3}{3} - \frac{R^4}{4R}\right) = 4\pi\rho_0 \left(\frac{R^3}{12}\right) = 4\pi \left(\frac{3Q}{\pi R^3}\right)\left(\frac{R^3}{12}\right) = Q, \text{ as was to be shown.}$$

b) Apply Gauss's Law, to a spherical surface of radius r, where $r > R$:

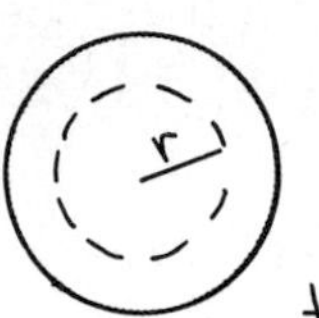

$$\epsilon_0 \oint \vec{E} \cdot d\vec{A} = Q_{enclosed}$$

$$\epsilon_0 E\, 4\pi r^2 = Q$$

$$E = \frac{Q}{4\pi\epsilon_0 r^2}, \text{ for } r \geq R \text{ ; same as for point charge of charge } Q$$

c) Apply Gauss's Law to a spherical surface of radius r, where $r < R$:

$$\oint \vec{E} \cdot d\vec{A} = E\, 4\pi r^2$$

To calculate the enclosed charge Q_{encl} use the same technique as in (a), except integrate dq out to r rather than R. (We want only that charge inside radius r.)

$$Q_{encl} = \int_0^r 4\pi r'^2 \rho_0 \left(1 - \frac{r'}{R}\right) dr' = 4\pi\rho_0 \int_0^r \left(r'^2 - \frac{r'^3}{R}\right) dr' = 4\pi\rho_0 \left(\frac{r'^3}{3} - \frac{r'^4}{4R}\right)\Big|_0^r$$

$$Q_{encl} = 4\pi\rho_0 \left(\frac{r^3}{3} - \frac{r^4}{4R}\right) = 4\pi\rho_0 r^3 \left(\frac{1}{3} - \frac{1}{4}\frac{r}{R}\right)$$

$$\rho_0 = \frac{3Q}{\pi R^3} \Rightarrow Q_{encl} = 12Q\frac{r^3}{R^3}\left(\frac{1}{3} - \frac{1}{4}\frac{r}{R}\right) = Q\left(\frac{r^3}{R^3}\right)\left(4 - 3\frac{r}{R}\right)$$

Thus

$$\epsilon_0 \oint \vec{E} \cdot d\vec{A} = Q_{encl} \Rightarrow \epsilon_0 E\, 4\pi r^2 = Q\left(\frac{r^3}{R^3}\right)\left(4 - 3\frac{r}{R}\right)$$

$$E = \frac{Qr}{4\pi\epsilon_0 R^3}\left(4 - 3\frac{r}{R}\right), \quad r \leq R$$

d) With $r = R$ in the $r \geq R$ expression (part (b)):

$$E(R) = \frac{Q}{4\pi\epsilon_0 R^2}$$

With $r = R$ in the $r \leq R$ expression (part (c)):

$$E(R) = \frac{Q}{4\pi\epsilon_0 R^2}(4 - 3) = \frac{Q}{4\pi\epsilon_0 R^2}$$

So these two results agree at the surface of the sphere.

CHAPTER 26

Exercises 1, 5, 7, 9, 15, 17, 19, 21, 25, 27

Problems 29, 33, 37, 39, 41, 45

Exercises

26-1

a) $U = \frac{1}{4\pi\epsilon_0}\frac{qQ}{r} = (9.0\times10^{9}\,\text{N}\cdot\text{m}^2\cdot\text{C}^{-2})\frac{(-0.5\times10^{-6}\,\text{C})(4.0\times10^{-6}\,\text{C})}{0.8\,\text{m}} = \underline{-0.0225\,\text{J}}$

(Note: The electrical potential energy for a pair of point charges can be either positive or negative, depending on the signs of the charges.)

b) Use conservation of energy. Only the coulomb force does work

$\Rightarrow K_1 + U_1 = K_2 + U_2$, with $U = \frac{1}{4\pi\epsilon_0}\frac{qQ}{r}$

released from rest $\Rightarrow K_1 = 0$

$K_2 = \frac{1}{2}mv_2^2$

$U_1 = -0.0225\,\text{J}$, from (a)

$U_2 = \frac{1}{4\pi\epsilon_0}\frac{qQ}{r_2} = (9\times10^{9}\,\text{N}\cdot\text{m}^2\cdot\text{C}^{-2})\frac{(-0.5\times10^{-6}\,\text{C})(4.0\times10^{-6}\,\text{C})}{0.4\,\text{m}} = -0.0450\,\text{J}$

$K_1 + U_1 = K_2 + U_2 \Rightarrow 0 - 0.0225\,\text{J} = \frac{1}{2}mv_2^2 - 0.0450\,\text{J}$

$\frac{1}{2}mv_2^2 = +0.0225\,\text{J}$

$v_2 = \sqrt{\frac{2(0.0225\,\text{J})}{3\times10^{-4}\,\text{kg}}} = \underline{12.2\,\text{m}\cdot\text{s}^{-1}}$

26-5

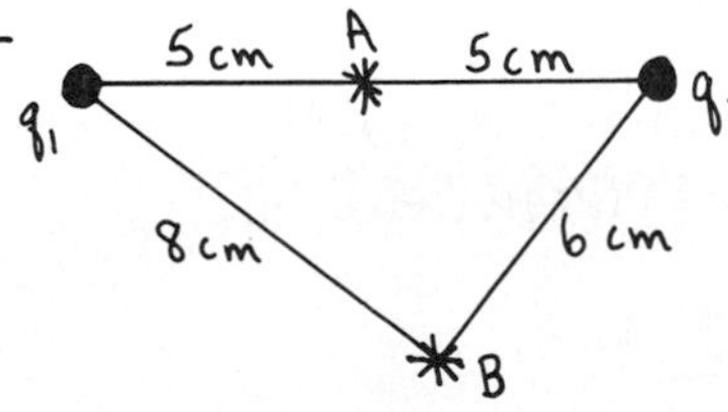

$V = \frac{1}{4\pi\epsilon_0}\sum_i \frac{q_i}{r_i}$

a) $V_A = \frac{1}{4\pi\epsilon_0}\left(\frac{q_1}{0.05\,\text{m}} + \frac{q_2}{0.05\,\text{m}}\right)$

$V_A = \frac{9\times10^{9}\,\text{N}\cdot\text{m}^2\cdot\text{C}^{-2}}{0.05\,\text{m}}(40\times10^{-9}\,\text{C} - 30\times10^{-9}\,\text{C}) = \underline{1.8\times10^{3}\,\text{V}}$

b) $V_B = \frac{1}{4\pi\epsilon_0}\left(\frac{q_1}{0.08\,\text{m}} + \frac{q_2}{0.06\,\text{m}}\right) = 9\times10^{9}\,\text{N}\cdot\text{m}^2\cdot\text{C}^{-2}\left(\frac{40\times10^{-9}\,\text{C}}{0.08\,\text{m}} + \frac{-30\times10^{-9}\,\text{C}}{0.06\,\text{m}}\right)$

$V_B = 4.5\times10^{3}\,\text{V} - 4.5\times10^{3}\,\text{V} = \underline{0}$

c) $W_{A\to B} = U_A - U_B = q'(V_A - V_B) = 25\times10^{-9}\,\text{C}\,(1.8\times10^{3}\,\text{V} - 0) = \underline{+4.5\times10^{-5}\,\text{J}}$

26-7

a)

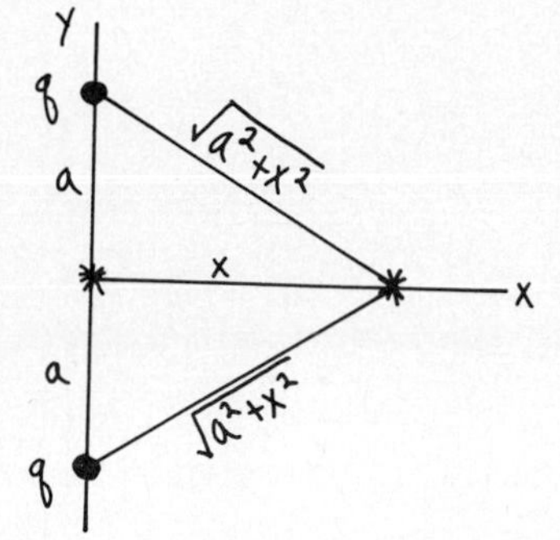

b) $V = \frac{1}{4\pi\epsilon_0} \sum_i \frac{q_i}{r_i}$

$V_0 = \frac{1}{4\pi\epsilon_0}\left(\frac{q}{a} + \frac{q}{a}\right) = \frac{q}{2\pi\epsilon_0 a}$

c) $V = \frac{1}{4\pi\epsilon_0} \sum_i \frac{q_i}{r_i}$

From the sketch in (a), $V = \frac{1}{4\pi\epsilon_0}\left(\frac{q}{\sqrt{a^2+x^2}} + \frac{q}{\sqrt{a^2+x^2}}\right)$

$V = \frac{2q}{4\pi\epsilon_0\sqrt{a^2+x^2}}$, as was to be shown.

d) V is maximum at $x=0$, and goes to zero as x gets large. Also V is symmetric about the y-axis: $V(x) = V(-x)$. Thus a sketch of V(x) is

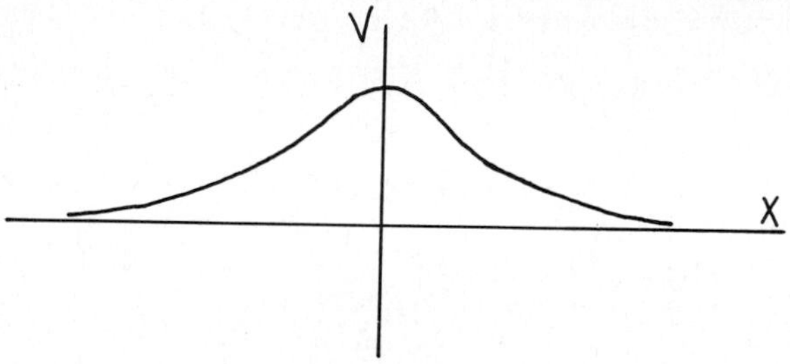

e) $V = \frac{1}{2}V_0 \Rightarrow \frac{2q}{4\pi\epsilon_0\sqrt{a^2+x^2}} = \frac{1}{2}\frac{q}{2\pi\epsilon_0 a} \Rightarrow \frac{2}{\sqrt{a^2+x^2}} = \frac{1}{a}$

$4a^2 = a^2+x^2$

$x^2 = 3a^2 \Rightarrow \underline{x = \pm\sqrt{3}\,a}$

26-9

a) From Eq(26-19), $E = \frac{V_{ab}}{d} = \frac{500\text{V}}{0.1\text{m}} = \underline{5000\ \text{V}\cdot\text{m}^{-1}}$

b)

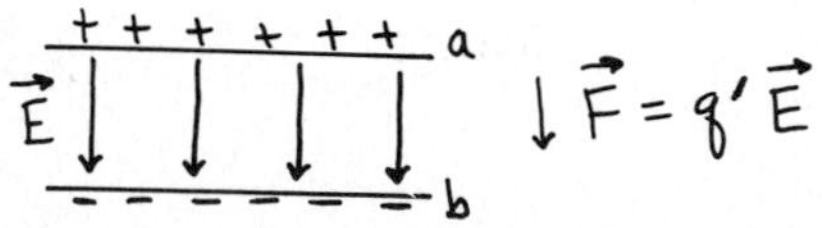

The positive plate a is at higher potential.

$W_{a\to b} = \int_a^b \vec{F}\cdot d\vec{l} = \int_a^b F\,dl = q'\int_a^b E\,dl = q'Ed$, where d is the separation between the plates.

$W_{a\to b} = (2\times10^{-9}\text{C})(5000\text{V}\cdot\text{m}^{-1})(0.1\text{m}) = \underline{1.0\times10^{-6}\text{J}}$

c) $W_{a\to b} = U_a - U_b = q'(V_a - V_b) = q'V_{ab} = (2.0\times10^{-9}\text{c})(500\text{V}) = \underline{1.0\times10^{-6}\text{J}}$

26-15

a)

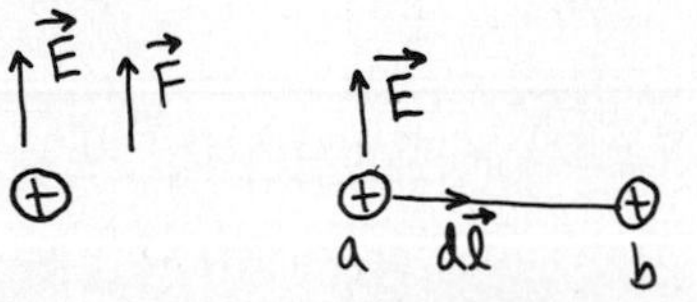

$W_{a\to b} = \int_a^b \vec{F}\cdot d\vec{l} = q\int_a^b \vec{E}\cdot d\vec{l}$

But $\vec{E}\cdot d\vec{l} = 0 \Rightarrow W_{a\to b} = \underline{0}$

26-15 (cont)

b)

$\vec{E} \cdot d\vec{l} = -E\,dl$

$W_{a\to b} = q\int_a^b \vec{E} \cdot d\vec{l} = -qE\int_a^b dl = -qE(0.8\text{m})$

Thus

$W_{a\to b} = -(2.5\times10^{-8}\text{C})(5\times10^{4}\text{N}\cdot\text{C}^{-1})(0.8\text{m}) = \underline{-1.0\times10^{-3}\text{J}}$

c)

$\vec{E} \cdot d\vec{l} = E\cos 45^\circ\, dl$

$W_{a\to b} = q\int_a^b \vec{E} \cdot d\vec{l} = qE\cos 45^\circ \int_a^b dl = qE\cos 45^\circ(2.60\text{m})$

$W_{a\to b} = (2.5\times10^{-8}\text{C})(5\times10^{4}\text{N}\cdot\text{C}^{-1})(\cos 45^\circ)(2.60\text{m}) = \underline{2.30\times10^{-3}\text{ J}}$

26-17

Outside the sphere, $r > R$, $E(r) = \dfrac{Q}{4\pi\epsilon_0 r^2}$

inside the sphere, $r < R$, $E(r) = \dfrac{Qr}{4\pi\epsilon_0 R^3}$

Let point B be at ∞; $V_B = 0$.

First take point A to be at a distance $r > R$ from the center of the sphere:

$V_A - V_B = \int_A^B \vec{E} \cdot d\vec{l}$

$V_r - 0 = \int_r^\infty \vec{E} \cdot d\vec{r}$; both $\vec{E}$ and $d\vec{r}$ are radially outward $\Rightarrow \vec{E} \cdot d\vec{r} = E\,dr$

$V_r = \int_r^\infty E\,dr$

In this range of r, $r > R$ to ∞, $E = \dfrac{Q}{4\pi\epsilon_0 r^2}$.

$V_r = \int_r^\infty \dfrac{Q}{4\pi\epsilon_0 r^2}\,dr = \dfrac{Q}{4\pi\epsilon_0}\int_r^\infty \dfrac{dr}{r^2} = \dfrac{Q}{4\pi\epsilon_0}\left(-\dfrac{1}{r}\Big|_r^\infty\right) = \underline{\dfrac{Q}{4\pi\epsilon_0 r}}$

From this expression, at the surface of the sphere $V_R = \dfrac{Q}{4\pi\epsilon_0 R}$.

To find V_r at a point for which $r < R$, take point B at R and point A at r:

$V_A - V_B = \int_A^B \vec{E} \cdot d\vec{l} \Rightarrow V_r - V_R = \int_r^R \vec{E} \cdot d\vec{r}$

Inside the sphere $\vec{E}$ is radially outward $\Rightarrow \vec{E} \cdot d\vec{r} = E\,dr$.

In this range of r, $r < R$, $E = \dfrac{Qr}{4\pi\epsilon_0 R^3}$.

Thus $V_r - V_R = \int_r^R E\,dr = \int_r^R \dfrac{Qr}{4\pi\epsilon_0 R^3}\,dr = \dfrac{Q}{4\pi\epsilon_0 R^3}\int_r^R r\,dr = \dfrac{Q}{4\pi\epsilon_0 R^3}\left(\dfrac{1}{2}r^2\Big|_r^R\right)$

26-17 (cont)

$$V_r - V_R = \frac{Q}{8\pi\epsilon_0 R^3}(R^2 - r^2)$$

$$V_r = V_R + \frac{Q}{8\pi\epsilon_0 R^3}(R^2 - r^2) = \frac{Q}{4\pi\epsilon_0 R} + \frac{Q}{8\pi\epsilon_0 R} - \frac{Qr^2}{8\pi\epsilon_0 R^3}$$

$$\underline{V_r = \frac{Q}{8\pi\epsilon_0 R^3}(3R^2 - r^2)}$$

26-19

From Example 26-14, $V = \frac{1}{4\pi\epsilon_0}\frac{Q}{2a}\ln\left[\frac{\sqrt{a^2+x^2}+a}{\sqrt{a^2+x^2}-a}\right]$

$$E_x = -\frac{dV}{dx} = -\frac{Q}{8\pi\epsilon_0 a}\frac{d}{dx}\left[\ln\left(\frac{\sqrt{a^2+x^2}+a}{\sqrt{a^2+x^2}-a}\right)\right]$$

$$\frac{d}{dx}\ln f(x) = \frac{1}{f(x)}\frac{df}{dx}$$

$$\Rightarrow \frac{d}{dx}\left[\ln\left(\frac{\sqrt{a^2+x^2}+a}{\sqrt{a^2+x^2}-a}\right)\right] = \left(\frac{\sqrt{a^2+x^2}-a}{\sqrt{a^2+x^2}+a}\right)\left(\frac{x}{\sqrt{a^2+x^2}}\right)\left(\frac{1}{\sqrt{a^2+x^2}-a}\right) + \left(\sqrt{a^2+x^2}-a\right)\left(-\frac{1}{\sqrt{a^2+x^2}-a}\right)^2\left(\frac{x}{\sqrt{a^2+x^2}}\right)$$

$$\frac{d}{dx}\left[\ln\left(\frac{\sqrt{a^2+x^2}+a}{\sqrt{a^2+x^2}-a}\right)\right] = \frac{x}{\sqrt{a^2+x^2}\,(\sqrt{a^2+x^2}+a)} - \frac{x}{\sqrt{a^2+x^2}\,(\sqrt{a^2+x^2}-a)}$$

$$= \frac{x}{\sqrt{a^2+x^2}}\left[\frac{1}{\sqrt{a^2+x^2}+a} - \frac{1}{\sqrt{a^2+x^2}-a}\right] = \frac{x}{\sqrt{a^2+x^2}}\left[\frac{\sqrt{a^2+x^2}-a-\sqrt{a^2+x^2}-a}{(\sqrt{a^2+x^2}+a)(\sqrt{a^2+x^2}-a)}\right]$$

$$= -\frac{2ax}{\sqrt{a^2+x^2}}\left(\frac{1}{a^2+x^2-a^2}\right) = -\frac{2a}{x\sqrt{a^2+x^2}}$$

Thus

$$E_x = -\left(\frac{Q}{8\pi\epsilon_0 a}\right)\left(-\frac{2a}{x\sqrt{a^2+x^2}}\right) = +\frac{Q}{4\pi\epsilon_0 x\sqrt{a^2+x^2}},$$

which agrees with E_x as calculated in Example 25-8.

26-21

a) No electric field ⇒ no electric force.

$6\pi\eta r v$ (viscous force) ↑ ; v ↓ ; (−10e) ; ↓ mg

$$v = v_T \Rightarrow a = 0$$

$$mg - 6\pi\eta r v_T = 0$$

$$v_T = \frac{mg}{6\pi\eta r} = \frac{(3\times10^{-14}\,\mathrm{kg})(9.8\,\mathrm{m\cdot s^{-2}})}{6\pi(180\times10^{-7}\,\mathrm{N\cdot s\cdot m^{-2}})(2\times10^{-6}\,\mathrm{m})} = \underline{4.33\times10^{-4}\,\mathrm{m\cdot s^{-1}}}$$

b) ↑ $\vec{F} = -10e\vec{E}$; (−10e) ; ↓ $\vec{E}$

($\vec{E}$ downward, charge negative ⇒ force is upward.)

Note: $mg = (3\times10^{-14}\,\mathrm{kg})(9.8\,\mathrm{m\cdot s^{-2}})$

$mg = 2.94\times10^{-13}\,\mathrm{N}$

26-21 (cont)

$$F = 10eE = 10(1.60\times10^{-19}\,C)(3\times10^{5}\,N\cdot C^{-1}) = 4.80\times10^{-13}\,N$$

$F > mg \Rightarrow$ the oil drop travels upward, so the viscous force is downward.

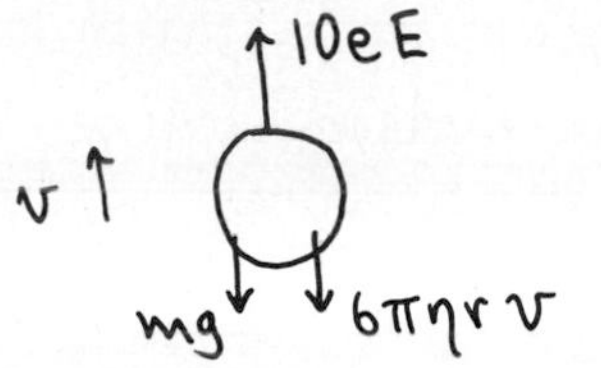

$$v = v_T \Rightarrow a = 0$$

$$10eE - mg - 6\pi\eta r v_T = 0$$

$$v_T = \frac{10eE - mg}{6\pi\eta r} = \frac{4.80\times10^{-13}\,N - 2.94\times10^{-13}\,N}{6\pi(180\times10^{-7}\,N\cdot s\cdot m^{-2})(2\times10^{-6}\,m)} = \underline{2.74\,m\cdot s^{-1}},$$

upward

26-25

a) $K_1 + U_1 = K_2 + U_2$

$K_1 = 0$

$K_2 = U_1 - U_2 = q'(V_1 - V_2) = -e(V_1 - V_2) \Rightarrow K_2 = e\Delta V$

rest energy $E_0 = mc^2 = (9.11\times10^{-31}\,kg)(3.0\times10^{8}\,m\cdot s^{-1})^2 = 8.20\times10^{-14}\,J$

$K_2 = 1\%\,E_0 \Rightarrow K_2 = 8.20\times10^{-16}\,J$

$$\Delta V = \frac{K_2}{e} = \frac{8.20\times10^{-16}\,J}{1.60\times10^{-19}\,C} = 5125\,J\cdot C^{-1} = \underline{5125\,V}$$

b) $K = \frac{1}{2}mv^2 \Rightarrow v = \sqrt{\frac{2K}{m}} = \sqrt{\frac{2(8.20\times10^{-16}\,J)}{9.11\times10^{-31}\,kg}} = 4.24\times10^{7}\,m\cdot s^{-1}$

$$\frac{v}{c} = \frac{4.24\times10^{7}\,m\cdot s^{-1}}{3.00\times10^{8}\,m\cdot s^{-1}} = 0.141 \Rightarrow v = \underline{0.141c}$$

c) Now repeat, for a proton:

$K_2 = e\Delta V$ as above.

But now $E_0 = mc^2 = (1.67\times10^{-27}\,kg)(3.00\times10^{8}\,m\cdot s^{-1})^2 = 1.50\times10^{-10}\,J$

$K_2 = 1\%\,E_0 \Rightarrow K_2 = 1.50\times10^{-12}\,J$

$$\Delta V = \frac{K_2}{e} = \frac{1.50\times10^{-12}\,J}{1.60\times10^{-19}\,C} = \underline{9.38\times10^{6}\,V}$$

$v = \sqrt{\frac{2K}{m}} = \sqrt{\frac{2(1.50\times10^{-12}\,J)}{1.67\times10^{-27}\,kg}} = 4.24\times10^{7}\,m\cdot s^{-1} \Rightarrow v = \underline{0.141c}$, as in (b).

Note: $K = 1\%\,E_0 \Rightarrow \frac{1}{2}mv^2 = (0.01)mc^2 \Rightarrow v = \sqrt{0.02}\,c = 0.141c$, independent of m.

26-27

a)

$\vec{F} = -e\vec{E} \Rightarrow a = \frac{eE}{m}$, downward

$$a = \frac{eE}{m} = \frac{(1.60\times10^{-19}\,C)(20{,}000\,N\cdot C^{-1})}{9.11\times10^{-31}\,kg}$$

$$a = 3.51\times10^{15}\,m\cdot s^{-2}$$

This acceleration is constant and downward, so the motion is like that of a projectile.

X-component

$v_{ox} = v_0 = 2\times10^{7}\,m\cdot s^{-1}$
$a_x = 0$
$x - x_0 = 0.04\,m$
$t = ?$

$$x - x_0 = v_{ox}t + \tfrac{1}{2}a_x t^2 \quad (a_x = 0)$$

$$t = \frac{x - x_0}{v_{ox}} = \frac{0.04\,m}{2\times10^{7}\,m\cdot s^{-1}} = 2.0\times10^{-9}\,s$$

Y-component

$v_{oy} = 0$
$a_y = 3.51\times10^{15}\,m\cdot s^{-2}$
$t = 2.0\times10^{-9}\,s$
$y - y_0 = ?$

$$y - y_0 = v_{oy}t + \tfrac{1}{2}a_y t^2 \quad (v_{oy} = 0)$$

$$y - y_0 = \tfrac{1}{2}(3.51\times10^{15}\,m\cdot s^{-2})(2.0\times10^{-9}\,s)^2 = 7.02\times10^{-3}\,m$$

$$y - y_0 = \underline{0.702\,cm}$$

b)

$v_x = 2\times10^{7}\,m\cdot s^{-1}$

$v_y = v_{oy} + a_y t = 0 + (3.51\times10^{15}\,m\cdot s^{-2})(2.0\times10^{-9}\,s) = 7.02\times10^{6}\,m\cdot s^{-1}$

$\tan\theta = \frac{v_y}{v_x} = \frac{7.02\times10^{6}\,m\cdot s^{-1}}{2.0\times10^{7}\,m\cdot s^{-1}} = 0.351 \Rightarrow \theta = \underline{19.3°}$, below the horizontal

c) After it leaves the electric field between the plates the electron travels in a straight line, with constant velocity components (if we neglect gravity):

X-component

$a_x = 0$
$v_{ox} = 2.0\times10^{7}\,m\cdot s^{-1}$
$x - x_0 = 0.12\,m$
$t = ?$

$$x - x_0 = v_{ox}t + \tfrac{1}{2}a_x t^2 \quad (a_x = 0)$$

$$t = \frac{x - x_0}{v_{ox}} = \frac{0.12\,m}{2.0\times10^{7}\,m\cdot s^{-1}} = 6.0\times10^{-9}\,s$$

Y-component

$v_{oy} = 7.02\times10^{6}\,m\cdot s^{-1}$ (from part (b))
$a_y = 0$
$t = 6.0\times10^{-9}\,s$
$y - y_0 = ?$

$$y - y_0 = v_{oy}t + \tfrac{1}{2}a_y t^2 \quad (a_y = 0)$$

$$y - y_0 = (7.02\times10^{6}\,m\cdot s^{-1})(6.0\times10^{-9}\,s)$$

$$y - y_0 = 0.0421\,m = 4.21\,cm$$

26-27 (cont)

Thus the electron travels downward a distance 4.21 cm after it leaves the plates. While traversing the plates it traveled downward 0.702 cm, so its total downward deflection is 4.21 cm + 0.70 cm = 4.91 cm

Note: Is the neglect of gravity justified? The acceleration due to gravity is $9.8\ m \cdot s^{-2}$. In $t = 6.0 \times 10^{-9}\ s$ the downward deflection due to gravity is thus

$y - y_0 = \frac{1}{2} a_y t^2 = \frac{1}{2}(9.8\ m \cdot s^{-2})(6.0 \times 10^{-9}\ s)^2 = 1.76 \times 10^{-16}\ m$, completely negligible!

Problems

26-29

a) $\overleftarrow{E}$; 1 ⊕ ← 5 cm → ⊕ 2 $\qquad \vec{F}_E \longleftarrow ⊕ \longrightarrow \vec{F}_{ext}$

$\vec{F}_E$ is the force due to the electric field and $\vec{F}_{ext}$ is the other force.

$$W_{other} = (K_2 + U_2) - (K_1 + U_1)$$

$W_{other} = W_{F_{ext}} = 9 \times 10^{-5}\ J$, positive since $\vec{F}_{ext}$ is to the right and the object moves to the right

$K_1 = 0$

$K_2 = 4.5 \times 10^{-5}\ J$

$$W_{other} = K_2 + U_2 - U_1 \Rightarrow U_1 - U_2 = K_2 - W_{other}$$

But $W_{F_E} = -(U_2 - U_1) = U_1 - U_2$

$$\Rightarrow W_{F_E} = K_2 - W_{other} = 4.5 \times 10^{-5}\ J - 9.0 \times 10^{-5}\ J = -4.5 \times 10^{-5}\ J$$

($W_{F_E} < 0$ since the force is to the left and the displacement is to the right.)

b) $W_{F_E} = -F_E d = -qEd \Rightarrow E = -\frac{W_{F_E}}{qd} = -\frac{-4.5 \times 10^{-5}\ J}{(3 \times 10^{-9}\ C)(0.05\ m)} = 3.0 \times 10^{5}\ N \cdot C^{-1}$

c) The problem asks for $V_1 - V_2$.

$$W_{1 \to 2} = U_1 - U_2 = q(V_1 - V_2)$$

$$V_1 - V_2 = \frac{W_{1 \to 2}}{q} = \frac{-4.5 \times 10^{-5}\ J}{3 \times 10^{-9}\ C} = -1.5 \times 10^{4}\ V$$

We know that $V_1 < V_2$ because the electric field points from high potential toward low potential.

26-33

a)

+ + + + + +

$\vec{E}$ $\vec{F}$ (+) $\vec{F}$ (−) $V_{ab} = 1600\,V$

− − − − − − −

Calculate the electric field E, from that the force, and therefore the acceleration of each object. Then we can use constant acceleration kinematic equations.

$$E = \frac{V_{ab}}{d} = \frac{1600\,V}{0.04\,m} = 4.0\times10^{4}\,V\cdot m^{-1} = 4.0\times10^{4}\,N\cdot C^{-1}$$

$$F = qE = eE = (1.60\times10^{-19}\,C)(4.0\times10^{4}\,N\cdot C^{-1}) = 6.4\times10^{-15}\,N,$$ same for proton and electron

for the electron $a_e = \frac{F}{m_e} = \frac{6.4\times10^{-15}\,N}{9.11\times10^{-31}\,kg} = 7.03\times10^{15}\,m\cdot s^{-2}$

for the proton $a_p = \frac{F}{m_p} = \frac{6.4\times10^{-15}\,N}{1.67\times10^{-27}\,kg} = 3.83\times10^{12}\,m\cdot s^{-2}$

Consider the motion of each object:

($\rightarrow x$, $\downarrow y$) Let $y=0$ at the positive plate. At $t=0$ the proton is at $y_0 = 0$ and the electron is at $y_0 = +0.04\,m$.

proton

$v_0 = 0$

$y - y_0 = y$

$a = a_p$

$y - y_0 = v_0 t + \frac{1}{2}at^2 \Rightarrow \boxed{y = \frac{1}{2}a_p t^2}$

electron

$v_0 = 0$

$y - y_0 = y - 0.04\,m$

$a = -a_e$ (is upward)

$y - y_0 = v_0 t + \frac{1}{2}at^2$

$y - 0.04\,m = -\frac{1}{2}a_e t^2$

$\boxed{(0.04\,m - y) = \frac{1}{2}a_e t^2}$

Combined, these equations give $\frac{1}{2}t^2 = \frac{y}{a_p} = \frac{(0.04\,m - y)}{a_e}$

$$y = \frac{(0.04\,m)\,a_p}{a_e + a_p} = \frac{(0.04\,m)(3.83\times10^{12}\,m\cdot s^{-2})}{3.83\times10^{12}\,m\cdot s^{-2} + 7.03\times10^{15}\,m\cdot s^{-2}} = 2.18\times10^{-5}\,m = \underline{2.18\times10^{-3}\,cm}$$

(distance from the positive plate)

b) $\Delta U = q\Delta V$

$\Delta U + \Delta K = 0 \Rightarrow \Delta K = -q\Delta V$

In terms of magnitudes, $K = qV \Rightarrow \frac{1}{2}mv^2 = qV \Rightarrow v = \sqrt{\frac{2qV}{m}}$

$v_p = \sqrt{\frac{2eV}{m_p}}$; $v_e = \sqrt{\frac{2eV}{m_e}} \Rightarrow \frac{v_p}{v_e} = \sqrt{\frac{m_e}{m_p}} \Rightarrow v_e = \sqrt{\frac{m_p}{m_e}}\,v_p$

$v_e = \sqrt{\frac{1.67\times10^{-27}\,kg}{9.11\times10^{-31}\,kg}}\,v_p = 42.8\,v_p$; The electron velocity is $\underline{42.8 \text{ times}}$ the proton velocity.

c) $K = qV = eV$; same magnitude of charge, same potential difference $\Rightarrow$ proton and electron have $\underline{\text{equal}}$ kinetic energies.

26-37

a) $V = Cx^{4/3}$

The problem says that $V = 160$ volts when $x = 8\times10^{-3}$ m.

$\Rightarrow C = \frac{V}{x^{4/3}} = \frac{160\text{ V}}{(8\times10^{-3}\text{ m})^{4/3}} = \underline{1.00\times10^{5}\text{ V}\cdot\text{m}^{-4/3}}$

b) $E_x = -\frac{\partial V}{\partial x} = -\frac{4}{3}Cx^{1/3}$.

(The minus sign means that E_x is in the negative x-direction, which means that $\vec{E}$ points from the positive anode toward the negative cathode.)

c) $F = qE = -eE = +\frac{4}{3}eCx^{1/3}$

Halfway between the electrodes $\Rightarrow x = 4\times10^{-3}$ m

Then

$$F = \frac{4}{3}(1.6\times10^{-19}\text{ C})(1.00\times10^{5}\text{ V}\cdot\text{m}^{-4/3})(4\times10^{-3}\text{ m})^{1/3}$$

$$F = 3.39\times10^{-15}\text{ C}\cdot\text{V}\cdot\text{m}^{-1} = \underline{3.39\times10^{-15}\text{ N}}$$

26-39

From problem 25-34, $E(r) = \frac{\rho r}{2\epsilon_0}$ for $r \le R$ (inside the cylindrical charge distribution)

$$E(r) = \frac{\lambda}{2\pi\epsilon_0 r} \quad \text{for } r \ge R$$

Also, $\rho\pi R^2 l = \lambda l \Rightarrow \rho = \frac{\lambda}{\pi R^2}$, so the result for $r \le R$ can be written also as

$$E(r) = \frac{\lambda r}{2\pi\epsilon_0 R^2} \quad (r \le R)$$

a) $V = 0$ at $r = R$

V_r for $r > R$ (outside the cylinder):

$$V_a - V_b = \int_a^b \vec{E}\cdot d\vec{l}$$

Take point a to be at R and point b to be at $r > R$. Let $d\vec{l} = d\vec{r}$.
$\vec{E}$ and $d\vec{r}$ both are radially outward $\Rightarrow \vec{E}\cdot d\vec{r} = E\,dr$.

Thus $V_R - V_r = \int_R^r E\,dr$. $(V_R = 0)$

In this interval $(r > R)$ $E(r) = \frac{\lambda}{2\pi\epsilon_0 r}$ and

$$V_R - V_r = \int_R^r \frac{\lambda}{2\pi\epsilon_0 r}dr = \frac{\lambda}{2\pi\epsilon_0}\int_R^r \frac{dr}{r} = \frac{\lambda}{2\pi\epsilon_0}\ln\left(\frac{r}{R}\right) \Rightarrow \boxed{V_r = -\frac{\lambda}{2\pi\epsilon_0}\ln\left(\frac{r}{R}\right)}$$

26-39 (cont)

(Note: $r = R \Rightarrow V_r = 0$. And the potential decreases with increasing distance from the cylinder; $V_r \to -\infty$ as $r \to \infty$.)

V_r for $r < R$ (inside the cylinder):

$$V_a - V_b = \int_a^b \vec{E} \cdot d\vec{l}$$

Take point a at r and b at R, where $r < R$. Let $d\vec{l} = d\vec{r}$. $\vec{E}$ and $d\vec{r}$ are both radially outward $\Rightarrow \vec{E} \cdot d\vec{r} = E\,dr$.

Thus $V_r - V_R = \int_r^R E\,dr$.

In this interval ($r < R$), $E(r) = \dfrac{\lambda r}{2\pi\epsilon_0 R^2}$ and

$$V_r - V_R = \int_r^R \frac{\lambda r}{2\pi\epsilon_0 R^2}\,dr = \frac{\lambda}{2\pi\epsilon_0 R^2}\int_r^R r\,dr = \frac{\lambda}{4\pi\epsilon_0 R^2}(R^2 - r^2) = \frac{\lambda}{4\pi\epsilon_0}\left(1 - (r/R)^2\right)$$

$$V_R = 0 \Rightarrow \boxed{V_r = \frac{\lambda}{4\pi\epsilon_0}\left(1 - (r/R)^2\right)}$$

(Note: $r = R \Rightarrow V_r = 0$. The potential is $\frac{\lambda}{4\pi\epsilon_0}$ at $r = 0$ and decreases with increasing r.)

b)

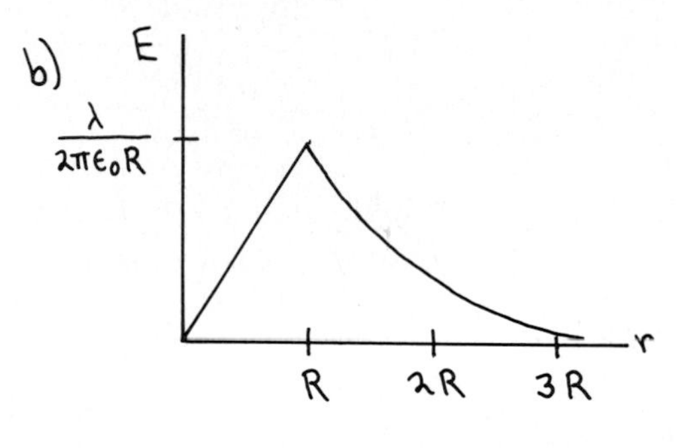

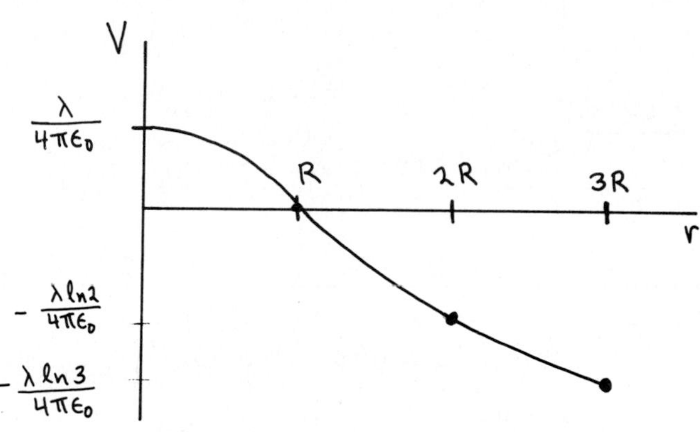

26-41

a)

a
dx'
x
P
x'
y'

(Call the coordinate axes x' and y' so as not to confuse them with the x and y distances given in the problem.)

Slice the charged rod up into thin slices of width dx'. The potential at P due to a slice having charge dq and at a distance x' from P is

$$dV = \frac{dq}{4\pi\epsilon_0 x'}$$

(This expression assumes $dV \to 0$ as $x' \to \infty$, as we were told to assume.)

26-41 (cont)

$$dq = \left(\frac{Q}{a}\right)dx' \Rightarrow dV = \frac{Q}{4\pi\epsilon_0 a}\frac{dx'}{x'}$$

Integrate over the length of the rod; x' ranges from x at the right-hand end to $x+a$ at the other end.

$$V = \int_x^{a+x} \frac{Q}{4\pi\epsilon_0 a}\frac{dx'}{x'} = \frac{Q}{4\pi\epsilon_0 a}\left(\ln x' \Big|_x^{a+x}\right) = \frac{Q}{4\pi\epsilon_0 a}\ln\left(\frac{a+x}{x}\right)$$

b)

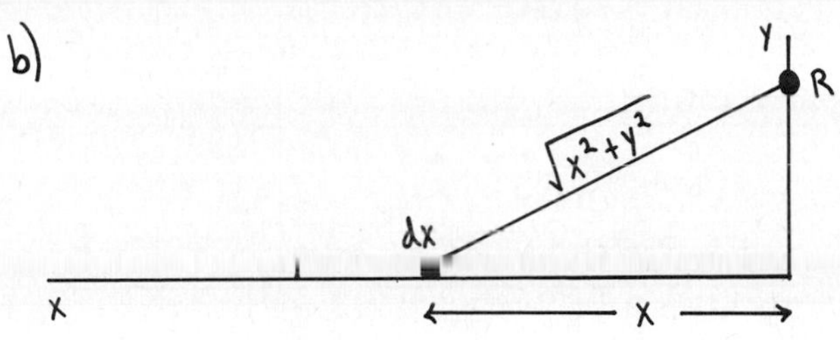

Use the coordinates shown in the sketch. The slice of width dx is a distance $\sqrt{x^2+y^2}$ from the point R.

$$dV = \frac{dq}{4\pi\epsilon_0\sqrt{x^2+y^2}}$$

$$dq = \frac{Q}{a}dx \Rightarrow dV = \frac{Q}{4\pi\epsilon_0 a}\frac{dx}{\sqrt{x^2+y^2}}$$

When we integrate over the length of the rod x ranges from $x=0$ to $x=a$.

$$V = \int_0^a \frac{Q}{4\pi\epsilon_0 a}\frac{dx}{\sqrt{x^2+y^2}} = \frac{Q}{4\pi\epsilon_0 a}\ln\left(x+\sqrt{x^2+y^2}\right)\Big|_0^a = \frac{Q}{4\pi\epsilon_0 a}\ln\left(\frac{a+\sqrt{a^2+y^2}}{y}\right)$$

(The expression for the integral was found in Appendix B.)

c) point P

$$V(x) = \frac{Q}{4\pi\epsilon_0 a}\ln\left(\frac{a+x}{x}\right) = \frac{Q}{4\pi\epsilon_0 a}\ln\left(1+\frac{a}{x}\right)$$

$\ln\left(1+\frac{a}{x}\right) = \frac{a}{x} - \frac{1}{2}\left(\frac{a}{x}\right)^2 + \ldots \rightarrow \frac{a}{x}$ for x large (Appendix B)

Thus $V(x) \longrightarrow \frac{Q}{4\pi\epsilon_0 a}\left(\frac{a}{x}\right) = \frac{Q}{4\pi\epsilon_0 x}$.

So for x large, V becomes that of a point charge (as it must).

point R

$$V(y) = \frac{Q}{4\pi\epsilon_0 a}\ln\left(\frac{a+\sqrt{a^2+y^2}}{y}\right) = \frac{Q}{4\pi\epsilon_0 a}\ln\left(\frac{a}{y}+\sqrt{1+a^2/y^2}\right)$$

$$\sqrt{1+a^2/y^2} = \left(1+a^2/y^2\right)^{1/2} = 1+\frac{a^2}{2y^2}+\ldots$$

$\frac{a}{y}+\sqrt{1+\frac{a^2}{y^2}} = 1+\frac{a}{y}+\frac{a^2}{2y^2}+\ldots \longrightarrow 1+\frac{a}{y}$ as y becomes large.

Thus $V(y) \longrightarrow \frac{Q}{4\pi\epsilon_0 a}\ln\left(1+\frac{a}{y}\right)$

26-41 (cont)

But $\ln\left(1+\frac{a}{y}\right) = \frac{a}{y} - \frac{a^2}{2y^2} + \dots \rightarrow \frac{a}{y}$

$$\Rightarrow V(y) \longrightarrow \frac{Q}{4\pi\epsilon_0 a}\left(\frac{a}{y}\right) = \frac{Q}{4\pi\epsilon_0 y}$$

Thus for y large V becomes that of a point charge.

26-45

Problem 25-35 dealt with the spherical charge distribution

$$\rho(r) = \rho_0(1 - r/R) \text{ for } r \le R$$

and $\rho(r) = 0$ for $r \ge R$; $\rho_0 = \frac{3Q}{\pi R^3}$

It was shown that $E(r) = \frac{Qr}{4\pi\epsilon_0 R^3}\left(4 - 3\frac{r}{R}\right)$ for $r \le R$

$$E(r) = \frac{Q}{4\pi\epsilon_0 r^2} \text{ for } r \ge R$$

a) $V_a - V_b = \int_a^b \vec{E}\cdot d\vec{\ell}$

Take a to be at r and b to be at ∞. Then $V_b = V_\infty = 0$. Take $d\vec{\ell} = d\vec{r}$, so $\vec{E}\cdot d\vec{r} = E\,dr$.

$V_r = \int_r^\infty E\,dr = \frac{Q}{4\pi\epsilon_0}\int_r^\infty \frac{dr}{r^2} = \frac{Q}{4\pi\epsilon_0}\left(-\frac{1}{r}\Big|_r^\infty\right) = \frac{Q}{4\pi\epsilon_0 r}$, the $V(r)$ for a point charge Q

b) $V_a - V_b = \int_a^b \vec{E}\cdot d\vec{\ell}$

Take b to be at R and a to be at r, where $r < R$.

From part (a), $V_R = \frac{Q}{4\pi\epsilon_0 R}$

$$V_r - V_R = \int_r^R \vec{E}\cdot d\vec{r} = \int_r^R E\,dr = \int_r^R \frac{Qr}{4\pi\epsilon_0 R^3}\left(4 - 3\frac{r}{R}\right)dr = \frac{Q}{4\pi\epsilon_0 R^3}\int_r^R \left(4r - 3\frac{r^2}{R}\right)dr$$

$$V_r - V_R = \frac{Q}{4\pi\epsilon_0 R^3}\left[2r^2 - \frac{r^3}{R}\right]\Big|_r^R = \frac{Q}{4\pi\epsilon_0 R^3}\left[2R^2 - R^2 - 2r^2 + \frac{r^3}{R}\right]$$

$$V_r - V_R = \frac{Q}{4\pi\epsilon_0 R^3}\left(R^2 - 2r^2 + \frac{r^3}{R}\right)$$

$$V_R = \frac{Q}{4\pi\epsilon_0 R} = \frac{QR^2}{4\pi\epsilon_0 R^3}$$

$$\Rightarrow V_r = \frac{Q}{4\pi\epsilon_0 R^3}\left(2R^2 - 2r^2 + \frac{r^3}{R}\right)$$

Note: $r = R \Rightarrow V_r = \frac{Q}{4\pi\epsilon_0 R^3}\left(2R^2 - 2R^2 + R^2\right) = \frac{Q}{4\pi\epsilon_0 R}$, as it should.

CHAPTER 27

Exercises 1, 7, 9, 13, 15, 19

Problems 21, 23, 27, 31, 33

Exercises

27-1

a) $V_{ab} = \dfrac{Q}{C} = \dfrac{0.2\times10^{-6}\,C}{500\times10^{-12}\,F} = \underline{400\,V}$

b) $C = \epsilon_0 \dfrac{A}{d} \Rightarrow A = \dfrac{dC}{\epsilon_0} = \dfrac{(0.2\times10^{-3}\,m)(500\times10^{-12}\,F)}{(8.85\times10^{-12}\,C^2\cdot N^{-1}\cdot m^{-2})} = 1.13\times10^{-2}\,m^2 = \underline{113\,cm^2}$

c) $E = \dfrac{V_{ab}}{d} = \dfrac{400\,V}{0.2\times10^{-3}\,m} = \underline{2.0\times10^{6}\,V\cdot m^{-1}}$

d) $\sigma = \dfrac{Q}{A} = \dfrac{0.2\times10^{-6}\,C}{1.13\times10^{-2}\,m^2} = \underline{1.77\times10^{-5}\,C\cdot m^{-2}}$

27-7

a) and b)

C_1, C_2, C_3, C_4; a, b, d; $V_{ab} = 48\,V$

$C_1 = C_2 = C_3 = C_4 = 2\,\mu F$

Replace the capacitor combinations by their equivalents:

C_1 C_2 = C_{12}: $\dfrac{1}{C_{12}} = \dfrac{1}{C_1} + \dfrac{1}{C_2} = \dfrac{1}{2\,\mu F} + \dfrac{1}{2\,\mu F} = \dfrac{2}{2\,\mu F} \Rightarrow C_{12} = 1\,\mu F$

C_{12}, C_3 = C_{123}: $C_{123} = C_{12} + C_3 = 1\,\mu F + 2\,\mu F = 3\,\mu F$

C_{123}, C_4 = C_{1234}: $\dfrac{1}{C_{1234}} = \dfrac{1}{C_{123}} + \dfrac{1}{C_4} = \dfrac{1}{3\,\mu F} + \dfrac{1}{2\,\mu F} = \dfrac{5}{6\,\mu F}$

$C_{1234} = \frac{6}{5}\,\mu F$

a, b; $V_{ab} = 48\,V$; $C_{1234} = \frac{6}{5}\,\mu F$

$C = \dfrac{Q}{V_{ab}} \rightarrow Q_{1234} = C_{1234} V_{ab} = (\tfrac{6}{5}\,\mu F)(48\,V)$

$Q_{1234} = 57.6\,\mu C$

27-7 (cont)

Now build back up the original circuit, piece by piece:

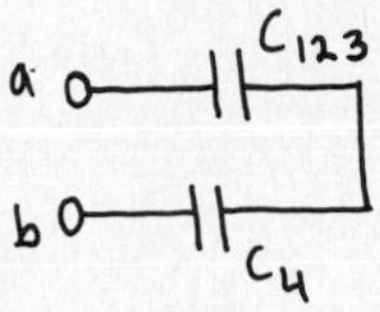

$Q_4 = Q_{123} = Q_{1234}$ (the amount of charge that comes out of the battery in the final equivalent circuit)

$\Rightarrow Q_4 = \underline{57.6\,\mu C}$

$V_4 = \frac{Q_4}{C_4} = \frac{57.6\,\mu C}{2\,\mu F} = \underline{28.8\,V}$

$V_{ab} = V_4 + V_{123} \Rightarrow V_{123} = V_{ab} - V_4 = 48.0\,V - 28.8\,V = 19.2\,V$

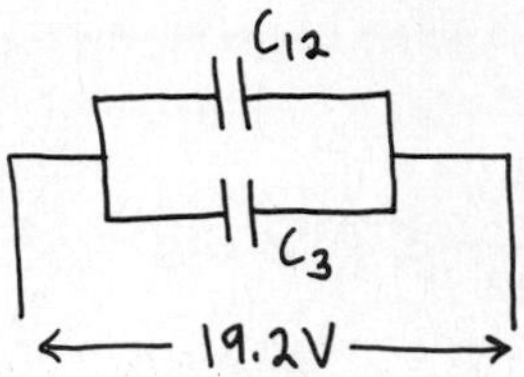

$V_3 = \underline{19.2\,V}$

$Q_3 = C_3 V_3 = (2\,\mu F)(19.2\,V) = \underline{38.4\,\mu C}$

$Q_{12} = C_{12} V_{12}$; $V_{12} = V_3 = 19.2\,V$

$\Rightarrow Q_{12} = (1\,\mu F)(19.2\,V) = 19.2\,\mu C$

$Q_1 = \underline{19.2\,\mu C}$; $V_1 = \frac{Q_1}{C_1} = \frac{19.2\,\mu C}{2\,\mu F} = \underline{9.6\,V}$

$Q_2 = \underline{19.2\,\mu C}$; $V_2 = \frac{Q_2}{C_2} = \frac{19.2\,\mu C}{2\,\mu F} = \underline{9.6\,V}$

c) $V_{ad} = V_3 = \underline{19.2\,V}$ (or, $V_{ad} = V_1 + V_2 = 9.6V + 9.6V = 19.2V$)

27-9

a) $C = \frac{Q}{V_{ab}} \Rightarrow V_{ab} = \frac{Q}{C} = \frac{0.5\,\mu C}{0.001\,\mu F} = \underline{500V}$

b) $U = \frac{1}{2} QV = \frac{1}{2}(0.5\,\mu C)(500V) = 125 \times 10^{-6}\,J = \underline{1.25 \times 10^{-4}\,J}$

c) $C = \epsilon_0 \frac{A}{d} \Rightarrow A = \frac{dC}{\epsilon_0} = \frac{(1.0 \times 10^{-3}\,m)(0.001 \times 10^{-6}\,F)}{8.85 \times 10^{-12}\,C^2 \cdot N^{-1} \cdot m^{-2}} = \underline{0.113\,m^2}$

d) dielectric breakdown for $E = 8.0 \times 10^5\,V \cdot m^{-1}$

$V_{ab} = Ed = (8.0 \times 10^5\,V \cdot m^{-1})(1.0 \times 10^{-3}\,m) = \underline{800V}$

27-13

a) $U = \frac{1}{2}\frac{q^2}{C}$; $C = \frac{\epsilon_0 A}{x} \Rightarrow U = \frac{x q^2}{2\epsilon_0 A}$

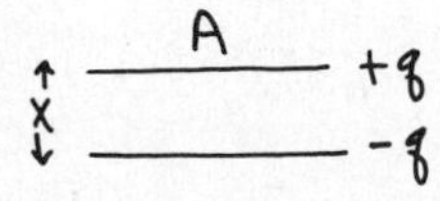

b) $x \rightarrow x + dx$

$U = \frac{(x+dx)q^2}{2\epsilon_0 A}$

27-13 (cont)

c) $dW = U(x+dx) - U(x) = \frac{q^2}{2\epsilon_0 A}[(x+dx)-x] = \frac{q^2 dx}{2\epsilon_0 A}$

But also $dW = F\,dx \Rightarrow F = \frac{dW}{dx} = \frac{q^2}{2\epsilon_0 A}$

d) 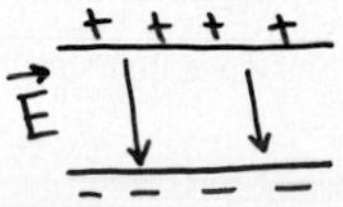

$E = \frac{\sigma}{\epsilon_0} = \frac{q}{\epsilon_0 A} \Rightarrow F = \frac{1}{2}qE$, not qE

The reason is that E is the field due to both plates.
If we consider the positive plate only,

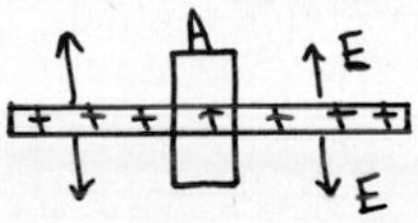

$\epsilon_0 \oint \vec{E}\cdot d\vec{A} = Q$

$\epsilon_0 2EA = \sigma A \Rightarrow E = \frac{\sigma}{2\epsilon_0} = \frac{q}{2\epsilon_0 A}$

The force this field exerts on the other plate, of charge $-q$, is $F = \frac{q^2}{2\epsilon_0 A}$.

27-15

$C = K\epsilon_0 \frac{A}{d}$. Minimum $A \Rightarrow$ smallest possible d.

$V_{ab} = Ed \Rightarrow d = \frac{V_{ab}}{E}$

V_{ab} must be able to be as large as 6000 V, but E must be smaller than $2\times10^7\ \mathrm{V\cdot m^{-1}}$

$\Rightarrow d = \frac{6000\ \mathrm{V}}{2\times10^7\ \mathrm{V\cdot m^{-1}}} = 3.0\times10^{-4}\ \mathrm{m}$

Then $A = \frac{dC}{K\epsilon_0} = \frac{(3.0\times10^{-4}\ \mathrm{m})(0.15\times10^{-6}\ \mathrm{F})}{3(8.85\times10^{-12}\ \mathrm{C^2\cdot N^{-1}\cdot m^{-2}})} = \underline{1.69\ \mathrm{m^2}}$

27-19

a) $K = \frac{E_0}{E}$

Without the dielectric the electric field would be

$E_0 = \frac{Q}{\epsilon_0 A} = \frac{1.0\times10^{-7}\ \mathrm{C}}{(8.85\times10^{-12}\ \mathrm{C^2\cdot N^{-1}\cdot m^{-2}})(100\times10^{-4}\ \mathrm{m^2})} = 1.13\times10^6\ \mathrm{V\cdot m^{-1}}$

$K = \frac{1.13\times10^6\ \mathrm{V\cdot m^{-1}}}{3.3\times10^5\ \mathrm{V\cdot m^{-1}}} = \underline{3.42}$

b) The surface charge density on the plates is

$\sigma = \frac{Q}{A} = \frac{1.0\times10^{-7}\ \mathrm{C}}{100\times10^{-4}\ \mathrm{m^2}} = 1.0\times10^{-5}\ \mathrm{C\cdot m^{-2}}$

eq. (27-13) $\sigma - \sigma_i = \frac{\sigma}{K} \Rightarrow \sigma_i = \sigma\left(1-\frac{1}{K}\right) = \sigma\left(\frac{K-1}{K}\right) = \frac{2.42}{3.42}(1.0\times10^{-5}\ \mathrm{C\cdot m^{-2}})$

27-19 (cont)

$$\sigma_i = 7.08 \times 10^{-6}\ C\cdot m^{-2}$$

This is the induced surface charge density.

Then $Q_i = \sigma_i A = (7.08\times10^{-6}\ C\cdot m^{-2})(100\times10^{-4}\ m^2) = \underline{7.08\times10^{-8}\ C}$

Problems

27-21

a) $C = \epsilon_0 \frac{A}{d} = (8.85\times10^{-12}\ C^2\cdot N^{-1}\cdot m^{-2}) \frac{(0.2\ m)^2}{2.0\times10^{-2}\ m} = \underline{1.77\times10^{-11}\ F}$

b) Remains connected to the battery $\Rightarrow V_{ab} = 50\ V$

$Q = CV_{ab} = (1.77\times10^{-11}\ F)(50\ V) = \underline{8.85\times10^{-10}\ C}$

c) $E = \frac{V_{ab}}{d} = \frac{50\ V}{2.0\times10^{-2}\ m} = \underline{2500\ V\cdot m^{-1}}$

d) $U = \frac{1}{2} Q V_{ab} = \frac{1}{2}(8.85\times10^{-10}\ C)(50\ V) = \underline{2.21\times10^{-8}\ J}$

27-23

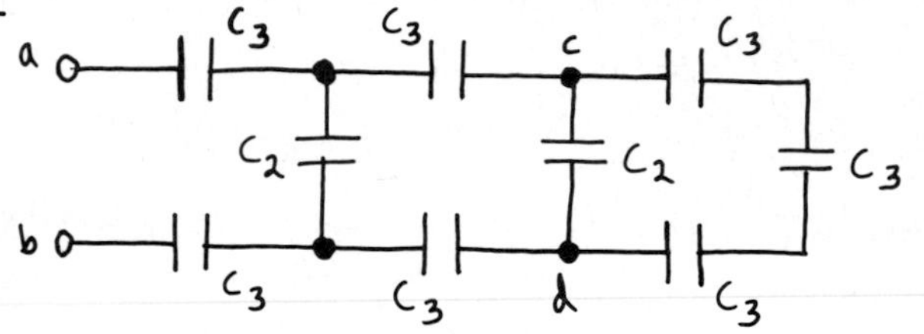

a) Replace the capacitors by their equivalent:

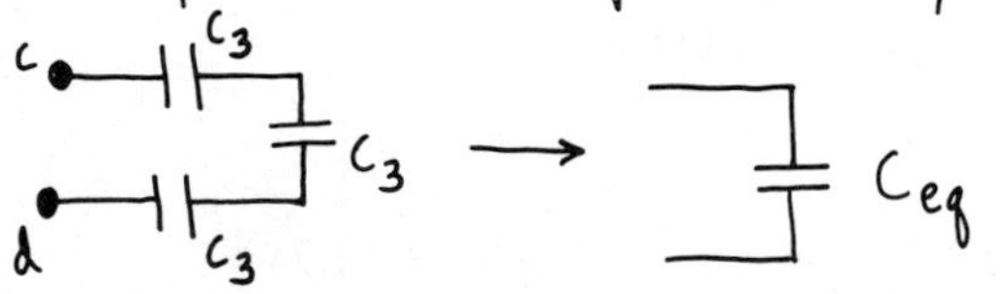

$$\frac{1}{C_{eq}} = \frac{1}{C_3} + \frac{1}{C_3} + \frac{1}{C_3} = \frac{3}{C_3}$$

$$C_{eq} = \frac{C_3}{3} = \frac{3\mu F}{3} = 1\mu F$$

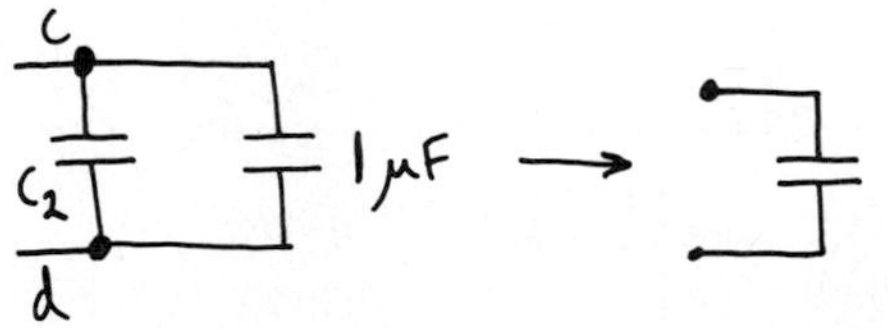

$$C_{eq} = 1\mu F + C_2 = 3\mu F$$

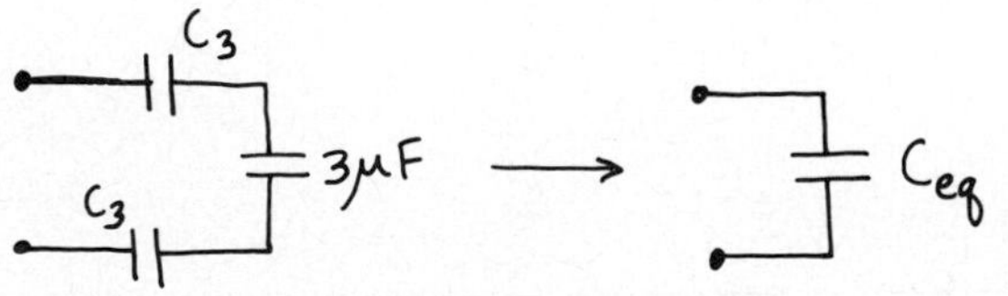

$$\frac{1}{C_{eq}} = \frac{1}{C_3} + \frac{1}{C_3} + \frac{1}{3\mu F} = \frac{3}{3\mu F}$$

$$C_{eq} = 1\mu F$$

27-23 (cont)

C_2, $C_{eq} = 1\mu F$ (in parallel) $\longrightarrow$ $C_{eq} = C_2 + 1\mu F = 3\mu F$

a, C_3, $3\mu F$, b, C_3 $\longrightarrow$ a, C_{eq}, b

$$\frac{1}{C_{eq}} = \frac{1}{C_3} + \frac{1}{C_3} + \frac{1}{3\mu F} = \frac{3}{3\mu F}$$

$$C_{eq} = \underline{\underline{1\mu F}}$$

b) a, C_3, $V_{ab} = 900V$, C_2, $1\mu F$ $\leftarrow$ This $1\mu F$ is the equivalent capacitance of the rest of the network. b, C_3

Replace C_2 and the $1\mu F$ capacitor by their $3\mu F$ equivalent:

a, $C_3 = 3\mu F$, $900V$, $3\mu F$, b, $C_3 = 3\mu F$

The potential is the same for each of these three capacitors, and the sum of these potentials is $V_{ab} = 900V \Rightarrow V = 300V$ for each capacitor.

$$Q_3 = C_3 V = (3\mu F)(300V) = \underline{900\mu C}$$

$$Q_2 = C_2 V = (2\mu F)(300V) = \underline{600\mu C}$$

c) a, C_3, C_3, c, C_2, $300V$, C_2, $1\mu F$ (equivalent capacitance for rest of network), b, C_3, C_3, d

Replace C_2 and the $1\mu F$ capacitors by their equivalent:

C_3, c, $300V$, $3\mu F$, C_3, d

$\Rightarrow V_{cd} = \underline{100V}$ (100V across each of these capacitors, so their sum will be 300V)

27-27

d) C_1, C_1, C_2, $12V$

$V_1 + V_2 + V_3 = 12V$ (series)

$Q_1 = Q_2 = Q_3$

$$V = \frac{Q}{C} \Rightarrow Q_1\left(\frac{1}{C_1} + \frac{1}{C_2} + \frac{1}{C_3}\right) = 12V$$

27-27 (cont)

$$Q_1 = \frac{12\,V}{\frac{1}{8\times10^{-6}F} + \frac{1}{8\times10^{-6}F} + \frac{1}{4\times10^{-6}F}} = 24\,\mu C$$

$$Q_1 = Q_2 = Q_3 = \underline{24\,\mu C}$$

b) $U = \frac{1}{2}Q_1V_1 + \frac{1}{2}Q_2V_2 + \frac{1}{2}Q_3V_3$

But $Q_1 = Q_2 = Q_3 \Rightarrow U = \frac{1}{2}Q_1(V_1+V_2+V_3) = \frac{1}{2}Q_1(12V)$

$$U = \frac{1}{2}(24\times10^{-6}C)(12V) = \underline{1.44\times10^{-4}J}$$

c) Q_1 Q_2 Q_3 $\longrightarrow$ q_1 v_1 q_2 v_2 q_3 v_3

$$v_1 = v_2 = v_3$$

$$q_1+q_2+q_3 = Q_1+Q_2+Q_3 = 72\times10^{-6}C$$

$$\frac{q_1}{C_1} = \frac{q_2}{C_2} = \frac{q_3}{C_3}$$

$$C_1 = C_2 \Rightarrow q_1 = q_2$$

$$q_1 = \left(\frac{C_1}{C_3}\right)q_3 = \left(\frac{8\,\mu F}{4\,\mu F}\right)q_3 = 2q_3 \;;\; q_2 = 2q_3$$

Thus $q_1+q_2+q_3 = 72\times10^{-6}C \Rightarrow 2q_3+2q_3+q_3 = 72\times10^{-6}C$

$$5q_3 = 72\times10^{-6}C \Rightarrow q_3 = 14.4\,\mu C$$

Then $v_3 = \frac{q_3}{C_3} = \frac{14.4\times10^{-6}C}{4\times10^{-6}F} = 3.6\,V \;;\; v_1 = v_2 = v_3 = \underline{3.6V}$

d) $U = \frac{1}{2}C_1v_1^2 + \frac{1}{2}C_2v_2^2 + \frac{1}{2}C_3v_3^2$

$v_1 = v_2 = v_3 \Rightarrow U = \frac{1}{2}(C_1+C_2+C_3)v_1^2 = \frac{1}{2}(20\times10^{-6}C)(3.6V)^2 = \underline{1.30\times10^{-4}J}$

Note: $U_i = 1.44\times10^{-4}J$; $U_f = 1.30\times10^{-4}J$

The stored potential energy <u>decreases</u> by $0.14\times10^{-4}J$.

27-31

a)

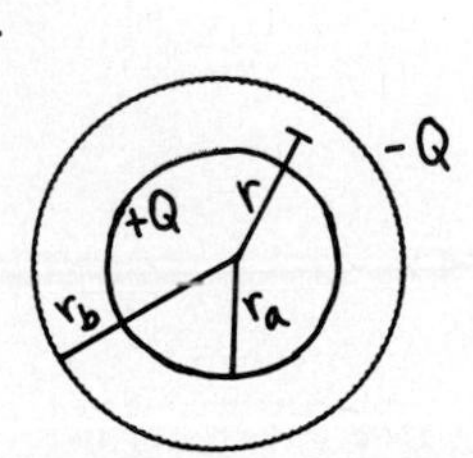

From chapter 26 the potential of a charged conducting sphere is

$V = \frac{q}{4\pi\epsilon_0 r}$ outside and $V = \frac{q}{4\pi\epsilon_0 R}$ inside.

27-31 (cont)

At distance r_a:

$$V_a = V_{inner} + V_{outer} = \frac{+Q}{4\pi\epsilon_0 r_a} + \frac{-Q}{4\pi\epsilon_0 r_b}$$ (r_a is outside the inner sphere and inside the outer sphere)

$$V_a = \frac{Q}{4\pi\epsilon_0}\left(\frac{1}{r_a} - \frac{1}{r_b}\right)$$

At distance r_b:

$$V_b = V_{inner} + V_{outer} = \frac{+Q}{4\pi\epsilon_0 r_b} + \frac{-Q}{4\pi\epsilon_0 r_b} = 0$$

(As we knew already, because $E=0$ outside the spheres.)

Thus $V_{ab} = V_a - V_b = \frac{Q}{4\pi\epsilon_0}\left(\frac{1}{r_a} - \frac{1}{r_b}\right)$

b) $C = \frac{Q}{V_{ab}} = \frac{Q}{\frac{Q}{4\pi\epsilon_0}\left(\frac{1}{r_a}-\frac{1}{r_b}\right)} = \frac{4\pi\epsilon_0}{\frac{r_b-r_a}{r_a r_b}} = 4\pi\epsilon_0\left(\frac{r_a r_b}{r_b - r_a}\right)$, as was to be shown

c) $r_b - r_a = d$

$r_a \simeq r_b \simeq r \Rightarrow A = 4\pi r^2 \simeq 4\pi r_a r_b \Rightarrow C = \epsilon_0 \frac{A}{d}$, which is eq. (27-2).

27-33

This capacitor can be viewed as follows. Consider inserting a thin metal plate between the dielectrics. $E=0$ inside the metal, so this does nothing to the potential between the capacitor plates. But then can separate this additional plate into two connected by a wire. The potential is constant throughout this conductor, so the potential between the capacitor plates still remains the same.

This final situation is easily analyzed as two capacitors in series:

$$C_1 = K_1 \frac{\epsilon_0 A}{d/2} = 2K_1 \frac{\epsilon_0 A}{d}$$

$$C_2 = K_2 \frac{\epsilon_0 A}{d/2} = 2K_2 \frac{\epsilon_0 A}{d}$$

The equivalent capacitance is given by

$$\frac{1}{C} = \frac{1}{C_1} + \frac{1}{C_2} = \frac{C_1 + C_2}{C_1 C_2} \Rightarrow C = \frac{C_1 C_2}{C_1 + C_2}$$

$$C = \frac{4K_1K_2\left(\frac{\epsilon_0 A}{d}\right)^2}{(K_1+K_2)\,2\frac{\epsilon_0 A}{d}} = \frac{2\epsilon_0 A}{d}\left(\frac{K_1K_2}{K_1+K_2}\right)$$, as was to be shown.

CHAPTER 28

Exercises 1, 7, 11, 13, 17, 19, 23, 25

Problems 31, 33, 35, 37, 39

Exercises

28-1

a) $I = \frac{\Delta Q}{\Delta t} = \frac{90\ \text{C}}{75\ \text{min}\left(\frac{60\ \text{s}}{1\ \text{min}}\right)} = \underline{0.02\ \text{A}}$

b) $J = \frac{I}{A} = \frac{0.02\ \text{A}}{\pi r^2} = \frac{0.02\ \text{A}}{\pi (0.5 \times 10^{-3}\ \text{m})^2} = 2.55 \times 10^4\ \text{A} \cdot \text{m}^{-2}$

$J = nqv \Rightarrow v = \frac{J}{nq} = \frac{2.55 \times 10^4\ \text{A} \cdot \text{m}^{-2}}{(5.8 \times 10^{28}\ \text{m}^{-3})(1.60 \times 10^{-19}\ \text{C})} = \underline{2.75 \times 10^{-6}\ \text{m} \cdot \text{s}^{-1}}$

28-7

a) $R = \frac{\rho l}{A}$

From Table 28-1, $\rho_{Al} = 2.63 \times 10^{-8}\ \Omega \cdot \text{m}$

$R = \frac{(2.63 \times 10^{-8}\ \Omega \cdot \text{m})(2.5\ \text{m})}{(0.01\ \text{m})(0.05\ \text{m})} = \underline{1.32 \times 10^{-4}\ \Omega}$

b) $l = \frac{AR}{\rho} = \frac{\pi (0.75 \times 10^{-3}\ \text{m})^2 (1.32 \times 10^{-4}\ \Omega)}{1.72 \times 10^{-8}\ \Omega \cdot \text{m}} = 0.0136\ \text{m} = \underline{1.36\ \text{cm}}$

28-11

$R_T = R_0 [1 + \alpha (T - T_0)]$

$T - T_0 = \frac{R_T - R_0}{R_0 \alpha} = \frac{214.2\ \Omega - 217.3\ \Omega}{217.3\ \Omega (-0.0005\ (\text{C}^\circ)^{-1})} = 28.5\ \text{C}^\circ$

($\alpha = -0.0005\ (\text{C}^\circ)^{-1}$ from Table 28-2)

$T_0 = 0^\circ\text{C} \Rightarrow T = \underline{28.5^\circ\text{C}}$

28-13

open circuit $\Rightarrow I = 0$

$V_{ab} = \mathcal{E} = \underline{1.52\ \text{V}}$

28-13 (cont)

closed circuit; $I = 1.5\,A$

a r $\mathcal{E}$ b, + −, i, R

$$V_b + \mathcal{E} - Ir = V_a$$

$$V_a - V_b = V_{ab} = \mathcal{E} - Ir$$

$$r = \frac{\mathcal{E} - V_{ab}}{I} = \frac{1.52\,V - 1.37\,V}{1.5\,A} = \underline{0.10\,\Omega}$$

28-17

a)

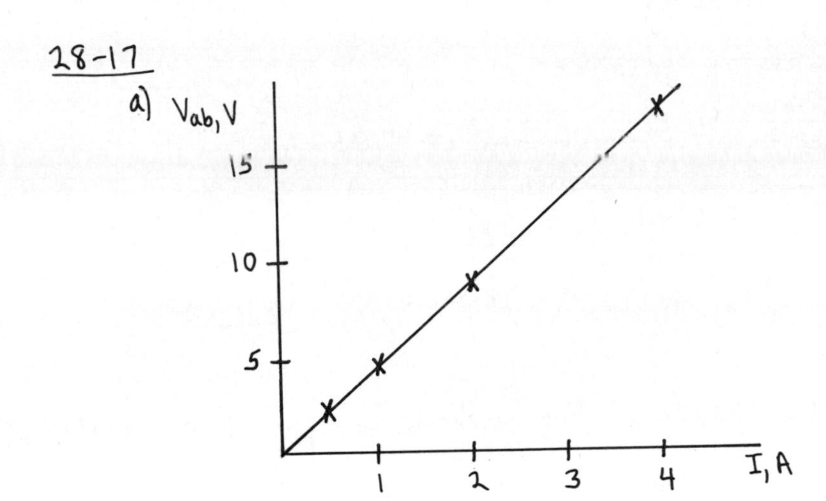

b) V versus I is a straight line $\Rightarrow V = IR$ where R is a constant, independent of I

Thus Ohm's Law is obeyed.

c) $R = \frac{V}{I}$

I	V	$\frac{V}{I} = R$
0.5 A	2.18 V	4.36 Ω
1.0 A	4.36 V	4.36 Ω
2.0 A	8.72 V	4.36 Ω
4.0 A	17.44 V	4.36 Ω

$\Rightarrow R = \underline{4.36\,\Omega}$

28-19

Eq. (28-21) $\quad P = I^2 R = \frac{V^2}{R} = IV \quad ; \quad p = \frac{P}{LA}$

a) E corresponds to V, and J to I, so use $P = IV$:

$$p = \frac{P}{LA} = \frac{IV}{LA} \; ; \; J = \frac{I}{A}, \; E = \frac{V}{L} \Rightarrow p = EJ$$

b) J corresponds to I and ρ to R, so use $P = I^2 R$:

$$p = \frac{P}{LA} = \frac{I^2 R}{LA}$$

$$R = \frac{\rho L}{A} \Rightarrow p = \frac{I^2}{LA}\left(\frac{\rho L}{A}\right) = \left(\frac{I}{A}\right)^2 \rho = J^2 \rho$$

c) Use $P = \frac{V^2}{R} \Rightarrow p = \frac{P}{LA} = \frac{V^2}{RLA}$

$$R = \frac{\rho L}{A} \Rightarrow p = \frac{V^2}{LA}\left(\frac{A}{\rho L}\right) = \left(\frac{V}{L}\right)^2 \frac{1}{\rho} = \frac{E^2}{\rho}$$

28-23

a) $P = VI = (12V)(50A) = 600W$

The battery can provide this for 1 hr. The total energy then supplied is $U = Pt = (600W)(3600s) = \underline{2.16\times10^6 J}$.

b) Heat of combustion of gasoline is $46\times10^6 J\cdot kg^{-1}$.

The mass of gasoline that supplies $2.16\times10^6 J$ is

$$m = \frac{2.16\times10^6 J}{46\times10^6 J\cdot kg^{-1}} = 4.70\times10^{-2} kg.$$

The volume of this mass of gasoline is

$$V = \frac{m}{\rho} = \frac{4.70\times10^{-2} kg}{900 kg\cdot m^{-3}} = 5.22\times10^{-5} m^3 = \underline{0.0522 \ell}$$

c) $t = \frac{U}{P} = \frac{2.16\times10^6 J}{300 W} = 7.2\times10^3 s = \underline{2.0 hr}$

28-25

a) $I = \frac{V}{R}$

The total resistance is the 1000Ω of the power supply plus the $10\times10^3\Omega$ of the person.

Thus

$$I = \frac{20\times10^3 V}{10\times10^3\Omega + 1000\Omega} = \underline{1.82 A} \quad \text{(lethal)}$$

b) $P = I^2 R = (1.82A)^2(10\times10^3\Omega) = \underline{3.31\times10^4 W}$

c) $I = 0.001 A \Rightarrow R = \frac{V}{I} = \frac{20\times10^3 V}{0.001 A} = 20\times10^6\Omega$

$R = R_{int.} + R_{person} \Rightarrow R_{int} = R - R_{person} = 20\times10^6\Omega - 10\times10^3\Omega = \underline{20\times10^6\Omega}$

Problems

28-31

+Q | K | -Q

If the plates have a separation d and area A, the resistance of the material between the plates is $R = \frac{\rho d}{A}$.

$$i = \frac{V}{R}, \quad V = \frac{Q}{C} \Rightarrow i = \frac{Q}{RC} = \frac{QA}{\rho d C}$$

$$C = K\frac{\epsilon_0 A}{d}$$

$$\Rightarrow i = \left(\frac{QA}{\rho d}\right)\left(\frac{d}{K\epsilon_0 A}\right) = \frac{Q}{K\rho\epsilon_0}$$, as was to be shown.

28-33

$R = \frac{\rho L}{A}$, for the original piece of wire.

Cut into pieces of length $l = L/3$

Each piece has resistance $R_p = \frac{\rho(L/3)}{A} = \frac{1}{3}\frac{\rho L}{A} = \frac{1}{3}R$

The three pieces are then connected in parallel.

The resistance R_{new} of this combination is given by

$$\frac{1}{R_{new}} = \frac{1}{R_p} + \frac{1}{R_p} + \frac{1}{R_p} = \frac{3}{R_p} \Rightarrow R_{new} = \frac{R_p}{3} = \frac{R}{9}$$

28-35

I, a, ε, r, b, $I = 3A$

$V_a = V_b - Ir + \varepsilon$

$\boxed{V_{ab} = \varepsilon - Ir}$

a, ε, r, b, I, $I = 2A$

$V_a = V_b + Ir + \varepsilon$

(The potential increases when go through a resistor opposite to the current.)

$\boxed{V_{ab} = \varepsilon + Ir}$

$8.5V = \varepsilon - (3A)r$

$11V = \varepsilon + (2A)r$

Subtract $\Rightarrow (5A)r = 2.5V \Rightarrow r = \underline{0.5\Omega}$

Then $\varepsilon = 8.5V + (3A)(0.5\Omega) = \underline{10V}$

or

$\varepsilon = 11V - (2A)(0.5\Omega) = 10V$

28-37

a) To charge a battery push current through the battery from + to −.

$\varepsilon = 12V$, I, a, $r = 0.2\Omega$, b, $I = 15A$

$V_a = V_b + Ir + \varepsilon$

$V_{ab} = \varepsilon + Ir = 12V + (15A)(0.2\Omega) = \underline{15V}$

b) $P = VI = (15V)(15A) = 225$ Watts

$U = Pt = (225W)(4hr)\left(\frac{3600s}{1hr}\right) = \underline{3.24 \times 10^6 J}$

28-37 (cont)

c) $P = I^2 r = (15\,A)^2 (0.2\,\Omega) = 45\,W$

$U = Pt = (45\,W)(4\,hr)\left(\frac{3600\,s}{1\,hr}\right) = \underline{6.48 \times 10^5\,J}$

d) $\mathcal{E} = 12\,V$, $r = 0.2\,\Omega$, R, $I = 15\,A$

$\mathcal{E} - IR - Ir = 0$

$R = \frac{\mathcal{E} - Ir}{I} = \frac{12\,V - (15\,A)(0.2\,\Omega)}{15\,A} = \underline{0.6\,\Omega}$

e) $P = I^2 R = (15\,A)^2 (0.6\,\Omega) = 135\,W$

The 60 A·hr battery can supply 15 A for 4 hr. Thus the total energy supplied to the external resistor is

$U = Pt = (135\,W)(4\,hr)\left(\frac{3600\,s}{1\,hr}\right) = \underline{1.94 \times 10^6\,J}$

f) $P = I^2 r = (15\,A)^2 (0.2\,\Omega) = 45\,W$

$U = Pt = (45\,W)(4\,hr)\left(\frac{3600\,s}{1\,hr}\right) = \underline{6.48 \times 10^5\,J}$

g) (b) ⇒ $3.24 \times 10^6\,J$ delivered to the battery during charging.
(e) ⇒ $1.94 \times 10^6\,J$ delivered to the external resistor during discharge

The difference is $1.30 \times 10^6\,J$, which is due to $6.5 \times 10^5\,J$ dissipated in the internal resistance during charging and $6.5 \times 10^5\,J$ during discharge.

28-39

$\mathcal{E}_1 = 12\,V$, $r_1 = 1\,\Omega$, $R = 8\,\Omega$, I, $\mathcal{E}_2 = 8\,V$, $r_2 = 1\,\Omega$

a) $-IR - \mathcal{E}_2 - Ir_2 - Ir_1 + \mathcal{E}_1 = 0$

$I = \frac{\mathcal{E}_1 - \mathcal{E}_2}{r_1 + r_2 + R} = \frac{12\,V - 8\,V}{10\,\Omega} = \underline{0.4\,A}$

b) $P = I^2 R + I^2 r_1 + I^2 r_2 = I^2 (R + r_1 + r_2)$

$P = (0.4\,A)^2 (8\,\Omega + 1\,\Omega + 1\,\Omega) = \underline{1.6\,\text{Watts}}$

c) chemical energy → electrical in the 12V battery, since the current is going from − to + in this battery.

$P = \mathcal{E}_1 I = (12\,V)(0.4\,A) = \underline{4.8\,W}$

d) electrical → chemical in the 8V battery, since the current is going from + to − in the battery.

$P = \mathcal{E}_2 I = (8\,V)(0.4\,A) = \underline{3.2\,W}$

28-39 (cont)

e) Rate of production of electrical energy = 4.8 W (part (c))

Rate of consumption of electrical energy
= 3.2 W (in the 8 V battery) + 1.6 W (in the resistors) = 4.8 W

Thus the rate of production does equal the rate of consumption.

CHAPTER 29

Exercises 3, 7, 9, 11, 13, 15, 17, 21, 25

Problems 29, 33, 35, 37, 39, 45, 47, 49

Exercises

29-3

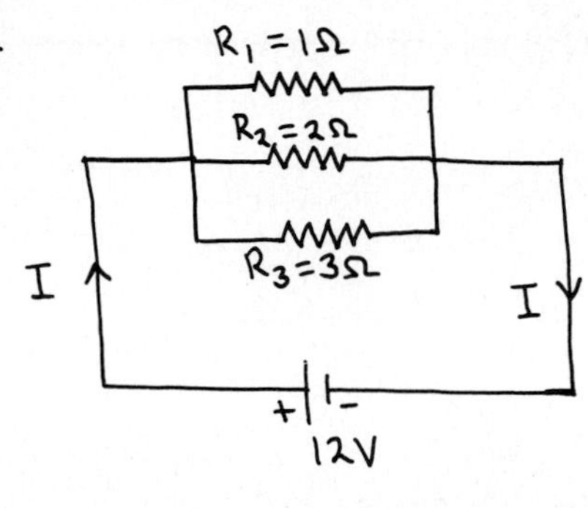

a) $\frac{1}{R_{eq}} = \frac{1}{R_1} + \frac{1}{R_2} + \frac{1}{R_3} = \frac{1}{1\Omega} + \frac{1}{2\Omega} + \frac{1}{3\Omega} = \frac{6+3+2}{6\Omega}$

$R_{eq} = \frac{6}{11}\Omega = \underline{0.545\Omega}$

b) For each resistor V = 12 volts.

$I = \frac{V}{R} \Rightarrow I_1 = \frac{V}{R_1} = \frac{12V}{1\Omega} = \underline{12A}$

$I_2 = \frac{V}{R_2} = \frac{12V}{2\Omega} = \underline{6A}$

$I_3 = \frac{V}{R_3} = \frac{12V}{3\Omega} = \underline{4A}$

c) $I = I_1 + I_2 + I_3 = 12A + 6A + 4A = \underline{22A}$

or

$I = \frac{V}{R_{eq}} = \frac{12V}{0.545\Omega} = 22A$

d) 12 V for each resistor

e) $P_1 = I_1^2 R_1 = (12A)^2(1\Omega) = \underline{144W}$

$P_2 = I_2^2 R_2 = (6A)^2(2\Omega) = \underline{72W}$

$P_3 = I_3^2 R_3 = (4A)^2(3\Omega) = \underline{48W}$

29-7

a)

28V, R, (1), I=2A, (2), I=2A, a, ε, X, 4A, 6Ω, 6A, (3), 3Ω, 6A

Apply the Point Rule to point a:

$\Sigma I = 0 \Rightarrow I + 4A = 6A$

$I = \underline{2A}$

29-7 (cont)

b) Apply loop rule to loop (1):

$$+28\,V - (6A)(3\Omega) - (2A)R = 0$$

$$R = \frac{28V - 18V}{2A} = \underline{5\Omega}$$

c) Apply loop rule to loop (2):

$$+28V - \mathcal{E} + (4A)(6\Omega) - (2A)(5\Omega) = 0$$

$$\mathcal{E} = 28V + 24V - 10V = \underline{42V}$$

Check by applying the loop rule to loop (3):

$$\mathcal{E} - (6A)(3\Omega) - (4A)(6\Omega) = 0$$

$$\mathcal{E} = 18V + 24V = 42V \checkmark$$

d) break at x ⇒ single loop

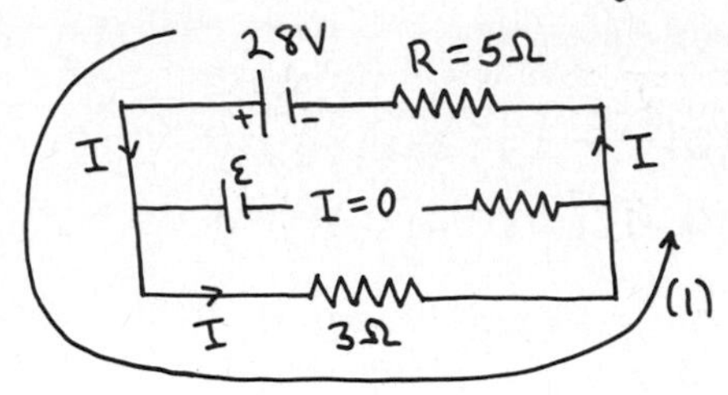

Loop Rule

$$\Rightarrow +28V - I(3\Omega) - I(5\Omega) = 0$$

$$I = \frac{28V}{8\Omega} = \underline{3.5\,A}$$

29-9

a) Break at a and b ⇒ no current in the branch containing the 3Ω resistor.

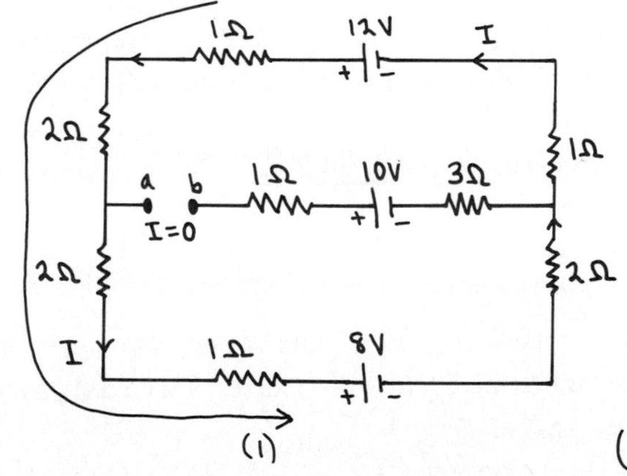

Apply Loop Rule to loop (1):

$$+12V - I(1\Omega + 2\Omega + 2\Omega + 1\Omega) - 8V - I(2\Omega + 1\Omega) = 0$$

$$I = \frac{12V - 8V}{9\Omega} = \frac{4}{9}A$$

Travel from point b to a through the 12 V battery:

$$V_b - 10V - (\tfrac{4}{9}A)(1\Omega) + 12V - (\tfrac{4}{9}A)(1\Omega) - (\tfrac{4}{9}A)(2\Omega) = V_a$$

(No potential drops for resistors having no current.)

$$V_a - V_b = 12V - 10V - (\tfrac{4}{9}A)(4\Omega) = 12V - 10V - \tfrac{16}{9}V = \underline{0.222V} \qquad (V_a > V_b)$$

As a check, travel from point b to point a through the 8V battery:

$$V_b - 10V + (\tfrac{4}{9}A)(2\Omega) + 8V + (\tfrac{4}{9}A)(1\Omega) + (\tfrac{4}{9}A)(2\Omega) = V_a$$

$$V_a - V_b = -10V + 8V + (\tfrac{4}{9}A)(5\Omega) = -10V + 8V + 2.22V = 0.222V \checkmark$$

29-9 (cont)

b) Connect a and b ⇒ current in that branch.

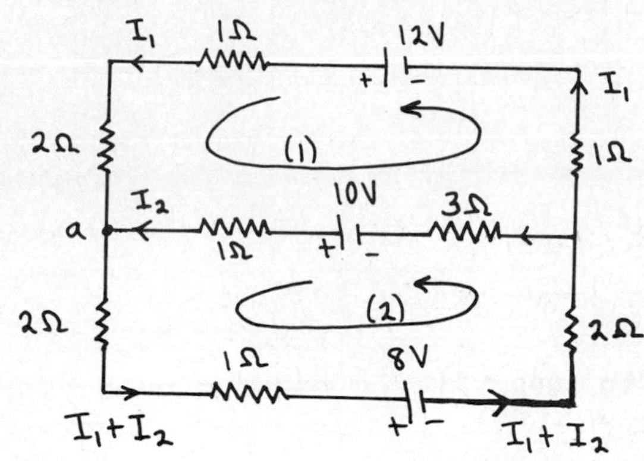

The currents as shown satisfy the point rule at point a:

$I_1 + I_2 - (I_1 + I_2) = 0$ ✓

Apply the loop rule to loop (1):

$$12V - I_1(1\Omega) - I_1(2\Omega) + I_2(1\Omega) - 10V + I_2(3\Omega) - I_1(1\Omega) = 0$$

$$2V - I_1(4\Omega) + I_2(4\Omega) = 0$$

$$1V - I_1(2\Omega) + I_2(2\Omega) = 0$$

$$\boxed{I_1(2\Omega) - I_2(2\Omega) = 1V} \quad (1)$$

Apply the loop rule to loop (2):

$$-(I_1+I_2)(2\Omega) - (I_1+I_2)(1\Omega) - 8V - (I_1+I_2)(2\Omega) - I_2(3\Omega) + 10V - I_2(1\Omega) = 0$$

$$10V - 8V + I_1(-2\Omega - 1\Omega - 2\Omega) + I_2(-2\Omega - 1\Omega - 2\Omega - 3\Omega - 1\Omega) = 0$$

$$2V + I_1(-5\Omega) + I_2(-9\Omega) = 0$$

$$\boxed{I_1(5\Omega) + I_2(9\Omega) = 2V} \quad (2)$$

eq.(1) ⇒ $I_1 = I_2 + \frac{1}{2}A$

Use in eq.(2) ⇒ $(I_2 + \frac{1}{2}A)(5\Omega) + I_2(9\Omega) = 2V$

$$I_2(14\Omega) + 2.5V = 2V$$

$I_2 = -\dfrac{0.5V}{14\Omega} = -0.0357A$ (I_2 is opposite to the direction assumed in the circuit diagram.)

Then $I_1 = I_2 + \frac{1}{2}A = -0.0357A + 0.500A = \underline{0.464A}$

29-11

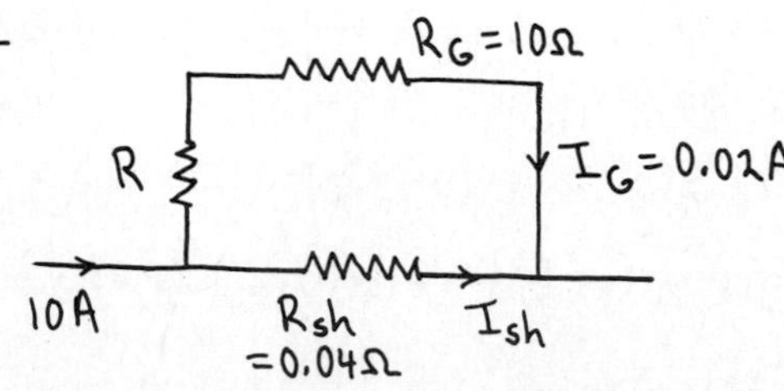

Want 10A in the external circuit to produce 0.02A through the galvanometer

⇒ $10A = I_G + I_{sh}$

$$I_{sh} = 10A - I_G = 10A - 0.02A = 9.98A$$

The potential drop across the shunt must equal the potential drop across the galvanometer branch

$$\Rightarrow I_{sh}R_{sh} = I_G(R + R_G)$$

$$R = \left(\frac{I_{sh}}{I_G}\right)R_{sh} - R_G = \left(\frac{9.98A}{0.02A}\right)(0.04\Omega) - 10\Omega = 19.96\Omega - 10\Omega = \underline{9.96\Omega}$$

29-13

For each range setting the circuit has the form

$R_G = 15\,\Omega$ R

$I = 1\times10^{-3}$ A I

V

For $V = 3$ volts, $R = R_1 \Rightarrow I(R_G + R_1) = V$

$$R_1 = \frac{V}{I} - R_G = \frac{3\,V}{1\times10^{-3}\,A} - 15\,\Omega = 3\times10^3\,\Omega - 15\,\Omega = \underline{2985\,\Omega}$$

The resistance of the meter is $R_1 + R_G = \underline{3.0\times10^3\,\Omega}$.

For $V = 15$ volts, $R = R_1 + R_2 \Rightarrow I(R_G + R_1 + R_2) = V$

$$R_2 = \frac{V}{I} - R_G - R_1 = \frac{15\,V}{1\times10^{-3}\,A} - 15\,\Omega - 2985\,\Omega = 15{,}000\,\Omega - 15\,\Omega - 2985\,\Omega = \underline{12{,}000\,\Omega}$$

The resistance of the meter is $R_1 + R_2 + R_G = \underline{1.5\times10^4\,\Omega}$

For $V = 150$ volts, $R = R_1 + R_2 + R_3 \Rightarrow I(R_G + R_1 + R_2 + R_3) = V$

$$R_3 = \frac{V}{I} - R_1 - R_2 - R_3 = \frac{150\,V}{1\times10^{-3}\,A} - 15\,\Omega - 2985\,\Omega - 12{,}000\,\Omega = 150{,}000\,\Omega - 15{,}000\,\Omega$$

$R_3 = \underline{135{,}000\,\Omega}$

The resistance of the meter is $R = \underline{1.5\times10^5\,\Omega}$.

29-15

$R_1 = 15{,}000\,\Omega$ $R_2 = 150{,}000\,\Omega$

V_1 V_2

I

$V = 120$ volts

$$V_1 + V_2 = V$$

$$I(R_1 + R_2) = V$$

$$I = \frac{V}{R_1 + R_2} = \frac{120\,V}{15{,}000\,\Omega + 150{,}000\,\Omega} = 7.27\times10^{-4}\,A$$

Then

$$V_1 = I_1 R_1 = (7.27\times10^{-4}\,A)(15{,}000\,\Omega) = \underline{10.9\,V}$$

$$V_2 = I_2 R_2 = (7.27\times10^{-4}\,A)(150{,}000\,\Omega) = \underline{109\,V}$$

(Note: $V_1 + V_2 = 120\,V$, as required.)

29-17

a)

$R = 1\times10^6\,\Omega$

$C = 10\times10^{-6}$ F

$V = 100$ volts

$$q = VC(1 - e^{-t/RC}) \quad \text{(eq. 29-9)}$$

$$Q_f = VC = (100\,V)(10\times10^{-6}\,F) = 1\times10^{-3}\,C$$

$$RC = (1\times10^6\,\Omega)(10\times10^{-6}\,F) = 10\,s$$

$t = 0 \Rightarrow q = (1\times10^{-3}\,C)(1 - 1) = \underline{0}$

$t = 5\,s \Rightarrow q = (1\times10^{-3}\,C)(1 - e^{-5/10}) = \underline{3.93\times10^{-4}\,C}$

$t = 10\,s \Rightarrow q = (1\times10^{-3}\,C)(1 - e^{-10/10}) = \underline{6.32\times10^{-4}\,C}$

$t = 20\,s \Rightarrow q = (1\times10^{-3}\,C)(1 - e^{-20/10}) = \underline{8.65\times10^{-4}\,C}$

$t = 100\,s \Rightarrow q = (1\times10^{-3}\,C)(1 - e^{-100/10}) = \underline{1.00\times10^{-3}\,C}$

29-17 (cont)

b) $i = \frac{V}{R} e^{-t/RC}$

$I_0 = \frac{V}{R} = \frac{100V}{1\times10^6\Omega}$

$I_0 = 1.0\times10^{-4}\text{ A}$

$t=0 \Rightarrow i = (1\times10^{-4}\text{ A})e^0 = \underline{1\times10^{-4}\text{ A}}$

$t=5\text{s} \Rightarrow i = (1\times10^{-4}\text{ A})e^{-5/10} = \underline{6.07\times10^{-5}\text{ A}}$

$t=10\text{s} \Rightarrow i = (1\times10^{-4}\text{ A})e^{-10/10} = \underline{3.68\times10^{-5}\text{ A}}$

$t=20\text{s} \Rightarrow i = (1\times10^{-4}\text{ A})e^{-20/10} = \underline{1.35\times10^{-5}\text{ A}}$

$t=100\text{s} \Rightarrow i = (1\times10^{-4}\text{ A})e^{-100/10} = \underline{4.54\times10^{-9}\text{ A}}$

c) $Q_i = 0,\ Q_f = 1.0\times10^{-3}\text{C}$

$I_0 = 1.0\times10^{-4}\text{ A}$

$I_0 = \frac{Q_f - Q_i}{t}$

$\Rightarrow t = \frac{Q_f - Q_i}{I_0} = \frac{1.0\times10^{-3}\text{C}}{1.0\times10^{-4}\text{A}} = \underline{10\text{s}}$

d) $q = VC(1 - e^{-t/RC})$

$q = 5\times10^{-4}\text{C};\ VC = 1.0\times10^{-3}\text{C};\ RC = 10\text{s}$

$\Rightarrow 5\times10^{-4}\text{C} = 10\times10^{-4}\text{C}(1 - e^{-t/10\text{s}})$

$\frac{1}{2} = 1 - e^{-t/10\text{s}} \Rightarrow e^{-t/10\text{s}} = \frac{1}{2} \Rightarrow -\frac{t}{10\text{s}} = \ln\left(\frac{1}{2}\right)$

$\frac{t}{10\text{s}} = \ln 2 \Rightarrow t = (10\text{s})\ln 2 = \underline{6.93\text{s}}$

e)

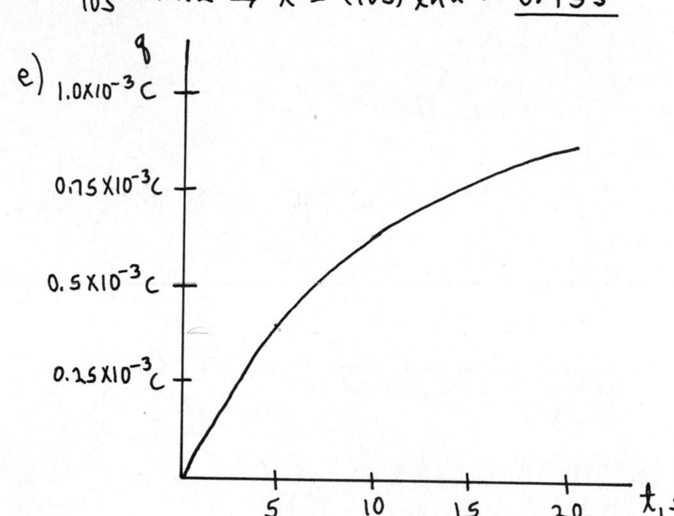

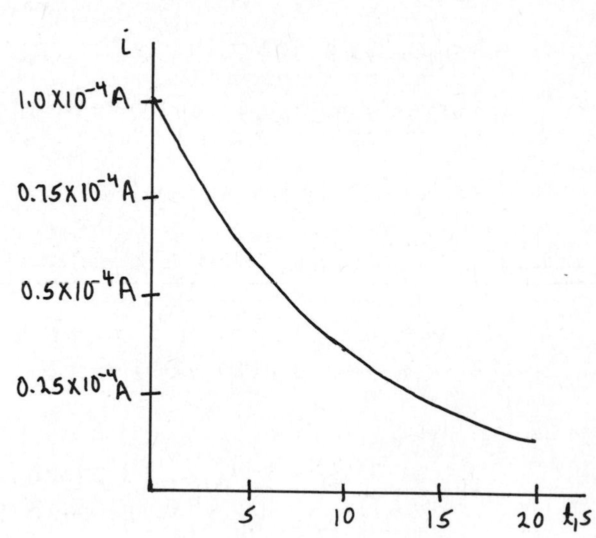

29-21

a) $I_c = 2\times10^{-3}\text{ A}$

$q = 0$ at $t = 0$.

The amount of charge brought to the plates by the charging current in time t is $Q = I_c t = (2\times10^{-3}\text{A})(5.0\times10^{-6}\text{s}) = \underline{1.0\times10^{-8}\text{C}}$

$E = \frac{\sigma}{\epsilon_0} = \frac{Q}{\epsilon_0 A} = \frac{1.0\times10^{-8}\text{C}}{(8.85\times10^{-12}\text{C}^2\cdot\text{N}^{-1}\cdot\text{m}^{-2})(4\times10^{-4}\text{m}^2)} = \underline{2.82\times10^6\text{ N}\cdot\text{C}^{-1}}$

$V_{ab} = Ed = (2.82\times10^6\text{ V}\cdot\text{m}^{-1})(3\times10^{-3}\text{m}) = \underline{8.46\times10^3\text{V}}$

29-21 (cont)

b) $E = \frac{Q}{\epsilon_0 A} \Rightarrow \frac{dE}{dt} = \frac{dQ/dt}{\epsilon_0 A}$

But $\frac{dQ}{dt} = I_c \Rightarrow \frac{dE}{dt} = \frac{I_c}{\epsilon_0 A} = \frac{2\times10^{-3}\,A}{(8.85\times10^{-12}\,C^2\cdot N^{-1}\cdot m^{-2})(4\times10^{-4}\,m^2)} = \underline{5.65\times10^{11}\,N\cdot C^{-1}\cdot s^{-1}}$

I_c constant $\Rightarrow \frac{dE}{dt} = \frac{I_c}{\epsilon_0 A}$ is constant in time.

c) $J_D = \epsilon_0 \frac{dE}{dt} = (8.85\times10^{-12}\,C^2\cdot N^{-1}\cdot m^{-2})(5.65\times10^{11}\,N\cdot C^{-1}\cdot s^{-1}) = \underline{5.00\,A\cdot m^{-2}}$

$I_D = J_D A = (5.00\,A\cdot m^{-2})(4\times10^{-4}\,m^2) = \underline{2.0\times10^{-3}\,A}$

Thus $I_D = I_c$.

29-25

a) $P = VI \Rightarrow I = \frac{P}{V}$

toaster: $I_t = \frac{P_t}{V} = \frac{1500\,W}{120\,V} = 12.5\,A$

frypan: $I_f = \frac{P_f}{V} = \frac{1200\,W}{120\,V} = 10.0\,A$

lamp: $I_\ell = \frac{P_\ell}{V} = \frac{100\,W}{120\,V} = 0.833\,A$

b) The total current drawn is $I_t + I_f + I_\ell = 12.5\,A + 10.0\,A + 0.833\,A = 23.3\,A$.
This will blow the 20 A fuse.

Problems

29-29

a) $P = I^2R$, so must have only part of the current go through each resistor
$\Rightarrow$ parallel combination.

A combination of the available resistors having the desired properties is two in series in parallel with two in series:

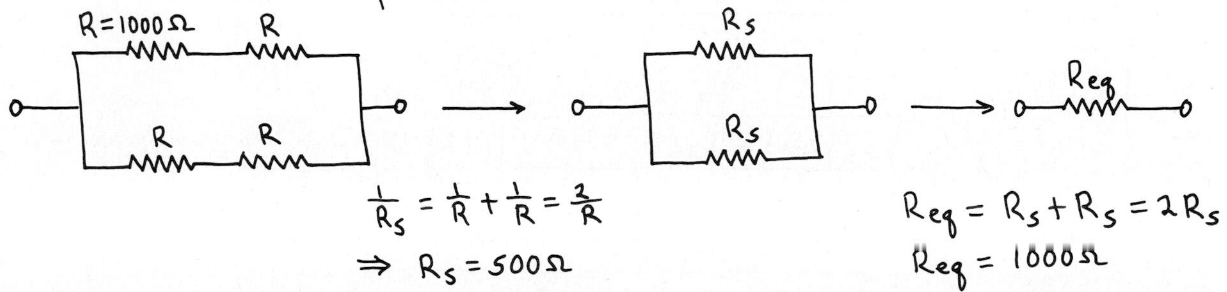

$\frac{1}{R_s} = \frac{1}{R} + \frac{1}{R} = \frac{2}{R}$

$\Rightarrow R_s = 500\,\Omega$

$R_{eq} = R_s + R_s = 2R_s$

$R_{eq} = 1000\,\Omega$

Thus the equivalent resistance is the required $1000\,\Omega$. Whether or not the combination has the desired power rating will be determined in part (b).

29-29 (cont)

b) The equivalent resistance must dissipate 2 W.

Thus $P = \frac{V^2}{R_{eq}} \Rightarrow V = \sqrt{P R_{eq}} = \sqrt{(2W)(1000\Omega)} = 44.7$ volts must be applied across the resistor network.

Thus there is 44.7 V across each parallel branch:

R R

⟵ 44.7V ⟶

This implies a potential of $\frac{1}{2}(44.7\,V) = 22.4\,V$ across each of the four resistors. The power dissipated in each one is thus
$P = \frac{V^2}{R} = \frac{(22.4V)^2}{1000\,\Omega} = \underline{0.50W}$. This is less than the 1 watt power rating of the individual resistors, so this network can dissipate the required 2 W.

29-33

Three unknown currents are shown in Fig. 29-26. Determine the currents in the other wires in terms of these, using the point rule. Note that no additional currents need to be introduced.

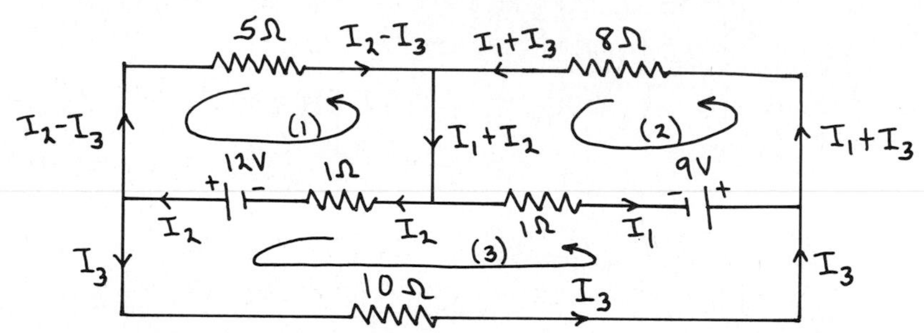

Apply the loop rule to the 3 loops indicated: (3 unknown currents ⇒ need 3 loop equations)

loop (1)

$$-12V + I_2(1\Omega) + (I_2 - I_3)(5\Omega) = 0$$

$$\boxed{I_2(6\Omega) - I_3(5\Omega) = 12V} \quad (1)$$

loop (2)

$$-I_1(1\Omega) + 9V - (I_1 + I_3)8\Omega = 0$$

$$\boxed{I_1(9\Omega) + I_3(8\Omega) = 9V} \quad (2)$$

loop (3)

$$-I_3(10\Omega) - 9V + I_1(1\Omega) - I_2(1\Omega) + 12V = 0$$

$$\boxed{-I_1(1\Omega) + I_2(1\Omega) + I_3(10\Omega) = 3V} \quad (3)$$

29-33 (cont)

eq.(1) $\Rightarrow I_2 = 2A + \frac{5}{6} I_3$

eq.(2) $\Rightarrow I_1 = 1A - \frac{8}{9} I_3$

Use these in eq.(3) $\Rightarrow -(1A - \frac{8}{9} I_3)(1\Omega) + (2A + \frac{5}{6} I_3)(1\Omega) + I_3(10\Omega) = 3V$

$-1V + I_3(\frac{8}{9}\Omega) + 2V + I_3(\frac{5}{6}\Omega) + I_3(10\Omega) = 3V$

$\left(\frac{16+15+180}{18}\right)\Omega\, I_3 = 2V \Rightarrow I_3 = 2V\left(\frac{18}{211\Omega}\right) = \underline{0.171\,A}$

$I_1 = 1A - \frac{8}{9} I_3 = 1A - \frac{8}{9}(0.171\,A) = \underline{0.848\,A}$

$I_2 = 2A + \frac{5}{6} I_3 = 2A + \frac{5}{6}(0.171A) = \underline{2.14A}$

29-35

a)

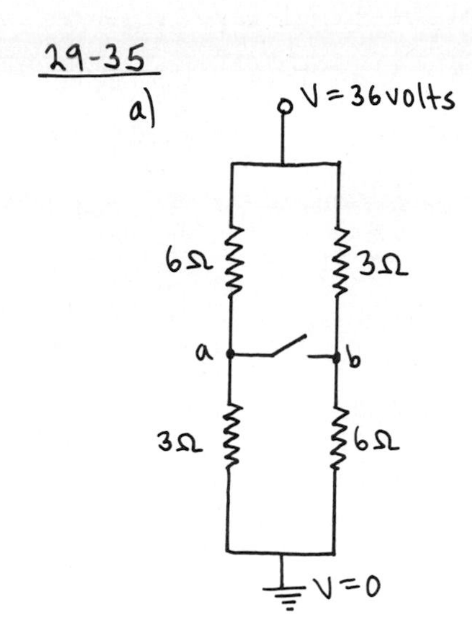

The 6Ω and 3Ω resistors on each side are in series. Thus, the equivalent circuit is:

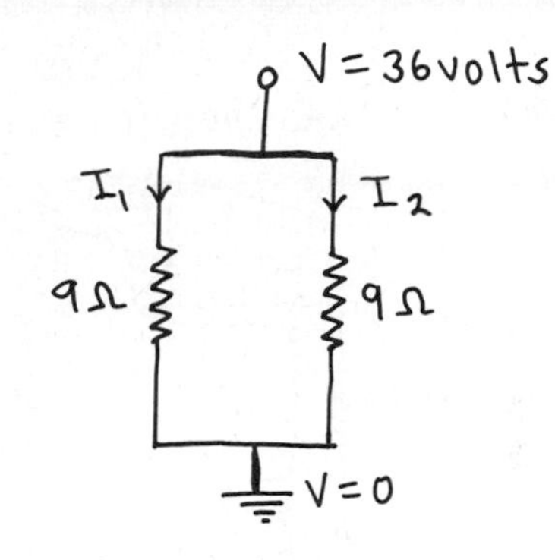

$I_1(9\Omega) = 36V \Rightarrow I_1 = 4A$

$I_2(9\Omega) = 36V \Rightarrow I_2 = 4A$

$V_a = V - I_1(6\Omega) = 36V - (4A)(6\Omega) = 12V$

$V_b = V - I_2(3\Omega) = 36V - (4A)(3\Omega) = 24V$

Thus

$V_{ab} = V_a - V_b = 12V - 24V = \underline{-12V}$ (point b is at higher potential)

b) close the switch $\Rightarrow$ the 6Ω and 3Ω resistors are in parallel

$\frac{1}{R_p} = \frac{1}{6\Omega} + \frac{1}{3\Omega} = \frac{3}{6\Omega} \Rightarrow R_p = 2\Omega$

Thus:

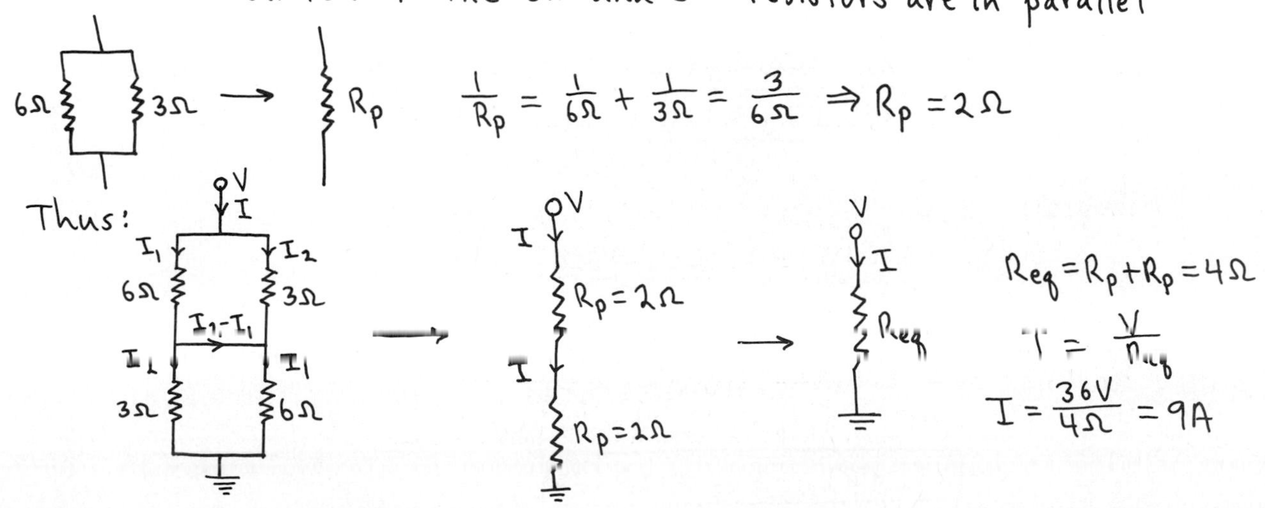

$R_{eq} = R_p + R_p = 4\Omega$

$I = \frac{V}{R_{eq}}$

$I = \frac{36V}{4\Omega} = 9A$

29-35 (cont)

The potential across each R_p is thus $V_p = I R_p = (9A)(2\Omega) = 18\text{ V}$.
The potential across each individual resistor is therefore 18 V.

$$I_1 = \frac{18\text{V}}{6\Omega} = 3\text{A}$$
$$I_2 = \frac{18\text{V}}{3\Omega} = 6\text{A}$$
$$\Rightarrow I_2 - I_1 = 6\text{A} - 3\text{A} = \underline{3\text{A}}$$

c) When the switch is open there is no current through the additional 3Ω resistor in Fig. 26-28(b). This circuit then is identical to that of Fig. 29-28(a).

Thus $V_{ab} = \underline{-12\text{V}}$, as in part (a).

d) When the switch S is closed in Fig. 29-28(b) one has the following circuit; (We have drawn in the 36 V battery, to make it easier to apply the loop rule.)

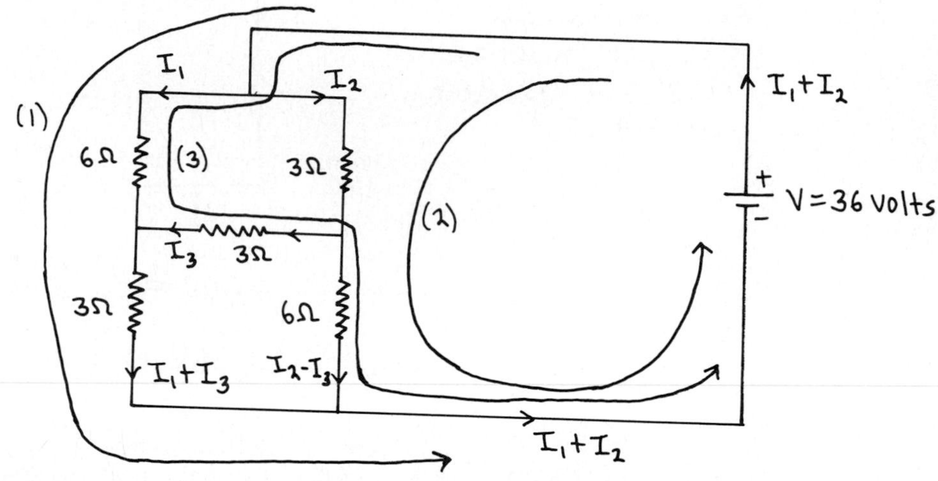

Three unknown currents are indicated, and the currents in the remaining branches have been determined by the point rule. Apply the loop rule to the three loops indicated:

loop (1) : $36\text{V} - I_1(6\Omega) - (I_1 + I_3)(3\Omega) = 0$

$$I_1(9\Omega) + I_3(3\Omega) = 36\text{V}$$
$$\boxed{3I_1 + I_3 = 12\text{A}} \quad (1)$$

loop (2): $36\text{V} - I_2(3\Omega) - (I_2 - I_3)6\Omega = 0$

$$I_2(9\Omega) - I_3(6\Omega) = 36\text{V}$$
$$\boxed{3I_2 - 2I_3 = 12\text{A}} \quad (2)$$

loop (3): $36\text{V} - I_1(6\Omega) + I_3(3\Omega) - (I_2 - I_3)(6\Omega) = 0$

$$I_1(6\Omega) + I_2(6\Omega) - I_3(9\Omega) = 36\text{V}$$
$$\boxed{2I_1 + 2I_2 - 3I_3 = 12\text{A}} \quad (3)$$

29-35 (cont)

$$eq(1) \Rightarrow I_1 = 4A - \tfrac{1}{3} I_3 \; ; \; eq(2) \Rightarrow I_2 = 4A + \tfrac{2}{3} I_3$$

Use these in eq. (3) $\Rightarrow 8A - \tfrac{2}{3} I_3 + 8A + \tfrac{4}{3} I_3 - 3I_3 = 12A$

$$\tfrac{7}{3} I_3 = 4A \Rightarrow I_3 = \tfrac{3}{7}(4A) = \underline{\tfrac{12}{7} A}$$, current through switch (from b to a)

Then

$$I_2 = 4A + \tfrac{2}{3}(I_3) = 4A + \tfrac{2}{3}\left(\tfrac{12}{7} A\right) = \tfrac{36}{7} A$$

$$I_1 = 4A - \tfrac{1}{3}(I_3) = 4A - \tfrac{1}{3}\left(\tfrac{12}{7} A\right) = \tfrac{24}{7} A$$

e) switch open (part (a)):

$I = I_1 + I_2$; $I_1 = 4A$; $I_2 = 4A$

$I = 8A$ (total current through circuit)

$$\Rightarrow R_{eq} = \frac{V}{I} = \frac{36V}{8A} = \underline{4.5\,\Omega}$$

f) switch closed:

$I = I_1 + I_2$; $I_1 = \tfrac{24}{7} A$; $I_2 = \tfrac{36}{7} A$

$I = \tfrac{24}{7} A + \tfrac{36}{7} A = \tfrac{60}{7} A = 8.57A$ (total current through circuit)

$$R_{eq} = \frac{V}{I} = \frac{36V}{8.57A} = \underline{4.20\,\Omega}$$

29-37

V=18V; 6Ω, 3Ω resistors; 6μF, 3μF capacitors; points a, b

a) switch open:

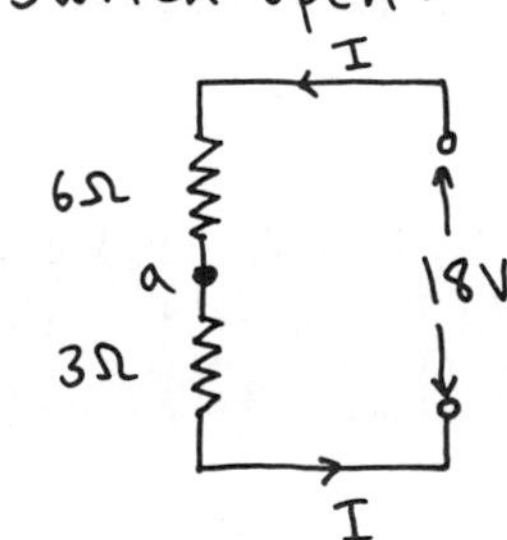

$$I = \frac{18V}{6\Omega + 3\Omega} = 2A$$

$$V_a = V - I(6\Omega)$$

$$V_a = 18V - (2A)(6\Omega) = 6V$$

V_6, C_6, Q_6; V_3, C_3, Q_3; 18V

$$Q_6 = Q_3 = Q$$

$$V_6 + V_3 = V$$

$$\frac{Q}{C_6} + \frac{Q}{C_3} = V$$

$$Q\left(\frac{1}{C_6} + \frac{1}{C_3}\right) = V$$

$$Q = \frac{V}{\frac{1}{C_6} + \frac{1}{C_3}} = \frac{18V}{\frac{1}{6\times10^{-6}F} + \frac{1}{3\times10^{-6}F}} = 3.6\times10^{-5}C = 36\mu C$$

$$V_6 = \frac{Q}{C_6} = \frac{36\mu C}{6\mu F} = 6V \; ; \; V_3 = \frac{Q}{C_3} = \frac{36\mu C}{3\mu F} = 12V$$

29-37 (cont)

Thus $V_b = 18V - V_a = 18V - 6V = 12V$

$V_a - V_b = 6V - 12V = \underline{-6\,V}$ (point b is at higher potential)

b) point b

c) switch closed:

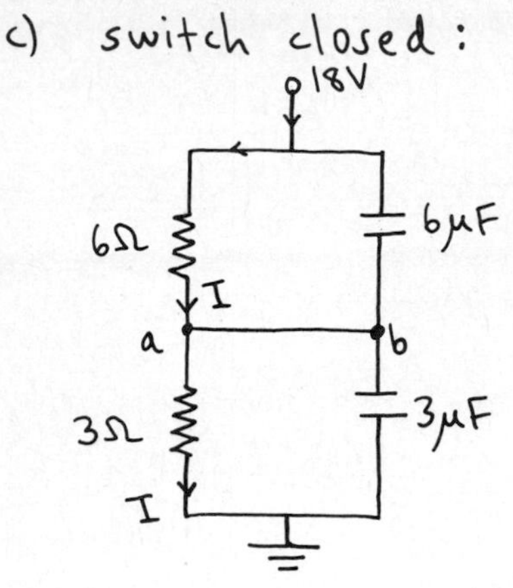

The only current path is still through the two resistors. (No current through the capacitors after they have received their equilibrium charges.)

$$\Rightarrow I = \frac{18V}{6\Omega + 3\Omega} = 2A$$

But now $V_6 = I(6\Omega) = 12V$

(The potential drop across the 6μF capacitor must equal the potential drop across the 6Ω resistor.)

$$\Rightarrow V_b = 18V - 12V = \underline{6\,V}$$

d) switch open ⇒ $Q_6 = 36\mu C$ (charge on the 6μF capacitor)

$Q_3 = 36\mu C$ (charge on the 3μF capacitor)

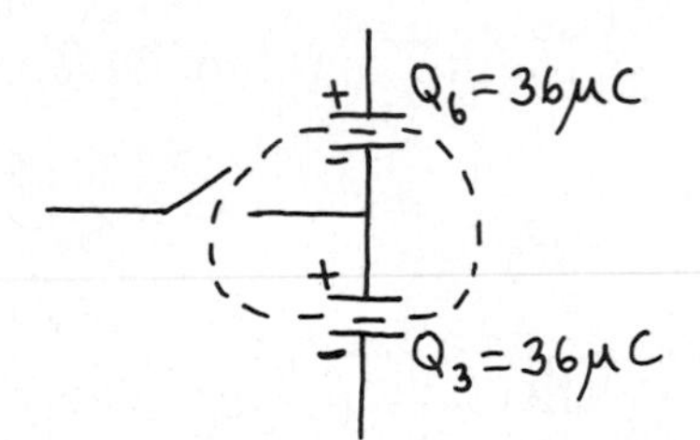

The net charge on the conductor enclosed by the dashed line is zero.

switch closed ⇒ $V_6 = 12V$, $Q_6 = C_6 V_6 = (6\mu F)(12V) = 72\mu C$

$V_3 = 6V$, $Q_3 = C_3 V_3 = (3\mu F)(6V) = 18\mu C$

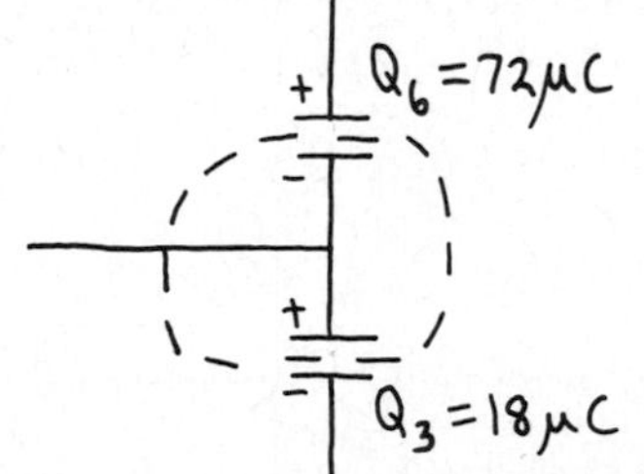

The net charge on the conductor enclosed by the dashed line is now

$$-72\mu C + 18\mu C = -54\mu C$$

To get onto these capacitor plates this charge must have passed through the switch ⇒ $-54\mu C$ of charge flowed through the switch S when it was closed.

29-39

a)

$I_1 + I_2 = I$

$I(400\,\Omega) = 45\,V \Rightarrow I = 0.1125\,A$

$I_2(600\,\Omega) = 45\,V \Rightarrow I_2 = 0.075\,A$

Thus $I_1 = I - I_2 = 0.1125\,A - 0.075\,A = 0.0375\,A$

But $I_1 R_V = 45\,V \Rightarrow R_V = \dfrac{45\,V}{0.0375\,A} = \underline{1200\,\Omega}$

b)

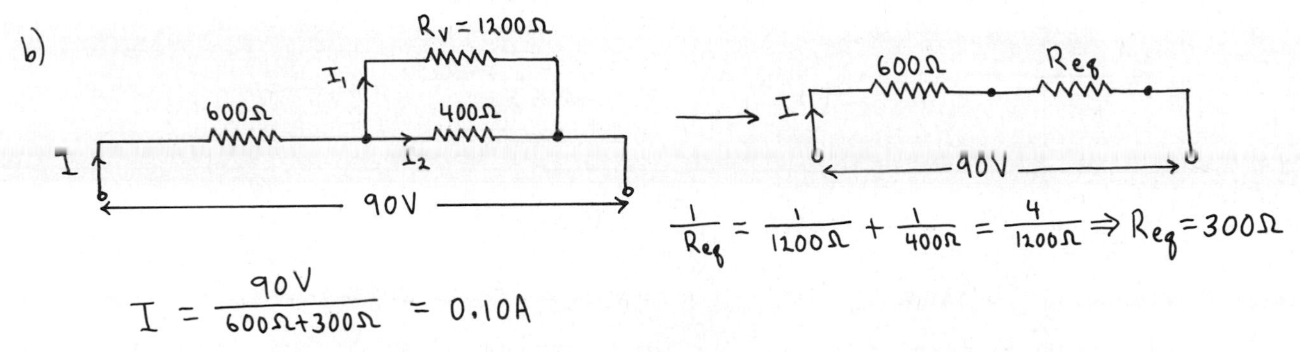

$\dfrac{1}{R_{eq}} = \dfrac{1}{1200\,\Omega} + \dfrac{1}{400\,\Omega} = \dfrac{4}{1200\,\Omega} \Rightarrow R_{eq} = 300\,\Omega$

$$I = \frac{90\,V}{600\,\Omega + 300\,\Omega} = 0.10\,A$$

The potential drop across the $600\,\Omega$ resistor is $I(600\,\Omega) = 60\,V$
The potential drop across the voltmeter resistance is thus $90\,V - 60\,V = \underline{30\,V}$; this is what the voltmeter reads.

29-45

a) The potentiometer circuit is equivalent to:

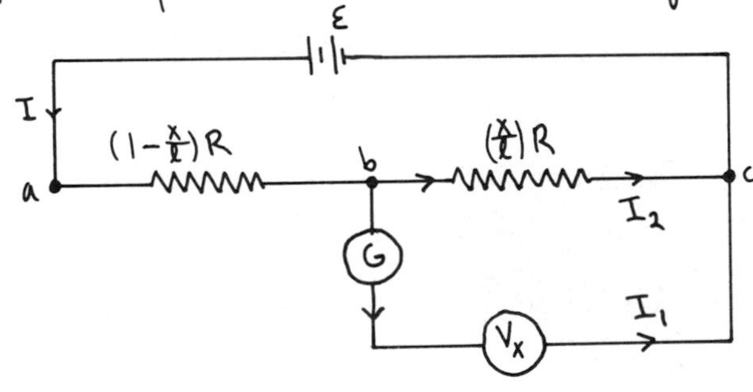

The resistance between points b and c is proportional to the distance between b and c.

The unknown potential difference V_x equals $I_2(\frac{x}{\ell}R)$.

(G) reads zero $\Rightarrow I_1 = 0,\ I_2 = I$

$$I = \frac{\mathcal{E}}{(1-\frac{x}{\ell})R + (\frac{x}{\ell})R} = \frac{\mathcal{E}}{R}$$

Thus $V_x = \frac{\mathcal{E}}{R}(\frac{x}{\ell}R) = \frac{x}{\ell}\mathcal{E}$

b) The resistance of the galvanometer is unimportant because the reading is taken with zero current through the galvanometer.

c) From (a), $V_x = \frac{x}{\ell}\mathcal{E} \Rightarrow V_x = \dfrac{0.193\,m}{1.000\,m}(12.00\,V) = \underline{4.52\,V}$

29-45 (cont)

No current through V_x when reading is taken.

$V_{ab} = \mathcal{E} - Ir$

$I = 0 \Rightarrow V_{ab} = \mathcal{E}$

Thus the terminal voltage of the battery equals the emf $\mathcal{E}$ of the battery; the potentiometer measures the emf.

29-47

eq. (29-15), $i = I_0 e^{-t/RC}$; $I_0 = \frac{V}{R}$

a) $P_V = Vi = V\left(\frac{V}{R}\right)e^{-t/RC} = \frac{V^2}{R} e^{-t/RC}$

$dU_V = P_V\, dt$

$U_V = \int_0^\infty P_V\, dt = \frac{V^2}{R}\int_0^\infty e^{-t/RC}\, dt = \frac{V^2}{R}(-RC)\left(e^{-t/RC}\Big|_0^\infty\right) = \underline{CV^2}$

b) $P_R = i^2 R = R\left(\frac{V^2}{R^2}\right) e^{-2t/RC} = \frac{V^2}{R} e^{-2t/RC}$

$dU_R = P_R\, dt$

$U_R = \int_0^\infty P_R\, dt = \frac{V^2}{R}\int_0^\infty e^{-2t/RC}\, dt = \frac{V^2}{R}\left(-\frac{RC}{2}\right)\left(e^{-2t/RC}\Big|_0^\infty\right) = \underline{\tfrac{1}{2}CV^2}$

c) $U_C = \frac{1}{2}CV^2$

Thus $U_C = U_V - U_R$, as was to be shown.

d) $\frac{U_C}{U_V} = \frac{\frac{1}{2}CV^2}{CV^2} = \frac{1}{2}$; this ratio is independent of R.

29-49

a) $V = IR$, or $E = J_C \rho$

$J_{C,max} = \frac{E_{max}}{\rho} = \frac{E_0}{\rho} = \frac{0.1\,V\cdot m^{-1}}{1.72\times10^{-8}\,\Omega\cdot m} = \underline{5.81\times10^{6}\,A\cdot m^{-2}}$

b) $J_D = \epsilon_0 \frac{dE}{dt}$

$E = E_0 \sin\omega t$

$\frac{dE}{dt} = \omega E_0 \cos\omega t$; $\left(\frac{dE}{dt}\right)_{max} = \omega E_0 = 2\pi f E_0$

$J_{D,max} = \epsilon_0\left(\frac{dE}{dt}\right)_{max} = \epsilon_0 2\pi f E_0 = (8.85\times10^{-12}\,C^2\cdot N^{-1}\cdot m^{-2})(2\pi)(60\,s^{-1})(0.1\,V\cdot m^{-1})$

$J_{D,max} = \underline{3.34\times10^{-10}\,A\cdot m^{-2}}$

[Note that at this low frequency (that of household current in the United States) $J_{D,max} \ll J_{C,max}$.]

CHAPTER 30

Exercises 1, 3, 7, 9, 13, 19, 23, 25, 27

Problems 29, 31, 37, 41, 43, 45

Exercises

30-1 $\vec{F} = q\vec{v} \times \vec{B}$

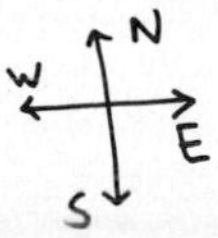

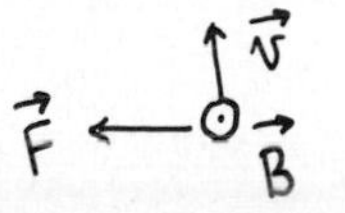

$\vec{v} \times \vec{B} \longrightarrow$, which is opposite to $\vec{F}$

Thus q must be negative.

30-3

$$\vec{v} = (-3\times10^4\,\text{m}\cdot\text{s}^{-1})\vec{i} + (5\times10^4\,\text{m}\cdot\text{s}^{-1})\vec{j} \;;\; q = -2.5\times10^{-8}\,\text{C}$$

a) $\vec{B} = (2\text{T})\vec{i}$

$$\vec{F} = q\,\vec{v}\times\vec{B} = (-2.5\times10^{-8}\,\text{C})(2\text{T})\left[(-3\times10^4\,\text{m}\cdot\text{s}^{-1})\vec{i} + (5\times10^4\,\text{m}\cdot\text{s}^{-1})\vec{j}\,\right]\times\vec{i}$$

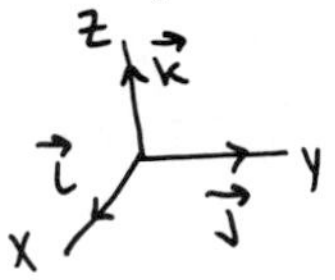

$$\vec{i}\times\vec{i} = \vec{j}\times\vec{j} = \vec{k}\times\vec{k} = 0$$

$$\vec{i}\times\vec{j} = \vec{k} \;;\; \vec{j}\times\vec{k} = \vec{i} \;;\; \vec{k}\times\vec{i} = \vec{j}$$

$$\vec{F} = -(2.5\times10^{-8}\,\text{C})(2\text{T})\left[(-3\times10^4\,\text{m}\cdot\text{s}^{-1})\,\underbrace{\vec{i}\times\vec{i}}_{0} + (5\times10^4\,\text{m}\cdot\text{s}^{-1})\,\underbrace{\vec{j}\times\vec{i}}_{-\vec{k}}\,\right]$$

$$\vec{F} = +(2.5\times10^{-8}\,\text{C})(2\text{T})(5\times10^4\,\text{m}\cdot\text{s}^{-1})\vec{k} = +\underline{(2.5\times10^{-3}\,\text{N})\vec{k}}$$

b) $\vec{B} = (2\text{T})\vec{k}$

$$\vec{F} = q\vec{v}\times\vec{B} = (-2.5\times10^{-8}\,\text{C})(2\text{T})\left[(-3\times10^4\,\text{m}\cdot\text{s}^{-1})\vec{i} + (5\times10^4\,\text{m}\cdot\text{s}^{-1})\vec{j}\,\right]\times\vec{k}$$

$$\vec{F} = -(2.5\times10^{-8}\,\text{C})(2\text{T})\left[(-3\times10^4\,\text{m}\cdot\text{s}^{-1})\,\underbrace{\vec{i}\times\vec{k}}_{-\vec{j}} + (5\times10^4\,\text{m}\cdot\text{s}^{-1})\,\underbrace{\vec{j}\times\vec{k}}_{\vec{i}}\,\right]$$

$$\vec{F} = -(2.5\times10^{-8}\,\text{C})(2\text{T})\left[(3\times10^4\,\text{m}\cdot\text{s}^{-1})\vec{j} + (5\times10^4\,\text{m}\cdot\text{s}^{-1})\vec{i}\,\right] = \underline{-(2.5\times10^{-3}\,\text{N})\vec{i} - (1.5\times10^{-3}\,\text{N})\vec{j}}$$

30-7

a)

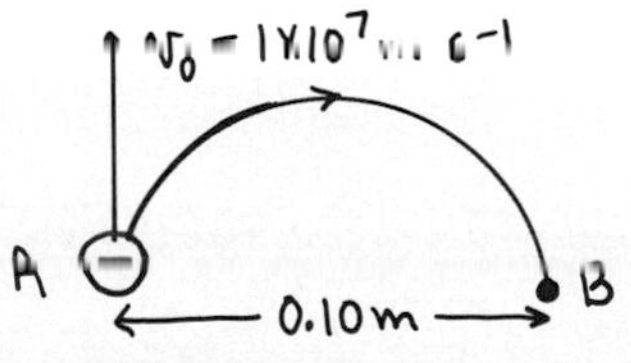

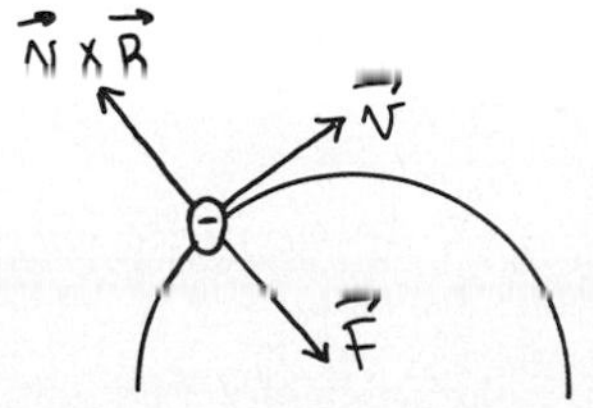

Semicircular path $\Rightarrow$ the force must be radially inward.
$\vec{v}$ is tangential.
Negative charge $\Rightarrow$ $\vec{F}$ and $\vec{v}\times\vec{B}$ are opposite.

30-7 (cont)

For $\vec{v} \times \vec{B}$ to be in the direction shown in the sketch, $\vec{B}$ must be directed into the plane of the paper ($\vec{B} \otimes$).

$F = ma$

$$\Rightarrow qvB = m\frac{v^2}{R} \Rightarrow B = \frac{mv}{qR} = \frac{(9.11\times10^{-31}\,kg)(1\times10^{7}\,m\cdot s^{-1})}{(1.60\times10^{-19}\,C)(0.05\,m)} = \underline{1.14\times10^{-3}\,T}$$

b) $v = 1\times10^{7}\,m\cdot s^{-1}$

The distance, along the semicircular path, from A to B is $s = \pi r$.

$$t = \frac{s}{v} = \frac{\pi R}{v} = \frac{\pi\,(0.05\,m)}{1\times10^{7}\,m\cdot s^{-1}} = \underline{1.57\times10^{-8}\,s}$$

30-9

a) $F = ma \Rightarrow qvB = m\frac{v^2}{R}$; $v = \frac{qBR}{m}$

$m = 2u = 2(1.66\times10^{-27}\,kg)$

A deuteron has a charge of $+e = 1.60\times10^{-19}\,C$

$$v = \frac{(1.60\times10^{-19}\,C)(1.5\,T)(0.40\,m)}{2(1.66\times10^{-27}\,kg)} = \underline{2.89\times10^{7}\,m\cdot s^{-1}}$$

b) $$s = \pi R \Rightarrow t = \frac{s}{v} = \frac{\pi R}{v} = \frac{\pi\,(0.40\,m)}{2.89\times10^{7}\,m\cdot s^{-1}} = \underline{4.35\times10^{-8}\,s}$$

c) conservation of energy $\Rightarrow \Delta U + \Delta K = 0$

$\Delta K = \frac{1}{2}mv^2 - 0$

$\Delta U = q\,\Delta V$ $\quad\Rightarrow \frac{1}{2}mv^2 = -q\,\Delta V$

$$\Delta V = -\frac{mv^2}{2q} = -\frac{2(1.66\times10^{-27}\,kg)(2.89\times10^{7}\,m\cdot s^{-1})^2}{2(1.60\times10^{-19}\,C)} = -8.67\times10^{6}\,\text{volts}$$

(The deuteron must be accelerated through a potential <u>drop</u> of 8.67×10^{6} volts.)

30-13

a) $\uparrow F_E = qE$, $\ominus$, $\downarrow F_B = qvB$

no deflection $\Rightarrow F_E = F_B$

$qE = qvB$

$$v = \frac{E}{B} = \frac{34\times10^{4}\,V\cdot m^{-1}}{2\times10^{-2}\,T} = \underline{1.7\times10^{7}\,m\cdot s^{-1}}$$

b) $\uparrow \vec{F}_E$, $\ominus \rightarrow \vec{v}$, $\downarrow \vec{F}_B$

$\uparrow \vec{F}_E \;\; \downarrow \vec{E}$

$\downarrow \vec{F}_B \;\; \uparrow \vec{v}\times\vec{B} \Rightarrow \vec{B} \otimes$

30-13 (cont)

c) $F_B = ma = m\frac{v^2}{R} \Rightarrow R = \frac{mv^2}{F_B} = \frac{mv^2}{qvB} = \frac{mv}{qB} = \frac{(9.11\times10^{-31}\,kg)(1.7\times10^{7}\,m\cdot s^{-1})}{(1.60\times10^{-19}\,C)(2\times10^{-2}\,T)}$

$R = \underline{4.84\times10^{-3}\,m}$

30-19

$\vec{l} = (0.01m)\,\vec{i},\quad I = 5A,\quad \vec{F} = I\,\vec{l}\times\vec{B}$

$\vec{i}\times\vec{i} = \vec{j}\times\vec{j} = \vec{k}\times\vec{k} = 0$
$\vec{i}\times\vec{j} = \vec{k}$
$\vec{j}\times\vec{k} = \vec{i}$
$\vec{k}\times\vec{i} = \vec{j}$

a) $\vec{B} = -(0.6T)\,\vec{j}$
$\vec{F} = (5A)(0.01m)(0.6T)(-\vec{i}\times\vec{j}) = \underline{-(0.03N)\,\vec{k}}$

b) $\vec{B} = +(0.5T)\,\vec{k}$
$\vec{F} = (5A)(0.01m)(0.5T)\,\vec{i}\times\vec{k} = \underline{-(0.025N)\,\vec{j}}$

c) $\vec{B} = -(0.3T)\,\vec{i}$
$\vec{F} = (5A)(0.01m)(0.3T)(-\vec{i}\times\vec{i}) = \underline{0}$

d) $\vec{B} = (0.2T)\,\vec{i} - (0.3T)\,\vec{k}$
$\vec{F} = (5A)(0.01m)[(0.2T)\underbrace{\vec{i}\times\vec{i}}_{0} - (0.3T)\underbrace{\vec{i}\times\vec{k}}_{-\vec{j}}] = (5A)(0.01m)(0.3T)\,\vec{j} = \underline{+(0.015N)\,\vec{j}}$

e) $\vec{B} = (0.9T)\,\vec{j} - (0.4T)\,\vec{k}$
$\vec{F} = (5A)(0.01m)[(0.9T)\underbrace{\vec{i}\times\vec{j}}_{\vec{k}} - (0.4T)\underbrace{\vec{i}\times\vec{k}}_{-\vec{j}}] = (5A)(0.01m)(0.9T)\,\vec{k} + (5A)(0.01m)(0.4T)\,\vec{j}$
$\underline{\vec{F} = (0.020N)\,\vec{j} + (0.045N)\,\vec{k}}$

30-23

a) $\uparrow\vec{B}$

$\vec{m} = NI\vec{A}$; direction is $\odot$ $(+z)$ $\Rightarrow \vec{m} = NIA\,\vec{k}$
$\vec{B} = B\vec{j}$

$\vec{i}\times\vec{i} = \vec{j}\times\vec{j} = \vec{k}\times\vec{k} = 0$
$\vec{i}\times\vec{j} = \vec{k},\ \vec{j}\times\vec{k} = \vec{i},\ \vec{k}\times\vec{i} = \vec{j}$
$\vec{i}\cdot\vec{i} = \vec{j}\cdot\vec{j} = \vec{k}\cdot\vec{k} = 1$
$\vec{i}\cdot\vec{j} = \vec{j}\cdot\vec{k} = \vec{k}\cdot\vec{i} = 0$

$\vec{\Gamma} = \vec{m}\times\vec{B} = NIAB\,\vec{k}\times\vec{j} = NIAB(-\vec{i}) = \underline{-NIAB\,\vec{i}}$
$U = -\vec{m}\cdot\vec{B} = -(NIA\vec{k})\cdot B\vec{j} = \underline{0}$

b) $\vec{B} = B\vec{j}$; $\vec{m} = NI\vec{A}$, $\vec{A}\otimes \Rightarrow \vec{m} = NIA\,\vec{j}$
$\vec{\Gamma} = \vec{m}\times\vec{B} = NIAB\,\vec{j}\times\vec{j} = \underline{0}$
$U = -\vec{m}\cdot\vec{B} = -(NIA)\vec{j}\cdot(B)\vec{j} = \underline{-NIAB}$

30-23 (cont)

c) $\vec{B}$, Y, X, Z

$\vec{B} = B\vec{j}\ \uparrow$; $\vec{m} = NI\vec{A}$, $\vec{A}\ \otimes \Rightarrow \vec{m} = -NIA\vec{k}$

$\vec{\Gamma} = \vec{m}\times\vec{B} = NIAB(-\vec{k}\times\vec{j}) = \underline{+NIAB\vec{i}}$

$U = -\vec{m}\cdot\vec{B} = -(-NIA\vec{k})\cdot(B\vec{j}) = \underline{0}$

d) Z, I, Y, X

$\vec{B} = B\vec{j}\ \otimes$; $\vec{m} = NI\vec{A}$, $\vec{A}\ \odot \Rightarrow \vec{m} = -NIA\vec{j}$

$\vec{\Gamma} = \vec{m}\times\vec{B} = NIAB(-\vec{j}\times\vec{j}) = \underline{0}$

$U = -\vec{m}\cdot\vec{B} = -(-NIA\vec{j})\cdot(B\vec{j}) = \underline{+NIAB}$

30-25

a) $I = 4.5A$, 120 V, I_f, $R_f = 150\Omega$, I_r, $R_r = 2\Omega$, $\mathcal{E}$

$I_f R_f = 120V$

$I_f = \frac{120V}{R_f} = \frac{120V}{150\Omega} = 0.80A$

b) $I_f + I_r = I \Rightarrow I_r = I - I_f = 4.5A - 0.8A = \underline{3.7A}$

c) $-I_r R_r - \mathcal{E} + 120V = 0$

$\mathcal{E} = 120V - I_r R_r = 120V - (3.7A)(2\Omega) = 120V - 7.4V = \underline{112.6V}$

d) The mechanical power is the electrical power input minus the rate of dissipation of electrical energy in the resistance of the motor.

electrical power input to the motor:

$P_{in} = I(120V) = (4.5A)(120V) = 540W$

electrical power loss in the two resistances:

$P_{loss} = I_f^2 R_f + I_r^2 R_r = (0.80A)^2(150\Omega) + (3.7A)^2(2\Omega) = 96W + 27.4W = 123W$

mechanical power:

$P_{mech} = P_{in} - P_{loss} = 540W - 123W = \underline{417W}$

Note: The mechanical power is the power associated with the induced emf $\mathcal{E}$.

$P_{mech} = P_{\mathcal{E}} = \mathcal{E} I_r = (112.6V)(3.7A) = 417W$ ✓

30-27

a)

Y, X, Z; 0.02m; $I = 200A$

$\vec{B}\odot$, $B = 1.5T$

thickness $= 1\times10^{-3}m$

$$J_x = nqv \Rightarrow v = \frac{J_x}{nq}$$

$$J_x = \frac{I}{A} = \frac{200A}{(0.02m)(1\times10^{-3}m)} = 1.0\times10^{7}\ A\cdot m^{-2}$$

$$n = 7.4\times10^{28}\ m^{-3}$$

$$v = \frac{1.0\times10^{7}\ A\cdot m^{-2}}{(7.4\times10^{28}m^{-3})(1.60\times10^{-19}C)} = \underline{8.45\times10^{-4}\ m\cdot s^{-1}}$$

b) magnitude of E:

$$F_E = F_B \Rightarrow qE = qvB$$

$$\Rightarrow E = vB = (8.45\times10^{-4}\ m\cdot s^{-1})(1.5T) = \underline{1.27\times10^{-3}\ V\cdot m^{-1}}$$

direction of $\vec{E}$:

$\odot\vec{B}$; $v_e \leftarrow \ominus \rightarrow I$

$\vec{v}\times\vec{B}\uparrow$; $\vec{F}_B = -e\vec{v}\times\vec{B}\downarrow$

$\vec{F}_E = -\vec{F}_B \Rightarrow \vec{F}_E\uparrow$

$\vec{E} = \frac{\vec{F}_E}{-e} \Rightarrow \vec{E}\downarrow$; $\vec{E}$ is in the $\underline{+z\text{-direction}}$

c) The Hall emf refers to the potential difference between the two edges of the strip (at $z=0$ and $z = 2cm$), that results from the electric field calculated in (b).

$V_{ab} = Ed \Rightarrow \mathcal{E}_{Hall} = Ew$, where w is the width of the strip.

$$\mathcal{E}_{Hall} = (1.27\times10^{-3}\ V\cdot m^{-1})(0.02m) = \underline{2.54\times10^{-5}\ \text{volts}}$$

Problems

30-29 $\quad \vec{v}_0 = (4\times10^{3}\ m\cdot s^{-1})\vec{i}$; $\vec{B} = -(0.3T)\vec{j}$

a) $q = +0.4\times10^{-8}C$

direction of $\vec{E}$

z, y, x; $\vec{B}$; $\oplus$; $\vec{v}_0$

$\otimes\ \vec{v}\times\vec{B}$

$\otimes\ \vec{F}_B$

$\vec{F}_E = -\vec{F}_B \Rightarrow \vec{F}_E\odot$

$q>0$, $\vec{E} = \frac{\vec{F}_E}{q} \Rightarrow \vec{E}\odot$; $\vec{E}$ must be in the $\underline{+z\text{-direction}}$

magnitude of E

$$F_E = F_B \Rightarrow qE = qvB$$

$$F = vB = (4\times10^{3}\ m\cdot s^{-1})(0.3T) = \underline{1.2\times10^{3}\ V\cdot m^{-1}}$$

30-29 (cont)

b) $q = -0.4 \times 10^{-8}\,C$

direction of $\vec{E}$

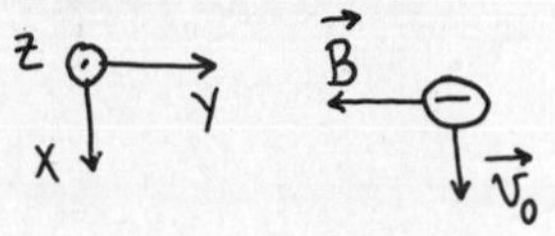

$\otimes\ \vec{v} \times \vec{B}$ $\qquad \vec{F}_E = -\vec{F}_B \Rightarrow \vec{F}_E \otimes$

$\odot\ \vec{F}_B$ $\qquad q<0,\ \vec{E} = \frac{\vec{F}_E}{q} \Rightarrow \vec{E}\odot$; $\vec{E}$ must be in the $+z$-direction

magnitude of E

As in (a), $E = vB = \underline{1.2 \times 10^{3}\, V\cdot m^{-1}}$

30-31

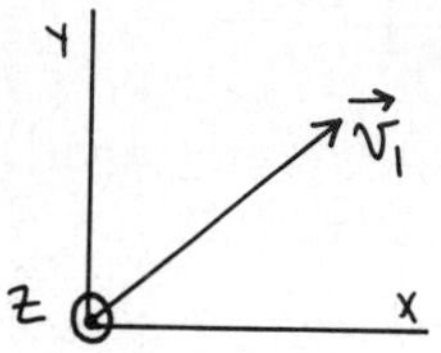

$\vec{F}_1 \otimes$

$\vec{F} = q\vec{v}\times\vec{B} \Rightarrow \vec{F}$ is perpendicular to $\vec{v}$ and $\vec{B}$

$\Rightarrow \vec{B}$ must have no z-component

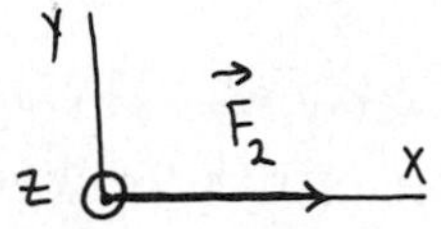

$\odot\ \vec{v}_2$

$\vec{F}$ perpendicular to $\vec{v}$ and $\vec{B}$

$\Rightarrow \vec{B}$ must have no x-component

Therefore $\vec{B}$ must lie in the -y-direction:

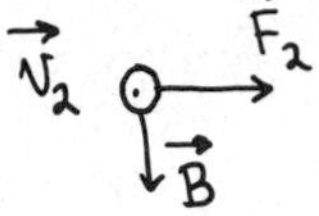

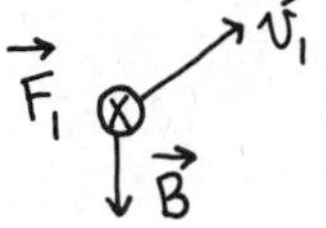

, $\vec{B} = -B\vec{j}$

magnitude of $\vec{B}$:

$$F_2 = qv_2B \Rightarrow B = \frac{F_2}{qv_2} = \frac{4\times10^{-5}\,N}{(4\times10^{-9}\,C)(2\times10^{4}\,m\cdot s^{-1})} = \underline{0.50\,T}$$

30-37

$\vec{B} = -(0.1\,T)\vec{k}$; $\vec{v} = (4\vec{i} - 3\vec{j} + 12\vec{k})\times10^{6}\,m\cdot s^{-1}$; $F = 2\,N$

a) $\vec{F} = q\vec{v}\times\vec{B}$

$$\vec{v}\times\vec{B} = [(4\vec{i}-3\vec{j}+12\vec{k})\times10^{6}\,m\cdot s^{-1}] \times (-0.1\,T)\vec{k}$$

$$= (-0.4\times10^{6}\,N\cdot C^{-1})\vec{i}\times\vec{k} + (0.3\times10^{6}\,N\cdot C^{-1})\vec{j}\times\vec{k} - (1.2\times10^{6}\,N\cdot C^{-1})\vec{k}\times\vec{k}$$

$\vec{i}\times\vec{k} = -\vec{j}$, $\vec{j}\times\vec{k} = \vec{i}$, $\vec{k}\times\vec{k} = 0$

$$\Rightarrow \vec{v}\times\vec{B} = +(0.4\times10^{6}\,N\cdot C^{-1})\vec{j} + (0.3\times10^{6}\,N\cdot C^{-1})\vec{i}$$

$$\vec{F} = q\vec{v}\times\vec{B} = q[(0.3\times10^{6}\,N\cdot C^{-1})\vec{i} + (0.4\times10^{6}\,N\cdot C^{-1})\vec{j}]$$

$$F = \sqrt{F_x^2 + F_y^2} = q\sqrt{(0.3\times10^{6}\,N\cdot C^{-1})^2 + (0.4\times10^{6}\,N\cdot C^{-1})^2} = q(0.5\times10^{6}\,N\cdot C^{-1})$$

$$F = 2\,N \Rightarrow q = \frac{2\,N}{0.5\times10^{6}\,N\cdot C^{-1}} = \underline{4.0\times10^{-6}\,C}$$

30-37 (cont)

b) $\vec{F} = m\vec{a} \Rightarrow \vec{a} = \frac{\vec{F}}{m}$

$\vec{F} = q[(0.3\times10^6\,N\cdot C^{-1})\,\vec{i} + (0.4\times10^6\,N\cdot C^{-1})\,\vec{j}]$

$q = 4.0\times10^{-6}\,C \Rightarrow \vec{F} = (1.2N)\,\vec{i} + (1.6N)\,\vec{j}$

$\vec{a} = \frac{\vec{F}}{m} = \frac{(1.2N)}{(1\times10^{-15}kg)}\,\vec{i} + \frac{(1.6N)}{(1\times10^{-15}kg)}\,\vec{j} = (1.2\times10^{15}\,m\cdot s^{-2})\,\vec{i} + (1.6\times10^{15}\,m\cdot s^{-2})\,\vec{j}$

$a = \sqrt{a_x^2 + a_y^2} = \sqrt{(1.2\times10^{15}\,m\cdot s^{-2})^2 + (1.6\times10^{15}\,m\cdot s^{-2})^2} = 2.0\times10^{15}\,m\cdot s^{-2}$

c) $\vec{F}$ is in the xy-plane $\Rightarrow$ in the z-direction the particle moves with constant velocity $12\times10^6\,m\cdot s^{-1}$.

In the xy-plane the force $\vec{F}$ causes the particle to move in a circle, with $\vec{F}$ directed in toward the center of the circle.

$F = ma$

$F = m\frac{v^2}{R} \Rightarrow R = \frac{mv^2}{F}$

$v^2 = v_x^2 + v_y^2 = (4\times10^6\,m\cdot s^{-1})^2 + (-3\times10^6\,m\cdot s^{-1})^2 = 25\times10^{12}\,m^2\cdot s^{-2}$

$R = \frac{(1\times10^{-15}kg)(25\times10^{12}\,m^2\cdot s^{-2})}{2N} = \underline{0.0125m}$

d) period $\tau = \frac{\text{distance}}{\text{velocity}} = \frac{2\pi R}{\text{velocity in xy-plane}}$

$v = \sqrt{25\times10^{12}}\,m\cdot s^{-1} = 5.0\times10^6\,m\cdot s^{-1} \Rightarrow \tau = \frac{2\pi(0.0125\,m)}{5\times10^6\,m\cdot s^{-1}} = 1.57\times10^{-8}s$

cyclotron frequency $f = \frac{1}{\tau} = \frac{1}{1.57\times10^{-8}s} = \underline{6.37\times10^7\,Hz}$

e) $t = 2\tau \Rightarrow$ is back to same x, y point

In the z-direction the particle moves with a constant velocity of $v_z = 12\times10^6\,m\cdot s^{-1}$.

In time $t = 2\tau$ it travels a distance $z = v_z 2\tau = (12\times10^6\,m\cdot s^{-1})(2)(1.57\times10^{-8}s) = 0.377m$

Thus at $t = 2\tau$ the coordinates of the particle are $(R, 0, 0.377m)$.

30-41

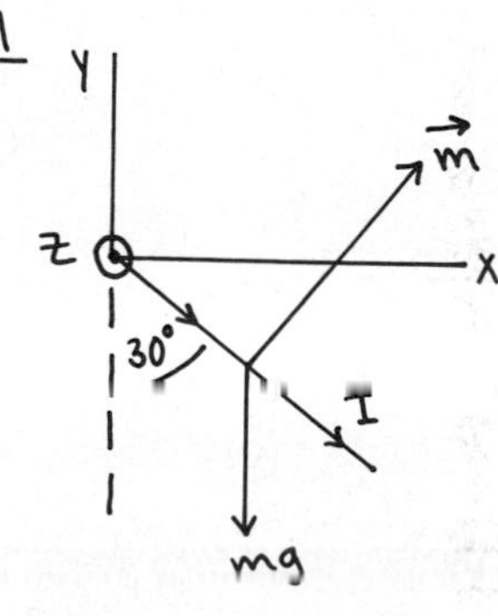

Take the pivot to be the z-axis. Then $\vec{\Gamma}_{mg}$ is ↻ ($\vec{\Gamma}_{mg}$ ⊗)

$\vec{\Gamma}_{mg} + \vec{\Gamma}_B = 0 \Rightarrow \vec{\Gamma}_B$ ↺ ($\vec{\Gamma}_B$ ⊙)

$\Rightarrow \vec{B}$ needs to be in the $\underline{+y\text{-direction}}$.

$\vec{B}$ 60° $\vec{m}$ 30°

$\vec{m}\times\vec{B}$ ⊙

30-41 (cont)

equilibrium $\Rightarrow \Gamma_{mg} = \Gamma_B$

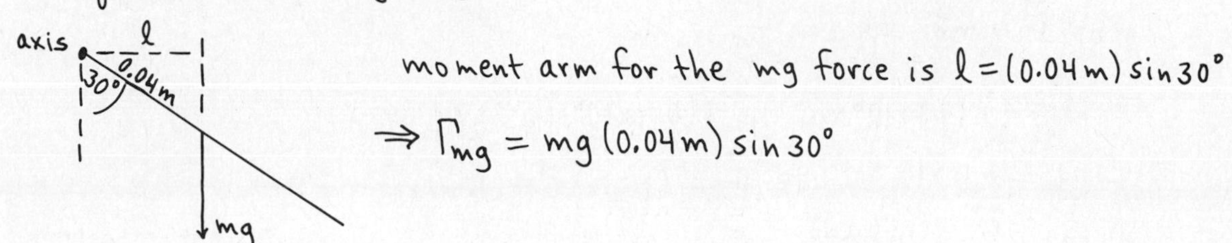

moment arm for the mg force is $l = (0.04\,m)\sin 30°$

$$\Rightarrow \Gamma_{mg} = mg(0.04\,m)\sin 30°$$

$$\Gamma_B = mB\sin 60° = IAB\sin 60°$$

$$\Gamma_{mg} = \Gamma_B \Rightarrow mg(0.04\,m)\sin 30° = IAB\sin 60° \Rightarrow B = \frac{mg(0.04\,m)\sin 30°}{IA\sin 60°}$$

$$m = 2(8\,cm + 6\,cm)(0.1\,g\cdot cm^{-1}) = 2.8\,g = 2.8\times 10^{-3}\,kg$$

$$A = (0.08\,m)(0.06\,m) = 4.8\times 10^{-3}\,m^2$$

$$B = \frac{(2.8\times 10^{-3}\,kg)(9.8\,m\cdot s^{-2})(0.04\,m)\sin 30°}{(10\,A)(4.8\times 10^{-3}\,m^2)\sin 60°} = \underline{0.0132\,T}$$

30-43

a) After the wire leaves the mercury, its acceleration is g, downward. The wire travels upward a total distance of 1.00m above its initial position. Its ends lose contact with the mercury after the wire has traveled 0.05m, so the wire travels upward 0.95m after it leaves the mercury.

(axes: y up, x right)

$v_{oy} = ?$

$y - y_o = 0.95\,m$

$a_y = -9.8\,m\cdot s^{-2}$

$v_y = 0$ (at the max height)

$$v_y^2 \to 0 = v_{oy}^2 + 2a_y(y - y_o)$$

$$v_{oy} = \sqrt{-2a_y(y-y_o)} = \sqrt{-2(-9.8\,m\cdot s^{-2})(0.95\,m)}$$

$$v_{oy} = \underline{4.32\,m\cdot s^{-1}}$$

b) $F_B = IlB$, $l = 0.25\,m$, $B = 0.01\,T$

$\longrightarrow I$ $\quad \vec{F}_B = I\vec{l}\times\vec{B}$ $\quad$ ($\vec{F}_B$ up; $\vec{B}$ into page ⊗; $\vec{l}$ to the right)

$$\sum F_y = F_B - mg = ma_y \Rightarrow a_y = \frac{F_B}{m} - g = \frac{IlB}{m} - g$$

This a_y causes the wire to attain an upward velocity of $4.32\,m\cdot s^{-1}$ (part (a)) after traveling 0.05m, starting from rest:

$$v_y^2 = v_{oy}^2 \to 0 + 2a_y(0.05\,m)$$

$$a_y = \frac{v_y^2}{0.10\,m} = \frac{(4.32\,m\cdot s^{-1})^2}{0.10\,m} = 187\,m\cdot s^{-2}$$

Then $\dfrac{IlB}{m} = a_y + g \Rightarrow I = \dfrac{m}{lB}(a_y + g) = \dfrac{9.79\times 10^{-5}\,kg}{(0.25\,m)(0.01\,T)}(187\,m\cdot s^{-2} + 9.8\,m\cdot s^{-2})$

$$I = \underline{7.71\,A}$$

c) $V = IR \Rightarrow R = \dfrac{V}{I} = \dfrac{1.5\,V}{7.71\,A} = \underline{0.195\,\Omega}$

30-45

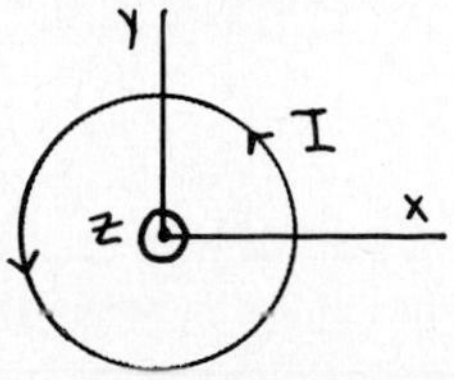

$A = 10\times10^{-4}\ m^2$
$i = 10A$
$B = 2.6T$
$\vec{m} = NIA\vec{k}$ (+z-direction)

a) $\vec{m}$ is in the +z-direction

$m = NIA = (1)(10A)(10\times10^{-4}\ m^2) = 1.0\times10^{-2}\ A\cdot m^{-2} = \underline{1.0\times10^{-2}\ J\cdot T^{-1}}$

b) $\vec{\Gamma} = [(-6\times10^{-3})\vec{i} + (8\times10^{-3})\vec{j}]\ N\cdot m$

$\vec{m} = m\vec{k}$

$\vec{B} = B_x\vec{i} + B_y\vec{j} + B_z\vec{k}$

$\vec{m}\times\vec{B} = mB_x \underbrace{\vec{k}\times\vec{i}}_{\vec{j}} + mB_y \underbrace{\vec{k}\times\vec{j}}_{-\vec{i}} + mB_z \underbrace{\vec{k}\times\vec{k}}_{0}$

Thus $\vec{\Gamma} = \vec{m}\times\vec{B} = -mB_y\vec{i} + mB_x\vec{j}$

Compare this to the expression given for $\vec{\Gamma}$

$\Rightarrow -6\times10^{-3}\ N\cdot m = -mB_y \Rightarrow B_y = \dfrac{6\times10^{-3}\ N\cdot m}{1.0\times10^{-2}\ J\cdot T^{-1}} = \underline{0.6T}$

and

$8\times10^{-3}\ N\cdot m = mB_x \Rightarrow B_x = \dfrac{8\times10^{-3}\ N\cdot m}{1.0\times10^{-2}\ J\cdot T^{-1}} = \underline{0.8T}$

The observed torque thus determines B_x and B_y, but tells us nothing about B_z.
($\vec{m}$ is in the z-direction, so the B_z component of $\vec{B}$ contributes no torque.)

But $U = -\vec{m}\cdot\vec{B} = -m\vec{k}\cdot(B_x\vec{i} + B_y\vec{j} + B_z\vec{k}) = -mB_z$.
$U_{negative} \Rightarrow B_z$ positive.

Then to get the magnitude of B_z we can use that

$B = \sqrt{B_x^2 + B_y^2 + B_z^2} = 2.6T$

$\Rightarrow B_z = \pm\sqrt{B^2 - B_x^2 - B_y^2} = \pm\sqrt{(2.6T)^2 - (0.8T)^2 - (0.6T)^2} = \pm 2.4T$

Thus $B_z = \underline{+2.4T}$.

CHAPTER 31

Exercises 1, 3, 5, 9, 11, 15, 21, 25

Problems 27, 31, 33, 37, 39, 41

Exercises

31-1

$\vec{B} = \frac{\mu_0}{4\pi} \frac{q\,\vec{v} \times \hat{r}}{r^2} = \frac{\mu_0}{4\pi} \frac{q\,\vec{v} \times \vec{r}}{r^3}$; $\vec{r}$ is the vector from the charge to the point P where the field is to be calculated.

a) $\vec{v} = (8 \times 10^6\ \text{m}\cdot\text{s}^{-1})\,\vec{i}$

$\vec{r} = (0.5\,\text{m})\,\vec{i}$; $r = 0.5\,\text{m}$

$\vec{v} \times \vec{r} = (4 \times 10^6\ \text{m}^2\cdot\text{s}^{-1})\,\vec{i} \times \vec{i} = 0$; $\underline{B = 0}$

b) $\vec{r} = -0.5\,\text{m}\,\vec{j}$; $r = 0.5\,\text{m}$

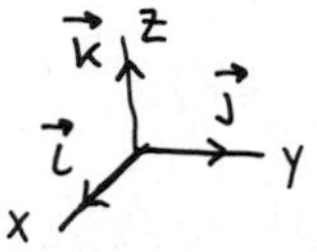

$\vec{v} \times \vec{r} = -(4 \times 10^6\,\text{m}^2\cdot\text{s}^{-1})\,\vec{i} \times \vec{j} = -(4 \times 10^6\,\text{m}^2\cdot\text{s}^{-1})\,\vec{k}$

$$\vec{B} = \left(\frac{4\pi \times 10^{-7}\,\text{T}\cdot\text{A}^{-1}\cdot\text{m}}{4\pi}\right) \frac{(5 \times 10^{-6}\,\text{C})(-4 \times 10^6\,\text{m}^2\cdot\text{s}^{-1})\,\vec{k}}{(0.5\,\text{m})^3} = -(1.60 \times 10^{-5}\,\text{T})\,\vec{k}$$

c) $\vec{r} = (0.5\,\text{m})\,\vec{k}$; $r = 0.5\,\text{m}$

$\vec{v} \times \vec{r} = (4 \times 10^6\,\text{m}^2\cdot\text{s}^{-1})\,\vec{i} \times \vec{k} = -(4 \times 10^6\,\text{m}^2\cdot\text{s}^{-1})\,\vec{j}$

Same as in (b) except for the direction

$\Rightarrow\ \underline{\vec{B} = -(1.60 \times 10^{-5}\,\text{T})\,\vec{j}}$

d) $\vec{r} = -(0.5\,\text{m})\,\vec{j} + (0.5\,\text{m})\,\vec{k}$; $r = \sqrt{(0.5\,\text{m})^2 + (0.5\,\text{m})^2} = 0.707\,\text{m}$

$\vec{v} \times \vec{r} = (4 \times 10^6\,\text{m}^2\cdot\text{s}^{-1})(-\vec{i} \times \vec{j} + \vec{i} \times \vec{k}) = 4 \times 10^6\,\text{m}^2\cdot\text{s}^{-1}(-\vec{k} - \vec{j}) = -4 \times 10^6\,\text{m}^2\cdot\text{s}^{-1}(\vec{j} + \vec{k})$

$$\vec{B} = \left(\frac{4\pi \times 10^{-7}\,\text{T}\cdot\text{A}^{-1}\cdot\text{m}}{4\pi}\right) \frac{(5 \times 10^{-6}\,\text{C})(-4 \times 10^6\,\text{m}^2\cdot\text{s}^{-1})(\vec{j} + \vec{k})}{(0.707\,\text{m})^3} = \underline{-(5.66 \times 10^{-6}\,\text{T})(\vec{j} + \vec{k})}$$

31-3

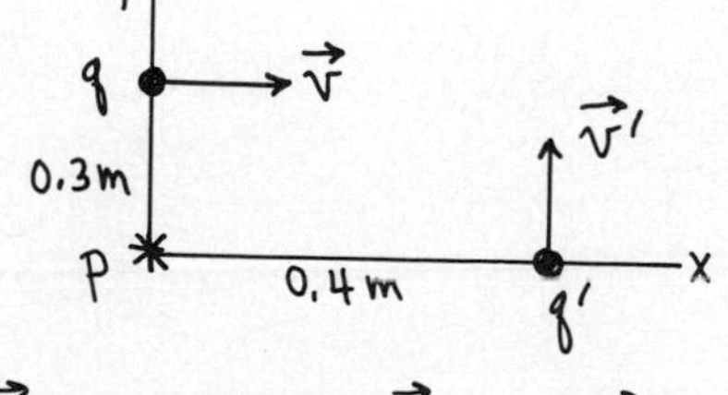

$q = 5 \times 10^{-6}\,\text{C}$, $q' = -3 \times 10^{-6}\,\text{C}$

$v = v' = 4 \times 10^5\,\text{m}\cdot\text{s}^{-1}$

$\vec{B} = \frac{\mu_0}{4\pi} \frac{q\,\vec{v} \times \vec{r}}{r^3}$; $\vec{B}_{tot} = \vec{B}_q + \vec{B}_{q'}$

B_q :

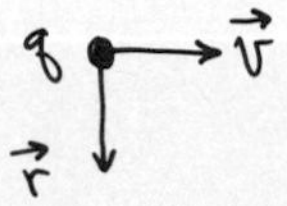

$\vec{v} = (4 \times 10^5\,\text{m}\cdot\text{s}^{-1})\,\vec{i}$

$\vec{r} = -(0.3\,\text{m})\,\vec{j}$

$\vec{v} \times \vec{r} = -(1.2 \times 10^5\,\text{m}^2\cdot\text{s}^{-1})\,\vec{i} \times \vec{j} = -(1.2 \times 10^5\,\text{m}^2\cdot\text{s}^{-1})\,\vec{k}$

31-3 (cont)

$$\vec{B}_q = \left(\frac{4\pi\times10^{-7}\,T\cdot A^{-1}\cdot m}{4\pi}\right)\frac{(5\times10^{-6}C)(-1.2\times10^{5}\,m^2\cdot s^{-1})\vec{k}}{(0.3\,m)^3} = -(2.22\times10^{-6}T)\vec{k}$$

$\vec{B}_{q'}$:

$\vec{v}' = (4\times10^{5}\,m\cdot s^{-1})\vec{j}$; $\vec{r} = -(0.4\,m)\vec{i}$

$\vec{v}'\times\vec{r} = -(1.6\times10^{5}\,m^2\cdot s^{-1})\vec{j}\times\vec{i} = +(1.6\times10^{5}\,m^2\cdot s^{-1})\vec{k}$

$$\vec{B}_{q'} = \left(\frac{4\pi\times10^{-7}\,T\cdot A^{-1}\cdot m}{4\pi}\right)\frac{(-3\times10^{-6}C)(1.6\times10^{5}\,m^2\cdot s^{-1})\vec{k}}{(0.4\,m)^3} = -(0.75\times10^{-6}T)\vec{k}$$

$$\vec{B}_{tot} = \vec{B}_q + \vec{B}_{q'} = [-2.22\times10^{-6}T - 0.75\times10^{-6}T]\vec{k} = \underline{-(2.97\times10^{-6}T)\vec{k}} \text{ (into page)}$$

31-5

$$d\vec{B} = \frac{\mu_0 I}{4\pi}\frac{d\vec{l}\times\vec{r}}{r^3}$$

straightline segment on left:

$d\vec{l}\times\vec{r} = 0 \Rightarrow dB = 0 \Rightarrow$ no contribution to B at P

straightline segment on right:

$d\vec{l}\times\vec{r} = 0 \Rightarrow dB = 0 \Rightarrow$ no contribution to B at P

semicircular section: Divide this section up into small segments of length dl

$d\vec{l}\times\vec{r}$ ⊗

$|d\vec{l}\times\vec{r}| = r\,dl$ (since $d\vec{l}$ is tangential and $\vec{r}$ is radial, and they therefore are perpendicular)

The magnitude of the magnetic field dB at P due to this infinitesimal segment is

$$dB = \frac{\mu_0 I}{4\pi}\frac{|d\vec{l}\times\vec{r}|}{r^3} = \frac{\mu_0 I}{4\pi}\frac{r\,dl}{r^3} = \frac{\mu_0 I\,dl}{4\pi r^2}$$

But $r = R \Rightarrow dB = \frac{\mu_0 I}{4\pi R^2}\,dl$.

The total B at P due to the current in the entire semicircle is $\vec{B} = \int d\vec{B}$, where the integral is over all small segments into which the semicircle is divided.

But $d\vec{B}$ for each small segment of the semicircle is ⊗, so $B = \int dB$

$$\Rightarrow B = \int \frac{\mu_0 I}{4\pi R^2}\,dl = \frac{\mu_0 I}{4\pi R^2}\int dl$$

31-5 (cont)

But $\int dl = \pi R$, the length of the semicircle

$$\Rightarrow B = \frac{\mu_0 I}{4\pi R^2}(\pi R) = \underline{\frac{\mu_0 I}{4R}}$$, directed into the plane of the paper.

31-9

eq (31-15) $B = \frac{\mu_0 I}{2\pi r}$

a) currents all in same direction $\Rightarrow$ the B's from the individual wires all add

$$\Rightarrow B = 6\frac{\mu_0 I}{2\pi r} = 6\frac{(4\pi \times 10^{-7} T\cdot A^{-1}\cdot m)}{2\pi}\left(\frac{0.5 A}{0.10 m}\right) = \underline{6.0\times 10^{-6} T}$$

b) The magnetic fields cancel for a pair of wires with equal and opposite currents $\Rightarrow$ only two wires contribute to B. (For the other four, two have current in one direction and two in the other direction.)

$$B = 2\frac{\mu_0 I}{2\pi r} = \frac{1}{3}(6.0\times 10^{-6} T) = \underline{2.0\times 10^{-6} T}$$

31-11

a) [diagram: wire #1 carrying I to the right, wire #2 carrying I to the left; P midway, distance a from each]

At P, $\vec{B}_1$ is ⊗, $\vec{B}_2$ is ⊗ $\Rightarrow B = B_1 + B_2$

$$B_1 = B_2 = \frac{\mu_0 I}{2\pi r} = \frac{\mu_0 I}{2\pi a}$$

Thus $B = \frac{\mu_0 I}{\pi a}$.

b) [diagram: P at distance a above wire #1 (I to the right); wire #2 (I to the left) 2a below wire #1]

At P, $\vec{B}_1$ is ⊙, $\vec{B}_2$ is ⊗ $\Rightarrow B = B_1 - B_2$.

$$B = \frac{\mu_0 I}{2\pi r} \Rightarrow B_1 = \frac{\mu_0 I}{2\pi a},\ B_2 = \frac{\mu_0 I}{2\pi(3a)}$$

$$B = \frac{\mu_0 I}{2\pi}\left(\frac{1}{a} - \frac{1}{3a}\right) = \frac{\mu_0 I}{3\pi a}$$

c) [diagram: wires #1 and #2 both carrying I to the right; P midway, distance a from each]

At point P, $B_1 = B_2$ and $\vec{B}_1$ is ⊗, $\vec{B}_2$ is ⊙

$$\Rightarrow B = B_1 - B_2 = 0$$

d) [diagram: P at distance a above wire #1; wire #2 2a below wire #1; both carry I to the right]

At point P, $\vec{B}_1$ is ⊙, $\vec{B}_2$ is ⊙

$$\Rightarrow B = B_1 + B_2.$$

$$B_1 = \frac{\mu_0 I}{2\pi a},\quad B_2 = \frac{\mu_0 I}{2\pi(3a)}$$

$$B = \frac{\mu_0 I}{2\pi}\left(\frac{1}{a} + \frac{1}{3a}\right) = \frac{2\mu_0 I}{3\pi a}$$

31-15

The wire CD rises until the upward force due to the currents balances the downward force of gravity.

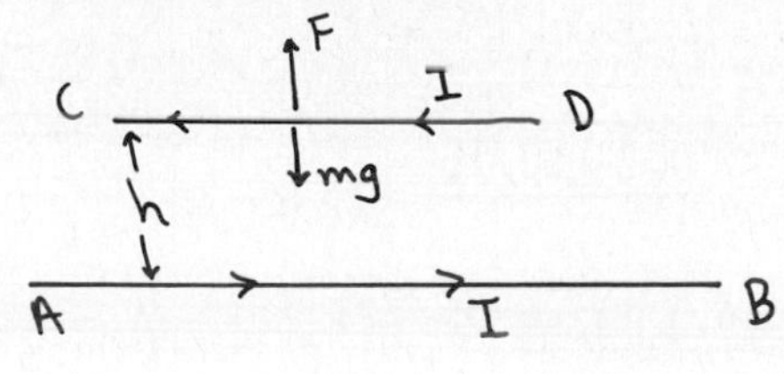

$m = (0.005\,\text{kg}\cdot\text{m}^{-1})(1.0\,\text{m}) = 0.005\,\text{kg}$

opposite currents $\Rightarrow$ force is repulsive

balanced $\Rightarrow$ $F = mg$

$$F = \frac{\mu_0 I I' l}{2\pi r} = \frac{\mu_0 I^2 l}{2\pi h}$$

Thus $\frac{\mu_0 I^2 l}{2\pi h} = mg \Rightarrow h = \frac{\mu_0}{2\pi}\frac{l I^2}{mg}$

$$h = \frac{4\pi\times10^{-7}\,\text{N}\cdot\text{A}^{-2}}{2\pi}\,\frac{(1\,\text{m})(50\,\text{A})^2}{(0.005\,\text{kg})(9.8\,\text{m}\cdot\text{s}^{-2})} = 0.0102\,\text{m} = \underline{1.02\,\text{cm}}$$

31-21

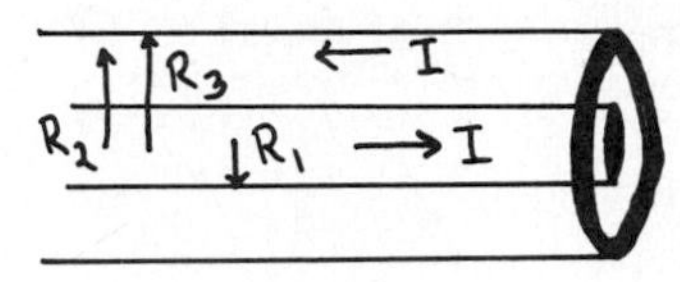

$\oint \vec{B}\cdot d\vec{l} = \mu_0 I_{encl}$

(I_{encl} = current enclosed by the closed path chosen for $\oint \vec{B}\cdot d\vec{l}$)

a)

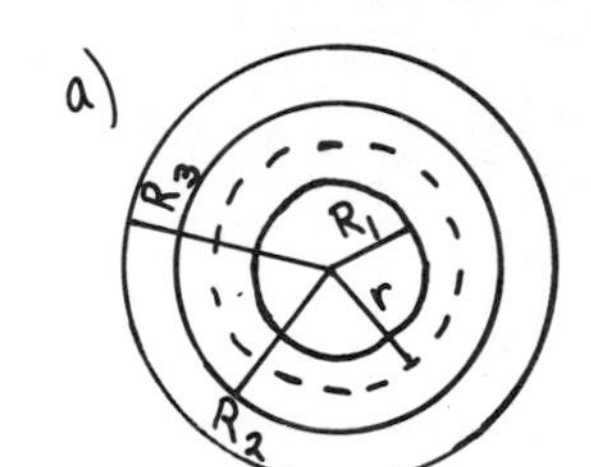

Apply Ampere's Law to a path that is a circle of radius r, where $R_1 < r < R_2$.

By symmetry, $\vec{B}$ is tangent to this path and constant around it $\Rightarrow \oint \vec{B}\cdot d\vec{l} = \oint B\,dl = B\oint dl = B(2\pi r)$

$I_{encl} = I$ (All of the current of the inner conductor but none of the outer conductor is enclosed by the path.)

$$\Rightarrow B(2\pi r) = \mu_0 I \Rightarrow B = \frac{\mu_0 I}{2\pi r}$$

b)

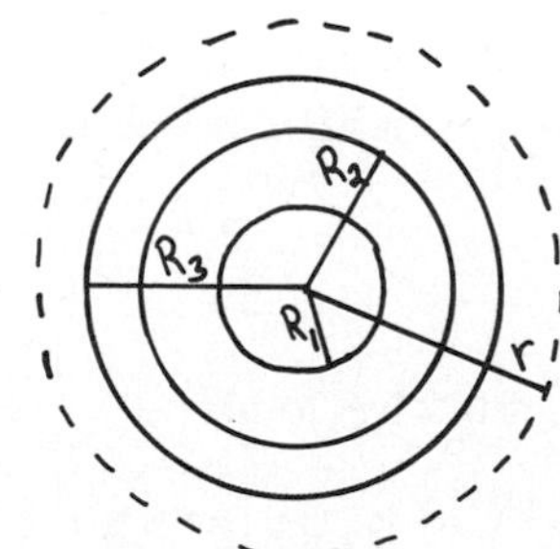

Apply Ampere's Law to a path that is a circle of radius r, where $r > R_3$.

As in (a) $\oint \vec{B}\cdot d\vec{l} = B(2\pi r)$

But now $I_{encl} = 0$, since all the current in each conductor is enclosed by the path and these currents are equal and opposite.

$$\Rightarrow B(2\pi r) = 0 \Rightarrow B = 0.$$

31-25

$$B = K_m \frac{\mu_0 N I}{2\pi r} \quad \text{(eq. 31-29, with } \mu_0 \to K_m \mu_0\text{)}$$

a) $2\pi r = 0.30\,\text{m}$; $K_m = 1400$

$$I = \frac{(2\pi r) B}{N \mu_0 K_m} = \frac{(0.30\,\text{m})(0.1\,\text{T})}{(400)(4\pi\times10^{-7}\,\text{T}\cdot\text{A}^{-1}\cdot\text{m})(1400)} = \underline{0.0426\,\text{A}}$$

$\mu_0 = \frac{\mu}{K_m}$

b) $K_m = 5200 \Rightarrow I = \frac{(2\pi r) B}{N\mu_0 K_m} = \frac{(0.30\,\text{m})(0.1\,\text{T})}{(400)(4\pi\times10^{-7}\,\text{T}\cdot\text{A}^{-1}\cdot\text{m})(5200)} = \underline{0.0115\,\text{A}}$

Problems

31-27

0.1m ⊖ → $v = 5\times10^{4}\,\text{m}\cdot\text{s}^{-1}$

$I = 1.5\,\text{A}$

At the electron's position,

$$B = \frac{\mu_0 I}{2\pi r} = \left(\frac{4\pi\times10^{-7}\,\text{T}\cdot\text{A}^{-1}\cdot\text{m}}{2\pi}\right)\left(\frac{1.5\,\text{A}}{0.1\,\text{m}}\right) = 3.0\times10^{-6}\,\text{T}$$

$\vec{B}$ is ⊙

$$\vec{F} = q\vec{v}\times\vec{B}$$

$\vec{B}$ ⊙ $\rightarrow \vec{v}$ $\quad \vec{v}\times\vec{B} \downarrow$ but $q = -e < 0 \Rightarrow \vec{F} \uparrow$ (away from wire)

$$F = qvB = evB = (1.6\times10^{-19}\,\text{C})(5\times10^{4}\,\text{m}\cdot\text{s}^{-1})(3.0\times10^{-6}\,\text{T}) = \underline{2.40\times10^{-20}\,\text{N}}$$

31-31

a)

⊗ $I_1 = 6\,\text{A}$; 1.0m; ⊙ I_2; 0.5m; $\vec{B}_1$ ← P → $\vec{B}_2$

$\vec{B}_1$ and $\vec{B}_2$ must be equal and opposite for the resultant field at P to be zero.

$\Rightarrow \vec{B}_2$ to the right $\Rightarrow I_2$ out of paper

$$B_1 = \frac{\mu_0 I_1}{2\pi r_1} = \frac{\mu_0}{2\pi}\left(\frac{6\,\text{A}}{1.5\,\text{m}}\right)$$

$$B_2 = \frac{\mu_0 I_2}{2\pi r_2} = \frac{\mu_0}{2\pi}\left(\frac{I_2}{0.5\,\text{m}}\right)$$

$$B_1 = B_2 \Rightarrow \frac{6\,\text{A}}{1.5\,\text{m}} = \frac{I_2}{0.5\,\text{m}} \Rightarrow I_2 = \left(\frac{0.5\,\text{m}}{1.5\,\text{m}}\right) 6\,\text{A} = \underline{2\,\text{A}}$$

b) $\vec{B}$ at Q:

$\vec{B}_2$ ← Q → $\vec{B}_1$; 0.5m; I_1 ⊗; 1.0m; I_2 ⊙

$$\vec{B} = \vec{B}_1 + \vec{B}_2 \Rightarrow B = B_1 - B_2, \text{ to the right.}$$

$$B_1 = \frac{\mu_0 I_1}{2\pi r_1} = \left(\frac{4\pi\times10^{-7}\,\text{T}\cdot\text{A}^{-1}\cdot\text{m}}{2\pi}\right)\left(\frac{6\,\text{A}}{0.5\,\text{m}}\right) = 2.4\times10^{-6}\,\text{T}$$

$$B_2 = \frac{\mu_0 I_2}{2\pi r_2} = \left(\frac{4\pi\times10^{-7}\,\text{T}\cdot\text{A}^{-1}\cdot\text{m}}{2\pi}\right)\left(\frac{2\,\text{A}}{1.5\,\text{m}}\right) = 2.67\times10^{-7}\,\text{T}$$

$$B = B_1 - B_2 = 2.40\times10^{-6}\,\text{T} - 2.67\times10^{-7}\,\text{T} = \underline{2.13\times10^{-6}\,\text{T}},$$

directed to the right

31-31 (cont)

c)

I_1, B_2, 0.6m, 90°, 1.0m, 0.8m, B_1, I_2

$\sin\theta = \frac{0.8\text{m}}{1.0\text{m}} = 0.8 \Rightarrow \theta = 53.1^\circ$; $\phi = 36.9^\circ$

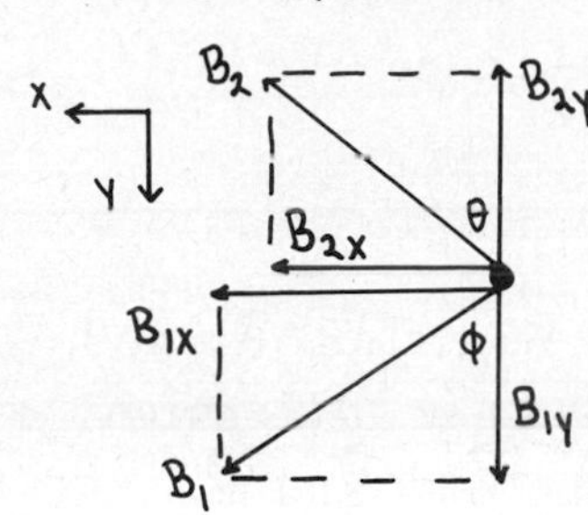

$$B_1 = \frac{\mu_0 I_1}{2\pi r_1} = \left(\frac{4\pi\times10^{-7}\,\text{T}\cdot\text{A}^{-1}\cdot\text{m}}{2\pi}\right)\left(\frac{6\,\text{A}}{0.6\text{m}}\right)$$

$$B_1 = 2.0\times10^{-6}\,\text{T}$$

$$B_{1y} = B_1\cos\phi = (2.0\times10^{-6}\,\text{T})\cos 36.9^\circ = 1.60\times10^{-6}\,\text{T}$$

$$B_{1x} = B_1\sin\phi = (2.0\times10^{-6}\,\text{T})\sin 36.9^\circ = 1.20\times10^{-6}\,\text{T}$$

$$B_2 = \frac{\mu_0 I_2}{2\pi r_2} = \left(\frac{4\pi\times10^{-7}\,\text{T}\cdot\text{A}^{-1}\cdot\text{m}}{2\pi}\right)\left(\frac{2\,\text{A}}{0.8\text{m}}\right) = 5.0\times10^{-7}\,\text{T}$$

$$B_{2y} = -B_2\cos\theta = -(5.0\times10^{-7}\,\text{T})\cos 53.1^\circ = -3.00\times10^{-7}\,\text{T}$$

$$B_{2x} = B_2\sin\theta = (5.0\times10^{-7}\,\text{T})\sin 53.1^\circ = 4.00\times10^{-7}\,\text{T}$$

Then

$$B_x = B_{1x} + B_{2x} = 1.20\times10^{-6}\,\text{T} + 4.0\times10^{-7}\,\text{T} = 1.60\times10^{-6}\,\text{T}$$

$$B_y = B_{1y} + B_{2y} = 1.60\times10^{-6}\,\text{T} - 3.0\times10^{-7}\,\text{T} = 1.30\times10^{-6}\,\text{T}$$

$$B = \sqrt{B_x^2 + B_y^2} = \underline{2.06\times10^{-6}\,\text{T}}$$

B_x, θ, B_y, B

$$\tan\theta = \frac{B_y}{B_x} = \frac{1.3\times10^{-6}\,\text{T}}{1.6\times10^{-6}\,\text{T}} = 0.812 \Rightarrow \theta = \underline{39.1^\circ}$$

(below the horizontal and to the left)

31-33

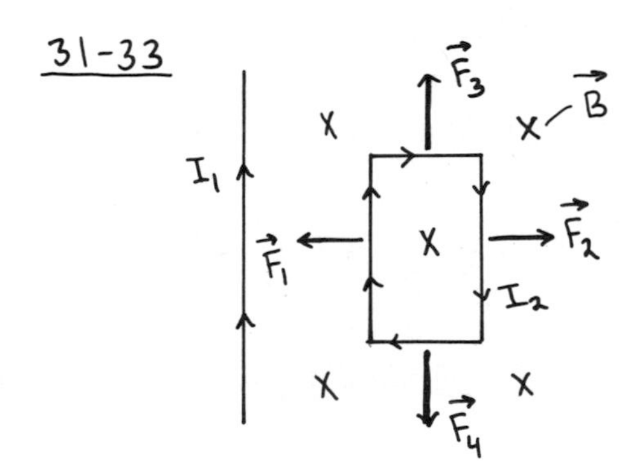

At the location of the rectangular loop the magnetic field due to the current I_1 in the long, straight wire is directed into the paper. The force on each side of the rectangle is thus as shown in the sketch ($\vec{F} = I\vec{\ell}\times\vec{B}$).

$F_3 = F_4$, but these forces are in opposite directions $\Rightarrow \vec{F}_3 + \vec{F}_4 = 0$.

$F_1 = I_2 \ell B$, where B is the magnetic field due to I_1, at a distance of 0.01m from the wire.

$$\Rightarrow B = \frac{\mu_0}{2\pi}\frac{I_1}{(0.01\text{m})} = \left(\frac{4\pi\times10^{-7}\,\text{T}\cdot\text{A}^{-1}\cdot\text{m}}{2\pi}\right)\left(\frac{20\,\text{A}}{0.01\text{m}}\right) = 4.0\times10^{-4}\,\text{T}$$

$$\Rightarrow F_1 = (10\text{A})(0.20\text{m})(4.0\times10^{-4}\,\text{T}) = 8.0\times10^{-4}\,\text{N}$$

$F_2 = I_2 \ell B$, but now r in the expression for B is 0.10m.

$$\Rightarrow B = \frac{\mu_0 I_1}{2\pi(0.10\text{m})} = 4.0\times10^{-5}\,\text{T}$$

$$F_2 = (10\text{A})(0.20\text{m})(4.0\times10^{-5}\,\text{T}) = 8.0\times10^{-5}\,\text{N}$$

31-33 (cont)

$$\vec{F} = \vec{F}_1 + \vec{F}_2 \Rightarrow F = F_1 - F_2 = 8.0\times10^{-4}\,N - 0.8\times10^{-4}\,N = \underline{7.2\times10^{-4}\,N},$$

directed to the left (toward the wire).

31-37

N turns N turns

a, $\frac{a}{2}$, P, $\frac{a}{2}$, *, x

coil #1 coil #2

The $\vec{B}$ fields $\vec{B}_1$ and $\vec{B}_2$ from the coils are in the same direction.
Thus $\vec{B} = \vec{B}_1 + \vec{B}_2 \Rightarrow B = B_1 + B_2$

a) The field of a coil of radius a, at a distance d out the axis from the coil is given by

$$B = \frac{N\mu_0 I a^2}{2(d^2+a^2)^{3/2}} \quad \text{(eq. 31-21)}$$

$$\Rightarrow B_1 = \frac{N\mu_0 I a^2}{2([\frac{a}{2}+x]^2+a^2)^{3/2}}, \quad B_2 = \frac{N\mu_0 I a^2}{2([\frac{a}{2}-x]^2+a^2)^{3/2}}$$

$$B = B_1 + B_2 = \frac{N\mu_0 I a^2}{2}\left(\frac{1}{([\frac{a}{2}+x]^2+a^2)^{3/2}} + \frac{1}{([\frac{a}{2}-x]^2+a^2)^{3/2}}\right)$$

b) point P $\Rightarrow x = 0$

$$\Rightarrow B = \frac{N\mu_0 I a^2}{2}\left(\frac{1}{(\frac{a^2}{4}+a^2)^{3/2}} + \frac{1}{(\frac{a^2}{4}+a^2)^{3/2}}\right) = \frac{N\mu_0 I}{a}\,\frac{1}{(1+\frac{1}{4})^{3/2}}$$

$$\frac{1}{(1+\frac{1}{4})^{3/2}} = \frac{1}{(\frac{5}{4})^{3/2}} = \frac{8}{\sqrt{125}} \Rightarrow B = \frac{8N\mu_0 I}{a\sqrt{125}}$$

c) From the expression in (b)

$$B = \frac{8(100)(4\pi\times10^{-7}\,T\cdot A^{-1}\cdot m)(5A)}{(0.30\,m)\sqrt{125}} = \underline{1.50\times10^{-3}\,T}$$

31-39

For a toroidal solenoid, with magnetic material of relative permeability K_m filling the space inside the coils, the magnetic field is

$$B = K_m \frac{\mu_0 N I}{2\pi r} \quad \text{(eq. 31-29, with } \mu_0 \text{ replaced by } K_m\mu_0)$$

a) $K_m = \dfrac{B(2\pi r)}{\mu_0 N I} = \dfrac{(1.0T)(0.40m)}{(4\pi\times10^{-7}T\cdot A^{-1}\cdot m)(400)(2.0A)} = 398$

b) $\chi_m = K_m - 1 = \underline{397}$

31-41

a)

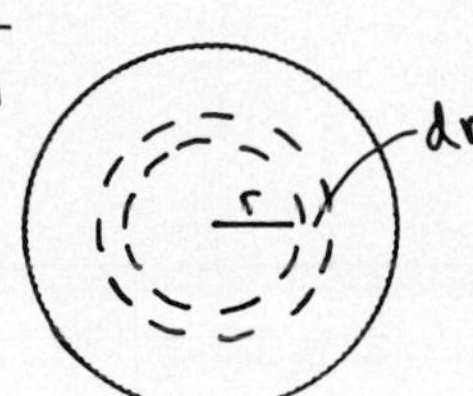

To integrate J over the cross section of the wire divide the wire cross section up into thin concentric rings of radius r and width dr. (We can't just take $I = JA = J\pi R^2$, since J varies across the cross section.)

The area of such a ring is dA, and the current through it is $dI = J\,dA$:

$$dA = 2\pi r\,dr \Rightarrow dI = J\,dA = \alpha r(2\pi r\,dr) = 2\pi\alpha r^2 dr$$

$$I = \int dI = 2\pi\alpha\int_0^R r^2 dr = 2\pi\alpha\left(\tfrac{1}{3}R^3\right) = \frac{2\pi\alpha R^3}{3} \Rightarrow \boxed{\alpha = \frac{3I}{2\pi R^3}}$$

b) (i) $r \le R$

Apply Ampere's Law to a circle of radius $r < R$:

$\oint \vec{B}\cdot d\vec{l} = \oint B\,dl = B\oint dl = B(2\pi r)$, by the symmetry and direction of $\vec{B}$

The current passing through the path is $I_{encl} = \int dI$, where in the integral r goes from 0 to r.

$$I_{encl} = 2\pi\alpha\int_0^r r^2 dr = \frac{2\pi\alpha r^3}{3} = \frac{2\pi}{3}\left(\frac{3I}{2\pi R^3}\right)r^3 = \frac{Ir^3}{R^3}$$

Thus $\oint \vec{B}\cdot d\vec{l} = \mu_0 I_{encl} \Rightarrow B(2\pi r) = \mu_0\dfrac{Ir^3}{R^3} \Rightarrow \boxed{B = \dfrac{\mu_0 I r^2}{2\pi R^3}}$.

(ii) $r \ge R$

Apply Ampere's Law to a circle of radius $r > R$:

$\oint \vec{B}\cdot d\vec{l} = \oint B\,dl = B\oint dl = B(2\pi r)$

$I_{encl} = I$; all the current in the wire passes through the path

Thus $\oint \vec{B}\cdot d\vec{l} = \mu_0 I_{encl} \Rightarrow B(2\pi r) = \mu_0 I$

$$\boxed{B = \frac{\mu_0 I}{2\pi r}}$$

Note: At $r = R$ the expression in (i) ($r \le R$) gives $B = \dfrac{\mu_0 I}{2\pi R}$

At $r = R$ the expression in (ii) ($r \ge R$) gives $B = \dfrac{\mu_0 I}{2\pi R}$, which is the same.

CHAPTER 32

Exercises 1, 5, 7, 11, 13, 15

Problems 21, 23, 27, 31, 33

<u>Exercises</u>

<u>32-1</u>

a) $\mathcal{E} = Bl v \Rightarrow v = \frac{\mathcal{E}}{Bl} = \frac{2.40\,V}{(1.2\,T)(0.40\,m)} = \underline{5.0\,m\cdot s^{-1}}$

b) $I = \frac{\mathcal{E}}{R} = \frac{2.40\,V}{1.2\,\Omega} = \underline{2.0\,A}$

c) $\vec{v} \times \vec{B} \uparrow \Rightarrow$ current is counterclockwise

$\vec{F} = I\vec{l} \times \vec{B}$

$F = IlB = (2.0\,A)(0.40\,m)(1.2\,T) = \underline{0.96\,N}$, <u>to the left</u>

<u>32-5</u>

a) When the loop is completely outside the field region

$$\mathcal{E} = 0 \Rightarrow I = 0 \Rightarrow \vec{F} = I\vec{l} \times \vec{B} = 0$$

$x = -(l + \frac{l}{2}) = -\frac{3l}{2}$

The loop is completely outside the field region for

$$x < -\frac{3l}{2} \text{ and } x > \frac{3l}{2}.$$

Loop entering the field region (but not completely inside); $-\frac{3l}{2} < x < -\frac{l}{2}$

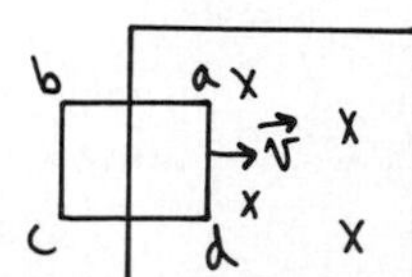

In side da, $\vec{v} \times \vec{B}$ is $\uparrow$, so the induced emf pushes current from d to a.

In sides ab and cd, $\vec{v} \times \vec{B}$ is $\uparrow$, so the force does not push current along the conductor and these sides contribute nothing to the induced emf. Side bc is out of the field, so contributes nothing to the induced emf. The induced current hence is counterclockwise.

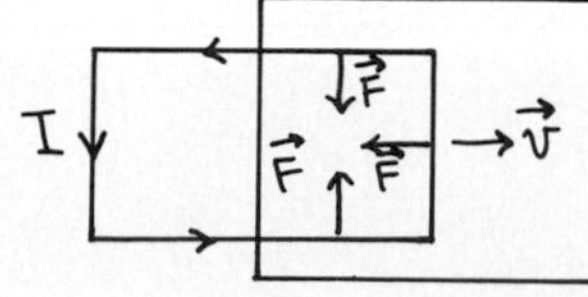

$\vec{F} = I\vec{l} \times \vec{B}$; the net force exerted on the loop by the magnetic field is to the left.

32-5 (cont)

$$I = \frac{\mathcal{E}}{R} \;;\; \mathcal{E} = B\ell v \Rightarrow I = \frac{B\ell v}{R}$$

$$F = I\ell B \Rightarrow F = \left(\frac{B\ell v}{R}\right)\ell B = \frac{B^2\ell^2 v}{R}$$

Loop completely inside the field region; $-\frac{\ell}{2} < x < \frac{\ell}{2}$

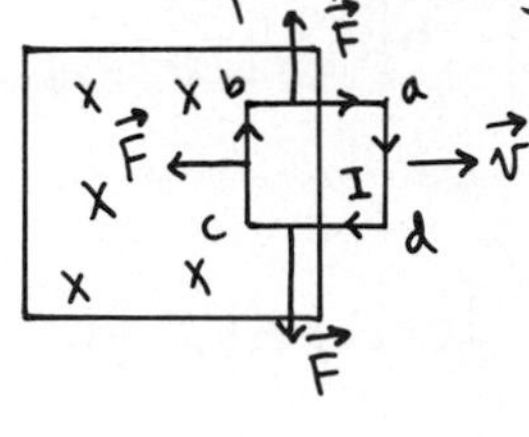

In side ad, $\vec{v} \times \vec{B}$ is ↑.

In side bc, $\vec{v} \times \vec{B}$ is ↑.

The induced emf's in these two sides cancel.

As in the above, for sides ab and cd $\vec{v} \times \vec{B}$ is perpendicular to the wire so there is no emf.

Thus the net emf $= 0 \Rightarrow I = 0 \Rightarrow F = 0$.

Loop leaving the field region; $\frac{\ell}{2} < x < \frac{3\ell}{2}$

In side bc $\vec{v} \times \vec{B}$ is ↑. The induced current therefore is clockwise. The net force exerted on the loop by the magnetic field is to the left.

$$I = \frac{B\ell v}{R} \;,\; F = \frac{B^2\ell^2 v}{R}$$

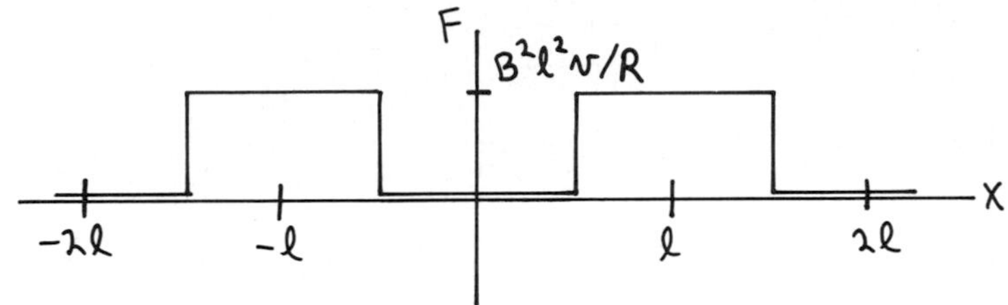

b) The current I was determined in (a). The graph of I versus x, for clockwise current positive, is:

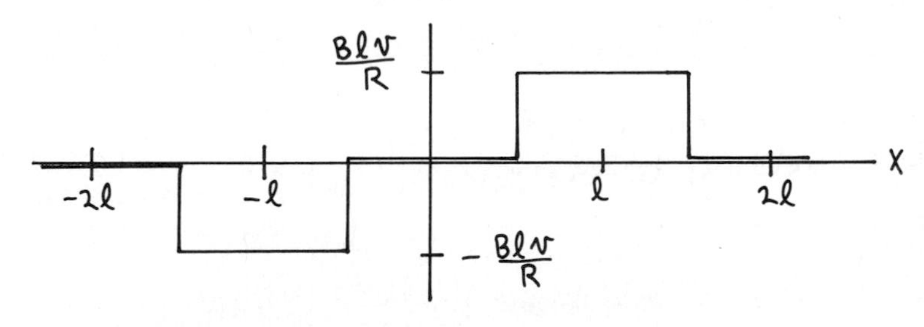

32-7

$$\mathcal{E} = -N\frac{d\Phi}{dt} \;;\; \mathcal{E}_{av} = -N\frac{\Delta\Phi}{\Delta t}$$

$\Delta t = 0.1\ s$

$\Delta\Phi = \Phi_f - \Phi_i$

32-7 (cont)

initial

$$\Phi_i = \int \vec{B}\cdot d\vec{A} = \int B\cos 45^\circ dA = B\cos 45^\circ \int dA = BA\cos 45^\circ$$

final

$$\Phi_f = \int \vec{B}\cdot d\vec{A} = \int B\,dA = B\int dA = BA$$

$$\Delta\Phi = BA - BA\cos 45^\circ = BA(1-\cos 45^\circ) = (2T)(0.12m)(0.25m)(1-\cos 45^\circ)$$
$$\Delta\Phi = +0.0176 \text{ wb}$$

$$\mathcal{E}_{av} = -N\frac{\Delta\Phi}{\Delta t} = -(50)\frac{0.0176 \text{ wb}}{0.1 s} = \underline{-8.8 V}$$

(The minus sign means the $\mathcal{E}_{av}$ is opposite to the direction associated with $\vec{A}$ (as defined in the sketch) by the right-hand rule.)

32-11

eq. (32-12) $B = \frac{RQ}{NA}$ (Be sure to study carefully the derivation of this equation in Example 32-7.)

$$B = \frac{(50\Omega + 30\Omega)(4\times10^{-5}C)}{(160)(4\times10^{-4} m^2)} = \underline{5.0\times10^{-2} T}$$

(Note that R is the total resistance of the circuit, coil plus meter.)

32-13

$$\mathcal{E}_{av} = -\frac{\Delta\Phi}{\Delta t}$$

The situation is similar to that of Fig. 32-10, except that here the current is decreasing.

$\Delta t = 0.05 s$

$\Phi_i = NBA$, where N is the number of turns in the secondary winding, $B = \mu_0 n I$ is the field inside the solenoid, and A is the area of the solenoid.

Thus $\Phi_i = N\mu_0 n I A = 2(4\pi\times10^{-7} T\cdot A^{-1}\cdot m)(10 cm^{-1})(\frac{100 cm}{1 m})(0.25A)(6\times10^{-4} m^2)$

$\Phi_i = 3.77\times10^{-7}$ wb

$\Phi_f = 0$, since the field in the solenoid has become zero.

$$\Delta\Phi = \Phi_f - \Phi_i = 0 - 3.77\times10^{-7} \text{ wb} = -3.77\times10^{-7} \text{ wb}$$

$$\mathcal{E}_{av} = -\frac{-3.77\times10^{-7} \text{ wb}}{0.05 s} = \underline{+7.54\times10^{-6} V}$$

32-15

a)

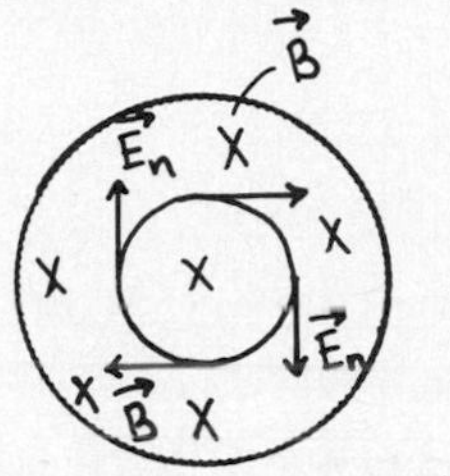

$\vec{E}_n$ is tangent to concentric circles. The direction of $\vec{E}_n$ (clockwise or counterclockwise) is the direction in which current would be induced in a circular concentric conducting loop.

Determine this direction as follows:

Take $\vec{A}$ to be into the paper, the same as the direction of $\vec{B}$. Then Φ is positive. B is decreasing $\Rightarrow \frac{d\Phi}{dt}$ is negative. $\mathcal{E} = -\frac{d\Phi}{dt}$ is then positive, so the associated current, by the right hand rule, is clockwise.

b) $\oint \vec{E}_n \cdot d\vec{l} = -\frac{d\Phi}{dt}$

$\oint \vec{E}_n \cdot d\vec{l} = \oint E_n dl = E_n \oint dl = E_n(2\pi r)$, for a path clockwise around the ring.

$\Phi = \int \vec{B} \cdot d\vec{A} = \int B\,dA = B\int dA = BA = B\pi r^2$

$\frac{d\Phi}{dt} = \pi r^2 \frac{dB}{dt}$

Thus $E_n(2\pi r) = -\pi r^2 \frac{dB}{dt}$

$E_n = -\frac{1}{2} r \frac{dB}{dt} = -\frac{1}{2}(0.10\text{m})(-0.1\,\text{T}\cdot\text{s}^{-1}) = +0.005\,\text{V}\cdot\text{m}^{-1}$, tangent to the ring in the clockwise direction.

$\mathcal{E} = \oint \vec{E}_n \cdot d\vec{l} = E_n 2\pi r = (+0.005\,\text{V}\cdot\text{m}^{-1})(2\pi)(0.10\text{m}) = 0.00314\,\text{V}$

c) $I = \frac{\mathcal{E}}{R} = \frac{0.00314\,\text{V}}{2\,\Omega} = 0.00157\,\text{A}$

d) $\mathcal{E} = IR$ for any segment of the ring. The potential difference between points a and b, or between any other two points of the ring, is zero.

Problems

32-21

a) $\mathcal{E} = Blv$

But a complication here is that different points of the rod have different velocities. Consider slicing the rod up into thin slices. Compute the emf $d\mathcal{E}$ for each slice, and add up (by integrating) all the $d\mathcal{E}$. (The thin slices of the rod act in series, so their emf's add.)

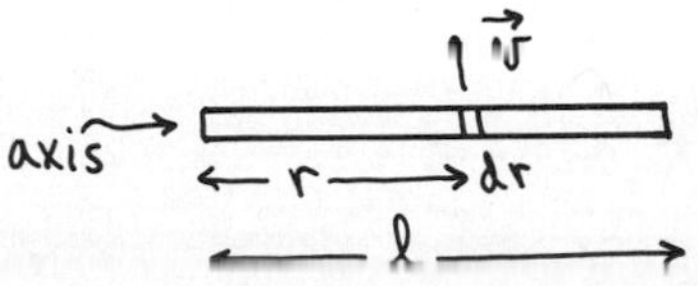

$d\mathcal{E} = Bv\,dr$

$v = r\omega \Rightarrow d\mathcal{E} = B\omega r\,dr$

$\mathcal{E} = \int d\mathcal{E} = B\omega \int_0^l r\,dr = \frac{1}{2} B\omega l^2$

32-21 (cont)

$$\omega = (2\,\text{rev}\cdot s^{-1})\left(\frac{2\pi\,\text{rad}}{1\,\text{rev}}\right) = 4\pi\,\text{rad}\cdot s^{-1}$$

$$\Rightarrow \mathcal{E} = \tfrac{1}{2}(0.5T)(4\pi\,\text{rad}\cdot s^{-1})(1m)^2 = \underline{3.14V}$$

b) No current flows so there is no IR potential drop. Thus the potential difference between the ends equals the emf of 3.14V calculated in (a).

32-23

a) $\Phi_{max} = BA = (0.5T)(400\times10^{-4}m^2) = \underline{2.0\times10^{-2}Wb}$

b) $\mathcal{E} = -\frac{d\Phi}{dt} \Rightarrow \mathcal{E}_{max} = \left(\frac{d\Phi}{dt}\right)_{max}$

$\Phi = BA\cos\theta = BA\cos\omega t$ (θ is the angle between $\vec{B}$ and the normal to the loop)

$\frac{d\Phi}{dt} = -\omega BA\sin\omega t \Rightarrow \left(\frac{d\Phi}{dt}\right)_{max} = \omega BA$

$\mathcal{E}_{max} = \omega BA = (10\,\text{rad}\cdot s^{-1})(0.5T)(400\times10^{-4}m^2) = \underline{0.20V}$

c) $\vec{\Gamma} = \vec{m}\times\vec{B} = I\vec{A}\times\vec{B}$

$\Gamma = IAB\sin\theta = IAB\sin\omega t$

$\mathcal{E} = -\frac{d\Phi}{dt} = \omega BA\sin\omega t$; $I = \frac{\mathcal{E}}{R} = \frac{\omega BA\sin\omega t}{R}$

$\Rightarrow \Gamma = \left(\frac{\omega BA\sin\omega t}{R}\right)AB\sin\omega t = \frac{B^2A^2\omega}{R}\sin^2\omega t$

$\Gamma_{max} = \frac{B^2A^2\omega}{R} = \frac{(0.5T)^2(400\times10^{-4}m^2)^2\,10\,\text{rad}\cdot s^{-1}}{2\,\Omega} = \underline{2.0\times10^{-3}\,N\cdot m}$

d) $\Gamma = \frac{B^2A^2\omega}{R}\sin^2\omega t$

For a small angular displacement

$$dW = \Gamma\,d\theta = \Gamma\omega\,dt = \frac{B^2A^2\omega^2}{R}\sin^2\omega t\,dt$$

one revolution: $t = 0$ to $t = \frac{2\pi}{\omega}$

$$W = \int dW = \frac{B^2A^2\omega^2}{R}\int_0^{2\pi/\omega}\sin^2\omega t\,dt$$

Electrical energy dissipated during 1 revolution:

$$P = I^2R = \frac{\omega^2B^2A^2\sin^2\omega t}{R}.$$

$dU = P\,dt$ is the energy dissipated in time dt.
The energy dissipated in 1 revolution thus is

$$U = \int dU = \int_0^{2\pi/\omega} P\,dt = \frac{\omega^2B^2A^2}{R}\int_0^{2\pi/\omega}\sin^2\omega t\,dt$$

The expressions for U and W are the same, so the work of the

32-23 (cont)

external torque in one revolution is equal to the electrical energy dissipated during 1 revolution.

(Note: $\int \sin^2 x\,dx = \frac{1}{2}x - \frac{1}{4}\sin 2x \Rightarrow \int_0^{2\pi/\omega} \sin^2 \omega t\,dt = \frac{1}{\omega}\int_0^{2\pi} \sin^2 y\,dy$
$= \frac{1}{\omega}\left[\pi - \frac{1}{4}\sin\pi\right] = \frac{\pi}{\omega}$

Thus $W = U = \frac{\omega B^2 A^2 \pi}{R}$.)

32-27

Calculate $\vec{E}_n$ at each point, and then use $\vec{F} = q\vec{E}_n$:

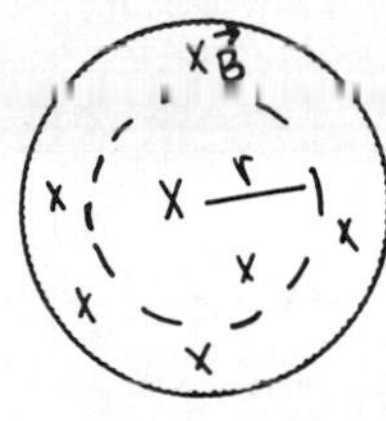

Apply $\oint \vec{E}_n \cdot d\vec{l} = -\frac{d\Phi}{dt}$ to a concentric circle of radius r.
Take $\vec{A}$ to be into the paper, in the direction of $\vec{B}$. Then $d\vec{l}$ is clockwise around the circle.

$$\frac{d\Phi}{dt} = \frac{d}{dt}(BA) = \frac{dB}{dt}A = \pi r^2 \frac{dB}{dt}$$

$\frac{dB}{dt} > 0 \Rightarrow \frac{d\Phi}{dt} > 0 \Rightarrow \oint \vec{E}_n \cdot d\vec{l}$ is negative $\Rightarrow \vec{E}_n$ is tangent to the circle in the counterclockwise direction.

$\oint \vec{E}_n \cdot d\vec{l} = -\oint E_n\,dl = -E_n \oint dl = -E_n 2\pi r$

Then

$\oint \vec{E}_n \cdot d\vec{l} = -\frac{d\Phi}{dt} \Rightarrow -E_n(2\pi r) = -\pi r^2 \frac{dB}{dt} \Rightarrow E_n = \frac{1}{2}r\frac{dB}{dt}$

point a

$\vec{E}_n \longleftarrow \bullet$ $\longleftarrow \vec{F}$

($\vec{F}$ is in the same direction as $\vec{E}_n$ since q is positive)

$F = qE_n = \frac{1}{2}qr\frac{dB}{dt}$, to the left

point b

$\uparrow \vec{E}_n$ $\uparrow \vec{F}$

$F = qE_n = \frac{1}{2}qr\frac{dB}{dt}$, upward

point c

$r = 0 \Rightarrow E_n = 0 \Rightarrow F = 0$

32-31

a) $\mathcal{E} = Blv$

But a complication is that $B = \frac{\mu_0 I}{2\pi r}$, where r is the distance from the long wire, varies over the length of the rod.

Divide the bar up into short pieces of length dr, calculate $d\mathcal{E}$ for each piece, and integrate over the length of the bar to get the total $\mathcal{E}$:

32-31 (cont)

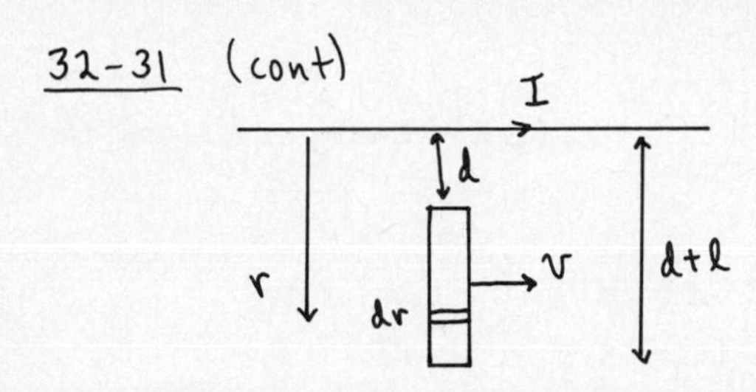

$$d\mathcal{E} = Bv\,dr = \frac{\mu_0 I v}{2\pi r}\,dr$$

$$\mathcal{E} = \int d\mathcal{E} = \frac{\mu_0 I v}{2\pi}\int_d^{d+l}\frac{dr}{r} = \frac{\mu_0 I v}{2\pi}\ln\left(\frac{d+l}{d}\right)$$

b)

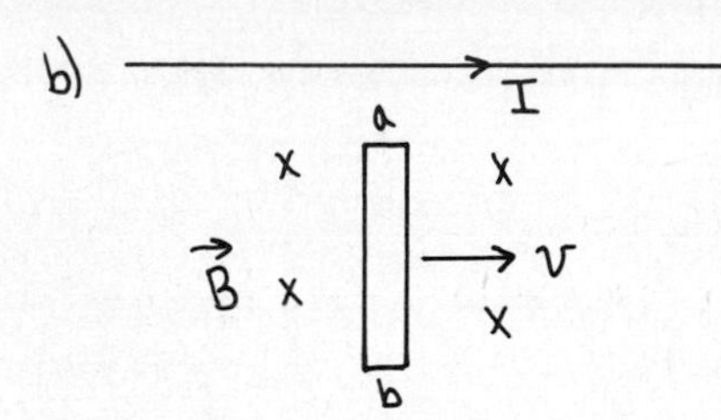

At the location of the bar the magnetic field $\vec{B}$ due to the current in the wire is into the plane of the figure.

$\uparrow \vec{v}\times\vec{B}$

$\vec{B}\ \otimes \longrightarrow \vec{v}$

The force on positive charge carriers in the bar hence would be directed from b to a $\Rightarrow$ a will be at higher potential.

c)

For sides ab and cd $\vec{E}_n = \vec{v}\times\vec{B}$ is perpendicular to the wire $\Rightarrow$ no emf.

Side ad: $\vec{E}_n = \vec{v}\times\vec{B}$ is $\uparrow$

Side bc: $\vec{E}_n = \vec{v}\times\vec{B}$ is $\uparrow$

These two $\vec{E}_n$ give rise to emf's equal in magnitude but oppositely directed (one tending to produce counterclockwise currents and the other clockwise currents), so the net $\mathcal{E} = 0$ and $\underline{I = 0}$.

32-33

a) $\Phi = B(t)\,A\cos\theta = B(t)\,A\cos\omega t$

(Note: $t = 0 \Rightarrow \Phi = B(0)\,A$, as it should since at $t=0$ the plane of the ring is perpendicular to the magnetic field.)

$B(t) = B_0 e^{-t/\tau} \Rightarrow \Phi = B_0 A e^{-t/\tau}\cos\omega t$

b) $\mathcal{E} = -\frac{d\Phi}{dt} = -B_0 A \frac{d}{dt}\left(e^{-t/\tau}\cos\omega t\right) = -B_0 A\left[-\frac{1}{\tau}e^{-t/\tau}\cos\omega t - \omega e^{-t/\tau}\sin\omega t\right]$

$\mathcal{E} = B_0 A e^{-t/\tau}\left[\frac{1}{\tau}\cos\omega t + \omega\sin\omega t\right]$

c) $I = \frac{\mathcal{E}}{R}$

At $t = 0$, $\mathcal{E} = B_0 A\left(\frac{1}{\tau}\right) \Rightarrow I = \frac{B_0 A}{R\tau} = \frac{(0.1\,T)\,\pi\,(0.20\,m)^2}{(0.01\,\Omega)(0.02\,s)} = \underline{62.8\,A}$

d) $\mathcal{E} = 0$ at $t = t_0 \Rightarrow B_0 A e^{-t_0/\tau}\left[\frac{1}{\tau}\cos\omega t_0 + \omega\sin\omega t_0\right] = 0$

$\Rightarrow \cos\omega t_0 + \tau\omega\sin\omega t_0 = 0$

$\tan\omega t_0 = -\frac{1}{\tau\omega} = -\frac{1}{(0.02\,s)(1000\,rad\cdot s^{-1})} = -0.05$

$\Rightarrow \omega t_0 = -0.0500\,rad$, or $(\pi - 0.0500)\,rad$, or $(2\pi - 0.0500)\,rad$, etc.

32-33 (cont)

Note: $\tan\theta$ is a periodic function with period π:

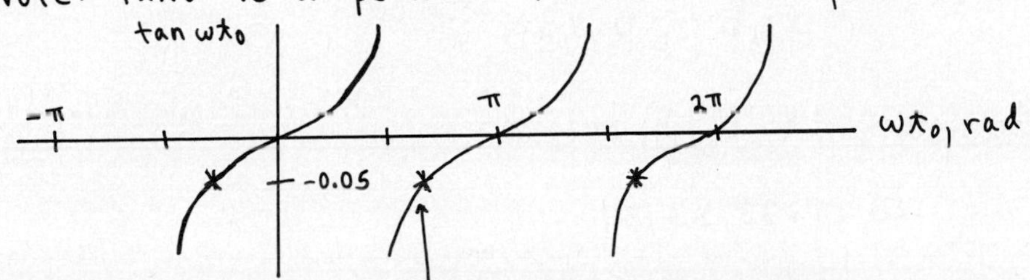

$$\Rightarrow \omega t_0 = (\pi - 0.0500)\text{rad} = 3.09\,\text{rad};\ t_0 = \frac{3.09\,\text{rad}}{1000\,\text{rad}\cdot s^{-1}} = \underline{3.02\times10^{-3}\,s}$$

e) $\mathcal{E}$ is maximum at $t = t_m \Rightarrow \frac{d\mathcal{E}}{dt} = 0$ at t_m

$$\frac{d\mathcal{E}}{dt} = 0 \Rightarrow \frac{d}{dt} e^{-t/\tau}\left[\frac{1}{\tau}\cos\omega t + \omega\sin\omega t\right] = 0 \text{ at } t = t_m$$

$$\Rightarrow -\frac{1}{\tau}e^{-t_m/\tau}\left[\frac{1}{\tau}\cos\omega t_m + \omega\sin\omega t_m\right] + e^{-t_m/\tau}\left[-\frac{\omega}{\tau}\sin\omega t_m + \omega^2\cos\omega t_m\right] = 0$$

$$\cos\omega t_m\left[\omega^2 - \frac{1}{\tau^2}\right] = \sin\omega t_m\left[2\frac{\omega}{\tau}\right]$$

$$\tan\omega t_m = \frac{\omega^2 - 1/\tau^2}{2\omega/\tau} = \frac{1}{2}\left[\tau\omega - \frac{1}{\tau\omega}\right]$$

$$\tan\omega t_m = \frac{1}{2}\left[(0.02s)(1000\,s^{-1}) - \frac{1}{(0.02s)(1000\,s^{-1})}\right] = 10.02\,s$$

$$\Rightarrow \omega t_m = 1.471\,\text{rad, or } (1.471 + \pi)\,\text{rad, etc}$$

Earliest time $\Rightarrow \omega t_m = 1.471$ rad

$$t_m = \frac{1.471\,\text{rad}}{1000\,\text{rad}\cdot s^{-1}} = \underline{1.471\times10^{-3}\,s}$$

At this t,

$$\mathcal{E} = (0.1T)\,\pi\,(0.2m)^2\, e^{-1.47\times10^{-3}s/0.02s}\left[\frac{1}{0.02s}\cos 1.47 + (1000s^{-1})\sin 1.47\right]$$

$$\mathcal{E} = 0.01257\,(0.9291)(5.031 + 994.9)\,V = \underline{11.7V}$$

CHAPTER 33

Exercises 3, 5, 7, 9, 13, 19, 21

Problems 23, 27, 33, 35, 37

Exercises

33-3

a) $\mathcal{E}_2 = M\frac{di_1}{dt} = (0.01\,H)(0.05\,A\cdot s^{-1}) = \underline{5.0\times10^{-4}\,V}$

$\frac{di_1}{dt}$ constant (uniform rate) $\Rightarrow$ $\mathcal{E}_2$ is constant

b) $\mathcal{E}_1 = M\frac{di_2}{dt} = (0.01\,H)(0.05\,A\cdot s^{-1}) = 5.0\times10^{-4}\,V$; same as in (a)

33-5

$$\mathcal{E} = L\frac{di}{dt} = (5\,H)(-0.02\,A\cdot s^{-1}) = \underline{-0.10\,V}$$

By Lenz's Law the induced emf is in the same direction as the current, to oppose its decrease.

33-7

a) $L = \frac{N\Phi}{i} = \frac{NBA}{i}$

$B = \frac{\mu_0 N i}{2\pi r} \Rightarrow L = \frac{\mu_0 N^2 A}{2\pi r} = \frac{(4\pi\times10^{-7}\,T\cdot A^{-1}\cdot m)(1000)^2(5\times10^{-4}\,m^2)}{2\pi\,(0.10\,m)} = \underline{1.0\times10^{-3}\,H}$

b) If the coils are both wound in the same sense then the combination acts like a single coil of 1500 turns. (The magnetic field is due to 1500 turns, and the flux due to this field links 1500 turns all with the same sign for Φ. Also, series $\Rightarrow$ same current in both coils.)

Thus

$$L = \frac{\mu_0 N^2 A}{2\pi r} = \left(\frac{1500}{1000}\right)^2(1.0\times10^{-3}\,H) = \underline{2.25\times10^{-3}\,H}$$

If one coil is wound opposite to the other:

The magnetic fields of the two coils are opposite to each other. Thus the field of the 500 turn coil cancels that due to 500 turns of the 1000 turn coil. The resultant magnetic field is therefore that of a 1000 − 500 = 500 turn coil. The fluxes for this resultant $\vec{B}$ field similarly cancel. (Positive flux for 500 turns of the 1000 turn coil, negative flux for the 500 turn coil.) Thus the net flux is due to the 500 extra turns in the larger coil.

$$\Rightarrow L = \left(\frac{500}{1000}\right)^2(1.0\times10^{-3}\,H) = \underline{2.5\times10^{-4}\,H}$$

33-9

$$U = \frac{1}{2}LI^2 \Rightarrow L = \frac{2U}{I^2} = \frac{2(0.1\,J)}{(20\,A)^2} = 5.0\times10^{-4}\,H$$

toroidal solenoid $\Rightarrow L = \frac{\mu_0 N^2 A}{2\pi r}$ (Example 33-2)

$$\Rightarrow N = \sqrt{\frac{2\pi r L}{\mu_0 A}} = \sqrt{\frac{2\pi\,(0.12\,m)(5.0\times10^{-4}\,H)}{(4\pi\times10^{-7}\,T\cdot A^{-1}\cdot m)(20\times10^{-4}\,m^2)}} = \underline{387 \text{ turns}}$$

33-13

$V_R = iR$ (drop), $V_L = L\frac{di}{dt}$

$$V = V_R + V_L$$

$$V = iR + L\frac{di}{dt} \Rightarrow i(t) = \frac{V}{R}\left(1 - e^{-\left(\frac{R}{L}\right)t}\right) \quad \text{(eq. 33-14)}$$

$$\frac{di}{dt} = \frac{V}{L}\, e^{(R/L)\,t} \quad \text{(eq. 33-15)}$$

a) at $t = 0$, $\frac{di}{dt} = \frac{V}{L} = \frac{12\,V}{3\,H} = \underline{4.0\,A\cdot s^{-1}}$

b) The sensible way to proceed here is to use $V = iR + L\frac{di}{dt}$

$$\Rightarrow \frac{di}{dt} = \frac{V - iR}{L} = \frac{12\,V - (1\,A)(6\,\Omega)}{3\,H} = \underline{2.0\,A\cdot s^{-1}}$$

(Note: $\frac{di}{dt}$ is largest at $t = 0$, and decreases exponentially as t increases.)

c) $t = 0.25$

$$i = \frac{V}{R}\left(1 - e^{-\left(\frac{R}{L}\right)t}\right) = \frac{12\,V}{6\,\Omega}\left(1 - e^{-\left(\frac{6\,\Omega}{3\,H}\right)(0.2\,s)}\right) = 2\,A\,(1 - e^{-0.4}) = \underline{0.659\,A}$$

d) final steady-state $\Rightarrow t \to \infty$

Then $i = \frac{V}{R} = \frac{12\,V}{6\,\Omega} = \underline{2.0\,A}$.

(Steady-state $\Rightarrow \frac{di}{dt} = 0 \Rightarrow V_L = 0 \Rightarrow V = V_R$; when the current is no longer changing the current is as if the inductor weren't there.)

33-19

$q = Q\cos\omega t$

eq. (33-25) $\Rightarrow \frac{d^2q}{dt^2} + \frac{1}{LC}q = 0$

$\frac{dq}{dt} = -\omega Q\sin\omega t$; $\frac{d^2q}{dt^2} = -\omega^2 Q\cos\omega t$

Substitute into eq. (33-25) $\Rightarrow -\omega^2 Q\cancel{\cos\omega t} + \frac{1}{LC}Q\cancel{\cos\omega t} = 0$

Thus this $q(t)$ does satisfy eq. (33-25) if $\omega = \frac{1}{\sqrt{LC}}$, and it does (eq. 33-24).

33-21

a) $\omega' = \sqrt{\frac{1}{LC} - \frac{R^2}{4L^2}}$

$R = 0 \Rightarrow \omega' = \omega = \frac{1}{\sqrt{LC}} = \frac{1}{\sqrt{(0.5H)(0.1\times10^{-3}F)}} = \underline{141\ rad\cdot s^{-1}}$

b) 10% decrease $\Rightarrow \omega' = 0.90\omega$

$\sqrt{\frac{1}{LC} - \frac{R^2}{4L^2}} = 0.90\frac{1}{\sqrt{LC}}$

Square both sides of the equation:

$\frac{1}{LC} - \frac{R^2}{4L^2} = 0.81\frac{1}{LC}$

$\frac{R^2}{4L^2} = \frac{0.19}{LC} \Rightarrow R = \sqrt{0.76\frac{L}{C}} = \sqrt{0.76\left(\frac{0.5H}{0.1\times10^{-3}F}\right)} = \underline{61.6\Omega}$

Problems

33-23

a) $\frac{di}{dt} = \frac{\Delta i}{\Delta t}$ (for constant $\frac{di}{dt}$) $\Rightarrow \frac{di}{dt} = +\frac{50A}{10s} = 5.0\ A\cdot s^{-1}$

$\mathcal{E} = L\frac{di}{dt} \Rightarrow L = \frac{\mathcal{E}}{di/dt} = \frac{25V}{5.0\ A\cdot s^{-1}} = \underline{5.0H}$

b) $L = \frac{N\Phi}{i} \Rightarrow N\Phi = Li = (5.0H)(50A) = \underline{250\ Wb}$

c) Energy stored in L is $U_L = \frac{1}{2}Li^2$, when the current is i. The rate at which energy is being stored is

$\frac{dU_L}{dt} = \frac{1}{2}L\frac{d(i^2)}{dt} = \frac{1}{2}L\left(2i\frac{di}{dt}\right) = Li\frac{di}{dt} = (5.0H)(50A)(5.0A\cdot s^{-1}) = 1250\ J\cdot s^{-1}$

The rate at which energy is being dissipated in the resistor is $P_R = i^2R = (50A)^2(25\Omega) = 6.25\times10^4\ J\cdot s^{-1}$.

The ratio is $\frac{dU_L/dt}{P_R} = \frac{1250\ J\cdot s^{-1}}{6.25\times10^4\ J\cdot s^{-1}} = \underline{0.020}$

33-27

$u_E = \frac{1}{2}\epsilon_0 E^2$ (eq. 27-9)

$u_B = \frac{B^2}{2\mu_0}$ (eq. 33-10)

$u_E = u_B \Rightarrow \frac{1}{2}\epsilon_0 E^2 = \frac{B^2}{2\mu_0} \Rightarrow B = \sqrt{\epsilon_0\mu_0}\ E$

$B = \sqrt{(8.85\times10^{-12}\ C^2\cdot N^{-1}\cdot m^{-2})(4\pi\times10^{-7}\ T\cdot A^{-1}\cdot m)}\ (400V\cdot m^{-1}) = \underline{1.33\times10^{-6}T}$

33-33

a) Series $\Rightarrow i_1 = i_2 = i$

Two coils rather than one $\Rightarrow$ twice the B field and twice the number of turns through which this field passes $\Rightarrow$ four times the total flux for a given i. (Since the coils are wound in the same sense.)

$$\Rightarrow \boxed{L_{series} = 4L}$$

b) Parallel $\Rightarrow i = i_1 + i_2$ (i = current in external circuit; i_1 and i_2 are the currents through the individual coils)

$i_1 = i_2$ (coils are identical) $\Rightarrow i_1 = i_2 = i/2$

Let N be the number of turns in each single coil

$$B_1 = \frac{\mu_0 N i_1}{l}, \quad B_2 = \frac{\mu_0 N i_2}{l} \Rightarrow B = B_1 + B_2 = \frac{\mu_0 N}{l}(i_1 + i_2) = \frac{\mu_0 N}{l} i$$

($B = B_1 + B_2$ since the coils are wound in the same sense. Note that B is the same as for a single coil alone.)

But this B field now has twice the turns to link (this B links both coils) $\Rightarrow$ twice the total flux as for a single coil

$$\Rightarrow \boxed{L_{parallel} = 2L}.$$

c) $\omega = \frac{1}{\sqrt{LC}}$

$$\omega_{series} = \frac{1}{\sqrt{L_{series} C}} = \frac{1}{\sqrt{4LC}} = \frac{1}{2}\frac{1}{\sqrt{LC}} = \underline{\frac{1}{2}\omega}.$$

33-35

The equation in question is $iR + L\frac{di}{dt} + \frac{q}{C} = 0$.

Multiply by $i \Rightarrow i^2R + Li\frac{di}{dt} + \frac{qi}{C} = 0$.

$\frac{d}{dt}U_L = \frac{d}{dt}\left(\frac{1}{2}Li^2\right) = \frac{1}{2}L\frac{d}{dt}(i^2) = \frac{1}{2}L\left(2i\frac{di}{dt}\right) = Li\frac{di}{dt}$, the second term.

$\frac{d}{dt}U_C = \frac{d}{dt}\left(\frac{q^2}{2C}\right) = \frac{1}{2C}\frac{d}{dt}(q^2) = \frac{1}{2C}2q\frac{dq}{dt} = \frac{qi}{C}$, the third term.

$i^2R = P_R$, rate at which electrical energy $\rightarrow$ heat in the resistor; P_R is positive

$\frac{d}{dt}U_L = P_L$, the rate at which the amount of energy stored in the inductor is changing.

$\frac{d}{dt}U_C = P_C$, the rate at which the amount of energy stored in the capacitor is changing.

Thus the equation says that $P_R + P_L + P_C = 0$.

33-35 (cont)

Note: At any given time one of P_c or P_L is negative. If the current is increasing the charge on the capacitor is decreasing, and vice versa.

33-37

a) The initial value of the current, after S_1 has been closed and S_2 open for a long time is $I = \frac{V}{R_0+R} = \frac{20\,V}{50\Omega+150\Omega} = 0.10\,A$

With both S_1 and S_2 closed the circuit becomes

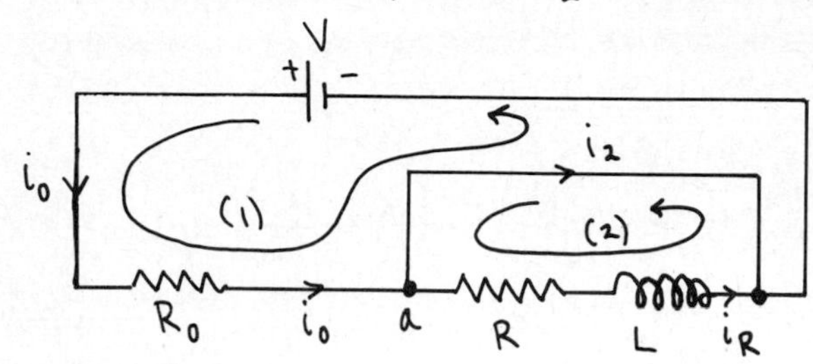

The current i_0 through R_0 can be calculated by considering loop (1), that goes through S_2:

$$V - i_0 R_0 = 0 \Rightarrow i_0 = \frac{V}{R_0} = \frac{20\,V}{50\Omega} = 0.4\,A$$

Consider loop (2): $-i_R R - L\frac{di_R}{dt} = 0$

(The emf induced in L is directed so as to try to maintain the current, that is initially 0.10A from left to right.

$\frac{di_R}{dt}$ is negative, so the $-L\frac{di_R}{dt}$ term is positive.

$i_R R + L\frac{di_R}{dt} = 0 \Rightarrow i_R = I e^{-Rt/L}$ (from eqs. (33-19) and (33-18))

$i_R = (0.1A)\,e^{-(150\Omega)t/5H} = (0.1A)\,e^{-(30s^{-1})t}$

Apply the point rule to the junction at a: $i_0 = i_2 + i_R$

Thus $i_0 = 0.40\,A$ (independent of time)

$i_R = (0.1A)\,e^{-(30s^{-1})t}$

$i_2 = i_0 - i_R = 0.4A - (0.1A)\,e^{-(30s^{-1})t}$

b) $t = 0.01s \Rightarrow i_2 = 0.4A - (0.1A)\,e^{-(30s^{-1})(0.01s)} = 0.4A - (0.1A)\,e^{-0.3} = \underline{0.326\,A}$

i_2 is calculated to be positive, so it is in the direction assumed in the sketch (from left to right through S_2)

CHAPTER 34

Exercises 1, 9, 11, 13, 15, 17, 19

Problems 23, 27, 29, 31

Exercises

34-1

a) $X_L = \omega L = 2\pi f L = 2\pi\,(60\,s^{-1})(1\,H) = \underline{377\,\Omega}$

b) $X_L = \omega L = 2\pi f L \Rightarrow L = \dfrac{X_L}{2\pi f} = \dfrac{1\,\Omega}{2\pi\,(60\,s^{-1})} = \underline{2.65\times10^{-3}\,H}$

c) $X_c = \dfrac{1}{\omega C} = \dfrac{1}{2\pi f C} = \dfrac{1}{2\pi\,(60\,Hz)(1\times10^{-6}\,F)} = \underline{2.65\times10^{3}\,\Omega}$

d) $X_c = \dfrac{1}{\omega C} = \dfrac{1}{2\pi f C} \Rightarrow C = \dfrac{1}{2\pi f X_c} = \dfrac{1}{2\pi(60\,s^{-1})(1\,\Omega)} = \underline{2.65\times10^{-3}\,F}$

34-9

a) (circuit: source ω, resistor R, capacitor C in series)

$$Z = \sqrt{R^2 + (X_L - X_c)^2} = \sqrt{R^2 + X_c^2}$$

(No inductor $\Rightarrow X_L = 0$)

$$X_c = \frac{1}{\omega C} = \frac{1}{(1000\,rad\cdot s^{-1})(2\times10^{-6}\,F)} = 500\,\Omega$$

$$Z = \sqrt{(300\,\Omega)^2 + (500\,\Omega)^2} = \underline{583\,\Omega}$$

b) $I = \dfrac{V}{Z} = \dfrac{50\,V}{583\,\Omega} = \underline{0.0858\,A}$

c) $V_R = IR = (0.0858\,A)(300\,\Omega) = \underline{25.7\,V}$

$V_c = IX_c = (0.0858\,A)(500\,\Omega) = \underline{42.9\,V}$

d) $\tan\phi = \dfrac{X_L - X_c}{R} = -\dfrac{X_c}{R} = -\dfrac{500\,\Omega}{300\,\Omega} = -1.67 \Rightarrow \phi = \underline{-59.1^\circ}$

$\phi < 0 \Rightarrow$ the source voltage lags the current

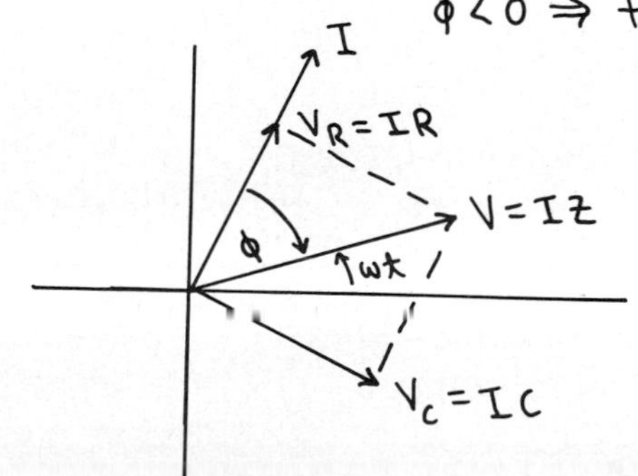

34-11

a) $f = 500\,Hz$

$$X_L = \omega L = 2\pi f L = 2\pi(500\,Hz)(0.1\,H) = 314\,\Omega$$

$$X_C = \frac{1}{\omega C} = \frac{1}{2\pi f C} = \frac{1}{2\pi(500\,Hz)(0.5\times10^{-6}C)} = 637\,\Omega$$

$R = 400\,\Omega$

$$\Rightarrow Z = \sqrt{R^2 + (X_L - X_C)^2} = \sqrt{(400\,\Omega)^2 + (314\,\Omega - 637\,\Omega)^2} = \underline{514\,\Omega}$$

phasor diagram ($X_C > X_L$)

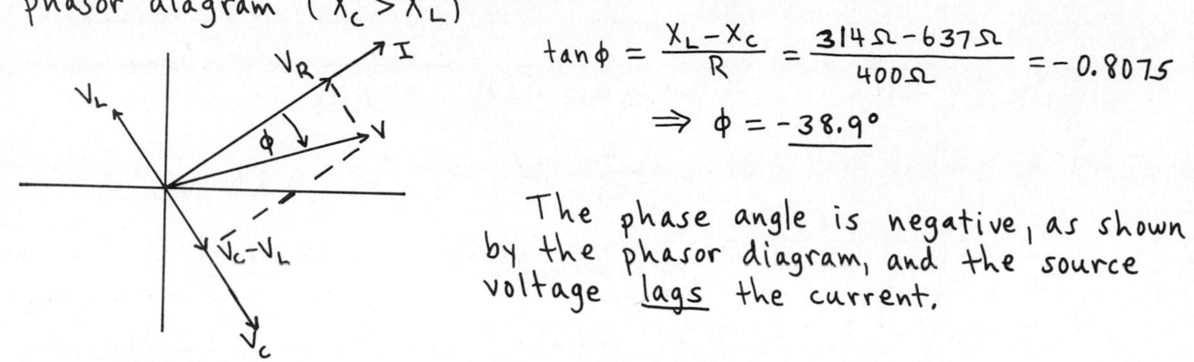

$$\tan\phi = \frac{X_L - X_C}{R} = \frac{314\,\Omega - 637\,\Omega}{400\,\Omega} = -0.8075$$

$$\Rightarrow \phi = \underline{-38.9°}$$

The phase angle is negative, as shown by the phasor diagram, and the source voltage lags the current.

b) $f = 1000\,Hz$

$$X_L = \omega L = 2\pi f L = 2\pi(1000\,Hz)(0.1\,H) = 628\,\Omega$$

$$X_C = \frac{1}{\omega C} = \frac{1}{2\pi f C} = \frac{1}{2\pi(1000\,Hz)(0.5\times10^{-6}F)} = 318\,\Omega$$

$R = 400\,\Omega$

$$\Rightarrow Z = \sqrt{R^2 + (X_L - X_C)^2} = \sqrt{(400\,\Omega)^2 + (628\,\Omega - 318\,\Omega)^2} = \underline{506\,\Omega}$$

phasor diagram ($X_L > X_C$)

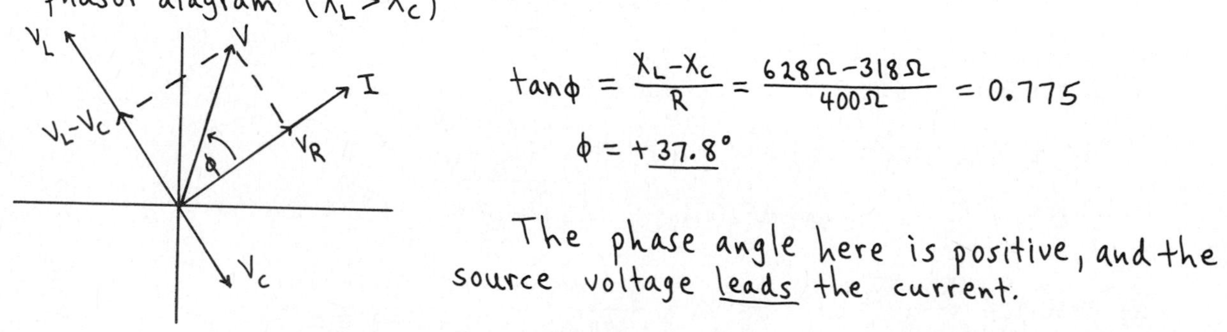

$$\tan\phi = \frac{X_L - X_C}{R} = \frac{628\,\Omega - 318\,\Omega}{400\,\Omega} = 0.775$$

$$\phi = +\underline{37.8°}$$

The phase angle here is positive, and the source voltage leads the current.

34-13

$R = 400\,\Omega$

$L = 0.1\,H$

$C = 0.5\times10^{-6}\,F$

$f = 100\,Hz$

$$X_L = \omega L = 2\pi f L = 2\pi(100\,Hz)(0.1\,H) = 62.8\,\Omega$$

$$X_C = \frac{1}{\omega C} = \frac{1}{2\pi f C} = \frac{1}{2\pi(100\,Hz)(0.5\times10^{-6}F)} = 3183\,\Omega$$

$$Z = \sqrt{R^2 + (X_L - X_C)^2} = \sqrt{(400\,\Omega)^2 + (62.8\,\Omega - 3183\,\Omega)^2} = 3146\,\Omega$$

$$I_{rms} = 0.25\,A \Rightarrow V_{rms} = I_{rms} Z = (0.25\,A)(3146\,\Omega) = 786\,V$$

$$\tan\phi = \frac{X_L - X_C}{R} = \frac{62.8\,\Omega - 3183\,\Omega}{400\,\Omega} = -7.80 \Rightarrow \phi = -82.7°$$

34-13 (cont)

a) $P_S = V_{rms} I_{rms} \cos\phi = (786\,V)(0.25\,A)\cos(-82.7°) = \underline{25.0\,W}$

b) $P_R = I_{rms}^2 R = (0.25\,A)^2(400\,\Omega) = \underline{25.0\,W}$

c) V_C is 90° out of phase with $i \Rightarrow P_C = 0$, for the capacitor. (Electrical energy is periodically stored and released in the capacitor, but not dissipated.)

d) V_L is 90° out of phase with $i \Rightarrow P_L = 0$, for the inductor. (Electrical energy is periodically stored and released in the magnetic field of the inductor, but not dissipated.)

e) $P_R + P_C + P_L = 25\,W + 0 + 0 = 25\,W \Rightarrow P_S = P_R + P_C + P_L$

The average power delivered by the source equals the average power consumed by the circuit, and all the power consumption occurs in the resistor.

34-15

$R = 300\,\Omega$, $L = 0.9\,H$, $C = 2.0\,\mu F$, $V = 50\,V$

a) $\omega = \frac{1}{\sqrt{LC}} = \frac{1}{\sqrt{(0.9\,H)(2.0\times10^{-6}\,F)}} = \underline{745\ rad\cdot s^{-1}}$

b) power factor $= \cos\phi$

At resonance $X_L = X_C \Rightarrow \tan\phi = \frac{X_L - X_C}{R} = 0 \Rightarrow \phi = 0$

Thus $\cos\phi = 1$ at resonance.

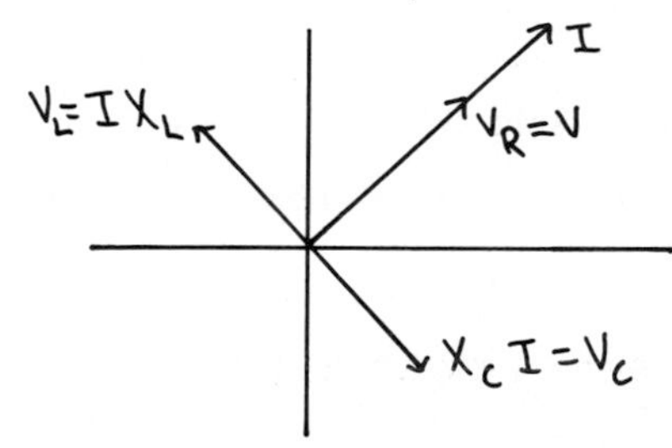

$X_L - X_C = 0 \Rightarrow V_L - V_C = 0$

I and V are in phase ; $\phi = 0$

c) $X_L = \omega L = (745\,rad\cdot s^{-1})(0.9\,H) = 671\,\Omega$

$X_C = \frac{1}{\omega C} = \frac{1}{(745\,rad\cdot s^{-1})(2\times10^{-6}\,F)} = 671\,\Omega$

$Z = \sqrt{R^2 + (X_L - X_C)^2} = R = 300\,\Omega$ (with $X_L - X_C$ crossed out, marked 0)

$V_{rms} = \frac{V}{\sqrt{2}} = \frac{50\,V}{\sqrt{2}} = 35.4\,V$

$I_{rms} = \frac{V_{rms}}{R} = \frac{35.4\,V}{300\,\Omega} = 0.118\,A$

34-15 (cont)

V_1 reads $V_{R,rms} = I_{rms} R = (0.118\,A)(300\,\Omega) = 35.4\,V$

V_2 reads $V_{L,rms} = I_{rms} X_L = (0.118\,A)(671\,\Omega) = 79.1\,V$

V_3 reads $V_{C,rms} = I_{rms} X_C = (0.118\,A)(671\,\Omega) = 79.1\,V$

V_4 reads $(V_L - V_C)_{rms} = 0$

V_5 reads V_{rms} (the rms source voltage) $= 35.4\,V$ (The same as $V_{R,rms}$, since at any time $v_L + v_C = 0$.)

d) $\omega = \frac{1}{\sqrt{LC}}$, independent of $R \Rightarrow$ remains $\underline{745\ rad\cdot s^{-1}}$

e) $I_{rms} = \frac{V_{rms}}{R} = \frac{35.4\,V}{100\,\Omega} = \underline{0.354\,A}$

34-17

$V = 120\,V$, $R = 200\,\Omega$, $L = 0.5\,H$, $C = 0.2\,\mu F$

a) $\omega = \frac{1}{\sqrt{LC}} = \frac{1}{\sqrt{(0.5\,H)(0.2\times10^{-6}\,F)}} = \underline{3162\ rad\cdot s^{-1}}$

b) The phasor diagram is as in Fig. 34-13, except that at resonance $I_C = I_L$:

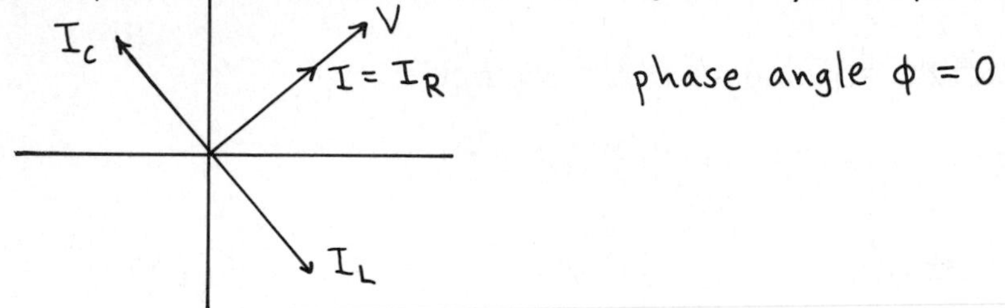

phase angle $\phi = 0$

c) resonance $\Rightarrow \omega C = \frac{1}{\omega L} \Rightarrow Z = R$

$I = \frac{V}{Z} = \frac{V}{R} = \frac{120\,V}{200\,\Omega} = \underline{0.60\,A}$

d) $I_R = \frac{V}{R} = 0.60\,A$; at resonance $I = I_R$. (At any time $i_C + i_L = 0$.)

34-19

a) $\frac{N_1}{N_2} = \frac{V_1}{V_2} = \frac{120\,V}{12\,V} = \underline{10}$

b) $I_2 = \frac{V_2}{R} = \frac{12\,V}{2\,\Omega} = \underline{6\,A}$

c) $P_2 = I_2^2 R = (6\,A)^2(2\,\Omega) = \underline{72\,W}$ or $P_2 = I_2 V_2 = (6\,A)(12\,V) = \underline{72\,W}$

d) transformer draws 72 W

$P = \frac{V^2}{R} \Rightarrow R = \frac{V^2}{P} = \frac{(120\,V)^2}{72\,W} = \underline{200\,\Omega}$

Problems

34-23

Voltage of source leads current by $30° \Rightarrow \phi = +30°$

$$\tan\phi = \frac{X_L - X_C}{R}$$

Only a coil (with resistance and inductance) in the circuit $\Rightarrow X_C = 0$

Thus
$$\tan\phi = \frac{X_L}{R} \Rightarrow X_L = R\tan\phi = (20\Omega)\tan 30° = 11.5\Omega$$

$$X_L = \omega L = 2\pi f L \Rightarrow L = \frac{X_L}{2\pi f} = \frac{11.5\,\Omega}{2\pi(100\,\text{Hz})} = \underline{0.0183\,\text{H}}$$

34-27

a) Voltage lags current $\Rightarrow X_C > X_L \Rightarrow$ add an inductor in series with the circuit

b) power factor $\cos\phi$ equals $1 \Rightarrow \phi = 0 \Rightarrow X_C = X_L$

Calculate the present value of $X_C - X_L$, to see how much more X_L is needed:

$\cos\phi = 0.6 \Rightarrow \phi = \pm 53.1°$

voltage lags $\Rightarrow \phi < 0 \Rightarrow \phi = -53.1°$, so $X_C > X_L$

$$\tan\phi = \frac{X_L - X_C}{R} \Rightarrow X_L - X_C = R\tan\phi$$

But we are given Z rather than R.

$$Z = \sqrt{R^2 + (X_L - X_C)^2} \Rightarrow Z^2 = R^2 + (X_L - X_C)^2 \Rightarrow R^2 = Z^2 - (X_L - X_C)^2$$

Then
$$\tan^2\phi = \frac{(X_L - X_C)^2}{R^2} = \frac{(X_L - X_C)^2}{Z^2 - (X_L - X_C)^2} \Rightarrow Z^2\tan^2\phi - (X_L - X_C)^2\tan^2\phi = (X_L - X_C)^2$$

$$(X_L - X_C)^2 = \frac{Z^2\tan^2\phi}{1 + \tan^2\phi} \Rightarrow X_L - X_C = \pm\sqrt{\frac{Z^2\tan^2\phi}{1 + \tan^2\phi}}$$

$$X_C > X_L \Rightarrow X_C - X_L = \sqrt{\frac{Z^2\tan^2\phi}{1 + \tan^2\phi}} = \sqrt{\frac{(50\Omega)^2\tan^2(-53.1°)}{1 + \tan^2(-53.1°)}} = 40\Omega$$

Therefore need to add 40Ω of X_L.

$$X_L = \omega L = 2\pi f L \Rightarrow L = \frac{X_L}{2\pi f} = \frac{40\,\Omega}{2\pi(60\,\text{Hz})} = \underline{0.106\,\text{H}}$$; amount of inductance to add

34-29

$R = 100\Omega$, $C = 0.1 \times 10^{-6}\,\text{F}$, $L = 0.1\,\text{H}$, $V = 100$ volts

a) resonance $\rightarrow X_L - X_L$

$$\frac{1}{\omega C} = \omega L \Rightarrow \omega = \frac{1}{\sqrt{LC}} = \frac{1}{\sqrt{(0.1\,\text{H})(0.1 \times 10^{-6}\,\text{F})}} = \underline{1.0 \times 10^4\,\text{rad}\cdot\text{s}^{-1}}$$

34-29 (cont)

$$f = \frac{\omega}{2\pi} = \frac{1.0\times10^4\ s^{-1}}{2\pi} = \underline{1592\ Hz}$$

b) At resonance $Z = R = 100\Omega$.
The current amplitude is $I = \frac{V}{Z} = \frac{100\,V}{100\Omega} = \underline{1.0\,A}$

c) $I = \frac{V}{R} = \frac{100\,V}{100\Omega} = \underline{1.0\,A}$

d) $X_L = \omega L = (1.0\times10^4\ rad\cdot s^{-1})(0.1\,H) = 1000\Omega$

$I = \frac{V}{X_L} = \frac{100\,V}{1000\,\Omega} = \underline{0.1\,A}$

e) $X_C = \frac{1}{\omega C} = \frac{1}{(1.0\times10^4\ rad\cdot s^{-1})(0.1\times10^{-6}F)} = 1000\,\Omega$

$I = \frac{V}{X_C} = \frac{100\,V}{1000\Omega} = \underline{0.1\,A}$

f) $U_L = \frac{1}{2} L I_L^2 = \frac{1}{2}(0.1H)(0.1A)^2 = \underline{5.0\times10^{-4}\,J}$

$U_C = \frac{1}{2} C V_C^2$
parallel circuit $\Rightarrow V_C = V = 100V$
So $U_C = \frac{1}{2}(0.1\times10^{-6}F)(100V)^2 = \underline{5.0\times10^{-4}\,J}$

34-31

$$I_{av} = \frac{1}{t_2 - t_1}\int_{t_1}^{t_2} I(t)\,dt$$

Average over one period of the alternating current $\Rightarrow t_1 = 0,\ t_2 = T$.
$I_{av} = \frac{1}{T}\int_0^T I(t)\,dt$

Need to determine the functional form of $I(t)$, for $t = 0$ to $t = T$:
$I = \left(\frac{2I_0}{T}\right)t$ for $t = 0$ to $\frac{T}{2}$.
(This gives I as a linear function of t; $I = 0$ at $t = 0$ and $I = I_0$ at $t = \frac{T}{2}$.)

$I = -I_0 + \left(\frac{2I_0}{T}\right)\left(t - \frac{T}{2}\right) = \left(\frac{2I_0}{T}\right)t - 2I_0$ for $t = \frac{T}{2}$ to T.
(this also gives I as a linear function of t, with $I = I_0$ at $t = \frac{T}{2}$ and $I = 0$ at $t = T$.)

$I(t)$ thus has one functional form for $t = 0$ to $\frac{T}{2}$ and another for $t = \frac{T}{2}$ to T.
So break $\int_0^T I\,dt$ into $\int_0^{T/2} I\,dt + \int_{T/2}^{T} I\,dt$:

34-31 (cont)

$$I_{av} = \frac{1}{T}\left[\int_0^{T/2} I\,dt + \int_{T/2}^{T} I\,dt\right] = \frac{1}{T}\left[\int_0^{T/2}\left(\frac{2I_0}{T}\right)t\,dt + \int_{T/2}^{T}\left[\left(\frac{2I_0}{T}\right)t - 2I_0\right]dt\right]$$

$$I_{av} = \frac{1}{T}\left[\frac{2I_0}{T}\int_0^{T/2} t\,dt + \frac{2I_0}{T}\int_{T/2}^{T} t\,dt - 2I_0\int_{T/2}^{T} dt\right]$$

$$I_{av} = \frac{1}{T}\left[\frac{2I_0}{T}\int_0^{T} t\,dt - 2I_0\int_{T/2}^{T} dt\right] = \frac{1}{T}\left[\frac{2I_0}{T}\left(\frac{T^2}{2}\right) - 2I_0\left(\frac{T}{2}\right)\right]$$

Thus $I_{av} = I_0 - I_0 = 0.$

(That this is indeed correct is easily verified by looking at the graph of $I(t)$.)

$$I_{rms}^2 = (I^2)_{av}$$

Thus $I_{rms}^2 = \frac{1}{T}\int_0^T I^2\,dt = \frac{1}{T}\left[\int_0^{T/2} I^2\,dt + \int_{T/2}^{T} I^2\,dt\right]$

$$I_{rms}^2 = \frac{1}{T}\left[\int_0^{T/2}\left(\frac{4I_0^2}{T^2}\right)t^2\,dt + \int_{T/2}^{T}\left[\frac{2I_0}{T}t - 2I_0\right]^2 dt\right.$$

$$I_{rms}^2 = \frac{1}{T}\left[\int_0^{T/2}\left(\frac{4I_0^2}{T^2}\right)t^2\,dt + \int_{T/2}^{T}\left(\frac{4I_0^2}{T^2}\right)t^2\,dt - \int_{T/2}^{T}\left(\frac{8I_0^2}{T}\right)t\,dt + \int_{T/2}^{T} 4I_0^2\,dt\right]$$

$$I_{rms}^2 = \frac{1}{T}\left[\frac{4I_0^2}{T^2}\int_0^T t^2\,dt - \frac{8I_0^2}{T}\int_{T/2}^{T} t\,dt + 4I_0^2\int_{T/2}^{T} dt\right]$$

$$I_{rms}^2 = \frac{1}{T}\left[\frac{4}{3}I_0^2 T - \frac{4I_0^2}{T}\left(T^2 - \frac{T^2}{4}\right) + 4I_0^2\left(\frac{T}{2}\right)\right]$$

$$I_{rms}^2 = \frac{1}{T}\left[\frac{4}{3}I_0^2 T - 3I_0^2 T + 2I_0^2 T\right] = I_0^2\left(\frac{4}{3} - 3 + 2\right) = \frac{1}{3}I_0^2$$

$$I_{rms} = \sqrt{I_{rms}^2} = \frac{I_0}{\sqrt{3}}$$

CHAPTER 35

Exercises 3, 7, 9, 11

Problems 17, 19, 21

Exercises

35-3

a) $c = f\lambda \Rightarrow \lambda = \frac{c}{f} = \frac{3.00 \times 10^8 \, m \cdot s^{-1}}{1020 \times 10^3 \, Hz} = \underline{294 \, m}$

b) eq. (35-2) $E = cB = (3.00 \times 10^8 \, m \cdot s^{-1})(1.6 \times 10^{-11} \, T) = \underline{4.8 \times 10^{-3} \, V \cdot m^{-1}}$

35-7

a) $E_{max} = c B_{max} \Rightarrow B_{max} = \frac{E_{max}}{c} = \frac{2 \times 10^{-2} \, V \cdot m^{-1}}{3.00 \times 10^8 \, m \cdot s^{-1}} = \underline{6.67 \times 10^{-11} \, T}$

b) $I = S_{av} = \frac{E_{max}^2}{2\mu_0 c} = \frac{(2 \times 10^{-2} \, V \cdot m^{-1})^2}{2(4\pi \times 10^{-7} \, T \cdot A^{-1} \cdot m)(3.00 \times 10^8 \, m \cdot s^{-1})} = 5.31 \times 10^{-7} \, W \cdot m^{-2},$

at a distance of $r = 50 \times 10^3 \, m$.

I gives the power per unit area. As in Example 35-3 surround the antenna with an imaginary sphere of radius r. The area of this sphere is $A = 4\pi r^2$. All the power radiated passes through this surface, so $P = IA$ (assuming I is the same at all points on this sphere)

$\Rightarrow P = (5.31 \times 10^{-7} \, W \cdot m^2)(4\pi)(50 \times 10^3 \, m)^2 = 1.67 \times 10^4 \, W = \underline{16.7 \, kW}$

c) $IA = \text{constant}$

$\Rightarrow \frac{E^2}{2\mu_0 c}(4\pi r^2) = \text{constant} \Rightarrow Er = \text{constant} \Rightarrow E_1 r_1 = E_2 r_2$

$r_2 = \left(\frac{E_1}{E_2}\right) r_1 = \left(\frac{2 \times 10^{-2}}{1 \times 10^{-2}}\right)(50 \, km) = \underline{100 \, km}$

35-9

a) $I = S_{av} = 1.4 \times 10^3 \, W \cdot m^{-2}$

By eq. (35-12) the momentum density is

$\frac{p}{V} = \frac{S}{c^2} = \frac{1.4 \times 10^3 \, W \cdot m^{-2}}{(3.0 \times 10^8 \, m \cdot s^{-1})^2} = 1.56 \times 10^{-14} \, kg \cdot m \cdot s^{-1}/m^3 = 1.56 \times 10^{-14} \, kg \cdot m^{-2} \cdot s^{-1}$

b) By eq. (35-13) the momentum flow rate is

$\frac{1}{A}\frac{dp}{dt} = \frac{S}{c} = \frac{1.4 \times 10^3 \, W \cdot m^{-2}}{3.0 \times 10^8 \, m \cdot s^{-1}} = \underline{4.67 \times 10^{-6} \, N \cdot m^{-2}}$

(Note: $\frac{dp}{dt}$ has units of force.)

35-11

a) By eq. (35-14), $v = \frac{c}{\sqrt{K K_m}} = \frac{3.00 \times 10^8 \text{ m}\cdot\text{s}^{-1}}{\sqrt{10(1000)}} = \underline{3.0 \times 10^6 \text{ m}\cdot\text{s}^{-1}}$

b) $v = f\lambda \Rightarrow \lambda = \frac{v}{f} = \frac{3.0\times10^6 \text{ m}\cdot\text{s}^{-1}}{100\times10^6 \text{ Hz}} = \underline{0.030 \text{ m}}$

Problems

35-17

a) For a solenoid $B = \mu_0 n i$ (eq. 31-28), and is uniform over the cross section of the solenoid.

i changing $\Rightarrow \frac{dB}{dt} = \mu_0 n \frac{di}{dt}$

Calculate the induced electric field from Faraday's Law:

$$\oint \vec{E}_n \cdot d\vec{l} = -\frac{d\Phi}{dt}$$

Take the path to be a circle of radius r concentric with the solenoid axis.

$$\oint \vec{E}_n \cdot d\vec{l} = E_n(2\pi r)$$

$$\Phi = BA; \quad \frac{d\Phi}{dt} = \mu_0 n A \frac{di}{dt} = \mu_0 n \pi r^2 \frac{di}{dt}$$

$$E_n(2\pi r) = -\mu_0 n \pi r^2 \frac{di}{dt} \Rightarrow E_n = -\tfrac{1}{2}\mu_0 n r \frac{di}{dt}$$

b) Take the positive direction for Φ to be in the direction of $\vec{B}$ (into the paper). Then the positive direction around the circle is clockwise (right-hand rule).

But $\frac{d\Phi}{dt}$ is positive, so Faraday's Law gives E_n negative

$\Rightarrow E_n$ is counterclockwise.

$\vec{S} = \frac{1}{\mu_0} \vec{E} \times \vec{B}$; $\vec{E} \times \vec{B}$ and therefore $\vec{S}$ are <u>radially inward</u>.

The magnitude of S is given by

$$S = \frac{1}{\mu_0} EB = \frac{1}{\mu_0}\left(\tfrac{1}{2}\mu_0 n r \frac{di}{dt}\right)(\mu_0 n i) = \tfrac{1}{2}\mu_0 n^2 r i \frac{di}{dt}$$

35-19

$\mathcal{E} = -\frac{d\Phi}{dt} = -\frac{d}{dt}(BA) = -\pi R^2 \frac{dB}{dt}$, where R is the radius of the loop.

$B = B_{max} \sin(\omega t - kx) \Rightarrow \frac{dB}{dt} = B_{max}\, \omega \cos(\omega t - kx)$

$\mathcal{E} = -\pi R^2 B_{max}\, \omega \cos(\omega t - kx)$

$\Rightarrow \mathcal{E}_{max} = \pi R^2 B_{max}\, \omega = \pi R^2 B_{max}(2\pi f) = 2\pi^2 f R^2 B_{max}$

35-19 (cont)

Need to calculate B_{max} at the antenna loop:

$P = IA = I(4\pi r^2)$, where r is the distance from the source to the loop and I is the intensity at that point. (See Example 35-3).

$$I = \frac{P}{4\pi r^2} = \frac{1\times10^6\,W}{4\pi\,(100\,m)^2} = 7.86\,W\cdot m^{-2}$$

$$I = S_{av} = \frac{E_{max}^2}{2\mu_0 c} \Rightarrow E_{max} = \sqrt{2\mu_0 c I} = \sqrt{2(4\pi\times10^{-7}\,T\cdot A^{-1}\cdot m)(3.00\times10^8\,m\cdot s^{-1})(7.86\,W\cdot m^{-2})}$$

$$E_{max} = 77.0\,V\cdot m^{-1}$$

$$B_{max} = \frac{E_{max}}{c} = \frac{77.0\,V\cdot m^{-1}}{3.00\times10^8\,m\cdot s^{-1}} = 2.57\times10^{-7}\,T$$

Thus

$$\mathcal{E}_{max} = 2\pi^2 f R^2 B_{max} = 2\pi^2(10\times10^6\,Hz)(0.25\,m)^2(2.57\times10^{-7}\,T)$$

$$\mathcal{E}_{max} = \underline{3.17\,V}$$

35-21

For modern transmission lines,

$$P = VI = (500\times10^3\,V)(1000\,A) = 5.00\times10^8\,W$$

The power in a beam of radius A is $P = IA = S_{av}A$, where $I = S_{av}$ is the intensity.

$$\text{intensity } I = \frac{P}{A} = \frac{5.0\times10^8\,W}{100\,m^2} = 5.0\times10^6\,W\cdot m^{-2}$$

$$\text{Then } I = \frac{E_{max}^2}{2\mu_0 c} \Rightarrow E_{max} = \sqrt{2\mu_0 c I} = \sqrt{2(4\pi\times10^{-7}\,T\cdot A^{-1}\cdot m)(3.0\times10^8\,m\cdot s^{-1})(5.0\times10^6\,W\cdot m^{-2})}$$

$$E_{max} = \underline{6.14\times10^4\,V\cdot m^{-1}}$$

$$B_{max} = \frac{E_{max}}{c} = \frac{6.14\times10^4\,V\cdot m^{-1}}{3.0\times10^8\,m\cdot s^{-1}} = \underline{2.05\times10^{-4}\,T}$$

CHAPTER 36

Exercises 1, 5, 9, 11, 13, 17, 21

Problems 27, 31, 33, 35, 39

Exercises

36-1

a) From Appendix F the orbital radius of the earth is $r = 1.49 \times 10^{11}$ m and the orbital period is 365.3 days. The orbital velocity thus is

$$v = \frac{2\pi r}{T} = \frac{2\pi\,(1.49 \times 10^{11}\,\text{m})}{(365.3\,\text{da})\left(\frac{24\,\text{hr}}{1\,\text{da}}\right)} = 1.07 \times 10^{8}\,\text{m}\cdot\text{hr}^{-1}$$

In 42 hr the earth travels a distance $d = vt = (1.07 \times 10^{8}\,\text{m}\cdot\text{hr}^{-1})(42\,\text{hr}) = \underline{4.49 \times 10^{9}\,\text{m}}$

b) $t = \frac{d}{c} = \frac{4.49 \times 10^{9}\,\text{m}}{3.00 \times 10^{8}\,\text{m}\cdot\text{s}^{-1}} = \underline{15.0\,\text{s}}$

36-5

a) [diagram: ray incident at 30° to surface, ϕ_a from normal, $n_a = 1.0$ (air); refracted ray ϕ_b, θ_b, $n_b = 1.5$ (glass)]

$n_a = 1.0$ (air)

$n_b = 1.5$ (glass)

$$\phi_a = 60^\circ$$

$$n_a \sin\phi_a = n_b \sin\phi_b$$

$$\sin\phi_b = \left(\frac{n_a}{n_b}\right)\sin\phi_a = \left(\frac{1}{1.5}\right)\sin 60^\circ = 0.577$$

$$\Rightarrow \phi_b = 35.3^\circ$$

Thus the angle θ_b between the refracted ray and the glass surface is

$$\theta_b = 90^\circ - 35.3^\circ = \underline{54.7^\circ}$$

b) [diagram: ϕ_a, $n_a = 1.0$ (air); ϕ_b, $n_b = 1.5$ (glass)]

$n_a = 1.0$ (air)

$n_b = 1.5$ (glass)

Want $\phi_a = \phi$ and $\phi_b = \frac{\phi}{2} = \frac{1}{2}\phi_a$

$$n_a \sin\phi_a = n_b \sin\phi_b$$

$$(1.0)\sin\phi = (1.5)\sin\left(\frac{\phi}{2}\right)$$

$$\sin\phi = (1.5)\sin(\phi/2)$$

To enable us to solve for ϕ use the trig identity (Appendix B)

$$\sin\phi = 2\sin\frac{\phi}{2}\cos\frac{\phi}{2}$$

$$\Rightarrow 2\sin\left(\frac{\phi}{2}\right)\cos\left(\frac{\phi}{2}\right) = 1.5\sin\left(\frac{\phi}{2}\right) \Rightarrow \cos\left(\frac{\phi}{2}\right) = 0.75$$

$$\frac{\phi}{2} = 41.4^\circ \Rightarrow \phi = \underline{82.8^\circ}$$

36-9

a) $v = \frac{c}{n} = \frac{3.0 \times 10^{8}\,\text{m}\cdot\text{s}^{-1}}{1.5} = \underline{2.0 \times 10^{8}\,\text{m}\cdot\text{s}^{-1}}$

b) $\lambda = \frac{\lambda_0}{n} = \frac{500\,\text{nm}}{1.5} = \underline{333\,\text{nm}}$

36-11

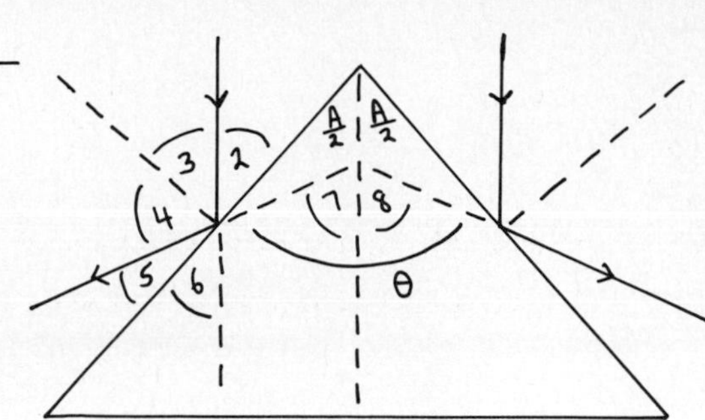

We are asked to show that θ = angle 7 + angle 8 = 2A.
By symmetry angle 7 = angle 8, so we need to show that angle 7 = A.

From the sketch it is apparent that angle 5 + angle 6 = angle 7 and that angle 2 = $\frac{A}{2}$.

But angle 2 = angle 6 $\Rightarrow$ angle 6 = $\frac{A}{2}$

The law of reflection says $\phi_a = \phi_r \Rightarrow$ angle 3 = angle 4. But then angle 2 = angle 5. Thus angle 5 = A/2.

Then angle 7 = angle 5 + angle 6 = $\frac{A}{2} + \frac{A}{2}$ = A, which completes the proof.

36-13

a) $n_a \sin\phi_a = n_b \sin\phi_b$

$$n = \frac{c}{v} \Rightarrow \frac{c}{v_a}\sin\phi_a = \frac{c}{v_b}\sin\phi_b \Rightarrow \sin\phi_a = \left(\frac{v_a}{v_b}\right)\sin\phi_b$$

$$\phi_a = \phi_{crit} \text{ when } \sin\phi_b = 1 \Rightarrow \sin\phi_{crit} = \left(\frac{v_a}{v_b}\right)$$

a → air (incident medium)
b → water

$$\sin\phi_{crit} = \frac{330\ m\cdot s^{-1}}{1320\ m\cdot s^{-1}} = 0.25 \Rightarrow \phi_{crit} = \underline{14.5°}$$

b) $n = \frac{c}{v} \Rightarrow$ air with the smaller v has the larger n for sound waves.

36-17

a) $I = I_{max}\cos^2\theta$

After first filter $I_1 = \frac{1}{2}I_0$; linearly polarized, along the polarizer axis

After second filter $I_2 = I_1\cos^2 45° = \left(\frac{1}{2}I_0\right)\left(\frac{1}{2}\right) = \frac{1}{4}I_0$; linearly polarized, along the polarizer axis

After third filter $I_3 = I_2\cos^2 45° = \left(\frac{1}{4}I_0\right)\left(\frac{1}{2}\right) = \frac{1}{8}I_0$; linearly polarized, along the polarizer axis

b) Remove second filter $\Rightarrow$ after first filter $I_1 = \frac{1}{2}I_0$ linearly polarized along the polarizer axis.
But now the next filter's axis is at 90°, so zero intensity is passed.

36-21

a) reflected beam completely linearly polarized
$\Rightarrow$ angle of incidence is the polarizing angle $\Rightarrow \phi_p = 58°$.

Brewster's Law: $\tan\phi_p = \frac{n'}{n}$; here n = index of incident medium $= 1.0$
$\Rightarrow n' = \tan\phi_p = \tan 58° = \underline{1.60}$, the refractive index of the glass.

b) $n_a \sin\phi_a = n_b \sin\phi_b$
$a \rightarrow$ air ; $n_a = 1.0$
$b \rightarrow$ glass ; $n_b = 1.6$

Thus $1.0 \sin 58° = (1.6) \sin\phi_b$
$\sin\phi_b = 0.53 \Rightarrow \phi_b = \underline{32.0°}$ (with respect to the normal)

Problems

36-27

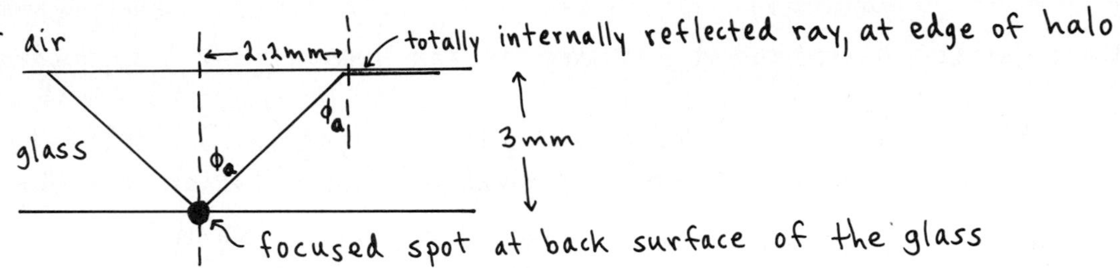

$n_a \sin\phi_a = n_b \sin\phi_b$

$n_a = n_{glass}$
$n_b = n_{air} = 1$
$\phi_a = \phi_{crit} \Rightarrow \phi_b = 90°$

$n_{glass} \sin\phi_{crit} = 1$

$n_{glass} = \frac{1}{\sin\phi_{crit}}$

From the distances given in the sketch, $\tan\phi_{crit} = \frac{2.2\ \text{mm}}{3\ \text{mm}} = 0.733$

$\phi_{crit} = 36.3°$

$n_{glass} = \frac{1}{\sin 36.3°} = \underline{1.69}$

36-31

$\phi_c = 60°$
$n_a \sin\phi_a = n_b \sin\phi_b$
$\phi_a = \phi_c \Rightarrow \phi_b = 90°$
$n_g \sin\phi_c = n_\ell \Rightarrow n_\ell = (1.5) \sin 60° = \underline{1.30}$

36-33

a) 1st filter: $I_1 = \frac{1}{2} I_0$

2nd filter: $I_2 = I_1 \cos^2\theta = \frac{1}{2} I_0 \cos^2\theta$

3rd filter: $I_3 = I_2 \cos^2(90° - \theta) = \frac{1}{2} I_0 \cos^2\theta \cos^2(90° - \theta)$

(Angle between 1st and 3rd is 90°, between 1st and 2nd is θ, so between 2nd and 3rd the angle is $(90° - \theta)$.)

$\cos(90° - \theta) = \cos(\theta - 90°) = \sin\theta$ (Appendix B)

$\Rightarrow I_3 = \frac{1}{2} I_0 (\cos\theta \sin\theta)^2$

But $\cos\theta \sin\theta = \frac{1}{2}\sin 2\theta$ (Appendix B again!)

$I_3 = \frac{1}{2} I_0 \left(\frac{1}{2}\sin 2\theta\right)^2 = \underline{\frac{1}{8} I_0 (\sin 2\theta)^2}$

b) I_3 maximum $\Rightarrow \sin 2\theta = 1 \Rightarrow 2\theta = 90° \Rightarrow \theta = \underline{45°}$

36-35

a) The polarizer passes $\frac{1}{2}$ of the unpolarized intensity $\Rightarrow \frac{1}{2} I_0$, independent of ϕ.

Out of the I_p polarized intensity the polarizer passes $I_p \cos^2(\phi - \theta)$, where $\phi - \theta$ is the angle between the plane of polarization and the axis of the polarizer.

ϕ ← polarization axis of polarized component

θ

$\phi - \theta$

axis of polarizer

The minimum total transmitted intensity is when the $I_p \cos^2(\phi - \theta)$ part is zero, which happens for $\phi - \theta = \pm 90°$.

From the table of data, this minimum transmitted intensity occurs for ϕ between 120° and 130°, say at 125°.

$\phi = 125°$ and $\phi - \theta = 90° \Rightarrow \underline{\theta = 35°}$.

b) For $\phi = 120°$ get $I = \frac{1}{2} I_0 + I_p \cos^2(\phi - \theta) = \frac{1}{2} I_0 + I_p \cos^2(120° - 35°)$

$I = \frac{1}{2} I_0 + I_p \cos^2(85°) = \frac{1}{2} I_0 + 0.00760\, I_p$

$\Rightarrow \boxed{5.2\ \mathrm{W \cdot m^{-2}} = \frac{1}{2} I_0 + 0.00760\, I_p}$

For $\phi = 0°$ get $I = \frac{1}{2} I_0 + I_p \cos^2(0° - 35°) = \frac{1}{2} I_0 + I_p \cos^2(35°)$

$\Rightarrow \boxed{18.4\ \mathrm{W \cdot m^{-2}} = \frac{1}{2} I_0 + 0.671\, I_p}$

Subtract these two eqs. $\Rightarrow 13.2\ \mathrm{W \cdot m^{-2}} = 0.663\, I_p$

$I_p = \underline{19.9\ \mathrm{W \cdot m^{-2}}}$

Then $I_0 = 2(5.2\ \mathrm{W \cdot m^{-2}} - 0.0076\, I_p) = 2(5.2\ \mathrm{W \cdot m^{-2}} - 0.15\ \mathrm{W \cdot m^{-2}}) = \underline{10.1\ \mathrm{W \cdot m^{-2}}}$

36-39

a) The number of full waves in the n_1 direction, for a plate of thickness d, is

$$\frac{d}{\lambda_1} = \frac{n_1 d}{\lambda_0}.$$

Similarly, the number of full waves in the n_2 direction is

$$\frac{d}{\lambda_2} = \frac{n_2 d}{\lambda_0}.$$

For a quarter-wave plate these must differ by $\frac{1}{4}$

$$\Rightarrow \frac{n_1 d}{\lambda_0} - \frac{n_2 d}{\lambda_0} = \frac{1}{4}$$

$$d\left(\frac{n_1 - n_2}{\lambda_0}\right) = \frac{1}{4} \Rightarrow d = \frac{\lambda_0}{4(n_1 - n_2)}$$

b) $\lambda_0 = 589\text{nm}$
$n_1 = 1.658$
$n_2 = 1.486$

$$\Rightarrow d = \frac{589\text{nm}}{4(1.658 - 1.486)} = \underline{856\text{nm}}$$

CHAPTER 37

Exercises 1, 5, 9, 11, 15, 17

Problems 19, 21, 23, 27

Exercises

37-1

plane mirror $\Rightarrow s = -s'$, $m = \frac{y'}{y} = +1$ $\left(m = -\frac{s'}{s}\right)$

$s = 80\text{cm} \Rightarrow s' = -80\text{cm}$
$y' = my = +1(6.0\text{cm}) = 6.0\text{cm}$ $\Rightarrow$ Image is 80 cm to right of mirror and is 6 cm tall.

37-5

a) concave $\Rightarrow R = +20\text{cm}$

$f = \frac{R}{2} = +10\text{cm}$

$s = 12\text{cm}$; object is outside the focal point

The principal rays are numbered as in the list in section 37-4.

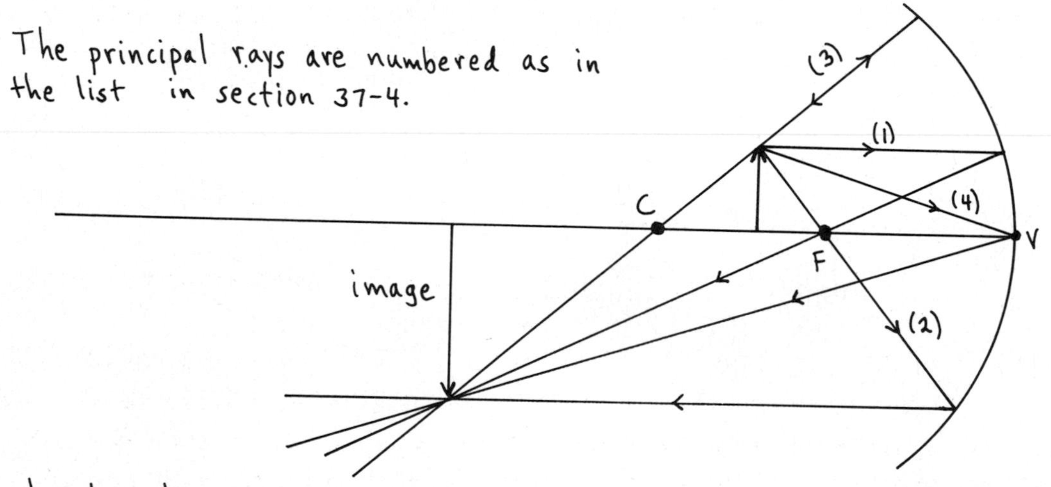

b) $\frac{1}{s} + \frac{1}{s'} = \frac{1}{f}$

$\frac{1}{s'} = \frac{1}{f} - \frac{1}{s} = \frac{s-f}{sf} \Rightarrow s' = \frac{sf}{s-f} = \frac{(12\text{cm})(10\text{cm})}{12\text{cm} - 10\text{cm}} = +60\text{cm}$

$m = -\frac{s'}{s} = -\frac{60\text{cm}}{12\text{cm}} = -5$; $y' = my = (-5)(2\text{cm}) = -10\text{cm}$

The image is calculated to be 60 cm in front of the mirror, is 10 cm tall, inverted ($m < 0$), and real ($s' > 0$). This all agrees qualitatively with the principal ray diagram.

37-9

a) convex $\Rightarrow R = -20\,cm$

$f = \frac{R}{2} = -10\,cm$

$s = 8\,cm \Rightarrow |s| < |f|$

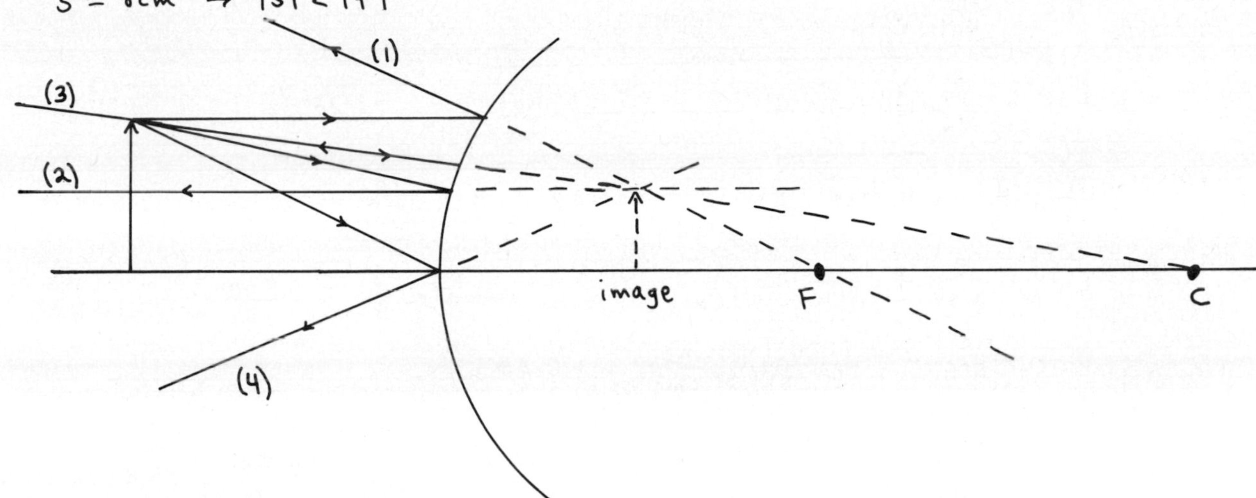

b) $\frac{1}{s} + \frac{1}{s'} = \frac{1}{f} \Rightarrow \frac{1}{s'} = \frac{1}{f} - \frac{1}{s} = \frac{s-f}{sf} \Rightarrow s' = \frac{sf}{s-f} = \frac{(8cm)(-10cm)}{8cm-(-10cm)} = \underline{-4.44cm}$

$m = -\frac{s'}{s} = -\frac{-4.44cm}{8cm} = \underline{+0.555}$

$y' = my = (0.555)(1.5cm) = \underline{0.832\,cm}$

The image is 4.44 cm behind the mirror, is 0.832 cm tall, is erect ($m>0$), and virtual ($s'<0$).

37-11

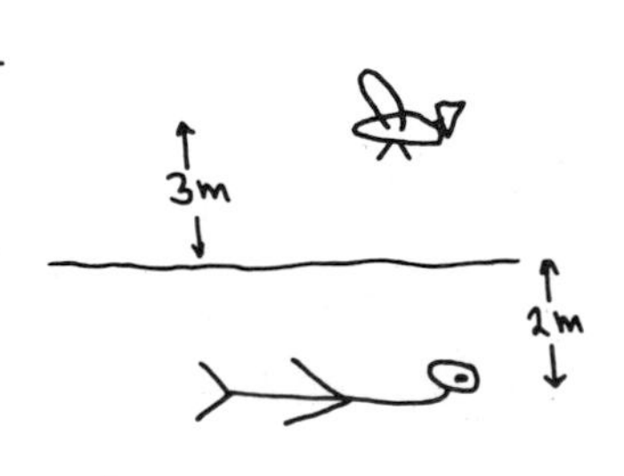

P'
P

$s' = -\frac{n'}{n} s$ (eq. 37-9)

$s' = -1.333(3m) = -4.00m$

The image point is 4.00 m above the water $\Rightarrow$ 6.00 m above the diver.

The ray sketch shows that the image is farther than the object from the surface because the light is bent toward the normal as it passes from air to water.

37-15

$n = n_\ell$ $\quad n' = 1.50$

$s = 60cm$ $\quad R = +4cm$

$s' = 100\,cm$

$\frac{n}{s} + \frac{n'}{s'} = \frac{n'-n}{R}$

$\frac{n_\ell}{60cm} + \frac{1.50}{100cm} = \frac{1.50 - n_\ell}{4cm}$

$\Rightarrow n_\ell\left(\frac{1}{60cm} + \frac{1}{4cm}\right) = 1.5\left(\frac{1}{4cm} - \frac{1}{100cm}\right)$

37-15 (cont)

$$n_\ell\left(\frac{10+150}{600\text{ cm}}\right) = \frac{3}{2}\left(\frac{150-6}{600\text{ cm}}\right) \Rightarrow n_\ell = \frac{3}{2}\left(\frac{144}{160}\right) = \underline{1.35}$$

37-17

$n=1.0$ $\quad n'=1.50$ $\quad s=20\text{ cm}$ $\quad R=-5\text{ cm}$

$$\frac{n}{s}+\frac{n'}{s'} = \frac{n'-n}{R} \Rightarrow \frac{1}{20\text{ cm}} + \frac{1.50}{s'} = \frac{1.50-1.00}{-5\text{ cm}}$$

$$\frac{3}{2s'} = -\frac{1}{10\text{ cm}} - \frac{1}{20\text{ cm}} = -\frac{3}{20\text{ cm}} \Rightarrow s' = \frac{3}{2}\left(-\frac{20\text{ cm}}{3}\right) = -10\text{ cm}$$

$$m = -\frac{ns'}{n's} = -\frac{(1.00)(-10\text{ cm})}{(1.50)(20\text{ cm})} = +0.333$$

$s'<0 \Rightarrow$ Image is 10 cm to the left of the vertex; the magnification is $m=+0.333$.

Problems

37-19

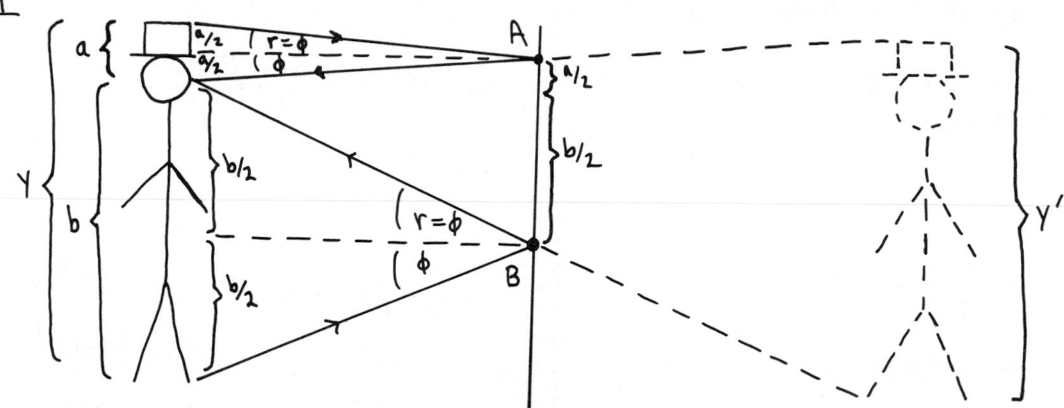

The image formation is given in the above sketch. The image height y' equals the object height y; the lateral magnification is unity. But the height of the mirror does not have to equal y for the man to see his full image.

Let a be the distance from the top of the man's head to his eyes, and b the distance from the man's feet to his eyes. Thus $y=a+b$. The light ray from the top of the man's head to his eyes strikes the mirror at point A. Since by the law of reflection $r=\phi$, point A is a distance $a/2$ above the man's eyes. Similarly the light ray from the man's feet to his eyes strikes the mirror at point B, which is a distance $b/2$ below his eyes.

Thus the total length of the mirror must equal the distance from A to B, which is $a/2+b/2 = \frac{1}{2}(a+b) = \frac{1}{2}y$. The mirror must be half the height of the man.

37-21

a) concave $\Rightarrow R>0,\ f>0$

$s' = +4\,m = 400\,cm$

$y = 0.5\,cm$; $|y'| = 40\,cm$

Image is to be formed on a screen $\Rightarrow$ real image $\Rightarrow s'>0$.

$m = -\frac{s'}{s} < 0$ since s' and s are both positive.

Thus $y' = -40\,cm$ (image is inverted), and $m = \frac{y'}{y} = \frac{-40\,cm}{0.5\,cm} = -80$

$$s = -\frac{s'}{m} = -\frac{400\,cm}{-80} = \underline{+5.0\,cm}$$

b) $\frac{1}{s} + \frac{1}{s'} = \frac{1}{f} \Rightarrow \frac{1}{f} = \frac{s+s'}{ss'} \Rightarrow f = \frac{ss'}{s+s'} = \frac{(5\,cm)(400\,cm)}{5\,cm + 400\,cm} = 4.94\,cm$

$f = \frac{R}{2} \Rightarrow R = 2f = 2(4.94\,cm) = \underline{9.88\,cm}$

37-23

a) convex $\Rightarrow R<0 \Rightarrow R = -10\,cm,\ f = -5\,cm$

$$\frac{1}{s} + \frac{1}{s'} = \frac{1}{f}$$

$$\frac{1}{s'} = \frac{1}{f} - \frac{1}{s} = \frac{s-f}{fs} \Rightarrow s' = \frac{fs}{s-f} = \frac{(-5\,cm)\,s}{s+5\,cm}$$

s is negative, so write $s = -|s| \Rightarrow s' = \frac{(5\,cm)|s|}{5\,cm - |s|}$

Thus $s'>0$ (real image) for $|s| < 5\,cm = |f|$

b) $m = -\frac{s'}{s}$

real image $\Rightarrow s'>0$
virtual object $\Rightarrow s<0$ $\Rightarrow m>0$; image is $\underline{\text{erect}}$

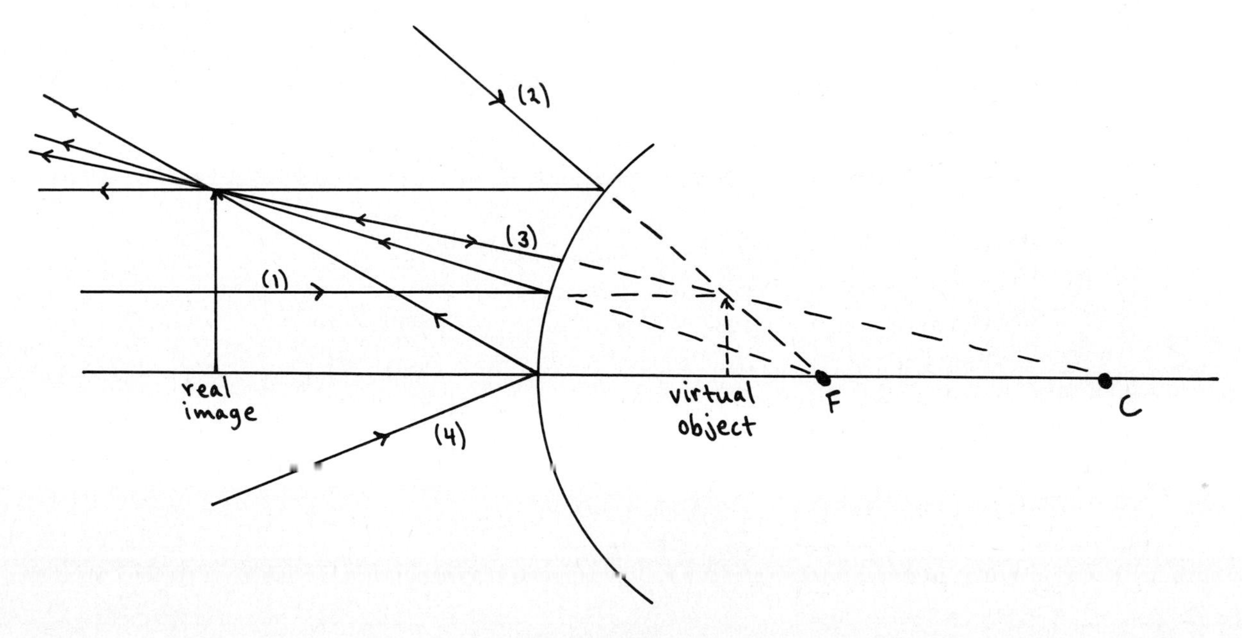

37-27

a)

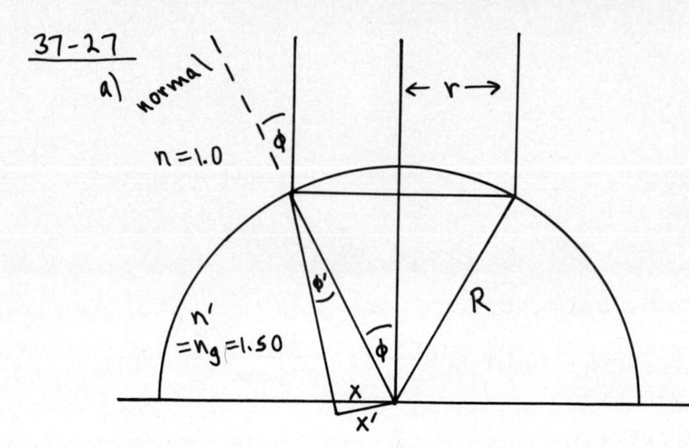

The width of the incident beam is exaggerated in the sketch, to make it easier to draw.

$r = 0.5\text{cm}$, the radius of the incident light beam

$R = 10\text{cm}$, the radius of the hemisphere

ϕ and ϕ' small $\Rightarrow x \approx x'$

$\sin\phi = \frac{r}{R}$, $\sin\phi' = \frac{x'}{R} \approx \frac{x}{R}$

Snell's Law $\Rightarrow n\sin\phi = n'\sin\phi'$

Use the above expressions for $\sin\phi$, $\sin\phi'$

$\Rightarrow (1.0)\frac{r}{R} = n_g \frac{x}{R} \Rightarrow x = \frac{r}{n_g} = \frac{0.50\text{ cm}}{1.50} = 0.333\text{ cm}$

The diameter of the circle is $d = 2x = \underline{0.666\text{ cm}}$.

b) R divides out of the expression; the result for the diameter of the spot is independent of the radius R of the hemisphere!

CHAPTER 38

Exercises 3, 5, 9, 11, 13, 15, 19, 23, 27, 31, 33, 37

Problems 39, 43, 45, 47, 49, 53, 57

Exercises

38-3

$f = -10\text{cm}$; $\frac{1}{s} + \frac{1}{s'} = \frac{1}{f} \Rightarrow \frac{1}{s'} = \frac{1}{f} - \frac{1}{s} = \frac{s-f}{sf} \Rightarrow s' = \frac{sf}{s-f}$

$s = 30\text{cm}$

a) $s' = \frac{sf}{s-f} = \frac{(30\text{cm})(-10\text{cm})}{30\text{cm} - (-10\text{cm})} = -\frac{300\text{ cm}^2}{40\text{ cm}} = \underline{-7.5\text{cm}}$

(The image is 7.5 cm from the lens, on same side as the object.)

b) $m = -\frac{s'}{s} = -\frac{-7.5\text{cm}}{30\text{ cm}} = \underline{+0.25}$

c) $s' < 0 \Rightarrow$ <u>virtual</u> image

d) $m > 0 \Rightarrow$ <u>erect</u> image

$s = 20\text{ cm}$

a) $s' = \frac{sf}{s-f} = \frac{(20\text{cm})(-10\text{cm})}{20\text{cm} - (-10\text{cm})} = -\frac{200\text{ cm}^2}{30\text{ cm}} = \underline{-6.67\text{ cm}}$

(The image is 6.67cm from the lens, on same side as the object.)

b) $m = -\frac{s'}{s} = -\frac{-6.67\text{cm}}{20\text{ cm}} = \underline{+0.333}$

c) $s' < 0 \Rightarrow$ <u>virtual</u> image

d) $m > 0 \Rightarrow$ <u>erect</u> image

$s = 15\text{ cm}$

a) $s' = \frac{sf}{s-f} = \frac{(15\text{cm})(-10\text{cm})}{15\text{cm} - (-10\text{cm})} = -\frac{150\text{ cm}^2}{25\text{ cm}} = \underline{-6.00\text{cm}}$

(The image is 6.00 cm from the lens, on same side as the object.)

b) $m = -\frac{s'}{s} = -\frac{-6.00\text{ cm}}{15\text{ cm}} = \underline{+0.400}$

c) $s' < 0 \Rightarrow$ <u>virtual</u> image

d) $m > 0 \Rightarrow$ <u>erect</u> image

38-3 (cont)

$s = 5\text{ cm}$

a) $s' = \frac{sf}{s-f} = \frac{(5\text{cm})(-10\text{cm})}{5\text{cm} - (-10\text{cm})} = -\frac{50\text{ cm}^2}{15\text{ cm}} = \underline{-3.33\text{ cm}}$

(The image is 3.33cm from the lens, on the same side as the object.)

b) $m = -\frac{s'}{s} = -\frac{-3.33\text{cm}}{5\text{cm}} = \underline{+0.666}$

c) $s' < 0 \Rightarrow$ virtual image

d) $m > 0 \Rightarrow$ erect image

38-5

a)

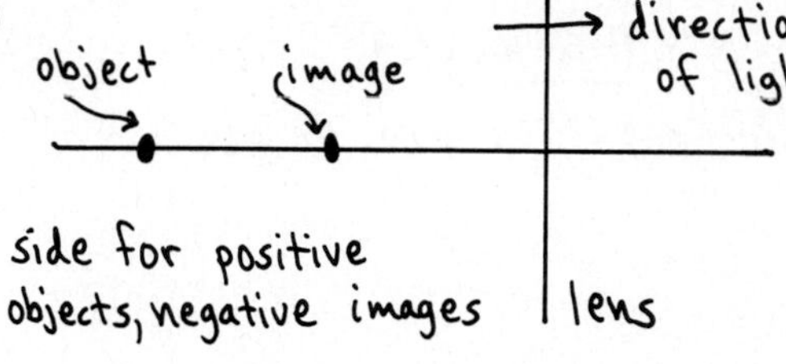

side for positive images, negative objects

$s = +20\text{cm}, \; s' = -4\text{cm}$

$\frac{1}{s} + \frac{1}{s'} = \frac{1}{f} \Rightarrow \frac{1}{f} = \frac{s'+s}{ss'} \Rightarrow f = \frac{ss'}{s+s'} = \frac{(+20\text{cm})(-4\text{cm})}{+20\text{cm} - 4\text{cm}} = -\frac{80\text{cm}^2}{16\text{cm}} = \underline{-5\text{ cm}}$

$f < 0 \Rightarrow$ lens is diverging

b) $m = -\frac{s'}{s} = -\frac{-4\text{cm}}{20\text{ cm}} = +0.200$

$m = \frac{y'}{y} \Rightarrow y' = ym = (2\text{cm})(+0.200) = \underline{+0.400\text{ cm tall}}$

$m > 0 \Rightarrow$ erect

c)

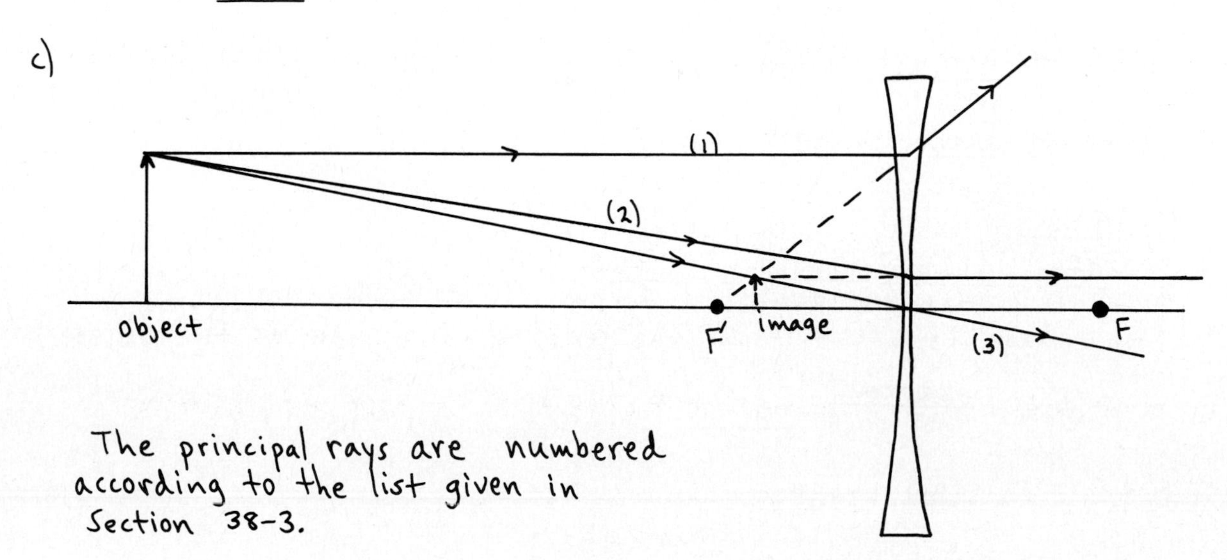

The principal rays are numbered according to the list given in Section 38-3.

38-9

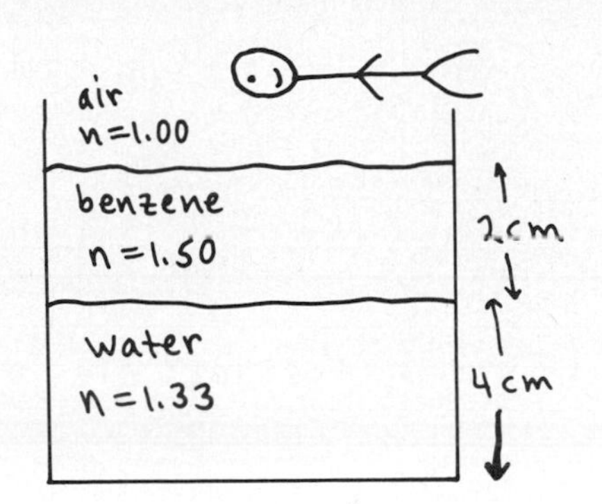

The bottom of the water serves as the object.

An image is formed by the water → benzene interface. This image serves as an object for the benzene → air interface, which forms the final image.

water → benzene interface

$n = 1.33$, $n' = 1.50$

$s = 4\text{cm}$ (distance of object [bottom of water] below the interface)

$$\frac{n}{s} + \frac{n'}{s'} = 0 \Rightarrow s' = -\frac{n'}{n}s = -\frac{1.50}{1.33}(4\text{cm}) = -4.51\text{ cm}$$

Thus the first image is 4.51 cm below the water-benzene interface and hence 4.51 cm + 2.00 cm = 6.51 cm below the benzene-air interface.

benzene → air interface

$n = 1.50$, $n' = 1.00$

$s = 6.51\text{cm}$

$$s' = -\left(\frac{n'}{n}\right)s = -\left(\frac{1.00}{1.50}\right)(6.51\text{cm}) = -4.34\text{ cm}$$

Thus this final image is 4.34 cm below the top surface of the benzene.

38-11

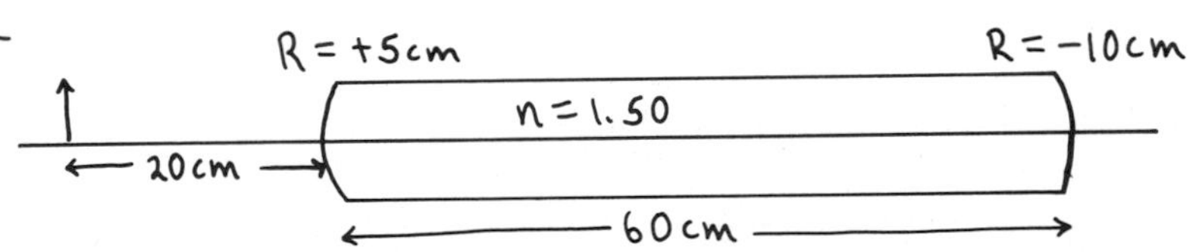

a) The object for the second surface is the image formed by the first surface.

b) Image formed by first surface:

$n = 1.00$

$n' = 1.50$

$s = 20\text{ cm}$

$R = +5\text{ cm}$

$$\frac{n}{s} + \frac{n'}{s'} = \frac{n'-n}{R} \Rightarrow \frac{1}{20\text{cm}} + \frac{1.5}{s'} = \frac{1.5-1.0}{5\text{cm}}$$

$$\frac{1.5}{s'} = \frac{1}{10\text{cm}} - \frac{1}{20\text{cm}} = \frac{1}{20\text{cm}}$$

$$s' = (1.5)(20\text{cm}) = 30\text{ cm}$$

This image is 30cm to the right of the first surface

$\Rightarrow$ 60 cm − 30 cm = 30cm to the left of the second surface.

Thus $s = +30\text{cm}$ for the second surface.

c) $s > 0 \Rightarrow$ real object (This "object" is on the side of the surface from which the rays actually come.)

38-11 (cont)

d) $n = 1.50$

$n' = 1.00$

$s = +30\text{ cm}$

$R = -10\text{ cm}$

$$\frac{n}{s} + \frac{n'}{s'} = \frac{n'-n}{R}$$

$$\frac{1.5}{30\text{cm}} + \frac{1.0}{s'} = \frac{1.0-1.5}{-10\text{cm}}$$

$$\frac{1}{s'} = \frac{1}{20\text{cm}} - \frac{3}{60\text{cm}} = 0 \Rightarrow s' = \infty$$

The final image is formed <u>at infinity</u>. (That is, the rays are parallel as they emerge from the rod.)

38-13

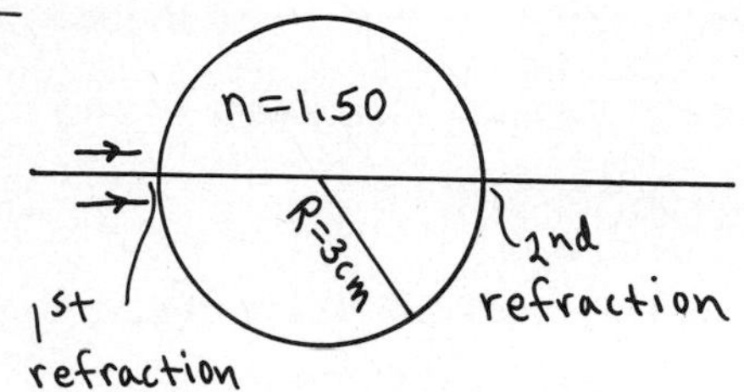

The image formed by the first refraction serves as the object for the second refraction. The two verticies are a distance of $2R = 6\text{ cm}$ apart.

1st refraction

$n = 1.00$

$n' = 1.50$

$R = +3\text{cm}$

$s = \infty$ (parallel rays)

$$\frac{n}{s} + \frac{n'}{s'} = \frac{n'-n}{R}$$

$$\frac{1.0}{\infty} + \frac{1.5}{s'} = \frac{1.5-1.0}{+3\text{cm}}$$

$$\frac{1.5}{s'} = \frac{1}{6\text{cm}} \Rightarrow s' = 1.5(6\text{cm}) = 9\text{cm}$$

$s' > 0 \Rightarrow$ the first image is 9 cm to the right of the first vertex $\Rightarrow$ 3 cm to the right of the second vertex.

2nd refraction

$n = 1.50$

$n' = 1.00$

$R = -3\text{cm}$

$s = -3\text{ cm}$ (virtual object)

$$\frac{n}{s} + \frac{n'}{s'} = \frac{n'-n}{R}$$

$$\frac{1.5}{-3\text{cm}} + \frac{1.0}{s'} = \frac{1.0-1.5}{-3\text{cm}}$$

$$\frac{1}{s'} = +\frac{1}{6\text{cm}} + \frac{3}{6\text{cm}} = \frac{4}{6\text{cm}} \Rightarrow s' = \frac{6}{4}\text{cm} = 1.5\text{cm}$$

$s' > 0 \Rightarrow$ this final image is 1.5 cm to the right of the 2nd vertex, or <u>4.5 cm from the center of the sphere.</u>

38-15

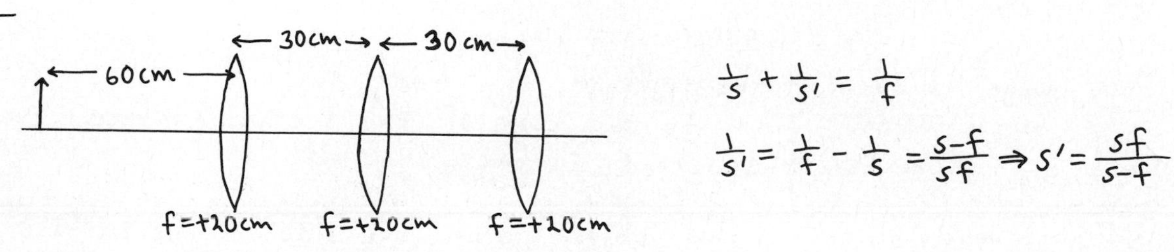

$$\frac{1}{s} + \frac{1}{s'} = \frac{1}{f}$$

$$\frac{1}{s'} = \frac{1}{f} - \frac{1}{s} = \frac{s-f}{sf} \Rightarrow s' = \frac{sf}{s-f}$$

38-15 (cont)

lens 1

$s = 60\text{cm}$ $\quad s' = \frac{sf}{s-f} = \frac{(60\text{cm})(20\text{cm})}{60\text{cm} - 20\text{cm}} = +30\text{cm}$

$f = 20\text{ cm}$

Thus the image is at the second lens.

lens 2

$s = 0$ $\quad s' = \frac{0}{0-20\text{cm}} = 0$

$f = 20\text{cm}$ The second lens does nothing.

lens 3

$s = 30\text{cm}$ $\quad s' = \frac{sf}{s-f} = \frac{(30\text{cm})(20\text{cm})}{30\text{cm} - 20\text{cm}} = 60\text{cm}$

$f = +20\text{ cm}$

The final image is 60cm to the right of the third lens.

38-19

a) $\frac{1}{f} = +2\text{m}^{-1} \Rightarrow f = +0.5\text{m} = +50\text{cm}$ (converging lens)

The purpose of the corrective lens is to take an object at 25cm from the eye and form a virtual image at the eye's near point. (The near point is the closest point on which the eye can focus.)

$s = 25\text{cm}$ $\quad \frac{1}{s} + \frac{1}{s'} = \frac{1}{f} \Rightarrow s' = \frac{sf}{s-f} = \frac{(25\text{cm})(50\text{cm})}{25\text{cm} - 50\text{cm}} = -50\text{cm}$

$f = 50\text{cm}$

The near point is 50 cm from this eye.

b) $\frac{1}{f} = -0.5\text{ m}^{-1} \Rightarrow f = -\frac{1}{0.5}\text{m} = -2.0\text{m} = -200\text{cm}$ (diverging lens)

The purpose of this corrective lens is to take an object at infinity and form a virtual image of it at the eye's far point. (The far point is the fartherest point on which the eye can focus.)

$s = \infty$ $\quad \frac{1}{s} + \frac{1}{s'} = \frac{1}{f}$

$f = -2.0\text{m}$ $\quad \cancel{\frac{1}{\infty}}^{\,0} + \frac{1}{s'} = -\frac{1}{2.0\text{m}} \Rightarrow s' = -2.0\text{m}$

The far point is 2.0m from this eye.

38-23

a)

image object eye
25 cm
s

$s' = -25\text{cm}$; $f = +10\text{cm}$

$\frac{1}{s} + \frac{1}{s'} = \frac{1}{f}$

$\frac{1}{s} = \frac{1}{f} - \frac{1}{s'} = \frac{s'-f}{fs'} \Rightarrow s = \frac{fs'}{s'-f} = \frac{(10\text{cm})(-25\text{cm})}{-25\text{cm} - 10\text{cm}} = 7.14\text{cm}$

b) $m = -\frac{s'}{s} = -\frac{-25\text{cm}}{7.14\text{cm}} = +3.50$

38-23 (cont)

$$m = \frac{y'}{y} \Rightarrow y' = my = (3.50)(1\,mm) = \underline{3.5\,mm} \text{ (image height)}$$

38-27

a) "f/2.8" ⇒ diameter of lens is $d = \frac{f}{2.8} = \frac{8\,cm}{2.8} = \underline{2.86\,cm}$

b) "f/5.6" ⇒ $d = \frac{8\,cm}{5.6} = 1.43\,cm$

d is smaller by a factor of 2 ⇒ the area of the lens opening is smaller by a factor of $2^2 = 4$ ⇒ need an exposure time 4 times longer.

$$\Rightarrow 4\left(\frac{1}{200}\,s\right) = \frac{1}{50}\,s$$

38-31

a)

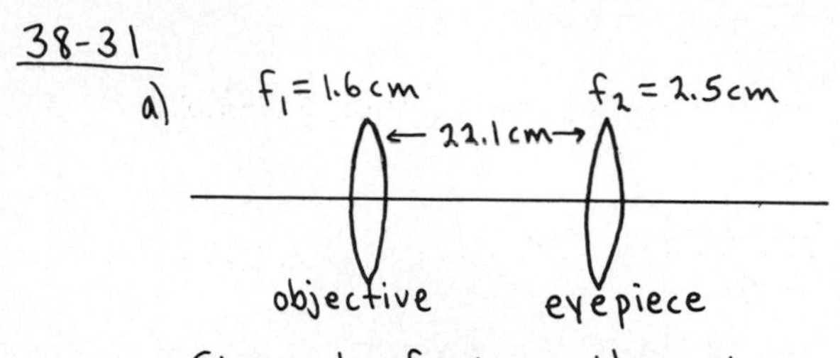

Start by finding the object distance for the eyepiece:

$\frac{1}{s} + \frac{1}{s'} = \frac{1}{f}$; $s' = \infty \Rightarrow s = f = 2.5\,cm$

The object for the eyepiece should be 2.5 cm to the left of the eyepiece.

But this object is the image of the objective. Thus the image formed by the objective should be 22.1 cm − 2.5 cm = 19.6 cm to the right of the objective.

$\Rightarrow s' = +19.6\,cm$ $\quad \frac{1}{s} + \frac{1}{s'} = \frac{1}{f}$

$f = 1.6\,cm$

$s = ?$ $\quad \frac{1}{s} = \frac{1}{f} - \frac{1}{s'} = \frac{s'-f}{fs'} \Rightarrow s = \frac{fs'}{s'-f} = \frac{(1.6\,cm)(19.6\,cm)}{19.6\,cm - 1.6\,cm} = \underline{1.74\,cm}$

b) for the objective

$$m_1 = -\frac{s'}{s} = -\frac{19.6\,cm}{1.74\,cm} = \underline{-11.3}$$

c) eq. (38-8) $M = m_1 M_2 = m_1 \frac{25\,cm}{f_2} = 11.3\left(\frac{25\,cm}{2.5\,cm}\right) = \underline{113}$

(We have disregarded the minus sign, as is customary.)

38-33

a) $M = -\frac{f_1}{f_2} = -\frac{100\,cm}{20\,cm} = \underline{-5}$ (using eq. 38-9)

b) $M = \frac{u'}{u} \Rightarrow u' = Mu$

u' = angular size of image $= \frac{y'}{f_2} \Rightarrow y' = f_2 M u$

38-33 (cont)

$u = $ angular size of object $= \dfrac{80\,\text{m}}{2\times10^3\,\text{m}} = 4.0\times10^{-2}$

$y' = f_2 M u = (20\,\text{cm})(-5)(4.0\times10^{-2}) = -4\,\text{cm}$

Height of image is 4 cm ($y' < 0$ since image is inverted).

38-37

For the primary mirror,

$$\frac{1}{s} + \frac{1}{s'} = \frac{1}{f}$$

$s = \infty \Rightarrow \frac{1}{s} = 0 \Rightarrow s' = f = 2.5\,\text{m}$ (image formed by primary is 2.5m to left of primary)

For the secondary mirror,

$s = 1.5\,\text{m} - 2.5\,\text{m} = -1.0\,\text{m}$ (virtual object)

$s' = 1.5\,\text{m} + 0.25\,\text{m} = 1.75\,\text{m}$, distance from the secondary mirror to the detector (where the final image is formed).

$$\frac{1}{f} = \frac{1}{s} + \frac{1}{s'} = \frac{s+s'}{ss'} \Rightarrow f = \frac{ss'}{s+s'} = \frac{(-1.0\,\text{m})(1.75\,\text{m})}{-1.0\,\text{m} + 1.75\,\text{m}} = -\frac{1.75\,\text{m}}{0.75} = -2.33\,\text{m}$$

$f < 0 \Rightarrow$ mirror is convex

$R = 2f = -4.66\,\text{m}$

Problems

38-39

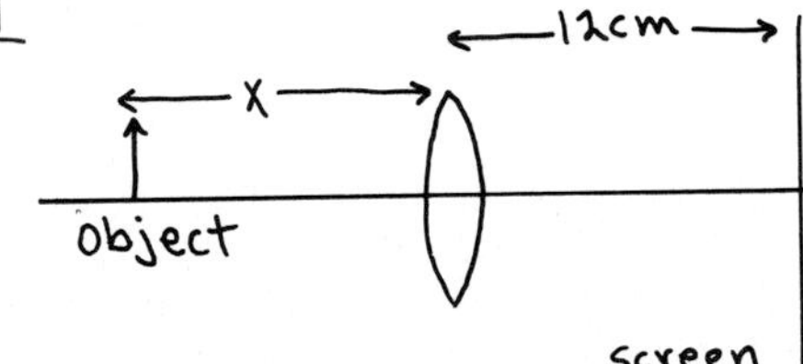

$s = x \Rightarrow s' = 12\,\text{cm}$

$$\frac{1}{s} + \frac{1}{s'} = \frac{1}{f} \Rightarrow \frac{1}{x} + \frac{1}{12\,\text{cm}} = \frac{1}{f}$$

$s = x + 2\,\text{cm} \Rightarrow s' = 12\,\text{cm} - 2\,\text{cm} - 2\,\text{cm} = 8\,\text{cm}$

(Lens is moved 2 cm to the right and screen is moved 2 cm to the left $\Rightarrow$ screen is then 8 cm from the lens.)

$$\frac{1}{s} + \frac{1}{s'} = \frac{1}{f} \Rightarrow \frac{1}{x+2\,\text{cm}} + \frac{1}{8\,\text{cm}} = \frac{1}{f}$$

Equate these two expressions for $\frac{1}{f} \Rightarrow \frac{1}{x} + \frac{1}{12\,\text{cm}} = \frac{1}{x+2\,\text{cm}} + \frac{1}{8\,\text{cm}}$

$$\frac{1}{x} - \frac{1}{x+2\,\text{cm}} = \frac{1}{8\,\text{cm}} - \frac{1}{12\,\text{cm}}$$

$$\frac{x + 2\,\text{cm} - x}{x(x+2\,\text{cm})} = \frac{12\,\text{cm} - 8\,\text{cm}}{(8\,\text{cm})(12\,\text{cm})} = \frac{4}{96\,\text{cm}} = \frac{1}{24\,\text{cm}}$$

$$\frac{2\,\text{cm}}{x(x+2\,\text{cm})} = \frac{1}{24\,\text{cm}} \Rightarrow 48\,\text{cm}^2 = x^2 + (2\,\text{cm})\,x$$

38-39 (cont)

$$x^2 + (2\text{cm})x - 48\text{cm}^2 = 0$$

$$(x+8\text{cm})(x-6\text{cm}) = 0 \; ; \; x \text{ positive} \Rightarrow x = 6\text{cm}$$

Then $\frac{1}{f} = \frac{1}{x} + \frac{1}{12\text{cm}} = \frac{1}{6\text{cm}} + \frac{1}{12\text{cm}} = \frac{3}{12\text{cm}} \Rightarrow f = \frac{12\text{cm}}{3} = \underline{4\text{cm}}$

38-43

a)

#1 — x — s — L−x — #2 — L

$L = 0.8\text{m}$

$|R| = 0.4\text{m}$

$\Rightarrow |f_1| = |f_2| = \frac{|R|}{2} = 0.2\text{m}$

Image formed by convex mirror (mirror #1):

convex $\Rightarrow f_1 = -0.2\text{m}$

$s_1 = L - x$

$$\frac{1}{s_1} + \frac{1}{s_1'} = \frac{1}{f_1} \Rightarrow \frac{1}{s_1'} = \frac{1}{f_1} - \frac{1}{s_1} = \frac{s_1 - f_1}{s_1 f_1} \Rightarrow s_1' = \frac{s_1 f_1}{s_1 - f_1} = \frac{(L-x)(-0.2\text{m})}{L - x + 0.2\text{m}}$$

$$L = 0.8\text{m} \Rightarrow s_1' = -(0.2\text{m})\left(\frac{0.8\text{m} - x}{1.0\text{m} - x}\right) < 0$$

$s_1' < 0 \Rightarrow$ image is $|s_1'| = 0.2\left(\frac{0.8\text{m}-x}{1.0\text{m}-x}\right)$ to left of mirror #1

$\Rightarrow |s_1'| + L$ to the left of mirror #2

Therefore $s_2 = |s_1'| + L = 0.8\text{m} + 0.2\text{m}\left(\frac{0.8\text{m}-x}{1.0\text{m}-x}\right) = \frac{0.8\text{m}^2 - (0.8\text{m})x + 0.16\text{m}^2 - (0.2\text{m})x}{1.0\text{m} - x}$

$$s_2 = \frac{0.96\text{m}^2 - (1.0\text{m})x}{1.0\text{m} - x}$$

Mirror #2 is concave $\Rightarrow f_2 = +0.2\text{m}$

final image is at the source $\Rightarrow s_2' = +x$

$$\frac{1}{s_2} + \frac{1}{s_2'} = \frac{1}{f_2} \Rightarrow \frac{1.0\text{m} - x}{0.96\text{m}^2 - (1.0\text{m})x} + \frac{1}{x} = \frac{1}{0.2\text{m}}$$

$$\frac{(1.0\text{m})x - x^2 + 0.96\text{m}^2 - (1.0\text{m})x}{x\left[0.96\text{m}^2 - (1.0\text{m})x\right]} = 5.0\text{m}^{-1}$$

$$-x^2 + 0.96\text{m}^2 = \left[(0.96\text{m}^2)x - (1.0\text{m})x^2\right]\left[5.0\text{m}^{-1}\right]$$

$$4x^2 - (4.8\text{m})x + 0.96\text{m}^2 = 0$$

quadratic formula

$$\Rightarrow x = \frac{1}{8}\left[4.8\text{m} \pm \sqrt{(-4.8\text{m})^2 - 4(4)(0.96\text{m}^2)}\right] = 0.6\text{m} \pm 0.346\text{m}$$

Thus $x = 0.946\text{m}$ (not possible, gives $x > L$) or $x = \underline{0.254\text{m}}$

38-43 (cont)

b) Repeat the above calculation, but now for the image first being formed by mirror #2 and then by mirror #1.

Image formed by the concave mirror (mirror #2):

$s_2 = X$

$f_2 = +0.2\,m$

$$\frac{1}{s_2} + \frac{1}{s_2'} = \frac{1}{f_2} \Rightarrow \frac{1}{s_2'} = \frac{1}{f_2} - \frac{1}{s_2} = \frac{s_2 - f_2}{s_2 f_2} \Rightarrow s_2' = \frac{s_2 f_2}{s_2 - f_2} = \frac{(0.2m)X}{X-(0.2m)}$$

$s_2' > 0 \Rightarrow$ image is formed a distance $|s_2'|$ to the left of the mirror.

$$\text{Thus } s_1 = 0.8m - \frac{(0.2m)X}{X-0.2m} = \frac{(0.8m)X - 0.16m^2 - (0.2m)X}{X-0.2m} = \frac{(0.6m)X - 0.16m^2}{X-0.2m}$$

$s_1' = L - X = 0.8m - X$

$f_1 = -0.2\,m$

$$\frac{1}{s_1} + \frac{1}{s_1'} = \frac{1}{f_1} \Rightarrow \frac{X-0.2m}{(0.6m)X - 0.16m^2} + \frac{1}{0.8m - X} = -\frac{1}{0.2m}$$

$$\frac{-[X-(0.2m)][X-(0.8m)] + (0.6m)X - 0.16m^2}{[(0.6m)X - 0.16m^2][0.8m - X]} \quad -5.0m^{-1}$$

$$-X^2 + (1.0m)X - 0.16m^2 + (0.6m)X - 0.16m^2 = [-0.128m^3 + (0.64m^2)X - (0.6m)X^2](-5.0m^{-1})$$

$$-X^2 + (1.6m)X - 0.32m^2 = +3X^2 - (3.2m)X + 0.64m^2$$

$$4X^2 - (4.8m)X + 0.96m^2 = 0$$

Same quadratic equation as in (a) $\Rightarrow X = \underline{0.254m}$

38-45

a) $f = 50\times10^{-3}m$

$$|m| = \left|\frac{y'}{y}\right| = \frac{\frac{2}{3}(30\times10^{-3}m)}{3m} = 6.67\times10^{-3}$$

$$|m| = \left|\frac{s'}{s}\right| \Rightarrow |s'| = |s||m|$$

Real object $\Rightarrow s > 0$

Image formed on film $\Rightarrow$ real image $\Rightarrow s' > 0$.

Thus $s' = |m|s = (6.67\times10^{-3})s$

The above gives one relation between s' and s. Another is given by

$$\frac{1}{s} + \frac{1}{s'} = \frac{1}{f} \Rightarrow \frac{1}{s} + \frac{1}{(6.67\times10^{-3})s} = \frac{1}{50\times10^{-3}m}$$

$$\frac{1}{s}(1+150) = 20m^{-1} \Rightarrow s = \frac{151}{20m^{-1}} = \underline{7.55m}$$

b) fill the viewfinder $\Rightarrow |y'| = 30\times10^{-3}m$

$$|m| = \frac{|y'|}{|y|} = \frac{30\times10^{-3}m}{3m} = 0.010$$

38-45 (cont)

$$|m| = \frac{s'}{s} \Rightarrow s' = |m|s = (0.010)\,s$$

Then

$$\frac{1}{s}+\frac{1}{s'}=\frac{1}{f} \Rightarrow \frac{1}{s} + \frac{1}{(0.010)s} = \frac{1}{50\times10^{-3}\text{m}}$$

$$\frac{1}{s}(1+100) = 20\text{m}^{-1} \Rightarrow s = \frac{101}{20\,\text{m}^{-1}} = \underline{5.05\,\text{m}}$$

38-47

Thin-walled glass ⇒ the glass has no effect on the light rays. The problem is that of refraction by a sphere of water surrounded by air.

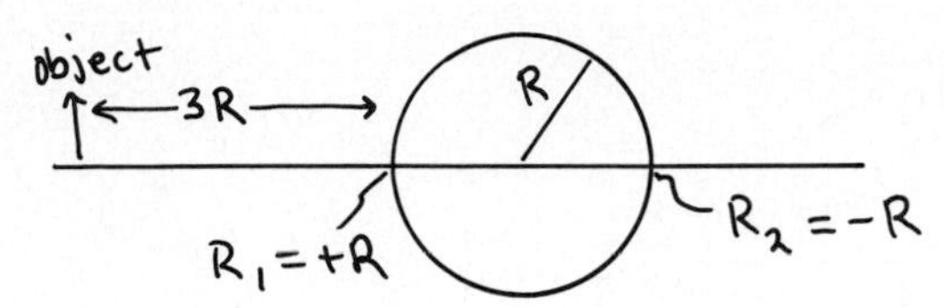

First refraction (air → water):

$n = 1.0$ (air)
$n' = 1.333$ (water)
$s_1 = 3R$
$R_1 = +R$

$$\frac{n}{s_1} + \frac{n'}{s_1'} = \frac{n'-n}{R_1}$$

$$\frac{1}{3R} + \frac{1.333}{s_1'} = \frac{1.333-1}{R}$$

$$\frac{1.333}{s_1'} = \frac{0.333}{R} - \frac{0.333}{R} = 0 \Rightarrow s_1' = \infty \text{ (parallel rays)}$$

Second refraction (water → air):

$n = 1.333$ (water)
$n' = 1.0$ (air)
$s_2 = -\infty$
$R_2 = -R$

$$\frac{n}{s_2} + \frac{n'}{s_2'} = \frac{n'-n}{R_2}$$

$$\frac{1.333}{-\infty} + \frac{1.0}{s_2'} = \frac{1.0-1.333}{-R}$$

$$\frac{1}{s_2'} = +\frac{1}{3R} \Rightarrow s_2' = +3R$$

The final image is 3R to the right of second surface ⇒ 4R from the center of the sphere, on the opposite side from the object.

38-49

air
$n=1.0$
glass
$n=1.5$
2cm
air
$n=1.0$
ray
8cm
object

First refraction (air → glass):

$n = 1.0$
$n' = 1.5$
$R_1 = \infty$ (plane surface)
$s_1 = 8\text{ cm}$

$$\frac{n}{s_1} + \frac{n'}{s_1'} = \frac{n'-n}{R_1} \to 0$$

$$s_1' = -\frac{n's_1}{n} = -\frac{(1.5)(8\text{cm})}{1.0} = -12\text{cm}$$

This image is 12cm below the lower surface ⇒ 14 cm below the upper surface.

38-49 (cont)

This image serves as the object for the second refraction (glass → air):

$n = 1.5$ $\qquad \dfrac{n}{s_2} + \dfrac{n'}{s_2'} = \dfrac{n' - n}{R_2} \rightarrow 0$

$n' = 1.0$

$R_2 = \infty$ $\qquad s_2' = -\dfrac{n' s_2}{n} = -\dfrac{(1.0)(14\text{cm})}{1.5} = -9.33\text{cm}$

$s_2 = 14\text{ cm}$

The final image is 9.33 cm below the top surface of the plate, so is 7.33 cm above the bottom surface and 8.0 cm − 7.33 cm = 0.67 cm above the page.

38-53

$n = 1.0$ (air) $\qquad n = 1.40$

$R = +0.70\text{cm}$

$s = 10\text{ cm}$; $s' = ?$

$$\frac{n}{s} + \frac{n'}{s'} = \frac{n' - n}{R}$$

$$\frac{1}{10\text{cm}} + \frac{1.4}{s'} = \frac{1.4 - 1.0}{0.70\text{cm}}$$

$$\frac{1.4}{s'} = 0.5714\text{cm}^{-1} - 0.10\text{cm}^{-1} = 0.4714\text{cm}^{-1}$$

$$s' = \frac{1.4}{0.4714\text{cm}^{-1}} = 2.97\text{cm}$$

The cornea-to-retina distance for this eye is therefore 2.97 cm. Exercise 38-20 says that this distance in a normal eye is 2.50 cm, so the nearsighted eye is elongated.

38-57

a)

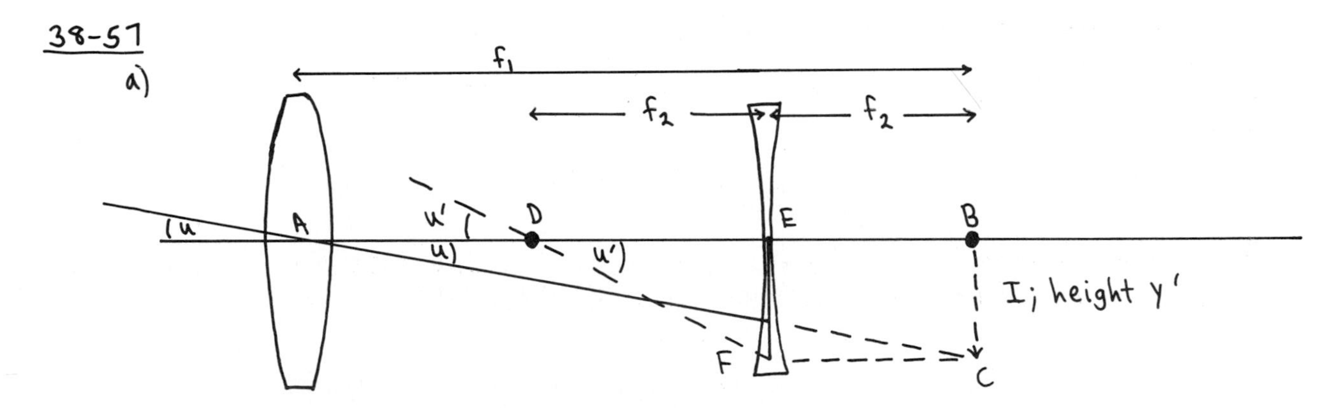

The angular size of the object is given by $\tan u = -\dfrac{y'}{f_1}$ (from triangle ABC).
u is small $\Rightarrow \tan u \approx u \Rightarrow u = -\dfrac{y'}{f_1}$.

The angular size of the image is given by $\tan u' = \dfrac{y'}{f_2}$. (From triangle DEF. Line EF has the same length as line BC.)
u' small $\Rightarrow \tan u' \approx u' \Rightarrow u' = \dfrac{y'}{f_2}$.

The angular magnification M is the ratio of the angular size of the image to the angular size of the object.

38-57 (cont)

$$\Rightarrow M = \frac{u'}{u} = \frac{y'/f_2}{-y'/f_1} = -\left(\frac{y'}{f_2}\right)\left(\frac{f_1}{y'}\right) = -\frac{f_1}{f_2}$$

Thus the angular magnification of the Galilean telescope is $M = -\frac{f_1}{f_2}$. Since f_2, the focal lens of the diverging lens, is negative, M is positive. This corresponds to an erect image.

b) In Exercise 38-33 the objective lens has focal length $f_1 = 100$ cm and the angular magnification is $M = -5 \Rightarrow |M| = 5$.

For a Galilean telescope $M = -\frac{f_1}{f_2}$ so $f_2 = -\frac{f_1}{M} = -\frac{100\text{cm}}{5} = \underline{-20\text{cm}}$

c) The telescope of Exercise 38-33:

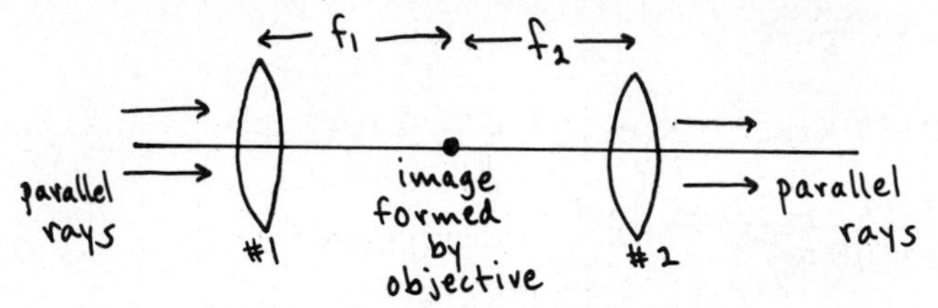

length $= f_1 + f_2 = 100\text{cm} + 20\text{cm} = \underline{120\text{cm}}$

Galilean telescope:

From Fig. (38-29), the length (separation between the lenses) is $|f_1| - |f_2|$.

Thus the length is $100\text{cm} - 20\text{cm} = \underline{80\text{cm}}$

The Galilean telescope is shorter.

CHAPTER 39

Exercises 1, 5, 7, 11, 13, 15, 19, 21

Problems 27, 29, 31, 37, 39

Exercises

39-1

The dark lines correspond to destructive interference and hence are located by

$$d \sin \theta_m = (m+\tfrac{1}{2})\lambda \Rightarrow \sin\theta_m = \frac{(m+\frac{1}{2})\lambda}{d}$$

1st dark line $\Rightarrow m=0$

2nd dark line $\Rightarrow m=1 \Rightarrow \sin\theta_1 = \frac{3\lambda}{2d} = \frac{3(600\times10^{-9}\,m)}{2(0.3\times10^{-3}\,m)} = 3.0\times10^{-3} \Rightarrow \theta_1 = 3.0\times10^{-3}\,rad$

3rd dark line $\Rightarrow m=2 \Rightarrow \sin\theta_2 = \frac{5\lambda}{2d} = \frac{5(600\times10^{-9}m)}{2(0.3\times10^{-3}\,m)} = 5.0\times10^{-3} \Rightarrow \theta_2 = 5.0\times10^{-3}\,rad$

Note that θ_1 and θ_2 are small, and that $\sin\theta_1 \simeq \theta_1$, $\sin\theta_2 \simeq \theta_2$.

The distances of these dark lines from the center of the central bright band are

$$y_1 = R\tan\theta_1 \simeq R\sin\theta_1 \simeq R\theta_1 = (0.50m)(3\times10^{-3}) = 1.5\times10^{-3}m$$

$$y_2 = R\tan\theta_2 \simeq R\sin\theta_2 \simeq R\theta_2 = (0.50m)(5\times10^{-3}) = 2.5\times10^{-3}\,m$$

The distance between these lines is therefore $\Delta y = y_2 - y_1 = 2.5\times10^{-3}m - 1.5\times10^{-3}m$

$$\Delta y = \underline{1.0\times10^{-3}\,m}$$

39-5

eq.(39-15) $I = I_0\cos^2\left(\frac{\pi d}{\lambda}\sin\theta\right)$

$$I = \tfrac{1}{2}I_0 \Rightarrow \tfrac{1}{2}I_0 = I_0\cos^2\left(\frac{\pi d}{\lambda}\sin\theta\right)$$

$$\cos\left(\frac{\pi d}{\lambda}\sin\theta\right) = \pm\frac{1}{\sqrt{2}} \Rightarrow \frac{\pi d}{\lambda}\sin\theta = \pm\frac{\pi}{4}, \pm\frac{3\pi}{4}, \pm\frac{5\pi}{4}, \pm\frac{7\pi}{4}, \ldots$$

θ small so $\sin\theta \simeq \theta \Rightarrow \frac{\pi d}{\lambda}\theta = \pm\frac{\pi}{4}, \pm\frac{3\pi}{4}, \pm\frac{5\pi}{4}, \pm\frac{7\pi}{4}, \ldots$

$$\Rightarrow \theta = \pm\frac{\lambda}{4d}, \pm\frac{3\lambda}{4d}, \pm\frac{5\lambda}{4d}, \pm\frac{7\lambda}{4d}, \ldots$$

The maxima are given by $\frac{\pi d}{\lambda}\sin\theta_m = 0, \pm\pi, \pm2\pi, \pm3\pi, \ldots$

$$\frac{\pi d}{\lambda}\theta_m \simeq 0, \pm\pi, \pm2\pi, \pm3\pi, \ldots \Rightarrow \theta_m = 0, \pm\frac{\lambda}{d}, \pm2\frac{\lambda}{d}, \pm\frac{3\lambda}{d}, \ldots$$

The angular separation between the θ_m^+ and θ_m^- points where $I = \frac{1}{2}I_0$ is therefore $\Delta\theta_m = \frac{2\lambda}{4d} = \frac{\lambda}{2d}$, independent of m.

39-5 (cont)

For example, for the m=1 maximum, $\theta_1 = \frac{\lambda}{d}$.

$\theta_1^+ = \frac{5\lambda}{4d}$ and $\theta_1^- = \frac{3\lambda}{4d}$ (θ_1 lies between θ_1^- and θ_1^+)

$\Delta\theta_1 = \theta_1^+ - \theta_1^- = \frac{5\lambda}{4d} - \frac{3\lambda}{4d} = \frac{2\lambda}{4d} = \frac{\lambda}{2d}$.

For the m=2 maximum, $\theta_2 = \frac{2\lambda}{d}$, $\theta_2^+ = \frac{9\lambda}{4d}$, $\theta_2^- = \frac{7\lambda}{4d} \Rightarrow \Delta\theta_2 = \theta_2^+ - \theta_1^- = \frac{\lambda}{2d}$.
etc.

39-7

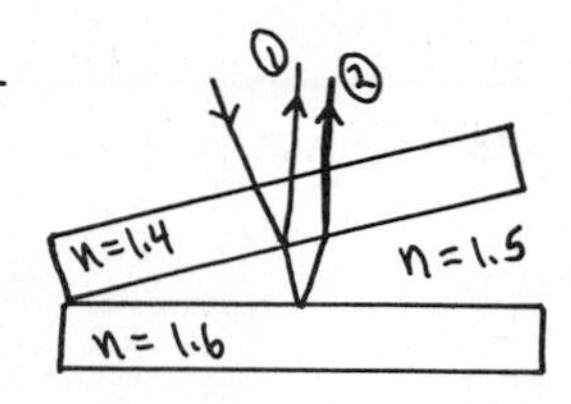

When ray ① reflects off the top of the wedge of silicone grease it undergoes a 180° phase change ($n=1.4 < n=1.5$).

When ray ② reflects off the top of the lower plate it undergoes a 180° phase change ($n=1.5 < n=1.6$).

This means then that the condition for an interference minimum is $2d = (m+\frac{1}{2})\lambda_n$, where λ_n is the wavelength in the silicone grease,

$$\lambda_n = \frac{\lambda}{n} = \frac{500\times10^{-9}\,m}{1.5} = 333\times10^{-9}\,m.$$

As in Example 39-4 (Fig. 39-7), $\frac{d}{x} = \frac{h}{\ell} \Rightarrow d = \frac{hx}{\ell}$.

$$\frac{2hx}{\ell} = (m+\tfrac{1}{2})\lambda_n$$

$$x = (m+\tfrac{1}{2})\frac{\ell\lambda_n}{2h} = (m+\tfrac{1}{2})\frac{(0.1m)(333\times10^{-9}m)}{2(0.02\times10^{-3}m)} = (m+\tfrac{1}{2})(8.33\times10^{-4}m)$$

$x_m = (m+\frac{1}{2})(8.33\times10^{-4}m)$

$x_{m+1} = (m+\frac{3}{2})(8.33\times10^{-4}m)$

$\Rightarrow \Delta x = x_{m+1} - x_m = 8.33\times10^{-4}m = \underline{0.833mm}$ is the spacing between adjacent fringes.

39-11

a) n=1.0 ① ② n=1.4]d n=1.5

Both rays ① and ② undergo a 180° phase change on reflection. Thus the condition for destructive interference is $2d = (m+\frac{1}{2})\lambda_n$, where λ_n is the wavelength in the coating.

Thus $d = \frac{(m+\frac{1}{2})\lambda_n}{2}$.

Thinnest coating $\Rightarrow m=0 \Rightarrow d = \frac{\lambda_n}{4} = \frac{(400nm/1.4)}{4} = 71.4\,nm = \underline{7.14\times10^{-8}m}$

b) See what other visible wavelengths have destructive interference:

$\lambda_n = \frac{2d}{m+\frac{1}{2}} \Rightarrow \lambda = \frac{2dn}{m+\frac{1}{2}}$ (wavelength in air)

$m=0 \Rightarrow \lambda = 4dn = 400nm$ (violet)

$m=1 \Rightarrow \lambda = 4dn/3 = \frac{400nm}{3} = 133\,nm$ (not visible), etc.

39-11 (cont)

Thus violet is the only color removed from the incident light.

See if any visible wavelengths have constructive interference:

$$2d = m\lambda_n = \frac{m\lambda}{n} \Rightarrow \lambda = \frac{2dn}{m}, \quad m = 1, 2, \ldots$$

$m=1 \Rightarrow \lambda = 2dn = \frac{1}{2}(4dn) = \frac{1}{2}(400\text{nm}) = 200\text{nm}$; not visible

$m=2 \Rightarrow \lambda = dn = \frac{1}{4}(4dn) = 100\text{nm}$; not visible

etc.

The effect of the coating is to remove the violet end of the white light color spectrum, so the residual color of the reflected light is reddish.

39-13

For the light at F:

One ray (call it ray 1) has traveled distance AB+BC+CD and has undergone 3 reflections (at A, B, and C).

The other ray (call it ray 2) has traveled distance AD and has undergone 1 reflection (at D).

The net result of the reflections is a 180° phase difference between the two rays. Thus destructive interference at F means that the path difference equals an integer number of wavelengths.

For the light at E:

The two rays (1 and 2) have the same path difference as above, so the path difference equals an integer number of wavelengths and introduces no phase difference. Ray 1, that travels from A to B to C to D to E, undergoes 4 reflections. (At A, B, C, and D.) Ray 2, that travels from A to D to E, undergoes no reflections. The net effect of the reflections is no phase difference. Thus the two rays arrive at E in phase ⇒ constructive interference.

39-15

θ, y_1, R

$$\sin\theta = \frac{n\lambda}{a}$$

First minimum ⇒ $n=1 \Rightarrow \sin\theta = \frac{\lambda}{a}$

$y_1 = R\tan\theta \simeq R\sin\theta$ (if θ is small)

Thus $y_1 = \frac{\lambda R}{a}$.

$$\lambda = \frac{a y_1}{R} = \frac{(2\times10^{-3}\text{m})(0.8\times10^{-3}\text{m})}{3.0\text{m}} = 5.33\times10^{-7}\text{m} = 533\text{nm}$$

39-19

4000 lines·cm^{-1} ⇒ the slit spacing is $d = \frac{1\times10^{-2}\,m}{4000} = 2.5\times10^{-6}\,m$

The line positions are given by $\sin\theta = m\frac{\lambda}{d}$.
second-order ⇒ $m=2$

$$\sin\theta_\alpha = 2\left(\frac{656\times10^{-9}\,m}{2.5\times10^{-6}\,m}\right) = 0.525 \Rightarrow \theta_\alpha = 31.7°$$

$$\sin\theta_\delta = 2\left(\frac{410\times10^{-9}\,m}{2.5\times10^{-6}\,m}\right) = 0.328 \Rightarrow \theta_\delta = 19.1°$$

The angular separation is $\theta_\alpha - \theta_\delta = 31.7° - 19.1° = \underline{12.6°}$

39-21

Rayleigh's criterion says that the two objects are resolved if the center of one diffraction pattern coincides with the first minimum of the other.

By eq. (39-20) the angular position of the first minimum relative to the center of the central maximum is $\sin\theta = \frac{\lambda}{a}$, where a is the slit width. Hence if the objects are resolved according to Rayleigh's criteria the angular separation between the centers of the images of the two objects must be at least $\frac{\lambda}{a}$.

But as discussed in Example 39-7, the angular separation of the image points equals the angular separation of the object points.
Thus $\frac{y}{s} = \frac{\lambda}{a}$, where $y = 1\,m$ is the separation of the two points and s is their distance from the observer.

$$\Rightarrow s = \frac{ya}{\lambda} = \frac{(1\,m)(1\times10^{-3}\,m)}{500\times10^{-9}\,m} = \underline{2000\,m}$$

Problems

39-27

The only effect of the water is to change the wavelength to
$$\lambda = \frac{600\,nm}{1.333} = 450\,nm$$

$$\sin\theta_m = \frac{(m+\frac{1}{2})\lambda}{d} \Rightarrow y_m = R\tan\theta_m \simeq R\sin\theta_m = \frac{(m+\frac{1}{2})\lambda R}{d}$$

$$m=1 \Rightarrow y_1 = \frac{3\lambda R}{2d} = \frac{3(450\times10^{-9}\,m)(0.5\,m)}{2(0.3\times10^{-3}\,m)} = 1.125\times10^{-3}\,m$$

$$m=2 \Rightarrow y_2 = \frac{5\lambda R}{2d} = \frac{5}{3}\left(\frac{3\lambda R}{2d}\right) = \frac{5}{3}(1.125\times10^{-3}\,m) = 1.875\times10^{-3}\,m$$

The separation between the two lines on the screen is
$$y_2 - y_1 = 1.875\times10^{-3}\,m - 1.125\times10^{-3}\,m = 7.50\times10^{-4}\,m = \underline{0.750\,mm}$$
(Equal to the answer in Exercise 39-1 divided by $n_{water} = 1.333$.)

39-29

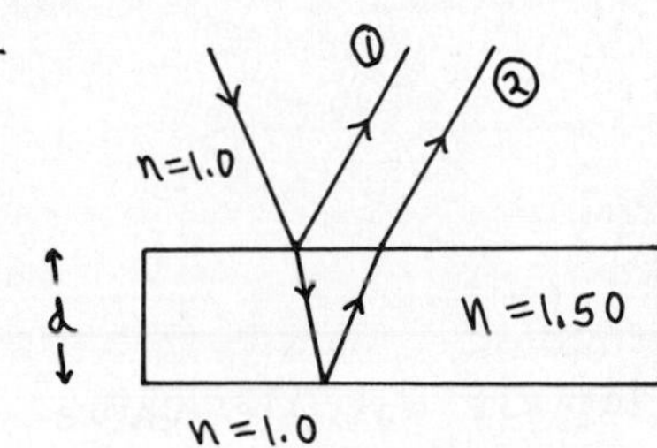

Ray ① undergoes a 180° phase change on reflection at the top surface of the glass.

Ray ② has no phase change on reflection from the lower surface of the glass.

The condition for constructive interference is thus

$$2d = (m+\tfrac{1}{2})\lambda_n, \quad m = 0, 1, 2, \ldots$$

$$2d = (m+\tfrac{1}{2})\frac{\lambda}{n} \Rightarrow \lambda = \frac{2dn}{m+\frac{1}{2}} = \frac{2(0.40\times10^{-6}\,\text{m})(1.50)}{m+\frac{1}{2}} = \frac{1200\,\text{nm}}{m+\frac{1}{2}}$$

m	λ
0	2400 nm
1	800 nm
2	480 nm
3	343 nm
⋮	⋮

Only $\lambda = 480\,\text{nm}$ is in the limits of the visible spectrum.

39-31

This problem deals with Newton's rings (section 39-4). The interference is between rays reflecting from the top and bottom edges of the air between the lens and plate.

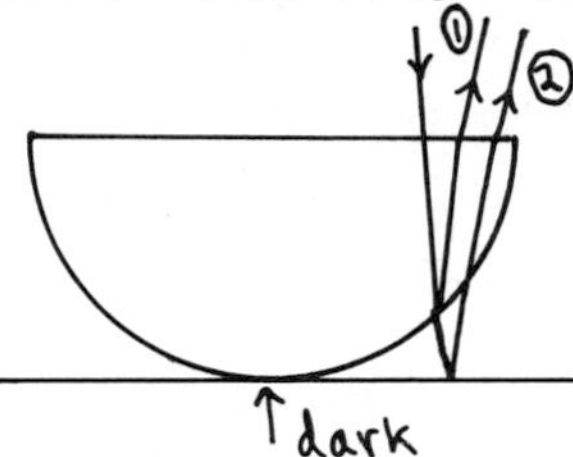

Ray ① does not undergo any phase change on reflection.

Ray ② does undergo a 180° phase change on reflection.

Hence the path difference $2d$, where d is the thickness of the air wedge, must satisfy $2d = (m+\tfrac{1}{2})\lambda$, $m = 0, 1, 2, \ldots$ for constructive interference.

Third bright ring $\Rightarrow m = 2$ and

$$d = \frac{5\lambda}{4} = \frac{5(650\times10^{-9}\,\text{m})}{4} = 8.125\times10^{-7}\,\text{m}$$

Now must relate this to the diameter of the ring:

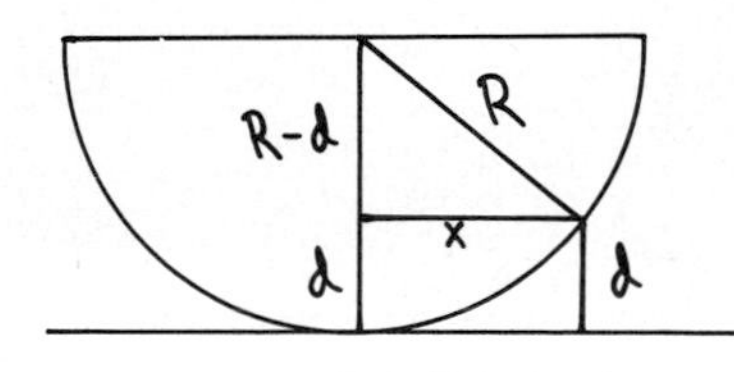

The radius of the ring is x.

$$x^2 + (R-d)^2 = R^2$$

$$\Rightarrow x = \sqrt{R^2-(R-d)^2} = \sqrt{R^2 - R^2 + 2Rd - d^2}$$

$$x = \sqrt{2Rd - d^2}$$

But $R = 1.20\,\text{m} >> d$, so can neglect d^2 relative to $2Rd$

$$\Rightarrow x = \sqrt{2Rd} = \sqrt{2(1.2\,\text{m})(8.125\times10^{-7}\,\text{m})} = 1.40\times10^{-3}\,\text{m}.$$

This is the radius. The diameter of the ring is $2x = 2.80\times10^{-3}\,\text{m} = \underline{2.80\,\text{mm}}$

39-37

$$d \sin\theta = m\lambda$$

fourth order $\Rightarrow m = 4$

$$d = \frac{1.0 \times 10^{-2}\,m}{5000} = 2.0 \times 10^{-6}\,m \qquad \Rightarrow \lambda = \frac{d \sin\theta}{4}$$

The longest wavelength corresponds to the largest possible value of $\sin\theta$, which is unity

$$\Rightarrow \lambda_{max} = \frac{d}{4} = \frac{2.0 \times 10^{-6}\,m}{4} = 5.0 \times 10^{-7}\,m = \underline{500\,nm}$$

39-39

Rayleigh's criterion $\Rightarrow \sin\theta = 1.22\frac{\lambda}{d}$

$\lambda \ll d \Rightarrow \sin\theta$ is small, so $\sin\theta \simeq \theta$ and $\theta = 1.22\frac{\lambda}{d}$

But, as discussed in Example 39-4, this θ is both the angular separation between the centers of the images and also the angular separation between the objects. The latter is y/s, where y is the distance between the objects and s is their distance from the observer.

Thus $\frac{y}{s} = 1.22\frac{\lambda}{d}$

$$s = \frac{yd}{1.22\lambda} = \frac{(20m)(4.0 \times 10^{-3}\,m)}{1.22(550 \times 10^{-9}\,m)} = 1.19 \times 10^{5}\,m = \underline{119\,km}$$

CHAPTER 40

Exercises 1, 3, 5, 7, 13, 15, 21, 25

Problems 27, 29, 33, 35, 37, 41

Exercises

40-1

Simultaneous to observer on train ⇒ light pulses from A′ and B′ arrive at O′ at the same time. In frame O light from A′ has a shorter distance to travel than light from B′, so O will conclude that the pulse at A (A′) started before the pulse at B (B′). To him bolt A appears to strike first.

40-3

a) $\Delta t = 1.6\times10^{-5}\ \text{s}$, $\Delta t' = 2.3\times10^{-6}\ \text{s}$

$$\Delta t = \frac{\Delta t'}{\sqrt{1-u^2/c^2}} \Rightarrow \sqrt{1-\frac{u^2}{c^2}} = \frac{\Delta t'}{\Delta t} = \frac{2.3\times10^{-6}\ \text{s}}{1.6\times10^{-5}\ \text{s}} = 0.144$$

$$1-u^2/c^2 = (0.144)^2 = 0.0207 \Rightarrow u = \sqrt{1.0-0.0207}\ c = \underline{0.990c}$$

b) "In the laboratory" ⇒ use Δt, the time measured in the laboratory

$$\Delta x = v\Delta t = (0.990c)(1.6\times10^{-5}\ \text{s}) = (0.990)(3.00\times10^{8}\ \text{m}\cdot\text{s}^{-1})(1.6\times10^{-5}\ \text{s}) = 4.75\times10^{3}\ \text{m} = \underline{4.75\ \text{km}}$$

40-5

$$l = l'\sqrt{1-\frac{u^2}{c^2}}$$

l = length measured by a stationary observer when the spacecraft is moving ⇒ $l = 200\ \text{m}$.

l' = length measured by a stationary observer when the spacecraft is at rest

$$l' = \frac{l}{\sqrt{1-u^2/c^2}} = \frac{200\ \text{m}}{\sqrt{1-\left(\frac{0.8c}{c}\right)^2}} = \frac{200\ \text{m}}{0.60} = \underline{333\ \text{m}}$$

40-7

eq. (40-15) $$x' = \frac{x-ut}{\sqrt{1-u^2/c^2}}$$

eq. (40-16) $$x' = -ut' + x\sqrt{1-u^2/c^2}$$

40-7 (cont)

Equate these two expressions for x' $\Rightarrow$ $\dfrac{x-ut}{\sqrt{1-u^2/c^2}} = -ut' + x\sqrt{1-u^2/c^2}$

Solve for t': $ut' = x\sqrt{1-u^2/c^2} - \dfrac{x-ut}{\sqrt{1-u^2/c^2}}$

$$ut' = \frac{x(1-u^2/c^2) - x + ut}{\sqrt{1-u^2/c^2}} = \frac{\cancel{x} - xu^2/c^2 - \cancel{x} + ut}{\sqrt{1-u^2/c^2}} = \frac{-u(xu/c^2 - t)}{\sqrt{1-u^2/c^2}}$$

$\Rightarrow t' = \dfrac{t - ux/c^2}{\sqrt{1-u^2/c^2}}$, which is eq. (40-17).

40-13

$$F = \frac{dp}{dt} = \frac{d}{dt}\left(\frac{mv}{\sqrt{1-v^2/c^2}}\right) \quad \text{(eq. 40-24)}$$

$$F = \frac{m}{\sqrt{1-v^2/c^2}}\frac{dv}{dt} + \frac{mv}{(1-v^2/c^2)^{3/2}}\left(-\frac{1}{2}\right)\left(-\frac{2v}{c^2}\right)\frac{dv}{dt}$$

$$F = \frac{dv}{dt}\left[\frac{m}{\sqrt{1-v^2/c^2}} + \frac{mv^2/c^2}{(1-v^2/c^2)^{3/2}}\right] = \frac{dv}{dt}\frac{m}{(1-v^2/c^2)^{3/2}}\left[1 - \cancel{\frac{v^2}{c^2}} + \cancel{\frac{v^2}{c^2}}\right]$$

$$F = \frac{dv}{dt}\frac{m}{(1-v^2/c^2)^{3/2}}$$

But $a = \dfrac{dv}{dt}$ $\Rightarrow$ $a = \dfrac{F}{m}(1-v^2/c^2)^{3/2}$, which is eq. (40-25).

40-15

a) $K = \dfrac{mc^2}{\sqrt{1-v^2/c^2}} - mc^2$ (eq. 40-31)

rest energy is mc^2

Thus $K = mc^2 \Rightarrow mc^2 = mc^2\left(\dfrac{1}{\sqrt{1-v^2/c^2}} - 1\right)$

$\Rightarrow \dfrac{1}{\sqrt{1-v^2/c^2}} = 2 \Rightarrow 1 - \dfrac{v^2}{c^2} = 0.25 \Rightarrow v = \sqrt{1-0.25}\; c = \underline{0.866c}$

b) $K = 10mc^2 \Rightarrow 10mc^2 = mc^2\left(\dfrac{1}{\sqrt{1-v^2/c^2}} - 1\right)$

$\Rightarrow \dfrac{1}{\sqrt{1-v^2/c^2}} = 11 \Rightarrow 1 - \dfrac{v^2}{c^2} = 0.00826 \Rightarrow v = \sqrt{1-0.00826}\; c = \underline{0.9959c}$

40-21

eq. (40-34) $E^2 = (mc^2)^2 + (pc)^2$

$$E = \sqrt{(mc^2)^2 + (pc)^2} = mc^2\sqrt{1 + \left(\frac{pc}{mc^2}\right)^2} = mc^2\sqrt{1 + \left(\frac{p}{mc}\right)^2}$$

40-21 (cont)

$pc \ll mc^2 \Rightarrow p \ll mc$, so can make a binomial expansion (Appendix B) of the square root and retain only the first two terms.

$$\sqrt{1+\left(\frac{p}{mc}\right)^2} = 1 + \frac{1}{2}\left(\frac{p}{mc}\right)^2 + \ldots$$

$$\Rightarrow E \simeq mc^2\left(1+\frac{1}{2}\left(\frac{p}{mc}\right)^2\right) = mc^2 + \frac{p^2}{2m}$$

$$p = \frac{mv}{\sqrt{1-v^2/c^2}} = mv\left(1-v^2/c^2\right)^{-1/2}$$

$p \ll mc \Rightarrow v \ll c$, so can expand $(1-v^2/c^2)^{-1/2}$ as well

$$(1-v^2/c^2)^{-1/2} = 1 + \frac{1}{2}\frac{v^2}{c^2} + \ldots$$

$p = mv\left(1+\frac{1}{2}\frac{v^2}{c^2}\right) = mv + \frac{1}{2}mv\frac{v^2}{c^2}$ → neglect $= mv$, the classical expression

Then $E = mc^2 + \frac{(mv)^2}{2m} = mc^2 + \frac{1}{2}mv^2$, as desired.

40-25

Approaching $\Rightarrow f = \sqrt{\frac{c+u}{c-u}}\, f'$ (eq. 40-41, but with the sign of u changed, as discussed in the text below the equation)

Convert into an equation relating λ and λ':

$$\lambda = \frac{c}{f} \Rightarrow f = \frac{c}{\lambda} \Rightarrow \frac{c}{\lambda} = \sqrt{\frac{c+u}{c-u}}\,\frac{c}{\lambda'} \Rightarrow \lambda = \sqrt{\frac{c-u}{c+u}}\,\lambda'$$

$\lambda' = 675\text{ nm}$, $\lambda = 525\text{ nm} \Rightarrow 525\text{ nm} = \sqrt{\frac{c-u}{c+u}}\,(675\text{ nm})$

$$0.6049(c+u) = c-u$$

$$1.6049u = 0.3951c \Rightarrow u = 0.246c = \underline{7.38\times10^7\ \text{m}\cdot\text{s}^{-1}}$$

Problems

40-27

a) $\Delta t' = 2.6\times10^{-8}\text{ s}$, the time measured in the rest frame of the π^+.

We can use $l = l'\sqrt{1-v^2/c^2}$ to relate the length l' of the tube in the lab frame to the length l of the tube in the rest frame of the π^+:

$$l = (3.0\times10^3\text{ m})\sqrt{1-v^2/c^2}.$$

Lives until it reaches the end $\Rightarrow l = v\Delta t' = (2.6\times10^{-8}\text{ s})\,v$

Equate these two expressions for $l \Rightarrow (3.0\times10^3\text{ m})\sqrt{1-v^2/c^2} = (2.6\times10^{-8}\text{ s})\,v$.

Write v as $(1-\Delta)c$, so that can solve for Δ rather than for v itself:

40-27 (cont) $\sqrt{1-\frac{v^2}{c^2}} = \sqrt{1-(1-2\Delta+\Delta^2)} = \sqrt{2\Delta}$ (Δ^2: neglect)

and

$$(3\times10^3\text{m})\sqrt{2\Delta} = (2.6\times10^{-8}\text{s})(1-\Delta)(3.00\times10^8\,\text{m}\cdot\text{s}^{-1})$$

$$(3\times10^3\text{m})\sqrt{2\Delta} = 7.8\text{m}(1-\Delta)$$

Square both sides of the equation $\Rightarrow (1.8\times10^7\text{m}^2)\Delta = (60.8\text{m}^2)(1-\Delta)^2$

But $60.8 \ll 1.8\times10^7$, so let $(1-\Delta)^2 \simeq 1$ (Δ will turn out to be $\ll 1$)

$\Rightarrow (1.8\times10^7\text{m}^2)\Delta = 60.8\text{m}^2 \Rightarrow \underline{\Delta = 3.38\times10^{-6}}$

b) $E = \frac{mc^2}{\sqrt{1-v^2/c^2}}$; $v=(1-\Delta)c \Rightarrow E = \frac{mc^2}{\sqrt{1-(1-\Delta)^2}} = \frac{mc^2}{\sqrt{1-1+2\Delta-\Delta^2}}$ (Δ^2: neglect)

$\Rightarrow E = \frac{mc^2}{\sqrt{2\Delta}}$

rest mass $mc^2 = 139.6\text{ MeV} \Rightarrow E = \frac{139.6\text{ MeV}}{\sqrt{2(3.38\times10^{-6})}} = \underline{5.37\times10^4\text{ MeV}}$

(Note: The velocity is very close to the velocity of light. In part (a), $l = (3\times10^3\text{m})\sqrt{2\Delta} = 7.8\text{m}$, so the particle makes it to the end of the tube before decaying because of the extreme relativistic length contraction of the tube in the particle's frame. Also, the particle's energy is much larger than its rest energy.)

40-29

a)

m $v=0$ — initial / $v \leftarrow$ m (Co nucleus) — final; photon $\rightsquigarrow E$

Apply conservation of momentum:

For the recoiling nucleus, $p = mv$ (to the left). (We will use the nonrelativistic expression. If we calculate that $v \ll c$ then this is ok, otherwise we will have to start over, using the relativistic expression for p.)

For the emitted photon, $p = \frac{E}{c}$ (eq. 40-34), to the right.

The initial momentum is zero, so the final momentum of the system must be zero

$$\Rightarrow mv = \frac{E}{c}$$

40-29 (cont)

$$v = \frac{E}{mc} = \frac{(1.33\times10^{6}\,\text{eV})\left(\frac{1.60\times10^{-19}\,\text{J}}{1\,\text{eV}}\right)}{(27+33)(1.66\times10^{-27}\,\text{kg})(3.00\times10^{8}\,\text{m·s}^{-1})} = \underline{7.12\times10^{3}\,\text{m·s}^{-1}}$$

b) $\frac{v}{c} = \frac{7.12\times10^{3}\,\text{m·s}^{-1}}{3.00\times10^{8}\,\text{m·s}^{-1}} = 2.4\times10^{-5}$, so $v \ll c$ and the nonrelativistic expression for the momentum of the recoiling nucleus is very accurate.

40-33

a) $K = eV \Rightarrow \frac{K}{e}$ (kinetic energy in electron volts) $= V = 1.80\times10^{5}\,\text{eV} = \underline{0.180\text{ MeV}}$

(An electron accelerated through a potential of 1V gains kinetic energy of 1eV; section 26-7.)

b) $E = K + mc^2$

rest energy $mc^2 = (9.11\times10^{-31}\,\text{kg})(3.00\times10^{8}\,\text{m·s}^{-1})^2 = 8.20\times10^{-14}\,\text{J}\left(\frac{1\,\text{eV}}{1.60\times10^{-19}\,\text{J}}\right)$

$= 5.12\times10^{5}\,\text{eV} = 0.512\text{ MeV}$

$\Rightarrow E = 0.180\text{ MeV} + 0.512\text{ MeV} = \underline{0.692\text{ MeV}}$

c) Obtain v from E, using the relativistic expression:

$$E = \frac{mc^2}{\sqrt{1-v^2/c^2}} \Rightarrow \sqrt{1-v^2/c^2} = \frac{mc^2}{E} = \frac{0.512\text{ MeV}}{0.692\text{ MeV}} = 0.740$$

$$1-\frac{v^2}{c^2} = (0.740)^2 = 0.548 \Rightarrow v = \sqrt{1-0.548}\;c = 0.672c = \underline{2.02\times10^{8}\,\text{m·s}^{-1}}$$

d) Use the classical expression for K.

$$K = \tfrac{1}{2}mv^2 \Rightarrow v = \sqrt{\frac{2K}{m}}\;;\; K = 1.80\times10^{5}\,\text{eV}\left(\frac{1.60\times10^{-19}\,\text{J}}{1\,\text{eV}}\right) = 2.88\times10^{-14}\,\text{J}$$

$$v = \sqrt{\frac{2(2.88\times10^{-14}\,\text{J})}{9.11\times10^{-31}\,\text{kg}}} = \underline{2.51\times10^{8}\,\text{m·s}^{-1}} \text{ (about 25\% too large)}$$

40-35

In crown glass the velocity of light is $v = \frac{c}{n} = \frac{3.00\times10^{8}\,\text{m·s}^{-1}}{1.52} = 1.97\times10^{8}\,\text{m·s}^{-1}$!

Calculate the kinetic energy of an electron that has this velocity. (Note: This velocity is close to the velocity of light, so we must use the relativistic expression for kinetic energy.)

$$K = \frac{mc^2}{\sqrt{1-v^2/c^2}} - mc^2 = mc^2\left(\frac{1}{\sqrt{1-v^2/c^2}} - 1\right) = mc^2\left(\frac{1}{\sqrt{1-(c/nc)^2}} - 1\right)$$

$$K = mc^2\left(\frac{1}{\sqrt{1-\left(\frac{1}{1.52}\right)^2}} - 1\right) = 0.328\,mc^2.$$

For an electron $mc^2 = (9.11\times10^{-31}\,\text{kg})(3.00\times10^{8}\,\text{m·s}^{-1})^2 = 8.20\times10^{-14}\,\text{J}\left(\frac{1\,\text{eV}}{1.60\times10^{-19}\,\text{J}}\right)$

$mc^2 = 5.12\times10^{5}\,\text{eV}$

40-35 (cont)

Thus $K = (0.328)(5.12\times10^5 \text{ eV}) = 1.68\times10^5 \text{ eV} = \underline{168 \text{ keV}}$

40-37

Observed frequency: $f = \frac{c}{\lambda} = \frac{3.00\times10^8 \text{ m}\cdot\text{s}^{-1}}{430.4\times10^{-9}\text{ m}} = 6.97\times10^{14}\text{ Hz}$

Frequency in the atom's rest frame: $f' = \frac{c}{\lambda'} = \frac{3.00\times10^8 \text{ m}\cdot\text{s}^{-1}}{121.6\times10^{-9}\text{ m}} = 2.47\times10^{15}\text{ Hz}$

$f < f' \Rightarrow$ the emitting atoms are receding from the earth.

eq. (40-41) $f = \sqrt{\frac{c-u}{c+u}}\, f' \Rightarrow \sqrt{\frac{c-u}{c+u}} = \frac{f}{f'} = \frac{6.97\times10^{14}\text{ Hz}}{2.47\times10^{15}\text{ Hz}} = 0.282$

$\frac{c-u}{c+u} = (0.282)^2 = 0.0795 \Rightarrow c - u = 0.0795c + 0.0795u$

$1.0795u = 0.9205c \Rightarrow u = 0.853c = \underline{2.56\times10^8 \text{ m}\cdot\text{s}^{-1}}$

40-41

By eq. (40-25), $a = \frac{dv}{dt} = \frac{F}{m}(1 - v^2/c^2)^{3/2}$

$$\Rightarrow \frac{dv}{(1-v^2/c^2)^{3/2}} = \left(\frac{F}{m}\right)dt$$

$$\int_0^v \frac{dv}{(1-v^2/c^2)^{3/2}} = \int_0^t \left(\frac{F}{m}\right)dt = \left(\frac{F}{m}\right)t$$

Let $x = \frac{v}{c} \Rightarrow dv = c\,dx$ and $x = 0$ when $v = 0$, $x = \frac{v}{c}$ when $v = v$.

$$\Rightarrow c\int_0^{v/c} \frac{dx}{(1-x^2)^{3/2}} = \left(\frac{F}{m}\right)t$$

$$\int_0^{v/c} \frac{dx}{(1-x^2)^{3/2}} = \left.\frac{x}{\sqrt{1-x^2}}\right|_0^{v/c} = \frac{v/c}{\sqrt{1-v^2/c^2}} \Rightarrow \frac{v}{\sqrt{1-v^2/c^2}} = \left(\frac{F}{m}\right)t$$

Solve this equation to get $v(t)$, and then look at the $t \to \infty$ behavior:

$$\frac{v^2}{1-v^2/c^2} = \left(\frac{F}{m}\right)^2 t^2 \Rightarrow v^2 = \left(\frac{F}{m}\right)^2 t^2 - \left(\frac{F}{mc}\right)^2 t^2 v^2$$

$$v^2\left(\frac{m^2c^2 + F^2t^2}{m^2c^2}\right) = \frac{F^2t^2}{m^2} \Rightarrow v = \frac{Fct}{\sqrt{m^2c^2 + F^2t^2}} = c\left(\frac{1}{\sqrt{1 + \left(\frac{mc}{Ft}\right)^2}}\right)$$

$\sqrt{1 + \left(\frac{mc}{Ft}\right)^2}$ is always > 1, so $\frac{1}{\sqrt{1+\left(\frac{mc}{Ft}\right)^2}}$ is always < 1, so $v < c$ always.

(The larger t is the closer v gets to c, but $\left(\frac{mc}{Ft}\right)^2$ is never zero, so v never reaches c.)

CHAPTER 41

Exercises 1, 5, 9, 11, 15, 19, 23

Problems 25, 27, 29, 35, 37

Exercises

41-1

a) $E = hf \Rightarrow f = \frac{E}{h} = \frac{(1.0\times10^{6}\,eV)\left(\frac{1.60\times10^{-19}\,J}{1\,eV}\right)}{6.626\times10^{-34}\,J\cdot s^{-1}} = \underline{2.41\times10^{20}\,Hz}$

b) $\lambda = \frac{c}{f} = \frac{3.00\times10^{8}\,m\cdot s^{-1}}{2.41\times10^{20}\,Hz} = \underline{1.24\times10^{-12}\,m}$

c) λ is a factor of 1240 times larger.

41-5

a) $E = hf = (6.626\times10^{-34}\,J\cdot s^{-1})(100\times10^{6}\,Hz) = \underline{6.63\times10^{-26}\,J}$

$E = (6.63\times10^{-26}\,J)\left(\frac{1\,eV}{1.60\times10^{-19}\,J}\right) = \underline{4.14\times10^{-7}\,eV}$

b) The number of photons emitted per second must be the total energy emitted per second (the power output of $50\times10^{3}\,W$) divided by the energy (in Joules) of one photon.

$\Rightarrow \frac{50\times10^{3}\,J\cdot s^{-1}}{6.63\times10^{-26}\,J\cdot photon^{-1}} = \underline{7.54\times10^{29}\,photons\cdot s^{-1}}$

41-9

eq (41-4) $eV_0 = hf - \phi$

The threshold frequency f_0 is the value of f when $eV_0 \to 0$:

$0 = hf_0 - \phi \Rightarrow hf_0 = \phi$

Thus eq. (41-4) can be written as $eV_0 = hf - hf_0$.
Or, in terms of wavelength, $f = \frac{c}{\lambda} \Rightarrow eV_0 = hc\left(\frac{1}{\lambda} - \frac{1}{\lambda_0}\right)$

$\frac{1}{\lambda_0} = \frac{1}{\lambda} - \frac{eV_0}{hc} = \frac{1}{2.54\times10^{-7}\,m} - \frac{(1.60\times10^{-19}\,C)(0.59\,V)}{(6.626\times10^{-34}\,J\cdot s^{-1})(3.0\times10^{8}\,m\cdot s^{-1})}$

$\frac{1}{\lambda_0} = 3.94\times10^{6}\,m^{-1} - 4.75\times10^{5}\,m^{-1} = 3.47\times10^{6}\,m^{-1}$

$\lambda_0 = 2.89\times10^{-7}\,m = \underline{289\,nm}$

41-11

a) eq. (41-9) $f = Rc\left(\frac{1}{2^2} - \frac{1}{n^2}\right)$

$H_\beta \Rightarrow n=4$

Thus

$$f = (1.097\times10^7\,m^{-1})(3.00\times10^8\,m\cdot s^{-1})\left(\frac{1}{4} - \frac{1}{16}\right) = \underline{6.17\times10^{14}\,Hz}$$

b) $\lambda = \frac{c}{f} = \frac{3.00\times10^8\,m\cdot s^{-1}}{6.17\times10^{14}\,Hz} = 4.86\times10^{-7}\,m = \underline{486\,nm}$

41-15

From Fig. (41-9), $\Delta E_{5s\to3p} = 20.66\,eV - 18.70\,eV = \underline{1.96\,eV}$

or $\Delta E = (1.96\,eV)\left(\frac{1.60\times10^{-19}\,J}{1\,eV}\right) = \underline{3.14\times10^{-19}\,J}$

The emitted photon has energy E equal to the transition energy ΔE of the atom.

$$\Delta E = E = hf = \frac{hc}{\lambda} \Rightarrow \lambda = \frac{hc}{E} = \frac{(6.626\times10^{-34}\,J\cdot s)(3.00\times10^8\,m\cdot s^{-1})}{3.14\times10^{-19}\,J} = 6.33\times10^{-7}\,m = \underline{633\,nm}$$

This agrees with the observed wavelength of 632.8 nm.

41-19

a) eq. (41-17) $F(\lambda) = \frac{2\pi hc^2}{\lambda^5}\left(\frac{1}{e^{hc/kT\lambda}-1}\right)$

Maximum in $F(\lambda)$ at $\lambda = \lambda_m \Rightarrow \left.\frac{dF}{d\lambda}\right|_{\lambda=\lambda_m} = 0.$

$\frac{dF}{d\lambda} = 0 \Rightarrow \left.\frac{d}{d\lambda}\left(\frac{\lambda^{-5}}{e^{\alpha/\lambda}-1}\right)\right|_{\lambda=\lambda_m} = 0$, where we have defined $\alpha = \frac{hc}{kT}$

$$-5\frac{\lambda_m^{-6}}{e^{\alpha/\lambda_m}-1} + \lambda_m^{-5}\left(-\frac{\alpha}{\lambda_m^2}\right)e^{\alpha/\lambda_m}\left(\frac{-1}{[e^{\alpha/\lambda_m}-1]^2}\right) = 0$$

$$-\frac{5}{\lambda_m^6}\left(e^{\alpha/\lambda_m}-1\right) + \frac{\alpha}{\lambda_m^7}e^{\alpha/\lambda_m} = 0$$

Let $\frac{\alpha}{\lambda_m} = x$; $x = \frac{hc}{\lambda_m kT}$.

Thus $-\frac{5x^6}{\alpha^6}(e^x - 1) + \frac{x^7}{\alpha^6}e^x = 0 \Rightarrow -5x^6e^x + 5x^6 + x^7e^x = 0$

$$-5e^x + 5 + xe^x = 0 \Rightarrow -5 + 5e^{-x} + x = 0 \Rightarrow \boxed{5e^{-x} = 5 - x}$$

This is eq. (41-19), and, as the text says, it gives that $x = 4.965$.

$x = 4.965$ and $x = \frac{hc}{\lambda_m kT} \Rightarrow \lambda_m = \frac{hc}{4.965\,kT}$, as was to be shown.

b) $\lambda_m T = \frac{hc}{4.965\,k} = \frac{(6.626\times10^{-34}\,J\cdot s)(3.00\times10^8\,m\cdot s^{-1})}{4.965\,(1.381\times10^{-23}\,J\cdot K^{-1})} = \underline{2.90\times10^{-3}\,m\cdot K}$

41-23

eq. (41-24) $\lambda' - \lambda = \frac{h}{mc}(1-\cos\phi)$

$$\frac{\lambda'-\lambda}{\lambda} = 0.01 \Rightarrow 0.01\lambda = \frac{h}{mc}(1-\cos\phi)$$

$$\frac{mc\lambda}{h}(0.01) = 1-\cos\phi \Rightarrow \cos\phi = 1 - \frac{mc\lambda}{h}(0.01)$$

$$\cos\phi = 1 - \frac{(9.11\times10^{-31}\,kg)(3.00\times10^{8}\,m\cdot s^{-1})(0.5\times10^{-10}\,m)}{6.626\times10^{-34}\,J\cdot s}(0.01) = 0.794 \Rightarrow \phi = \underline{37.4^\circ}$$

Problems

41-25

a) One photon dissociates one molecule of AgBr, so need to find the energy to dissociate a single molecule.

$$E = (1.00\times10^{5}\,J\cdot mol^{-1})\left(\frac{1\,mol}{6.02\times10^{23}\,molecules}\right) = 1.66\times10^{-19}\,J\left(\frac{1\,eV}{1.60\times10^{-19}\,J}\right) = \underline{1.04\,eV}$$

b) $E = hf = \frac{hc}{\lambda} \Rightarrow \lambda = \frac{hc}{E} = \frac{(6.626\times10^{-34}\,J\cdot s)(3.00\times10^{8}\,m\cdot s^{-1})}{1.66\times10^{-19}\,J} = 1.20\times10^{-6}\,m = \underline{1200\,nm}$

c) $f = \frac{c}{\lambda} = \frac{E}{h} = \frac{1.66\times10^{-19}\,J}{6.626\times10^{-34}\,J\cdot s} = \underline{2.51\times10^{14}\,Hz}$

d) $E = hf = (6.626\times10^{-34}\,J\cdot s)(100\times10^{6}\,Hz) = 6.63\times10^{-26}\,J = \underline{4.14\times10^{-7}\,eV}$

e) A photon of frequency $f = 100\,MHz$ has much too little energy (as calculated in (d)) to dissociate AgBr. The photons in the visible light from a firefly do individually have enough energy to dissociate AgBr. The huge number of 100 MHz TV radiation photons can't compensate for the fact that individually they have too little energy.

f) As explained in (e), the answer is no.

41-27

$$eV_0 = hf - \phi \Rightarrow eV_0 = \frac{hc}{\lambda} - \phi$$

Call the two wavelengths λ_1 and λ_2, and the corresponding stopping potentials V_{01} and V_{02}.

$$\Rightarrow eV_{01} = \frac{hc}{\lambda_1} - \phi \; ; \; eV_{02} = \frac{hc}{\lambda_2} - \phi .$$

Subtract these two equations $\Rightarrow e(V_{02}-V_{01}) = hc\left(\frac{1}{\lambda_2} - \frac{1}{\lambda_1}\right)$

$$\Delta V = V_{02} - V_{01} = \frac{hc}{e}\left(\frac{1}{\lambda_2} - \frac{1}{\lambda_1}\right) = \frac{(6.626\times10^{-34}\,J\cdot s)(3.00\times10^{8}\,m\cdot s^{-1})}{1.60\times10^{-19}\,C}\left(\frac{1}{360\times10^{-9}\,m} - \frac{1}{400\times10^{-9}\,m}\right)$$

$\Delta V = \underline{0.345\,V}$

41-29

$f = \frac{c}{\lambda}$; use to convert the λ entries in the table to f values:

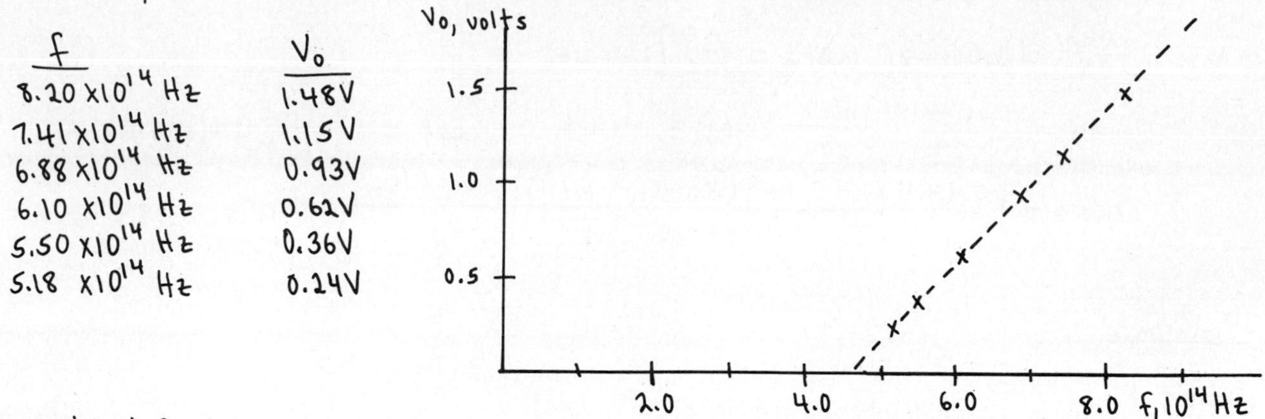

f	V_0
8.20×10^{14} Hz	1.48V
7.41×10^{14} Hz	1.15V
6.88×10^{14} Hz	0.93V
6.10×10^{14} Hz	0.62V
5.50×10^{14} Hz	0.36V
5.18×10^{14} Hz	0.24V

$eV_0 = hf - \phi$

$\boxed{V_0 = \frac{h}{e} f - \frac{\phi}{e}} \Rightarrow V_0$ versus f is a straight line.

a) f_{th} is f when $V_0 = 0$.

From the graph, $f = 4.5 \times 10^{14}$ Hz when $V_0 = 0 \Rightarrow f_{th} = \underline{4.5 \times 10^{14} \text{ Hz}}$.

b) $\lambda_{th} = \frac{c}{f_{th}} = \frac{3.00 \times 10^8 \text{ m}\cdot\text{s}^{-1}}{4.5 \times 10^{14} \text{ Hz}} = 6.7 \times 10^{-7} \text{ m} = \underline{670 \text{ nm}}$

c) $V_0 = 0 \Rightarrow \frac{h}{e} f_{th} = \frac{\phi}{e} \Rightarrow \phi = hf_{th} = (6.626 \times 10^{-34} \text{ J}\cdot\text{s})(4.5 \times 10^{14} \text{ Hz}) = 2.98 \times 10^{-19} \text{ J}$

$\phi = 2.98 \times 10^{-19} \text{ J} \left(\frac{1 \text{ eV}}{1.60 \times 10^{-19} \text{ J}}\right) = \underline{1.86 \text{ eV}}$

d) The slope of V_0 versus f is $\frac{h}{c}$.

$\text{slope} = \frac{1.48\text{V} - 0.24\text{V}}{8.20 \times 10^{14} \text{ Hz} - 5.18 \times 10^{14} \text{ Hz}} = \frac{1.24 \text{ V}}{3.02 \times 10^{14} \text{ Hz}} = 4.11 \times 10^{-15} \text{ V}\cdot\text{s}$

$1 \text{ V} = 1 \text{ N}\cdot\text{m}\cdot\text{C}^{-1} = 1 \text{ J}\cdot\text{C}^{-1} \Rightarrow \text{slope} = 4.11 \times 10^{-15} \text{ J}\cdot\text{s}\cdot\text{C}^{-1}$

Thus $\frac{h}{e} = 4.11 \times 10^{-15} \text{ J}\cdot\text{s}\cdot\text{C}^{-1}$ and $h = (4.11 \times 10^{-15} \text{ J}\cdot\text{s}\cdot\text{C}^{-1})(1.60 \times 10^{-19} \text{ C}) = \underline{6.58 \times 10^{-34} \text{ J}\cdot\text{s}}$.

This is in good agreement with the accurate value of 6.626×10^{-34} J.s. (It is too small by about 0.7%.)

41-35

a) eq. (41-24) $\lambda' - \lambda = \frac{h}{mc}(1 - \cos\phi) \Rightarrow \Delta\lambda = \frac{h}{mc}(1 - \cos\phi)$

largest $\Delta\lambda \Rightarrow \cos\phi = -1 \Rightarrow \Delta\lambda = \frac{2h}{mc} = \frac{2(6.626 \times 10^{-34} \text{ J}\cdot\text{s})}{(9.11 \times 10^{-31} \text{ kg})(3.00 \times 10^8 \text{ m}\cdot\text{s}^{-1})} = 4.85 \times 10^{-12} \text{ m}$

$\Delta\lambda = \underline{0.00485 \text{ nm}}$

41-35 (cont)

b) $\Delta\lambda = \dfrac{2h}{mc}$

Wavelength doubles $\Rightarrow \lambda' = 2\lambda \Rightarrow \Delta\lambda = \lambda' - \lambda = 2\lambda - \lambda = \lambda$

Thus $\lambda = \dfrac{2h}{mc}$.

$$E = \frac{hc}{\lambda} = hc\left(\frac{mc}{2h}\right) = \tfrac{1}{2}mc^2 = \tfrac{1}{2}(9.11\times10^{-31}\,\text{kg})(3.00\times10^{8}\,\text{m}\cdot\text{s}^{-1})^2 = 4.10\times10^{-14}\,\text{J}$$

$$E = 4.10\times10^{-14}\,\text{J}\left(\frac{1\,\text{eV}}{1.60\times10^{-19}\,\text{J}}\right) = \underline{2.56\times10^{5}\,\text{eV}}$$

41-37

a) Apply conservation of energy:

Initially, energy of photon is $\dfrac{hc}{\lambda_i}$

energy of electron is $m_e c^2$, since it is at rest.

After the collision, energy of photon is $\dfrac{hc}{\lambda_f}$

energy of electron is $m_e c^2 + K$.

initial energy = final energy

$$\Rightarrow \frac{hc}{\lambda_i} + m_e c^2 = \frac{hc}{\lambda_f} + m_e c^2 + K$$

$$K = hc\left(\frac{1}{\lambda_i} - \frac{1}{\lambda_f}\right) = (6.626\times10^{-34}\,\text{J}\cdot\text{s})(3.00\times10^{8}\,\text{m}\cdot\text{s}^{-1})\left(\frac{1}{0.100\times10^{-9}\,\text{m}} - \frac{1}{0.110\times10^{-9}\,\text{m}}\right)$$

$$K = (1.99\times10^{-25}\,\text{J}\cdot\text{m})(9.09\times10^{8}\,\text{m}^{-1}) = 1.81\times10^{-16}\,\text{J}$$

or

$$K = 1.81\times10^{-16}\,\text{J}\left(\frac{1\,\text{eV}}{1.60\times10^{-19}\,\text{J}}\right) = \underline{1.13\times10^{3}\,\text{eV}}$$

b) The photon has the energy lost by the electron

$$\Rightarrow \frac{hc}{\lambda} = K \Rightarrow \lambda = \frac{hc}{K} = \frac{(6.626\times10^{-34}\,\text{J}\cdot\text{s})(3.00\times10^{8}\,\text{m}\cdot\text{s}^{-1})}{1.81\times10^{-16}\,\text{J}} = 1.10\times10^{-9}\,\text{m}$$

$\lambda = \underline{1.10\,\text{nm}}$

CHAPTER 42

Exercises 1, 5, 9, 11, 13, 15

Problems 19, 21, 25, 27

Exercises

42-1

a) eq. (42-4) $v_n = \frac{1}{\epsilon_0}\frac{e^2}{2nh} = \frac{1}{n}\frac{(1.60\times10^{-19}\,\text{C})^2}{(8.854\times10^{-12}\,\text{C}^2\cdot\text{N}^{-1}\cdot\text{m}^{-2})(2)(6.626\times10^{-34}\,\text{J}\cdot\text{s})}$

$v_n = \frac{1}{n}(2.18\times10^6\,\text{m}\cdot\text{s}^{-1})$

$n=1 \Rightarrow v_1 = \underline{2.18\times10^6\,\text{m}\cdot\text{s}^{-1}}$

$n=2 \Rightarrow v_2 = \underline{1.09\times10^6\,\text{m}\cdot\text{s}^{-1}}$

$n=3 \Rightarrow v_3 = \underline{0.727\times10^6\,\text{m}\cdot\text{s}^{-1}}$

b) The orbital period is $\tau = \frac{2\pi r}{v}$.

$r_n = \epsilon_0\frac{n^2h^2}{\pi m e^2}$

$v_n = \frac{1}{\epsilon_0}\frac{e^2}{2nh}$ $\Rightarrow \tau = 2\pi\left(\epsilon_0\frac{n^2h^2}{\pi m e^2}\right)\left(\frac{\epsilon_0 2nh}{e^2}\right) = \frac{4n^3h^3\epsilon_0^2}{me^4} = n^3\left(\frac{4h}{m}\right)\left(\frac{h\epsilon_0}{e^2}\right)^2$

$\tau = n^3\left[\frac{4(6.626\times10^{-34}\,\text{J}\cdot\text{s})}{9.11\times10^{-31}\,\text{kg}}\right]\left[\frac{(6.626\times10^{-34}\,\text{J}\cdot\text{s})(8.854\times10^{-12}\,\text{C}^2\cdot\text{N}^{-1}\cdot\text{m}^{-2})}{(1.60\times10^{-19}\,\text{C})^2}\right]^2 = n^3(1.53\times10^{-16})\,\text{s}$

$n=1 \Rightarrow \tau_1 = \underline{1.52\times10^{-16}\,\text{s}}$

$n=2 \Rightarrow \tau_2 = \underline{1.22\times10^{-15}\,\text{s}}$

$n=3 \Rightarrow \tau_3 = \underline{4.13\times10^{-15}\,\text{s}}$

c) first excited state $\Rightarrow n=2$ (ground state is $n=1$)

number of orbits in 10^{-8} s is $\frac{10^{-8}\,\text{s}}{1.22\times10^{-15}\,\text{s}} = \underline{8.20\times10^6\ \text{orbits}}$

42-5

a) The force of attraction between an electron of charge of magnitude e and a nucleus of charge Ze is

$$F = \frac{1}{4\pi\epsilon_0}\frac{Ze^2}{r_n^2}$$

$$F = ma \Rightarrow \frac{1}{4\pi\epsilon_0}\frac{Ze^2}{r_n^2} = \frac{mv_n^2}{r_n}$$

The quantization condition is $mvr = n\frac{h}{2\pi}$, just as it is for hydrogen. Therefore in all the equations derived for hydrogen replace e^2 by Ze^2.

42-5 (cont)

In particular, eq. (42-6) becomes $E_n = -\frac{1}{\epsilon_0^2}\frac{Z^2 m e^4}{8n^2h^2}$.
Thus for $Z=2$ the energy of a level of given n is $Z^2 = 4$ times what it is for $Z=1$.

b) Eq. (42-7) becomes $f = \frac{1}{\epsilon_0^2}\frac{mZ^2e^4}{8h^3}\left(\frac{1}{n_2^2}-\frac{1}{n_1^2}\right)$

and $\frac{1}{\lambda} = \frac{f}{c} = \frac{1}{\epsilon_0^2}\frac{mZ^2e^4}{8h^3c}\left(\frac{1}{n_2^2}-\frac{1}{n_1^2}\right)$.

$R = \frac{me^4}{\epsilon_0^2 8h^3c} \Rightarrow \frac{1}{\lambda} = Z^2R\left(\frac{1}{n_2^2}-\frac{1}{n_1^2}\right)$ $(R = 1.097\times10^7\ m^{-1})$

For $Z=2$ this becomes $\frac{1}{\lambda} = 4R\left(\frac{1}{n_2^2}-\frac{1}{n_1^2}\right)$

Lyman series: $n_2=1,\ n_1=2,3,\ldots$

longest λ is for $n_1=2$: $\frac{1}{\lambda}=4R\left(\frac{1}{1}-\frac{1}{4}\right)=3R=3.29\times10^7 m^{-1} \Rightarrow \lambda=30.4nm$

shortest λ is for $n_1\to\infty$: $\frac{1}{\lambda}=4R\left(\frac{1}{1}-\frac{1}{\infty}\right)=4R=4.39\times10^7 m^{-1}\Rightarrow\lambda=22.8nm$

The Lyman series is not in the visible for He^+.

Balmer series: $n_2=2,\ n_1=3,4,\ldots$

longest λ is for $n_1=3$: $\frac{1}{\lambda}=4R\left(\frac{1}{4}-\frac{1}{9}\right)=0.556R=6.10\times10^6 m^{-1}\Rightarrow\lambda=164nm$

shortest λ is for $n_1\to\infty$: $\frac{1}{\lambda}=4R\left(\frac{1}{4}-\frac{1}{\infty}\right)=R=1.10\times10^7 m^{-1}\Rightarrow\lambda=90.9nm$

The Balmer series is not in the visible.

Paschen series: $n_2=3,\ n_1=4,5,\ldots$

longest λ is for $n_1=4$: $\frac{1}{\lambda}=4R\left(\frac{1}{9}-\frac{1}{16}\right)=0.194R=2.13\times10^6 m^{-1}\Rightarrow\lambda=469nm$

shortest λ is for $n_1\to\infty$: $\frac{1}{\lambda}=4R\left(\frac{1}{9}-\frac{1}{\infty}\right)=0.444R=4.87\times10^6 m^{-1}\Rightarrow\lambda=205nm$

According to Section 35-7 the visible spectrum is from 400nm to 700nm. There are therefore some lines in the Paschen series in the visible.

Brackett series: $n_2=4,\ n_1=5,6,\ldots$

longest λ is for $n_1=5$: $\frac{1}{\lambda}=4R\left(\frac{1}{16}-\frac{1}{25}\right)=0.0900R=9.87\times10^4 m^{-1}\Rightarrow\lambda=1013nm$

shortest λ is for $n_1\to\infty$: $\frac{1}{\lambda}=4R\left(\frac{1}{16}-\frac{1}{\infty}\right)=0.250R=2.74\times10^5 m^{-1}\Rightarrow\lambda=365nm$

The Brackett series has lines in the visible.

Pfund series: $n_2=5,\ n_1=6,7,\ldots$

longest λ is for $n_1=6$: $\frac{1}{\lambda}=4R\left(\frac{1}{25}-\frac{1}{36}\right)=0.0489R=5.36\times10^5 m^{-1}\Rightarrow\lambda=1864nm$

shortest λ is for $n_1\to\infty$: $\frac{1}{\lambda}=4R\left(\frac{1}{25}-\frac{1}{\infty}\right)=0.160R=1.76\times10^6 m^{-1}\Rightarrow\lambda=568nm$

The Pfund series has lines in the visible.

$n_2=6$ series:

The shortest λ is for $n_1\to\infty$: $\frac{1}{\lambda}=4R\left(\frac{1}{36}-\frac{1}{\infty}\right)=0.111R=1.22\times10^6 m^{-1}$

42-5 (cont)

$\Rightarrow \lambda = 820\,nm$

This wavelength is too long to be in the visible $\Rightarrow$ all of the $n_2 = 6$ (and higher) series is not in the visible.

The conclusion is that the Paschen, Brackett and Pfund series have lines in the visible for He^+.

c) With $e^2 \to Ze^2$, eq. (42-3) becomes $r_n(Z) = \frac{1}{\epsilon_0}\frac{n^2h^2}{\pi m Z e^2} = \frac{1}{Z} r_n(Z=1)$

The radius for He^+ is smaller by a factor of 2 compared to the radius for H and the same n.

42-9

a) eq. (42-11) $\lambda = \frac{h}{\sqrt{2meV}}$

electron, $V = 500V \Rightarrow \lambda = \frac{6.626\times10^{-34}\,J\cdot s}{\sqrt{2(9.11\times10^{-31}\,kg)(1.60\times10^{-19}C)(500V)}} = 5.49\times10^{-11}m = 0.0549\,nm$

b) proton, $V = 500V \Rightarrow \lambda = \frac{6.626\times10^{-34}\,J\cdot s}{\sqrt{2(1.67\times10^{-27}\,kg)(1.60\times10^{-19}C)(500V)}} = 1.28\times10^{-12}m = 0.00128\,nm$

(Note: $\lambda_p = \sqrt{\frac{m_e}{m_p}}\,\lambda_e$.)

42-11

a) $\Delta x \Delta p_x \sim \frac{h}{2\pi}$

$\Delta x \sim \frac{h}{p}$ (uncertainty in position is on order of de Broglie wavelength)

Thus $\frac{h}{p}\Delta p \sim \frac{h}{2\pi} \Rightarrow \Delta p \sim \frac{p}{2\pi} \sim p$

b) $\Delta x = 0.5\times10^{-10}m$

$\Delta x \Delta p \sim \frac{h}{2\pi} \Rightarrow \Delta p \sim \frac{h}{2\pi\,\Delta x} = \frac{6.626\times10^{-34}\,J\cdot s}{2\pi(0.5\times10^{-10}m)} = 2.11\times10^{-24}\,kg\cdot m\cdot s^{-1}$

In the Bohr model, for $n=1$ $v_1 = \frac{1}{\epsilon_0}\frac{e^2}{2h}$ (eq. 42-4)

$p_1 = mv_1 = \frac{me^2}{2\epsilon_0 h} = \frac{(9.11\times10^{-31}kg)(1.60\times10^{-19}C)^2}{2(8.85\times10^{-12}C^2\cdot N^{-1}\cdot m^{-2})(6.626\times10^{-34}J\cdot s)} = 1.99\times10^{-24}\,kg\cdot m\cdot s^{-1}$

Thus the uncertainty Δp is roughly equal to p.

42-13

$E = mc^2 \Rightarrow \Delta E = (\Delta m)c^2$

$m = 3(1.67\times10^{-27}\,kg) = 5.01\times10^{-27}\,kg$

42-13 (cont)

$$\Delta m = 1\% \text{ of } m = 0.01(5.01\times10^{-27}\text{ kg}) = 5.01\times10^{-29}\text{ kg}$$

$$\Delta E = (5.01\times10^{-29}\text{ kg})(3.00\times10^{8}\text{ m}\cdot\text{s}^{-1})^2 = 4.51\times10^{-12}\text{ J}$$

$$\text{eq. (42-17)} \quad \Delta E\,\Delta t \sim \frac{h}{2\pi} \Rightarrow \Delta t = \frac{h}{2\pi\,\Delta E} = \frac{6.626\times10^{-34}\text{ J}\cdot\text{s}}{2\pi(4.51\times10^{-12}\text{ J})} = \underline{2.34\times10^{-23}\text{ s}}$$

42-15

a) eq. (42-21) $L_z = m\hbar$, $m = -l, -l+1, \ldots, l-1, l$

Therefore, the largest m for $l=3$ is $m=3$, so the largest $L_z = \underline{3\hbar}$

b) eq. (42-20) $L = \sqrt{l(l+1)}\,\hbar = \sqrt{3(4)}\,\hbar = \underline{3.46\hbar}$

Note: $L > \max L_z$

c) $$\cos\theta = \frac{L_z}{L} = \frac{m\hbar}{3.46\hbar} = \frac{m}{3.46}$$

The allowed values of m range from $-l$ to $+l$.

Note: $L_z < 0$ (for $m<0$) $\Rightarrow \theta > 90°$

$m=3 \Rightarrow \cos\theta = \frac{3}{3.46} = 0.867 \Rightarrow \theta = 29.9°$

$m=-3 \Rightarrow \cos\theta = -0.867 \Rightarrow \theta = 150.1°$

$m=2 \Rightarrow \cos\theta = \frac{2}{3.46} = 0.578 \Rightarrow \theta = 54.7°$

$m=-2 \Rightarrow \cos\theta = -0.578 \Rightarrow \theta = 125.3°$

$m=1 \Rightarrow \cos\theta = \frac{1}{3.46} = 0.289 \Rightarrow \theta = 73.2°$

$m=-1 \Rightarrow \cos\theta = -0.289 \Rightarrow \theta = 106.8°$

$m=0 \Rightarrow \cos\theta = 0 \Rightarrow \theta = 90.0°$

Problems

42-19

The only change in the Bohr model formulas is that we use the muon rather than the electron mass.

a) eq. (42-6), $n=1 \Rightarrow E_1 = -\frac{1}{\epsilon_0^2}\frac{m_\mu e^4}{8h^2} = -\frac{m_\mu}{8}\left(\frac{e^2}{\epsilon_0 h}\right)^2$

$$E_1 = -\frac{(207)(9.11\times10^{-31}\text{ kg})}{8}\left[\frac{(1.60\times10^{-19}\text{ C})^2}{(8.854\times10^{-12}\text{ C}^2\cdot\text{N}^{-1}\cdot\text{m}^{-2})(6.626\times10^{-34}\text{ J}\cdot\text{s})}\right]^2 = -4.49\times10^{-16}\text{ J}$$

$$E_1 = -(4.49\times10^{-16}\text{ J})\left(\frac{1\text{ eV}}{1.60\times10^{-19}\text{ J}}\right) = \underline{-2810\text{ eV}}$$

(Note: This result equals (207)(−13.6 eV), where −13.6 eV is the ground state energy of hydrogen.)

42-19 (cont)

b) eq. (42-3) with $n=1 \Rightarrow r_1 = \epsilon_0 \frac{h^2}{\pi m_\mu e^2} = \left(\frac{m_e}{m_\mu}\right)\left(\epsilon_0 \frac{h^2}{\pi m_e e^2}\right)$

But $\epsilon_0 \frac{h^2}{\pi m_e e^2} = 0.53\times10^{-10}\,m$, the hydrogen atom $n=1$ Bohr radius and $\frac{m_e}{m_\mu} = \frac{1}{207}$

$\Rightarrow r_1 = \frac{0.53\times10^{-10}\,m}{207} = \underline{2.56\times10^{-13}\,m}$

c) $f = \frac{1}{\epsilon_0^2}\frac{me^4}{8h^3}\left(\frac{1}{n_2^2} - \frac{1}{n_1^2}\right)$

$\frac{1}{\lambda} = \frac{f}{c} = \frac{me^4}{\epsilon_0^2 8h^3 c}\left(\frac{1}{n_2^2} - \frac{1}{n_1^2}\right) = \left(\frac{m_\mu}{m_e}\right)\left(\frac{m_e e^4}{\epsilon_0^2 8h^3 c}\right)\left(\frac{1}{n_2^2} - \frac{1}{n_1^2}\right)$

$\frac{m_e e^4}{\epsilon_0^2 8h^3 c} = R = 1.097\times10^7\,m^{-1}$ and $\frac{m_\mu}{m_e} = 207$

$\Rightarrow \frac{1}{\lambda} = 207(1.097\times10^7\,m^{-1})\left(\frac{1}{n_2^2} - \frac{1}{n_1^2}\right)$

$n_1 = 2, n_2 = 1 \Rightarrow \frac{1}{\lambda} = 207(1.097\times10^7\,m^{-1})\left(\frac{3}{4}\right)$

$\lambda = \frac{1}{207}\left(\frac{4}{3(1.097\times10^7\,m^{-1})}\right) = \frac{1}{207}(1.215\times10^{-7}\,m) = 5.87\times10^{-10}\,m = \underline{0.587\,nm}$

(The wavelengths in the spectrum are all smaller by a factor of 207 compared to the wavelengths for hydrogen.)

42-21

a) $U = -\mu B$

Bohr theory $\Rightarrow mvr = n\hbar \Rightarrow vr = n\frac{\hbar}{m}$.

$\mu = \frac{1}{2}evr$ (eq. 42-23)

Thus $\mu = n\left(\frac{\hbar e}{2m}\right) = n(9.27\times10^{-24}\,A\cdot m^2)$.

$U = -\mu B = -n(9.27\times10^{-24}\,A\cdot m^2)(2T) = -n(1.854\times10^{-23}\,J)\left(\frac{1\,eV}{1.60\times10^{-19}\,J}\right) = -n(1.16\times10^{-4}\,eV)$

$n=3$ level is shifted by $U = -3(1.16\times10^{-4}\,eV) = -3.48\times10^{-4}\,eV$

$n=2$ level is shifted by $U = -2(1.16\times10^{-4}\,eV) = -2.32\times10^{-4}\,eV$

b) The shift is small, so let's derive an explicit formula for the shift in wavelength, instead of subtracting two large numbers that are almost equal.

For $B=0$, the separation between the levels is E_0, and the wavelength is λ_0:

$E_0 = \frac{hc}{\lambda_0}$.

With $B = 2T$, let the energy separation between the two levels be E, and the wavelength be λ:

$E = \frac{hc}{\lambda}$.

42-21 (cont)

$n=3$ ——— ↕ E_0 ——— ↕ E ——— -3.48×10^{-4} eV

$n=2$ ——— ——— -2.32×10^{-4} eV

$B=0$ $B=2T$

Let $\Delta E = -3.48\times10^{-4}\text{eV} - (-2.32\times10^{-4}\text{eV})$
$\Delta E = -1.16\times10^{-4}$ eV.

ΔE is the change in the transition energy (separation between the $n=3$ and $n=2$ levels). $E = E_0 + \Delta E$. Since ΔE is negative the transition energy is decreased. (The upper level is lowered more than the lower level by the interaction with the magnetic field.)

The change ΔE in transition energy corresponds to a change in wavelength: $\lambda = \lambda_0 + \Delta\lambda$.

We need to derive an equation that relates $\Delta\lambda$ to ΔE:

$$E = \frac{hc}{\lambda} = \frac{hc}{\lambda_0 + \Delta\lambda} = \left(\frac{hc}{\lambda_0}\right)\frac{1}{1+\left(\frac{\Delta\lambda}{\lambda_0}\right)} = \frac{hc}{\lambda_0}\left(1+\frac{\Delta\lambda}{\lambda_0}\right)^{-1} \approx \frac{hc}{\lambda_0}\left(1-\frac{\Delta\lambda}{\lambda_0}\right)$$

In the last step we made a binomial expansion (Appendix B) and retained only the first two terms. This will be accurate if $\Delta\lambda/\lambda_0$ is small.

But also $E = E_0 + \Delta E \Rightarrow E_0 + \Delta E = \frac{hc}{\lambda_0} - \frac{hc}{\lambda_0^2}\Delta\lambda = E_0 - \frac{hc}{\lambda_0^2}\Delta\lambda$

$$\Rightarrow \boxed{\Delta E = -\left(\frac{hc}{\lambda_0^2}\right)\Delta\lambda}$$

This is the desired equation relating ΔE to $\Delta\lambda$. Note that the minus sign means that ΔE negative means $\Delta\lambda$ positive; a decrease in transition energy corresponds to an increase in the wavelength of the emitted photon.

$$\Delta\lambda = -\frac{\lambda_0^2}{hc}\Delta E = -\frac{(656.3\times10^{-9}\text{m})^2}{(6.626\times10^{-34}\text{J}\cdot\text{s})(3.00\times10^8\text{m}\cdot\text{s}^{-1})}(-1.16\times10^{-4}\text{eV})\left(\frac{1.60\times10^{-19}\text{J}}{1\text{eV}}\right) = +4.02\times10^{-11}\text{m}$$

$\Delta\lambda = +0.0402$ nm

(Note: $\Delta\lambda$ is much less than λ_0, so the binomial expansion we made is accurate.)

42-25

a) $p_y = p_x\left(\frac{\lambda}{a}\right)$ (eq. 42-14) ; a is the aperature diameter

$p_y = m v_x\left(\frac{\lambda}{a}\right) \Rightarrow v_y = v_x\left(\frac{\lambda}{a}\right)$

The uncertainty in position of the point where the electrons strike the screen is $\Delta y = 2 v_y t$, where t is the time it takes the electrons to travel from the aperature to the screen. The factor of 2 is put in because the electrons could have either positive or negative v_y.

$t = \frac{d}{v_x}$, where $d = 0.3$m is the distance the electrons travel

42-25 (cont)

$$\Rightarrow \Delta y = 2\left(v_x \frac{\lambda}{a}\right)\frac{d}{v_x} = \frac{2\lambda d}{a}$$

$$\lambda = \frac{h}{p} = \frac{h}{mv} \Rightarrow \Delta y = \frac{2hd}{mva}$$

v is the velocity the electrons acquire by being accelerated through the potential of $\Delta V = 20{,}000$ volts:

$$K = e\Delta V$$

$$\frac{1}{2}mv^2 = e\Delta V \Rightarrow v = \sqrt{\frac{2e\Delta V}{m}} = \sqrt{\frac{2(1.60\times10^{-19}\text{C})(20{,}000\text{V})}{9.11\times10^{-31}\text{kg}}} = 8.38\times10^{7}\,\text{m}\cdot\text{s}^{-1}$$

(Note: $\frac{v}{c} = 0.28$, so using the classical expression for K shouldn't introduce much error.)

$$\text{Then } \Delta y = \frac{2(6.626\times10^{-34}\,\text{J}\cdot\text{s})(0.3\text{m})}{(9.11\times10^{-31}\text{kg})(8.38\times10^{7}\,\text{m}\cdot\text{s}^{-1})(0.5\times10^{-3}\text{m})} = \underline{1.0\times10^{-8}\,\text{m}}$$

b) Δy is only on the order of the size of a single atom; this uncertainty has no effect on the clarity of the picture!

42-27

The number of states N with principal quantum number n is given by summing over all m, l, and s quantum numbers:

For a given l, m takes on the values $-l, -l+1, \ldots, +l$, for a total of $2l+1$ different values of m.

Then since l takes the values $0, 1, \ldots, n-1$,

$$N = 2\sum_{l=0}^{n-1}(2l+1).$$

The factor of 2 accounts for the two possible values of s ($s = \pm\frac{1}{2}$).

$$N = 4\sum_{l=0}^{n-1} l + 2\sum_{l=0}^{n-1} 1 = 4\sum_{l=0}^{n-1} l + 2n,$$ since there are n terms in the sum.

The problem tells us that $\sum_{l=1}^{M} l = \frac{M(M+1)}{2}$.

Apply with $M = n-1 \Rightarrow \sum_{l=0}^{n-1} l = \sum_{l=1}^{n-1} l = \frac{(n-1)(n-1+1)}{2} = \frac{(n-1)n}{2}.$

Thus $N = 4\left(\frac{(n-1)n}{2}\right) + 2n = 2n(n-1) + 2n = 2n(n-1+1) = 2n^2$, as was to be shown.

CHAPTER 43

Exercises 1, 7, 9

Problems 13, 15

Exercises

43-1

a) $n=5 \Rightarrow l = 0, 1, 2, 3, 4$ (since $l = 0, 1, \ldots, n-1$).
For each l, $m = -l, -l+1, \ldots, l$.

		number of states
$l = 0$	$m = 0$	1
$l = 1$	$m = -1, 0, +1$	3
$l = 2$	$m = -2, -1, 0, +1, +2$	5
$l = 3$	$m = -3, -2, -1, 0, +1, +2, +3$	7
$l = 4$	$m = -4, -3, -2, -1, 0, +1, +2, +3, +4$	9
		25 $= n^2$ (see Problem 42-27.)

b) Two electrons (because of the two possible values of the spin quantum number) in each l-m state $\Rightarrow 2(25) =$ 50 electrons in the $n=5$ shell.

43-7

a) $I = \sum mr^2 = 2mr^2$

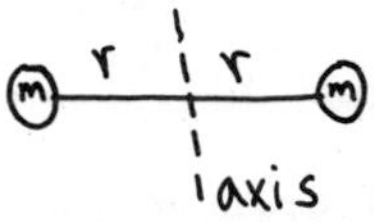

$2r = 0.074\text{ nm}$
$m = 1.67\times10^{-27}\text{ kg}$

$$I = 2(1.67\times10^{-27}\text{ kg})(0.037\times10^{-9}\text{ m})^2 = \underline{4.57\times10^{-48}\text{ kg}\cdot\text{m}^2}$$

b) $E_l = l(l+1)\dfrac{\hbar^2}{2I}$ (eq. 43-3)

$$\frac{\hbar^2}{2I} = \frac{(1.054\times10^{-34}\text{ J}\cdot\text{s})^2}{2(4.57\times10^{-48}\text{ kg}\cdot\text{m}^2)} = 1.22\times10^{-21}\text{ J} = 1.22\times10^{-21}\text{ J}\left(\frac{1\text{ eV}}{1.60\times10^{-19}\text{ J}}\right) = \underline{7.62\times10^{-3}\text{ eV}}$$

$l=0 \Rightarrow E_0 = 0$

$l=1 \Rightarrow E_1 = 2\left(\dfrac{\hbar^2}{2I}\right) = 1.52\times10^{-2}\text{ eV}$

$l=2 \Rightarrow E_2 = 6\left(\dfrac{\hbar^2}{2I}\right) = 4.57\times10^{-2}\text{ eV}$

c) $l=2 \rightarrow l=0$ transition $\Rightarrow \Delta E = E_2 - E_0 = 4.57\times10^{-2}\text{ eV}$.

Thus the energy of the photon emitted is $E = \Delta E = 4.57\times10^{-2}\text{ eV}\left(\dfrac{1.60\times10^{-19}\text{ J}}{1\text{ eV}}\right)$

$E = 7.31\times10^{-21}\text{ J}$

$$E = hf \Rightarrow f = \frac{E}{h} = \frac{7.31\times10^{-21}\text{ J}}{6.626\times10^{-34}\text{ J}\cdot\text{s}} = \underline{1.10\times10^{13}\text{ Hz}}$$

$$\lambda = \frac{c}{f} = \frac{3.00\times10^{8}\text{ m}\cdot\text{s}^{-1}}{1.10\times10^{13}\text{ Hz}} = 2.73\times10^{-5}\text{ m} = \underline{27.3\,\mu\text{m}}$$

43-9

Each atom occupies a cube of side 0.282 nm. Therefore the volume occupied by each atom is $V = (0.282 \times 10^{-9}\,m)^3 = 2.24 \times 10^{-29}\,m^3$.

From Appendix D the mass of a Na atom is $m_{Na} = \frac{M_{Na}}{N_A} \Rightarrow$
$m_{Na} = \frac{22.990 \times 10^{-3}\,kg \cdot mol^{-1}}{6.022 \times 10^{23}\,mol^{-1}} = 3.818 \times 10^{-26}\,kg$, and for a Cl atom it is
$m_{Cl} = \frac{M_{Cl}}{N_A} = \frac{35.453 \times 10^{-3}\,kg \cdot mol^{-1}}{6.022 \times 10^{23}\,mol^{-1}} = 5.887 \times 10^{-26}\,kg$.

The average mass (since in NaCl there are equal numbers of Na and Cl atoms) of the atoms in the crystal is

$$m = \tfrac{1}{2}(m_{Na} + m_{Cl}) = \tfrac{1}{2}(3.818 \times 10^{-26}\,kg + 5.887 \times 10^{-26}\,kg) = 4.85 \times 10^{-26}\,kg.$$

Thus $\rho = \frac{m}{V} = \frac{4.85 \times 10^{-26}\,kg}{2.24 \times 10^{-29}\,m^3} = \underline{2.16 \times 10^3\,kg \cdot m^{-3}}$.

Problems

43-13

a) $\omega = \sqrt{\frac{k}{m}}$; $f = \frac{\omega}{2\pi} = \frac{1}{2\pi}\sqrt{\frac{k}{m}} \Rightarrow k = m(2\pi f)^2$

The problem says to consider the Cl atom stationary, so it is the H atom that is moving back and forth on the end of the "spring".

$m_H = \frac{M_H}{N_A} = \frac{1.008 \times 10^{-3}\,kg \cdot mol^{-1}}{6.022 \times 10^{23}\,mol^{-1}} = 1.67 \times 10^{-27}\,kg$

$k = (1.67 \times 10^{-27}\,kg)(2\pi\,[8.6 \times 10^{13}\,Hz])^2 = \underline{488\,N \cdot m^{-1}}$

b) $\Delta E = E = hf$ (transition energy = photon energy)

$\Rightarrow \Delta E = (6.626 \times 10^{-34}\,J \cdot s)(8.6 \times 10^{13}\,Hz) = \underline{5.70 \times 10^{-20}\,J}$

$\Delta E = 5.70 \times 10^{-20}\,J\left(\frac{1\,eV}{1.60 \times 10^{-19}\,J}\right) = \underline{0.356\,eV}$

c) $\lambda = \frac{c}{f} = \frac{3.00 \times 10^8\,m \cdot s^{-1}}{8.6 \times 10^{13}\,Hz} = 3.49 \times 10^{-6}\,m = \underline{3490\,nm}$

(This wavelength corresponds to infra-red radiation.)

43-15

a) Define the zero of energy such that the energy of two separated Na and Cl atoms is zero. The energy of separated $Na^+ + Cl^-$ is then 1.3 eV (section 43-3).

The energy of $Na^+ + Cl^-$ at some finite separation r is just this energy plus the (negative) coulomb potential energy due to the attractive coulomb force between the + and − charges:

43-15 (cont)

$$E(Na^+ + Cl^-) = 1.3\,\text{eV}\left(\frac{1.60\times10^{-19}\,\text{J}}{1\,\text{eV}}\right) - \frac{1}{4\pi\epsilon_0}\frac{e^2}{r}.$$

Set this equal to 0, the energy of Na + Cl

$$\Rightarrow 2.08\times10^{-19}\,\text{J} - \frac{1}{4\pi\epsilon_0}\frac{e^2}{r} = 0.$$

$$r = \frac{1}{4\pi\epsilon_0}\frac{e^2}{2.08\times10^{-19}\,\text{J}} = (9.0\times10^9\,\text{N}\cdot\text{m}^2\cdot\text{C}^{-2})\frac{(1.60\times10^{-19}\,\text{C})^2}{2.08\times10^{-19}\,\text{J}} = 1.11\times10^{-9}\,\text{m} = \underline{1.11\,\text{nm}}$$

b) Let the energy of separated K + Br neutral atoms be zero.

To form separated $K^+ + Br^-$ requires energy $4.3\,\text{eV} - 3.5\,\text{eV} = 0.8\,\text{eV}$ (Exercise 43-4).

$$0.8\,\text{eV} = 0.8\,\text{eV}\left(\frac{1.60\times10^{-19}\,\text{J}}{1\,\text{eV}}\right) = 1.28\times10^{-19}\,\text{J}$$

Set the energy of $K^+ + Br^-$ at a separation r equal to zero

$$\Rightarrow 1.28\times10^{-19}\,\text{J} - \frac{1}{4\pi\epsilon_0}\frac{e^2}{r} = 0$$

$$r = \frac{1}{4\pi\epsilon_0}\frac{e^2}{1.28\times10^{-19}\,\text{J}} = (9.0\times10^9\,\text{N}\cdot\text{m}^2\cdot\text{C}^{-2})\frac{(1.60\times10^{-19}\,\text{C})^2}{1.28\times10^{-19}\,\text{J}} = 1.80\times10^{-9}\,\text{m}$$

$r = \underline{1.80\,\text{nm}}$

CHAPTER 44

Exercises 1, 5, 7, 9, 13, 17, 19, 21

Problems 25, 29

Exercises

44-1

a) If the gold nucleus is treated as a point charge,

$$U = \frac{1}{4\pi\epsilon_0}\frac{q_1 q_2}{r} = \frac{1}{4\pi\epsilon_0}\frac{Z_1 Z_2 e^2}{r}$$

$Z_1 = 2$ (alpha particle) ; $Z_2 = 79$ (gold nucleus)

$$U = (9.0\times10^9\ \text{N}\cdot\text{m}^2\cdot\text{C}^{-2})\frac{(2)(79)(1.60\times10^{-19}\text{C})^2}{1.0\times10^{-14}\text{m}} = \underline{3.64\times10^{-12}\text{J}}$$

$$U = (3.64\times10^{-12}\text{J})\left(\frac{1\text{eV}}{1.60\times10^{-19}\text{J}}\right) = 2.28\times10^7\text{eV} = \underline{22.8\text{ MeV}}$$

b) $K_i + U_i = K_f + U_f$

$K_f = 0$

alpha particle starts out far away from the gold nucleus $\Rightarrow r_i = \infty,\ U_i = 0.$

Thus $K_i = U_f = 3.64\times10^{-12}\text{J} = \underline{22.8\text{MeV}}$

c) An alpha particle is a bare helium nucleus, with 2 protons and 2 neutrons. Its mass is thus $4.00260\,u - 2(0.000549u) = 4.0015\,u = 4.0015\,u\left(\frac{1.661\times10^{-27}\text{kg}}{1u}\right)$

$m = 6.65\times10^{-27}$ kg

(4.00260 u from Table 44-2; the 2(0.000549u) is to subtract the mass of the two electrons in the neutral atom)

$$K_i = \tfrac{1}{2}mv_i^2 \Rightarrow v_i = \sqrt{\frac{2K_i}{m}} = \sqrt{\frac{2(3.64\times10^{-12}\text{J})}{6.65\times10^{-27}\text{kg}}} = \underline{3.31\times10^7\text{m}\cdot\text{s}^{-1}}$$

44-5

deuterium nucleus consists of 1 proton and 1 neutron

Its mass is

$m_D = 2.01410u - (1)(0.000549u) = 2.013551u$ (a neutral deuterium atom has one electron, whose mass must be subtracted to get the nuclear mass).

$m_n = 1.008665u$

$m_p = 1.007276u$

Mass defect therefore is $m_p + m_n - m_D = 1.007276u + 1.008665u - 2.013551u$
$= 2.39\times10^{-3}u$

$1u = 931.5\text{MeV} \Rightarrow (2.39\times10^{-3}u)(931.5\text{ MeV}\cdot u^{-1}) = \underline{2.23\text{ MeV}}$

44-7

a) The nuclear beta decay reaction is

$${}^{3}_{1}H \rightarrow {}^{3}_{2}He^{+} + e^{-}, \text{ where } e^{-} \text{ denotes an electron}$$

The ${}^{3}_{1}H$ atom has one electron. The ${}^{3}_{2}He^{+}$ atom has that same electron, so is a positive ion, and one electron is emitted from the nucleus.

The mass of the neutral ${}^{3}H$ atom is given in the problem as 3.01647u. The mass of ${}^{3}He^{+} + e^{-}$ is the same as of the ${}^{3}He$ neutral atom. According to Table 44-2 this mass is 3.01603u.

Therefore $m({}^{3}H) > m({}^{3}He^{+} + e^{-})$, so β-decay is energetically possible.

b) The mass defect is $3.01647u - 3.01603u = 4.4 \times 10^{-4} u$

The energy equivalent is $(4.4 \times 10^{-4} u)(931.5 \text{ MeV} \cdot u^{-1}) = \underline{0.410 \text{ MeV}}$

44-9

a) $t_{1/2} = \frac{0.693}{\lambda}$ (eq. 44-5)

$$\lambda = \frac{0.693}{t_{1/2}} = \frac{0.693}{4.50 \times 10^{9} \text{ yr}} = 1.54 \times 10^{-10} \text{ yr}^{-1} \left(\frac{1 \text{ yr}}{3.156 \times 10^{7} s} \right) = \underline{4.88 \times 10^{-18} s^{-1}}$$

b) activity = 1 curie $\Rightarrow \frac{dN}{dt} = 1 \text{ Ci} = 3.70 \times 10^{10} \text{ decays} \cdot s^{-1}$

$$\frac{dN}{dt} = -\lambda N \Rightarrow N = -\frac{dN/dt}{\lambda} = \frac{3.70 \times 10^{10} \text{ decays} \cdot s^{-1}}{4.88 \times 10^{-18} s^{-1}} = 7.58 \times 10^{27} \text{ nuclei}$$

$N = \frac{m}{m_u}$, where m is the mass of the sample and m_u is the mass of one uranium atom ; $m_u = 238u$

$$\Rightarrow m = N m_u = (7.58 \times 10^{27})(238)(1.661 \times 10^{-27} kg) = \underline{3.00 \times 10^{3} kg}$$

c) number emitted per second is $\left| \frac{dN}{dt} \right|$

$$\left| \frac{dN}{dt} \right| = \lambda N$$

$$m = 1g = 1 \times 10^{-3} kg \Rightarrow N = \frac{m}{m_u} = \frac{1 \times 10^{-3} kg}{238 (1.661 \times 10^{-27} kg)} = 2.53 \times 10^{21} \text{ nuclei}$$

$$\left| \frac{dN}{dt} \right| = (4.88 \times 10^{-18} s^{-1})(2.53 \times 10^{21}) = \underline{1.23 \times 10^{4} s^{-1}}$$

44-13

a) ${}^{2}_{1}H + {}^{9}_{4}Be \rightarrow {}^{7}_{3}Li + {}^{4}_{2}He$

There are 5 electrons on each side of the reaction. Their mass will cancel out, so we can use neutral atom masses.

${}^{2}_{1}H + {}^{9}_{4}Be$ has mass $2.01410u + 9.01219u = 11.02629u$

44-13 (cont)

$^7_3Li + ^4_2He$ has mass $7.01601\,u + 4.00260\,u = 11.01861\,u$

The mass decrease is $11.02629\,u - 11.01861\,u = 0.00768\,u$.
This corresponds to an energy release of $(0.00768\,u)(931.5\,MeV\cdot u^{-1}) = \underline{7.15\,MeV}$

b) Estimate the threshold energy by calculating the coulomb potential energy when the proton is at the radius r of the ^9_4Be.

$r = r_0 A^{1/3} \Rightarrow r = 1.2\times10^{-15}\,m\,(9)^{1/3} = 2.50\times10^{-15}\,m$ (^9_4Be nuclear radius)

$$U = \frac{1}{4\pi\epsilon_0}\frac{q_1 q_2}{r} = \frac{1}{4\pi\epsilon_0}\frac{(1)(4)e^2}{r} = (9.0\times10^9\,N\cdot m^2\cdot C^{-2})\frac{(4)(1.60\times10^{-19}\,C)^2}{2.50\times10^{-15}\,m} = 3.69\times10^{-13}\,J$$

$$U = (3.69\times10^{-13}\,J)\left(\frac{1\,eV}{1.60\times10^{-19}\,J}\right) = 2.31\times10^6\,eV = \underline{2.31\,MeV}$$

44-17

a) eq. (44-10) $\omega = \frac{qB}{m}$

$$f = \frac{\omega}{2\pi} = \frac{qB}{2\pi m} \Rightarrow B = \frac{2\pi m f}{q}$$

A deuteron is a deuterium nucleus, and thus has mass

$$2.01410\,u - 0.000549\,u = 2.01355\,u = 2.01355\,u\left(\frac{1.661\times10^{-27}\,kg}{1\,u}\right) = 3.34\times10^{-27}\,kg,$$

and charge $+e$.

$$B = \frac{2\pi(3.34\times10^{-27}\,kg)(10\times10^6\,Hz)}{1.60\times10^{-19}\,C} = \underline{1.31\,T}$$

b) $\omega = \frac{v}{r}$; $f = \frac{\omega}{2\pi} \Rightarrow v = r\omega = 2\pi r f = 2\pi(0.32\,m)(10\times10^6\,Hz) = \underline{2.01\times10^7\,m\cdot s^{-1}}$

$\frac{v}{c} = 0.067$, so non-relativistic expression for K will be ok.

$$K = \tfrac{1}{2}mv^2 = \tfrac{1}{2}(3.34\times10^{-27}\,kg)(2.01\times10^7\,m\cdot s^{-1})^2 = 6.75\times10^{-13}\,J$$

$$K = (6.75\times10^{-13}\,J)\left(\frac{1\,eV}{1.60\times10^{-19}\,J}\right) = 4.22\times10^6\,eV = \underline{4.22\,MeV}$$

44-19

$\pi^0 \rightarrow 2\gamma$

The rest mass energy of the π^0 is converted into the energy of the photons.

π^0: $v = 0$, $p_i = 0$ — initial

$p = \frac{h}{\lambda}$, $p = \frac{h}{\lambda}$, $p_f = 0$ — final

44-19 (cont)

The initial momentum of the π^0 at rest is zero. The momenta of the emitted photons must therefore be equal and opposite. $p = \frac{h}{\lambda}$ for a photon, so the photons must have equal wavelengths, and therefore equal frequencies and energies.

The rest mass energy of the π^0 is 135.0 MeV (Table 44-3). The energy of each photon therefore is $\frac{1}{2}(135.0\text{ MeV}) = \underline{67.5\text{ MeV}}$.

$$f = \frac{E}{h} = \frac{(67.5\times10^{6}\text{ eV})\left(\frac{1.60\times10^{-19}\text{ J}}{1\text{ eV}}\right)}{6.626\times10^{-34}\text{ J}\cdot\text{s}} = \underline{1.63\times10^{22}\text{ Hz}}$$

and

$$\lambda = \frac{c}{f} = \frac{3.00\times10^{8}\text{ m}\cdot\text{s}^{-1}}{1.63\times10^{22}\text{ Hz}} = 1.84\times10^{-14}\text{ m} = 1.84\times10^{-5}\text{ nm}$$

44-21

The quantum numbers of the quark combinations are sums of the corresponding quantum numbers of the individual quarks in the combination. The quark quantum numbers are listed in Table 44-4.

a) uus

$Q = (\frac{2}{3} + \frac{2}{3} - \frac{1}{3})e = +e$, $B = \frac{1}{3} + \frac{1}{3} + \frac{1}{3} = 1$

$S = 0 + 0 - 1 = -1$, $C = 0 + 0 + 0 = 0$

b) $c\bar{s}$

The quantum numbers of the $\bar{s}$ are opposite in sign to those of the s.

$Q = (\frac{2}{3} + \frac{1}{3})e = +e$, $B = \frac{1}{3} + (-\frac{1}{3}) = 0$

$S = 0 + 1 = 1$, $C = 1 + 0 = 1$

c) $\overline{ddu}$

$Q = [\frac{1}{3} + \frac{1}{3} + (-\frac{2}{3})]e = 0$, $B = -\frac{1}{3} - \frac{1}{3} - \frac{1}{3} = -1$

$S = 0$, $C = 0$

d) $c\bar{b}$

$Q = (\frac{2}{3} + \frac{1}{3})e = +e$, $B = \frac{1}{3} - \frac{1}{3} = 0$

$S = 0$, $C = 1$

Problems

44-25

For ^{14}C, $t_{1/2} = 5568$ yr (from the text, in the last paragraph of section 44-4)

$$\lambda = \frac{0.693}{t_{1/2}} = \frac{0.693}{5568\text{ yr}} = 1.24\times10^{-4}\text{ yr}^{-1}$$

$$N = N_0 e^{-\lambda t}\ ;\ \frac{N}{N_0} = e^{-\lambda t}$$

$$\ln\left(\frac{N}{N_0}\right) = -\lambda t \Rightarrow t = -\frac{1}{\lambda}\ln\left(\frac{N}{N_0}\right)$$

44-25 (cont)

For the speciman $\frac{N}{N_0} = \frac{1}{8}$.

$$\Rightarrow t = -\frac{1}{\lambda}\ln\left(\frac{1}{8}\right) = \frac{\ln 8}{\lambda} = \frac{\ln 8}{1.24\times10^{-4}\,yr^{-1}} = \underline{1.67\times10^{4}\,yr}$$, the age of the speciman

44-29

$$\Lambda^0 \rightarrow p + \pi^-$$

a) mass defect $= m_{\Lambda^0} - m_p - m_{\pi^-}$

The masses are listed in Table 44-3:

mass defect $= 1115\,MeV/c^2 - 938.3\,MeV/c^2 - 139.6\,MeV/c^2 = 37\,MeV/c^2$

The energy equivalent of this mass defect appears as kinetic energy of the products $\Rightarrow K = \underline{37\,MeV}$

b) conservation of linear momentum:

$$\vec{P}_{\Lambda^0} = \vec{P}_p + \vec{P}_{\pi^-}$$

Λ^0 at rest $\Rightarrow P_{\Lambda^0} = 0 \Rightarrow \vec{P}_p = -\vec{P}_{\pi^-}$

Thus $m_p v_p = m_{\pi^-} v_{\pi^-}$

$$(938.3\,MeV/c^2)\,v_p = (139.6\,MeV/c^2)\,v_{\pi^+}$$

$$v_{\pi^+} = 6.72\,v_p$$

$$K = \tfrac{1}{2}m_p v_p^2 + \tfrac{1}{2}m_{\pi^+} v_{\pi^+}^2$$

Work in mks units, to be sure not to get confused:

$$K = (37\times10^6\,eV)\left(\frac{1.60\times10^{-19}\,J}{1\,eV}\right) = 5.92\times10^{-12}\,J$$

$$m_p = \left(\frac{938.3\,MeV}{931.5\,MeV\cdot u^{-1}}\right)(1.661\times10^{-27}\,kg\cdot u^{-1}) = 1.673\times10^{-27}\,kg$$

$$m_{\pi^+} = \left(\frac{139.6\,MeV}{931.5\,MeV\cdot u^{-1}}\right)(1.661\times10^{-27}\,kg\cdot u^{-1}) = 2.489\times10^{-28}\,kg$$

Then we have

$$\tfrac{1}{2}(1.673\times10^{-27}\,kg)\,v_p^2 + \tfrac{1}{2}(2.489\times10^{-28}\,kg)\,v_{\pi^+}^2 = 5.92\times10^{-12}\,J$$

Use $v_{\pi^+} = 6.72\,v_p$

$$\Rightarrow (8.365\times10^{-28}\,kg)\,v_p^2 + (1.244\times10^{-28}\,kg)(6.72\,v_p)^2 = 5.92\times10^{-12}\,J$$

$$v_p = \sqrt{\frac{5.92\times10^{-12}\,J}{6.45\times10^{-27}\,kg}} = 3.03\times10^{7}\,m\cdot s^{-1}$$

$$v_{\pi^+} = 6.72\,v_p = 6.72(3.03\times10^{7}\,m\cdot s^{-1}) = 2.04\times10^{8}\,m\cdot s^{-1}$$

Then

$$K_p = \tfrac{1}{2}(1.67\times10^{-27}\,kg)(3.03\times10^{7}\,m\cdot s^{-1})^2 = 7.67\times10^{-13}\,J \quad (13\%\text{ of total})$$

44-29 (cont)

$$K_{\pi^+} = \frac{1}{2}(2.489 \times 10^{-28}\,\text{kg})(2.04 \times 10^{8}\,\text{m}\cdot\text{s}^{-1})^2 = 5.18 \times 10^{-12}\,\text{J} \quad (87\% \text{ of total})$$

(Note: $v_{\pi^+}/c = 0.68$, so use of the nonrelativistic momentum and kinetic energy expressions is not very accurate.)